EXCITATORY AMINO ACIDS

Excitatory Amino Acids

Edited by

Peter J. Roberts

Department of Physiology and Pharmacology
University of Southampton, UK

Jon Storm-Mathisen

Anatomical Institute
University of Oslo, Norway

and

Harry F. Bradford

Department of Biochemistry
Imperial College, London, UK

MACMILLAN

First published 1986

Published by
THE MACMILLAN PRESS LTD
Houndmills, Basingstoke, Hampshire RG21 2XS
and London
Companies and representatives
throughout the world

Distributed in North America by
SHERIDAN HOUSE PUBLISHERS
145 Palisade Street, Dobbs Ferry, NY 10522

**Printed and bound in Great Britain at
The Camelot Press Ltd, Southampton**

British Library Cataloguing in Publication Data
Excitatory amino acids.
1. Neurotransmitters   2. Amino acids
I. Roberts, P.J.   II. Storm-Mathisen, J.
III. Bradford, H.F.
612'.8   QP364.7
ISBN 0–333–40655–9

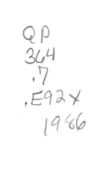

The Contributors

J.I. Addae
Department of Physiology
St George's Hospital Medical School
Cranmer Terrace
London SW17 0RE
UK

C. Aldinio
FIDIA Neurobiological Research
    Laboratories
Via Ponte della Fabbrica 3/A
35031 Abano Terme
Italy

R.A. Altschuler
Laboratory of Neuro-otolaryngology
NINCDS
National Institutes of Health
Bethesda
MD 20205
USA

J.A. Aram
Department of Physiology
Royal Veterinary College
London NW1 0TU
UK

R.N. Auer
Laboratory for Experimental Brain Research
University Hospital
University of Lund
Sweden

M. Baudry
Department of Psychobiology and Center for
    the Neurobiology of Learning and Memory
University of California at Irvine
California 92717
USA

M. Beccaro
FIDIA Neurobiological Research
    Laboratories
Via Ponte della Fabbrica 3/A
35031 Abano Terme
Italy

J. Bockaert
Centre CNRS–INSERM de Pharmacologie-
    Endocrinologie
B.P. 5055
34033 Montpellier
France

L. Brehm
Department of Chemistry BC
The Royal Danish School of Pharmacy
DK–2100 Copenhagen
Denmark

M.B. Bromberg
University of Michigan
Neuroscience Laboratory Building
1103 East Huron
Ann Arbor
Michigan 48104
USA

P.A. Brooks
Department of Pharmacology
St George's Hospital Medical School
Cranmer Terrace
London SW17 0RE
UK

S.P. Butcher
Institute of Neurobiology
University of Gothenburg
S–400 33 Gothenburg
Sweden

G. Le M. Campbell
Department of Neurology
The Graduate Hospital
Philadelphia
PA 19141
USA

V. Carlá
Department of Preclinical and Clinical
    Pharmacology
University of Florence
Viale G.B. Morgagni 65
50134 Firenze
Italy

D.O. Carpenter
Division of Child Psychiatry
Departments of Psychiatry, Neuroscience,
    Pharmacology and Pediatrics
The Johns Hopkins University School of
    Medicine
Baltimore
Maryland 21205
USA

H.H. Chang
Departments of Human Development and
    Biochemistry
and The Center for Biomedical Research
University of Kansas
Lawrence
Kansas
USA

A.G. Chapman
Department of Neurology
Institute of Psychiatry
De Crespigny Park
Denmark Hill
London SE5 8AF
UK

G.G.S. Collins
Department of Pharmacology
University of Sheffield
Western Bank
Sheffield S10 2TN
UK

J.F. Collins
Department of Chemistry
City of London Polytechnic
London EC3N 2EY
UK

J.H. Connick
Department of Physiology
St George's Hospital Medical School
Cranmer Terrace
London SW17 0RE
UK

C.W. Cotman
Department of Psychobiology
University of California at Irvine
California 92717
USA

J.T. Coyle
Division of Child Psychiatry
Departments of Psychiatry, Neuroscience,
    Pharmacology and Pediatrics
The Johns Hopkins University School of
    Medicine
Baltimore
Maryland 21205
USA

V. Crunelli
Department of Pharmacology
St George's Hospital Medical School
Cranmer Terrace
London SW17 0RE
UK

M. Cuénod
Brain Research Institute
University of Zurich
August-Forel-Strasse 1
8029 Zurich
Switzerland

K. Curry
Department of Physiology
University of British Columbia
Vancouver
British Columbia V6T 1W5
Canada

D.R. Curtis
Department of Pharmacology
The Australian National University
Canberra
ACT 2601
Australia

N.C. Danbolt
Anatomical Institute
University of Oslo
Karl Johansgate 47
N–0162 Oslo 1
Norway

N.H. Diemer
Institute of Neuropathology
University of Copenhagen
Denmark

J. Dingwall
Friedrich Miescher-Institut
P.O. Box 2543
CH–4002 Basel
Switzerland

K.Q. Do
Brain Research Institute
University of Zurich
August-Forel-Strasse 1
8029 Zurich
Switzerland

J. Drejer
A/S Ferrosan
Research Division
Sydmarken 5
DK–2860 Soeborg
Denmark

B. Engelsen
Department of Neurology
University of Bergen
N–5016 Haukeland Sykehus
Norway

M. Errami
CNRS INP 5
B.P. 71–13402 Marseille Cédex 9
France

G.E. Fagg
Friedrich Miescher-Institut
P.O. Box 2543
CH–4002 Basel
Switzerland

B.O. Fischer
Anatomical Institute
University of Oslo
Karl Johansgate 47
N–0162
Oslo 1
Norway

R. Fisher
Division of Child Psychiatry
Departments of Psychiatry, Neuroscience,
 Pharmacology and Pediatrics
The Johns Hopkins University School of
 Medicine
Baltimore
Maryland 21205
USA

F. Fonnum
Division for Environmental Toxicology
Norwegian Defence Research Establishment
P.O. Box 25
N–2007 Kjeller
Norway

V.M. Fosse
Division for Environmental Toxicology
Norwegian Defence Research Establishment
P.O. Box 25
N–2007 Kjeller
Norway

A.C. Foster
Merck Sharpe and Dohme
Neurosciences Research Centre
Terlings Park
Harlow
Essex CM20 2QR
UK

T. Fu-long
Anatomical Institute
University of Oslo
Karl Johansgate 47
N–0162 Oslo 1
Norway

J.M. ffrench-Mullen
Division of Child Psychiatry
Departments of Psychiatry, Neuroscience,
 Pharmacology and Pediatrics
The Johns Hopkins University School of
 Medicine
Baltimore
Maryland 21205
USA

V. Gallo
Laboratory of Organ and System
 Physiopathology
Istituto Superiore di Sanità
Viale Regina Elena 299
Rome
Italy

J.W. Geddes
Department of Biochemistry
University of Saskatchewan
Saskatoon S7N 0W0
Canada

O. Goldberg
Department of Organic Chemistry
The Weizmann Institute of Science
Rehovot 76100
Israel

J.T. Greenamyre
University of Michigan
Neuroscience Laboratory Building
1103 East Huron
Ann Arbor
Michigan 48104
USA

V. Gundersen
Anatomical Institute
University of Oslo
Karl Johansgate 47
N–0162 Oslo 1
Norway

H. Hagberg
Institute of Neurobiology
University of Gothenburg
S–400 33 Gothenburg
Sweden

A. Hamberger
Institute of Neurobiology
University of Gothenburg
S–400 33 Gothenburg
Sweden

S.H. Hansen
Department of Chemistry BC
The Royal Danish School of Pharmacy
DK–2100 Copenhagen
Denmark

P.L. Herrling
Wander Research Institute
(a Sandoz Research Unit)
P.O. Box 2747
CH–3001 Berne
Switzerland

T. Honoré
A/S Ferrosan
Research Division
Sydmarken 5
DK–2860 Soeborg
Denmark

I. Jacobson
Institute of Neurobiology
University of Gothenburg
S–400 33 Gothenburg
Sweden

F.F. Johansen
Institute of Neuropathology
University of Copenhagen
Denmark

P. Kalén
Biskopsgatan 5
S–223 62 Lund
Sweden

B.I. Kanner
Department of Biochemistry
Hadassah Medical School
Hebrew University
P.O. Box 1172
Jerusalem
Israel

J.S. Kelly
Department of Pharmacology
St George's Hospital Medical School
Cranmer Terrace
London SW17 0RE
UK

L. Kerkerian
CNRS INP 5
B.P. 71–13402 Marseille Cédex 9
France

A. King
Department of Pharmacology
St Bartholomew's Hospital Medical College
University of London
Charterhouse Square
London EC1M 6BQ
UK

T. Klockgether
Max-Planck-Institute for Experimental
    Medicine
Hermann-Rein-Strasse 3
D–3400 Göttingen
West Germany

C. Köhler
Department of Pharmacology
ASTRA Research Laboratories
Södertälje
Sweden

M. Koida
Department of Pharmacology
Setsunan University
Hirakata
Osaka 573–01
Japan

K. Koller
Division of Child Psychiatry
Departments of Psychiatry, Neuroscience,
    Pharmacology and Pediatrics
The Johns Hopkins University School of
    Medicine
Baltimore
Maryland 21205
USA

P. Krogsgaard-Larsen
Department of Chemistry BC
The Royal Danish School of Pharmacy
DK–2100 Copenhagen
Denmark

E. Kurylo
Department of Biochemistry
University of Saskatchewan
Saskatoon S7N 0W0
Canada

E. Kvamme
Neurochemical Laboratory
Preclinical Medicine
University of Oslo
P.O. Box 1115 Blindern
0317 Oslo 3
Norway

J.H. Laake
Anatomical Institute
University of Oslo
Karl Johansgate 47
N–0162 Oslo 1
Norway

J.D. Lane
Department of Pharmacology
Texas College of Osteopathic Medicine
Fort Worth
Texas 76107
USA

T.H. Lanthorn
Health Care Division
Monsanto Company
Chesterfield
MO 63198
USA

J. Lauridsen
Department of Chemistry BC
The Royal Danish School of Pharmacy
DK–2100 Copenhagen
Denmark

A. Lehmann
Institute of Neurobiology
University of Gothenburg
P.O. Box 33031
S–400 33 Gothenburg
Sweden

N. Leresche
Department of Pharmacology
St George's Hospital Medical School
Cranmer Terrace
London SW17 0RE
UK

G. Levi
Laboratory of Organ and System
  Physiopathology
Istituto Superiore di Sanità
Viale Regina Elena 299
Rome
Italy

D. Lodge
Department of Physiology
Royal Veterinary College
London NW1 0TU
UK

G. Lombardi
Department of Preclinical and Clinical
  Pharmacology
University of Florence
Viale G.B. Morgagni 65
50134 Firenze
Italy

A.M. López Colomé
Centro de Investigaciones en Fisiología
  Celular
UNAM
Apartado Postal 70–600
04510 Mexico City
Mexico

H. McLennan
Department of Physiology
University of British Columbia
Vancouver
British Columbia V6T 1W5
Canada

S. Madsen
Anatomical Institute
University of Oslo
Karl Johansgate 47
N–0162 Oslo 1
Norway

U. Madsen
Department of Chemistry BC
The Royal Danish School of Pharmacy
DK–2100 Copenhagen
Denmark

D.S. Magnuson
Department of Physiology
University of British Columbia
Vancouver
British Columbia V6T 1W5
Canada

L. Maier
Ciba-Geigy AG
CH–4002 Basel
Switzerland

A. Matus
Friedrich Miescher-Institut
P.O. Box 2543
CH–4002 Basel
Switzerland

C. Matute
Brain Research Institute
University of Zurich
August-Forel-Strasse 1
8029 Zurich
Switzerland

S. Mazzari
FIDIA Neurobiological Research
  Laboratories
Via Ponte della Fabbrica 3/A
35031 Abano Terme
Italy

B.S. Meldrum
Department of Neurology
Institute of Psychiatry
De Crespigny Park
Denmark Hill
London SE5 8AF
UK

L.M. Mello
Department of Neurology
Institute of Psychiatry
De Crespigny Park
Denmark Hill
London SE5 8AF
UK

E.K. Michaelis
Departments of Human Development and
  Biochemistry
and the Center for Biomedical Research
University of Kansas
Lawrence
Kansas
USA

M.H. Millan
Department of Neurology
Institute of Psychiatry
De Crespigny Park
Denmark Hill
London SE5 8AF
UK

D.T. Monaghan
Department of Psychobiology
University of California at Irvine
California 92717
USA

F. Moroni
Department of Preclinical and Clinical
  Pharmacology
University of Florence
Viale G.B. Morgagni 65
50134 Firenze
Italy

E. Myrseth
Department of Neurosurgery
University Hospital of Bergen
N–5016 Haukeland Sykehus
Norway

H. Nakamuta
Department of Pharmacology
Setsunan University
Hirakata
Osaka 573–01
Japan

W.J. Nicklas
Department of Neurology
UMDNJ–Rutgers Medical School
Piscataway
NJ 08854
USA

E.Ø. Nielsen
Department of Chemistry BC
The Royal Danish School of Pharmacy
DK–2100 Copenhagen
Denmark

M. Nielsen
Sct Hans Mental Hospital
DK–4000 Roskilde
Denmark

A. Nieoullon
CNRS INP 5
B.P. 71–13402 Marseille Cédex 9
France

A. Nistri
Department of Pharmacology
St Bartholomew's Hospital Medical College
University of London
Charterhouse Square
London EC1M 6BQ
UK

G. Nordbø
Anatomical Institute
University of Oslo
Karl Johansgate 47
N–0162 Oslo 1
Norway

K. Ogita
Department of Pharmacology
Setsunan University
Hirakata
Osaka 573–01
Japan

E. Okuno
Maryland Psychiatric Research Center
P.O. Box 21247
Baltimore
Maryland 21228
USA

F. Orrego
Department of Physiology and Biophysics
Faculty of Medicine
University of Chile
Santiago
Chile

O.P. Ottersen
Anatomical Institute
University of Oslo
Karl Johansgate 47
N–0162 Oslo 1
Norway

S. Patel
Department of Neurology
Institute of Psychiatry
De Crespigny Park
Denmark Hill
London SE5 8AF
UK

R. Paulsen
Division for Environmental Toxicology
Norwegian Defence Research Establishment
P.O. Box 25
N–2007
Kjeller
Norway

M.J. Peet
Department of Physiology
University of British Columbia
Vancouver
British Columbia V6T 1W5
Canada

J.B. Penney
University of Michigan
Neuroscience Laboratory Building
1103 East Huron
Ann Arbor
Michigan 48104
USA

J.-P. Pin
Centre CNRS–INSERM de Pharmacologie-
    Endocrinologie
B.P. 5055
34033 Montpellier
France

M. Pirchio
Department of Pharmacology
St George's Hospital Medical School
Cranmer Terrace
London SW17 0RE
UK

M. Pritzel
Biskopsgatan 5
S–223 62 Lund
Sweden

R. Radian
Department of Biochemistry
Hadassah Medical School
Hebrew University
P.O. Box 1172
Jerusalem
Israel

M. Recasens
Centre CNRS–INSERM de Pharmacologie-
    Endocrinologie
B.P. 5055
34033 Montpellier
France

N. Riveros
Department of Physiology and Biophysics
Faculty of Medicine
University of Chile
Santiago
Chile

P.J. Roberts
Department of Physiology and Pharmacology
University of Southampton
Southampton SO9 3TU
UK

S. Roy
Departments of Human Development and
    Biochemistry
and the Center for Biomedical Research
University of Kansas
Lawrence
Kansas
USA

J.-F. Rumigny
Laboratoires Delalande
92500 Rueil-Malmaison
France

T.E. Salt
Department of Physiology
University College
P.O. Box 78
Cardiff CF1 1XL
UK

M. Sandberg
Institute of Neurobiology
University of Gothenburg
S–400 33 Gothenburg
Sweden

M.S.P. Sansom
Department of Zoology
The University of Nottingham
Nottingham NG7 2RD
UK

A. Schousboe
Department of Biochemistry A
Panum Institute
Blegdamsvej 3
DK–2200 Copenhagen
Denmark

R. Schwarcz
Maryland Psychiatric Research Center
P.O. Box 21247
Baltimore
Maryland 21228
USA

M. Schwarz
Max-Planck-Institute for Experimental
    Medicine
Hermann-Rein-Strasse 3
D–3400 Göttingen
West Germany

R.P. Shank
Department of Biological Research
McNeil Pharmaceutical, Spring House
PA 19477–0776
USA

D.A.S. Smith
Department of Physiology
St George's Hospital Medical School
Cranmer Terrace
London SW17 0RE
UK

K.-H. Sontag
Max-Planck-Institute for Experimental
    Medicine
Hermann-Rein-Strasse 3
D–3400 Göttingen
West Germany

T.W. Stone
Department of Physiology
St George's Hospital Medical School
Cranmer Terrace
London SW17 0RE
UK

T.M. Stormann
Departments of Human Development and
    Biochemistry
and the Center for Biomedical Research
University of Kansas
Lawrence
Kansas
USA

J. Storm-Mathisen
Anatomical Institute
University of Oslo
Karl Johansgate 47
N–0162 Oslo 1
Norway

P. Streit
Brain Research Institute
University of Zurich
August-Forel-Strasse 1
8029 Zurich
Switzerland

V.I. Teichberg
Department of Neurobiology
The Weizmann Institute of Science
Rehovot 76100
Israel

U. Tossman
Department of Pharmacology
Karolinska Institute
P.B. 60400
S–10 401 Stockholm
Sweden

L. Turski
Department of Neurology
Institute of Psychiatry
De Crespigny Park
Denmark Hill
London SE5 8AF
UK

and

Max-Planck-Institute for Experimental
   Medicine
Hermann-Rein-Strasse 3
D–3400 Göttingen
West Germany

W.A. Turski
Max-Planck-Institute for Experimental
   Medicine
Hermann-Rein-Strasse 3
D–3400 Göttingen
West Germany

and

Wander Research Institute
(a Sandoz Research Unit)
P.O. Box 2747
CH–3001 Berne
Switzerland

T. Ueda
Mental Health Research Institute and the
   Departments of Pharmacology and
   Psychiatry
The University of Michigan
Ann Arbor
Michigan 48109
USA

U. Ungerstedt
Department of Pharmacology
Karolinska Institute
P.B. 60400
S–10 401 Stockholm
Sweden

P.N.R. Usherwood
Department of Zoology
The University of Nottingham
Nottingham NG7 2RD
UK

J.C. Watkins
Pharmacology Department
The Medical School
University Walk
Bristol BS8 1TD
UK

R.J. Wenthold
Laboratory of Neuro-otolaryngology
NINCDS
National Institutes of Health
Bethesda
MD 20205
USA

E. Westerberg
Laboratory for Experimental Brain Research
University Hospital
University of Lund
Sweden

T. Wieloch
Laboratory for Experimental Brain Research
University Hospital
University of Lund
Sweden

L. Wiklund
Department of Histology
Biskopsgatan 5
S–223 62 Lund
Sweden

J.D. Wood
Department of Biochemistry
University of Saskatchewan
Saskatoon S7N 0W0
Canada

Y. Yoneda
Department of Pharmacology
Setsunan University
Hirakata
Osaka 573–01
Japan

A.B. Young
University of Michigan
Neuroscience Laboratory Building
1103 East Huron
Ann Arbor
Michigan 48104
USA

R. Zaczek
Division of Child Psychiatry
Departments of Psychiatry, Neuroscience,
   Pharmacology and Pediatrics
The Johns Hopkins University School of
   Medicine
Baltimore
Maryland 21205
USA

Contents

Preface and Acknowledgements

The excitatory amino acids, and L-glutamate in particular, are now widely considered to play a dominant role in central synaptic transmission. From electrophysiological and binding studies, coupled with elaborate structural activity studies of receptor function, something of the complexities of the receptor(s) is beginning to be interpreted. Similarly, at the presynaptic end of excitatory amino acid transmitter function, immunocytochemistry, retrograde and anterograde labelling, studies of compartmentalisation and turnover of amino acids, mechanisms of uptake and release, etc are enabling us to build up a more complete understanding of these substances in excitatory synaptic transmission.

In addition to their physiological role as transmitters, there is growing evidence that they might also be implicated in a whole range of pathological states, spanning from epilepsy and ischaemia, to the neurodegenerative disorders such as Huntington's disease and senile dementia of the Alzheimer type. Although the crucial evidence for the direct involvement of excitatory transmitters in the aetiology of any of these disorders is still lacking, the discovery of selective antagonists at excitatory amino acid receptor subtypes, and their clear efficacy as anticonvulsants, and probably also as a valuable adjunct in the treatment of stroke, points the way to the impending novel therapeutic importance of drugs developed so as to interact specifically with these receptors. This book is a series of contributions based upon a satellite symposium of the International Society for Neurochemistry, held in Bournemouth, England during the summer of 1985. The chapters are intended to provide a review of the field that is likely to be of value both to non-experts and to workers who are just starting in the field.

We are greatly indebted to a number of organisations for their generous support: The International Society for Neurochemistry; Hoffmann-La Roche Inc, Nutley, N.J., USA; The Wellcome Trust, London; Ciba-Geigy, Basel, Switzerland; ICI plc, Alderley Park; Astra Läkemedel AB, Södertälje, Sweden; Wander Ltd, Berne, Switzerland; Merck Sharp & Dohme Neuroscience Research Centre, Harlow; Stuart Pharmaceuticals, Wilmington, USA; The Wellcome Research Laboratories, Beckenham; The Boots Co plc, Nottingham; Beecham Pharmaceuticals Ltd; Lilly Research Centre, Windlesham; Glaxo Group Research Ltd.

Southampton, Oslo and London, 1986

P.J.R.
J.S.-M.
H.F.B.

1

Twenty-five Years of Excitatory Amino Acid Research

J.C. Watkins

INTRODUCTION

Rather than attempt an overview of the whole field (which, in any case, is comprehensively and much more eloquently provided by the rest of this volume) I am taking the opportunity offered by the nature of the occasion to review those aspects of the subject with which my colleagues and I, over a period of twenty seven years, have been most closely associated. In chronicling our endeavours over these years, our failures as well as successes, some trails leading nowhere, others to completely unexpected and important findings, I hope to provide some indication, particularly to the young, of the true nature of scientific research, which rarely flows as smoothly as polished publications might suggest.

My personal involvement in neuroscience began in late 1956 when, soon after gaining my Ph.D. in organic chemistry, I decided against continuing a career in chemistry for chemistry's own sake, and wished instead to apply my specialist training in the biological sphere. Brain chemistry seemed to me to be a relatively untapped possibility. Indeed, the Journal of Neurochemistry was in its first year. My decision to enter this field was not taken lightly; it would inevitably constitute a major professional leap for me, since I had had absolutely no training in, or even minimal exposure to, any of the biological sciences, not even biochemistry, let alone electrophysiology. Undaunted, if somewhat apprehensive, I contacted Professor J.C. Eccles, at the Australian National University, Canberra (whose name and reputation were conveyed to me by medical acquaintances, I being ignorant at the time of practically all 'non-chemical' research), explaining my background and hopes, and was most fortunate to be rewarded by a letter of encouragement from him, and the offer of a place in his Department. This, of course, I accepted with alacrity and gratitude. It was a decision I have never had cause to regret.

I was to work with a certain D.R. Curtis who was at that time

developing for use in the central nervous system the new microelectrophoretic technique of pharmacological testing as pioneered
by Nastuk (1953) and Del Castillo & Katz (1955). My brief was to
extract ox brain in a search for new transmitters, isolating and
purifying as many constituents as I could, and then to supply them
to David Curtis who would test them microelectrophoretically on
single neurones of the cat spinal cord. At this time, acetylcholine, the biogenic amines and a crude peptoid extract termed
Substance P, were considered possible transmitters, but definitive
evidence was lacking.   The consensus was that we would probably
need to identify new candidates. Accepting this assignment with a
certain degree of feigned enthusiasm (for well I understood the
magnitude of the task), I ventured to suggest, that while I was engaged in this immensely laborious (and possibly futile) effort,
perhaps Dr. Curtis and his graduate student, John Phillis, might
start testing many of the *known* brain constituents that were already
available commercially or which could be obtained as gifts from
other investigators.   After all, these substances would certainly
be present in my extracts, and we ought therefore to learn of any
activity they might have before testing perhaps the same substances
as (possibly unidentified) components of my extracts.   This
suggestion accepted, the first electrophoretic tests with a range
of brain constituents, including L-glutamate, were made in January,
1958, by Curtis and Phillis, with myself in the role of keen ob-
server and occasional switch operator.

DISCOVERY OF "EXCITATORY AMINO ACIDS"

     There were several reasons for choosing L-glutamate as one of
the first substances to be tested.   Most importantly, it was known
to be one of the major constituents of brain, but also, it was
structurally and metabolically related to γ-aminobutyric acid (GABA)
which was already known at this time to have central depressant
activity.   At this stage we were unaware of the findings of Hayashi
(1954) who, several years before in work published in Japanese
journals, had reported that both L-glutamate and L-aspartate produced convulsions when directly injected into monkey and dog brains
or when administered via carotid injection.   To Hayashi, therefore,
must go the credit for first reporting an overt excitatory effect
of L-glutamate in the mammalian central nervous system, though at
the time, whether this was a direct or indirect effect was not
known.

     Our first results actually led us to the temporary conclusion
that L-glutamate and L-aspartate were both inhibitory.   Thus, Fig 1
shows the abolition of negative field potentials generated by the
activity of populations of spinal neurones following stimulation of
fibres in spinal roots.   This result, if somewhat more pronounced
than we had anticipated, was nevertheless not unexpected; after all,
glutamate (particularly) and aspartate can both be regarded as
GABA analogues!   Indeed, two papers were soon to appear describing
the action of glutamate and like compounds in producing spreading
depression in cerebral cortex (Van Harreveld, 1959; Purpura,

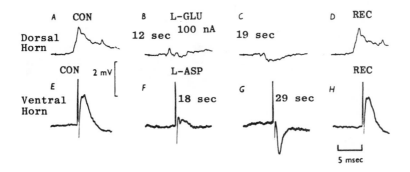

Fig 1.   Focal potentials recorded within the seventh lumbar segment of the cat spinal cord by double-barrel electrodes.   A-D recorded in the dorsal horn and evoked by orthodromic activation of neurones following maximum stimulation of the peroneal nerve.   B and C show depressant effects of L-glutamate (100 nA) at the times indicated after beginning the iontophoretic administration.   D shows recovery recorded 12 s later.   E-H, from another experiment, recorded in the ventral horn in response to maximum stimulation of the segmental ventral root. F & G show depression by L-aspartate (100 nA), at the times indicated, of the focal potential generated by the antidromic activation of motoneurones.   H, recovery, 20 s later. (Curtis *et al*, 1960).

Girado, Smith, Callan & Grundfest, 1959), though this effect was clearly different from the depression of evoked responses produced by GABA.

The (very different) truth of the matter was discovered in a somewhat round about way.   The electrophoretic method, though extremely elegant, was also very time consuming, and there was the added worry that testing brain constituents on just a few individual cells could result in activity being missed.   Moreover, the method was not suitable for testing the activity of crude brain extracts. I therefore asked David Curtis if he could suggest any preparation which could provide a rapid assessment of activity of either an excitatory or inhibitory kind, and which would be just as suitable for crude brain extracts as for pure substances.   He did indeed know of such a preparation - the isolated hemisected spinal cord of the frog or toad.   This preparation had first been used by Barron & Matthews (1938) and, more latterly, even by Eccles (1946). Ironically   (and unbeknownst to us), it had also been used to test the action of amino acids, and dramatic effects had even been observed, but could not be reliably reproduced (Brooks, Ransmeier & Gerard (1949).

John Phillis was mainly responsible for our experiments on the toad spinal cord *in vitro*.   In the first week of testing, during April 1958, we tested over 60 brain constituents, of which some 26 showed significant activity of one sort or another.   The excitatory

action of L-glutamate (the first substance tested) (Fig 2), and
also that of L-aspartate (tested the same day), and the inhibitory
actions of GABA, β-alanine and glycine, all tested within the first
week, were the most remarkable.   These substances therefore formed
the focus of our attention thereafter, unfortunately, in the process,
causing us to neglect many other active substances, including
dopamine (tested at the same time and shown to be moderately inhibitory).   My records of toad cord experiments conducted in 1958
indicate that of a total of around 120 substances tested (including
about 80 known or likely brain constituents), more than 50% showed
significant excitatory or inhibitory activity (Table 1).   We were
thus somewhat embarrassed by riches but decided to concentrate on the
amino acids in the first instance (as it turned out, quite a prolonged 'first instance' - now mid-way through its 28th year!).

Many of the substances of Table 1 have not, even yet, previously been reported as neuroactive.   Clearly, confirmation of their
action is required before any significance can be attached to these
results, many of which were from single experiments only.   They are
presented in the hope that, if, as seems likely, time does not permit
me ever to repeat the experiments, the results may nevertheless
stimulate a re-investigation by others.   Some of the comments in
Table 1 have, of course, benefitted from hindsight.

The results of the toad cord experiments (Curtis, Phillis &

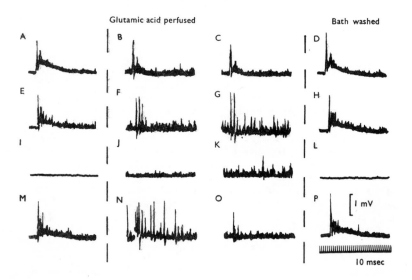

Fig. 2   Action of L-glutamic acid on potentials recorded from a ventral root following
stimulation of the corresponding dorsal root (A-H) and on spontaneous ventral root activity
(I-L) in the isolated hemisected spinal cord of the toad.   A, E, I, controls, B, F, J 10 sec
and C, G, K, 30 sec after addition of L-glutamate 0.5, 5 and 5 mM, respectively, to the superfusion medium.   Note general increase in reflex and spontaneous activity, but decrease in
initial phase of the DR-VRP induced by L-glutamate (Curtis *et al*, 1961).

Table 1   Results of screening tests in the toad spinal cord (1958)

A.   Substances producing excitation.

Date	No.	Substance	Action	Comment

1. __Amino acids__

Date	No.	Substance	Action	Comment
10.4.58	1	L-Glutamic acid	+ + +	Reported (Reference 1). Now well documented excitatory amino acid action; turns to inhibition if prolonged or severe ("overdepolarization")
10.4.58	6	L-Aspartic acid	+ + +	See 1 (this section)
15.4.58	31	DL-Asparagine	+ + +	Reported (ref. 1). Mechanism of action still unknown. Possibly NMDA receptor mediated, since less evident in *in vivo* studies (higher $[Mg^{2+}]_e$)
15.4.58	32	L-Glutamine	+	See 31 (this section)
15.4.58	34	DL-Serine	+ (+)	See 31 (this section)
17.4.58	68	L-Cysteic acid	+ + +	See 1 (this section)
5.5.58	69	D-Aspartic acid	+ + +	See 1 (this section)
18.6.58	89	D-Serine	+ +	See 31 (this section)
18.6.58	90	L-Serine	+	See 31 (this section)
18.6.58	94	L-Asparagine	+ (+)	See 31 (this section)
18.6.58	95	D-Asparagine	+ + +	See 31 (this section)
18.6.58	97	D-Glutamic acid	+ + +	See 1 (this section)
18.6.58	98	β-Hydroxyglutamic acid	+ + +	See 1 (this section)
18.6.58	99	O-Acetyl-DL-serine	(+)	*Not reported. Needs verification. Mechanism not apparent.
18.6.58		O-Acetyl-D-serine	+	*See 99 (this section
3.12.58	115	Aminomalonic acid	+ + +	See 1 (this section)
3.12.58	116	N-Acetylglutamic acid	+	*Not reported. Needs verification. May be related to recent observations of Coyle and colleagues (refs. 2, 3)

2. __Other substances.__

Date	No.	Substance	Action	Comment
10.4.58	21	Acetylcholine	+ (+)	Reported (ref. 4) Assumed activation of cholinergic synapses.
15.4.58	24	Citric acid	+ (+)	Not reported, later investigated as part of study on calcium chelators (ref. 5)
15.4.58	38	Succinic acid	+	See 99 (Section A1)
15.4.58	50	Nicotinamide	+	*See 99 (Section A1)
		DL-Citrulline	(+)	*See 99 (Section A1)
17.4.58	66	L-Malic acid	+	See 31 (Section A1)
5.5.58	70	O-Phosphocholine	+	*Not reported. ? Cholinomimetic action
5.5.58	71	DL-Pantothenic acid	(+)	*See 99 (Section A1)
5.5.58	82	Succinylcholine	+	See 21 (this section)
5.5.58	87	Carbachol	+ + +	See 21 (this section)
5.5.58	88	Acetyl-β-methyl choline	+ + +	See 21 (this section
18.6.58	91	D-Malic acid	+ + +	See 31 (Section A1)
11.7.58	102	DL-Kynurenine	+	*See 99 (Section A1). Convulsant action recently reported (ref.6)
22.7.58	103	Spermidine	+ +	*See 99 (Section A1)
22.7.58	105	L-Leucylglycine	+	See 99 (Section A1)
22.7.58	107	Isoniazid	+ +	*Not reported. Slowly developing action. Likely to be due to GAD inhibition (see ref. 7)
22.7.58	108	p-Chloromercuribenzoate	+ +	*Not reported. Slowly developing. Likely to be due to inhibition of amino acid uptake (ref. 8)
22.7.58	110	Nicotine	+ +	See 21 (this section)
18.8.58	114	Benziminizole	+	*See 99 (Section A1)

Table 1 (continued)

A.   Substances producing inhibition.

Date	No.	Substance	Action	Comment
1. Amino acids				
10.4.58	4	N-Phosphocreatine	– (–)	*Not reported. Mechanism unknown.
10.4.58	5	O-Phospho-DL-serine	– (–)	*Not reported. Action may be related to that of APB (ref. 9)
10.4.58	7	β-Alanine	– – –	Reported (ref. 1) Now well documented, inhibitory amino acid action.
10.4.58	8	(±)-β-Amino-n-butyric acid	– –	See 7 (this section)
10.4.58	9	Taurine	– – –	See 7 (this section)
10.4.58	10	L-Arginine	– (–)	*Not reported, needs verification. Mechanism uncertain
15.4.58	40	Glycine	– –	See 7 (this section)
15.4.58	41	Glycylglycine	– (–)	*Not reported. Needs verification. Probably GABA-like action
17.4.58	44	GABA	– – –	See 7 (this section)
17.4.58	47	Sarcosine	– (–)	See 7 (this section)
17.4.58	48	Glycocyamine	– –	Reported (ref. 1) Probably GABA-like action
17.4.58	58	DL+allo-cystathionine	– –	*Not reported. Needs verification. This type of structure now known to have excitatory amino acid antagonist activity.
5.5.58	79	N-Acetyl-DL-alanine	(–)	*Not reported. Needs verification, mechanism unknown.
5.5.58	80	ε-Aminocaproic acid	–	See 7 (this section)
5.5.58	86	δ-Aminolevulinic acid	– (–)	*Not reported. GABA-like action of this substance since established (ref. 10)
18.6.58	92	δ-Aminovaleric acid	– –	See 7 (this section)
18.6.58	93	(±)-2,4-Diaminobutyric acid	–	*Not reported. Probably GABA-like action, plus possible inhibition of GABA uptake.
18.6.58	96	D-Alanine	– (–)	See 7 (this section)
11.7.58	101	L-Alanine	–	See 7 (this section)
3.12.58	117	DL-α-Methylglutamic acid	–	*Not reported. Later reported to be a excitatory amino acid antagonist (ref. 11)
3.12.58	120	β-Phenyl-β-alanine	–	*Not reported. ?β-Alanine or baclofen-type action.
2. Other substances				
10.4.58	19	5-Hydroxytryptamine	– (–)	*Not reported. Inhibitory effects of biogenic amines now well documented. (see ref. 12)
15.4.58	22	Choline	– –	*Not reported. ?Anticholinergic action.
17.4.58	67	3-Hydroxytyramine (Dopamine)	– (–)	*Not reported. Inhibitory action of dopamine now well documented (see ref.12)
5.5.58	72	Thiamine	– –	*Not reported. Needs verification; mechanism unclear.
5.5.58	85	Tartaric acid	? –	*Not reported. Needs verification, doubtful importance.
22.7.58	109	Octan-2-ol	– –	*Not reported Needs verification. Tested because of enzyme inhibitory effects, but likely to be more a non-specific effect on neuronal membrane because of lipid solubility.
22.7.58	106	Phenylenediamine	– –	*Not reported, Needs verification. Mechanism not known, but tested because of inhibitory effects on GOT and GPT enzymes.

Table 1 (continued)

C. Substances without detected effect.

1. Amino acids

O-Phosphoethanolamine	DL-Threonine
Creatine	DL-Norleucine
L-Histidine	DL-Isoleucine
L-Tyrosine	DL-Norvaline
DL-DOPA[1]	L-Proline
DL-Ornithine	DL-Valine
L-Methionine	γ- Guanidinobutyric acid
Creatinine	Glycine ethyl ester
L-Lysine	β-Alanine ethyl ester
Glutathione	O-Acetyl-L-serine
L-Leucine	N-Acetyl-aspartic acid
DL-Phenylalanine	Ureidosuccinic acid
L-Cysteine	Ureidopropionic acid

2. Other substances.

Adenosine monophosphate	Nicotinic acid
Adenosine diphosphate	Lactic acid
Adenosine triphosphate	Ethanolamine
Noradrenaline[2]	Fumaric acid
Tyramine	Ketoglutaric acid
Histamine	Tricarballylic acid
Urea	β-Glycerophosphate
Biotin	Isocitrate
Glutaric acid	Nicotinamide
Betaine	Pyruvate
Guanidine	Carnosine
L-Ascorbic acid	Oxaloacetate
d-Glucosamine	Mandelic acid

* Significant action observed in initial screening test not reported at time and needs verification. Action may be indirect in some cases.

[1] L-DOPA now known to be a weak excitant (refs. 12,13,14)

[2] Result needs verification. Noradrenaline has depressant actions in other tissues (see ref. 12)

REFERENCES

1. Curtis, D.R., et al (1961) Br. J. Pharmac. 16 262-283
2. Zaczek, R., et al (1983) Proc. Natl. Acad. Sci (USA) 80 1116-1119
3. Koller, K.J. & Coyle, J.T. (1984) Eur. J. Pharmacol. 98 193-199
4. Kiraly, J.K. & Phillis, J.W. (1961) Br. J. Pharmac. 17 224-231
5. Curtis, D.R., et al (1960) J. Neurochem. 6 1-20
6. Lapin, I.P. (1981) Epilepsia 22 257-265
7. Watkins, J.C. (1967) Biochem. J. 102 14P
8. Curtis, D.R., et al (1970) Exp. Brain Res. 10 447-462
9. Foster, A.E., et al (1982) Brain Res. 242 374-377
10. Dichter, H.N. et al (1977) Brain Res. 126 189-195
11. Haldemann, S., et al (1972) Brain Res. 39 419-425
12. Krnjević, K. (1974) Physiol. Rev. 54 418-540
13. Krnjević, K. & Phillis, J.W. (1963) Br. J. Pharmac. 20 471-490
14. Davies, J., et al (1982) Comp. Biochem. Physiol. 72C 211-224

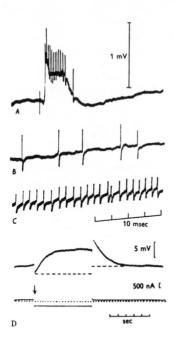

Fig. 3 A-C, Extracellular records showing re-
responses of an interneurone in the seventh
lumbar dorsal horn of the cat spinal cord
fired synaptically by a volley in the
peroneal nerve.   A, orthodromic response,
showing field potential on which are superimposed the repetitive spike discharges of a
single cell; B,C in the absence of afferent
stimulation, but 5 (B) and 10 (C) seconds
after beginning the iontophoretic administration of L-glutamate (120 nA).   D, Intracellular record from a gastrocnemius motoneurone
impaled by the central barrel of a co-axial
double-barrelled electrode showing depolarization (upper record) induced by L-aspartate
(520 nA) administered from the outer barrel
(Curtis *et al*, 1960).

Watkins, 1961) clearly indicated that the depressant actions of
glutamate and aspartate were preceded by, and, as soon realized,
actually caused by, a powerful initial excitatory effect.   When
tested by the ionophoretic[*] method on quiescent cells in the cat
spinal cord, L-glutamate and L-aspartate were clearly excitatory.
Intracellular studies by Curtis soon thereafter established the
membrane depolarizing effect of glutamate and aspartate in spinal
neurones and indicated the direct nature of the excitatory effect
(Fig 3).   The results of these experiments were published in a
letter to Nature in early 1959 (Curtis, Phillis & Watkins, 1959) and
full results a year later (Curtis, Phillis & Watkins, 1960).

It soon became apparent that glutamate and aspartate excited
all the different neuronal types in the spinal cord on which they
were tested.   This 'non-specificity', coupled with a) their extremely
high concentrations in the brain when only low concentrations of
transmitters were considered necessary, b) the need to administer
high concentrations (10^{-4} to 10^{-3}M) of the amino acids to show their
effect and c) our inability to find evidence for enzymic systems
capable of rapidly terminating their effects led us initially to
doubt that the acidic amino acids exercised a transmitter role.   We

_____

* It may be of interest that the term 'iontophoresis' rather than the more chemically acceptable 'ionophoresis' was apparently invented by editorial staff of an eminent Journal, who
returned the proofs of one of our early papers with the extra t consisently inserted.   The
term subsequently became adopted throughout the world, but now, 27 years later, the same
journal has begun to delete the t in manuscripts submitted to it.   Whether present staff of
that Journal are aware of the origin of the t I do not know.

did, however, consider that their action may underlie certain convulsive disorders, as indeed, did Hayashi, several years before.

While doubting a transmitter role of glutamate or aspartate we
thought is possible that a related substance may exercise that
function.   We therefore undertook extensive structure--activity
investigations to elucidate the molecular features responsible for
the effect.

EXCITATORY AMINO ACID RECEPTORS - FIRST CONCEPTS

The experimental work of the structure-activity study was
initially conducted by John Phillis using the isolated toad spinal
cord and then continued by David Curtis on single neurones of the
cat spinal cord using the ionophoretic method.   We first concentrated on the structural requirements necessary for effective
activation of the receptors, which we considered,through lack of
evidence to the contrary,to be of only one type.   The first
question that we addressed was the necessity or otherwise for the
presence of three charged groups (two acidic and one basic) in the
agonist molecule, and, by corollary, the probability or otherwise of
there being oppositely charged sites on the receptor.

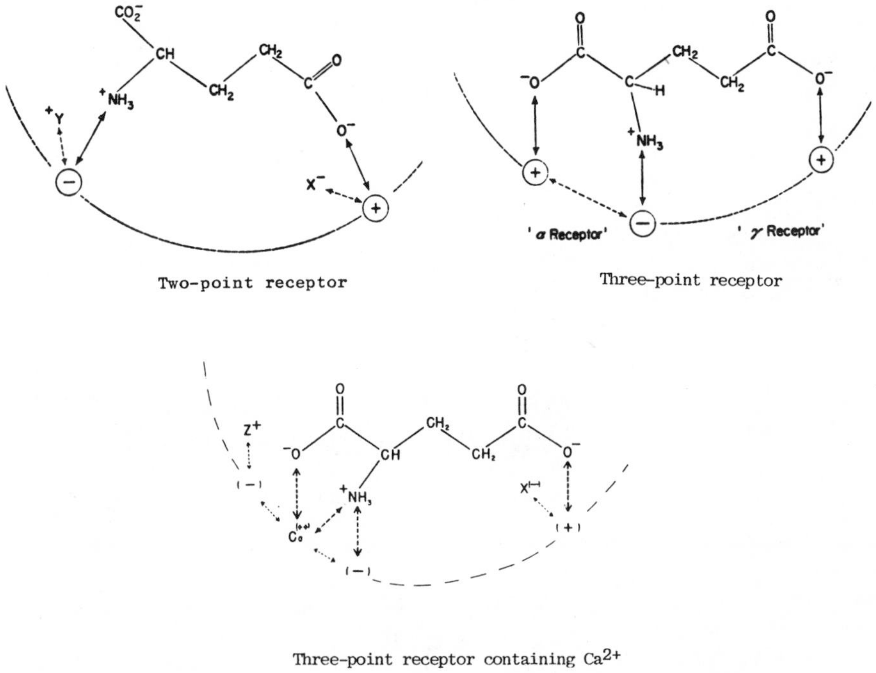

Two-point receptor   Three-point receptor

Three-point receptor containing Ca^{2+}

Fig. 4   Early receptor models (Curtis & Watkins, 1960, Curtis et al, 1961).

The actions in the toad spinal cord (Curtis, *et al*, 1961) of some substances which did not have two ionizable groups, for example, D-asparagine, and others which did not have an amino group, but an hydroxyl group instead (such as D-malic acid)* prompted us to consider a receptor structure that contained one or more divalent metal ions. Such metal ions could form co-ordinate bonds with electron donating groups of the excitant (these groups not necessarily capable of ionization in solution and perhaps including hydroxyl and amido groups) and one version of our model receptor actually contained Ca^{2+} (Fig 4c), We thought the interaction of the amino acid with the metal ion may have weakened the attachment of this ion to the receptor, and thus produced an effect similar (though not necessarily identical to) that which we found to be produced by Ca^{2+}-chelators (Curtis, Perrin & Watkins, 1960). In this latter case, our concept was that removal of free Ca^{2+} from the neuronal membrane by the Ca^{2+}-chelator resulted in a mass action release of Ca^{2+} from its sites of attachment, causing depolarization by increasing Na^+ permeability. It is still not known whether or not the same sites of the membrane are affected by Ca-chelation and amino acid excitants though the consensus of opinion is probably that different sites are involved, since calcium is considered to be an essential component of conducting membrane (e.g. Shanes, 1958).

With or without metal ions in the receptor, if the receptor had three active sites complementary to the three charged sites of the agonist (Fig 4b) one would expect the agonist to show stereoselectivity between D and L forms. Yet any stereoselectivity shown by D and L forms of the initial amino acids we tested in the spinal cord, namely, glutamate, aspartate and cysteate, was relatively slight. To accommodate this fact, we wondered if the receptor was of a "two-point" rather than "three-point" character, perhaps even the same receptor as that with which GABA interacted (Fig 4a) (Curtis & Watkins, 1960). In this case the difference in effect between GABA and, say, glutamate, would lie in the function of the additional free CO_2^- group of glutamate, which, not being bound to the receptor and perhaps protruding into a membrane pore, could conceivably mediate additional cationic transport through the membrane, thus causing depolarization. I would ask the reader to remember that, at this stage, no generally acceptable antagonists for GABA, which could easily have differentiated glutamate and GABA receptors, were yet recognized, though around this time it was reported that picrotoxin was a GABA antagonist in crustacean muscle, and had no action on glutamate-induced contractions in this tissue (Robbins, 1959).

Giving free rein to my speculative nature, at this stage I refined my 'general' model (Fig 4) into a 'special' model (Fig 5) (Watkins, 1965) which was, however, still general enough to accommodate both excitatory and inhibitory amino acids! I still

---

* Such actions are not evident in the cat spinal cord (Curtis & Watkins, 1960). In retrospect this may well be due to an action of the substances mainly on the NMDA receptor, the activity of which may be lower *in vivo* due to the presence of Mg^{2+} in extracellular fluid.

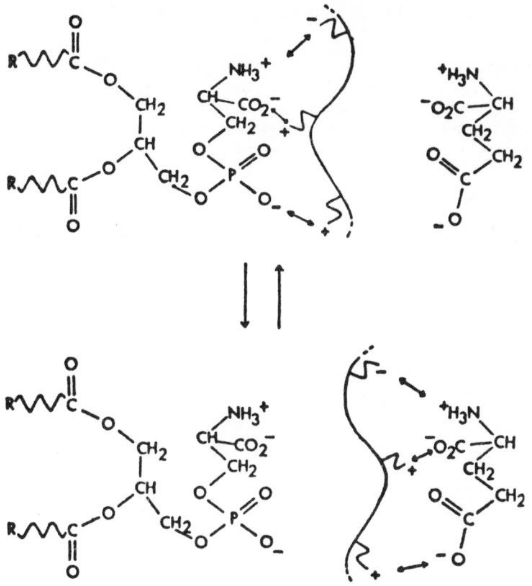

Fig. 5 Dissociation of an hypothetical phosphatidylserine-protein complex by glutamic acid. The charged phospholipid groups freed by the protein-amino acid complex formation were postulated to mediate ionic movements across the membrane (Watkins, 1965).

look upon this kind of receptor with some favour, especially in view of recent findings that the phosphonate group is an important moiety in the most active excitatory amino acid antagonists now known (e.g. Davies, Evans, Jones, Smith & Watkins, 1982) and the modicum of experimental support for the idea which has arisen over the years (Giambalvo & Rosenberg, 1976; Johnston & Kennedy, 1978; Foster, Fagg, Harris & Cotman, 1982).   Briefly, the idea was that the charge distribution in the molecules of GABA and glutamate are similar to those of the terminal residues of phosphatidylethanol-amine and phosphatidylserine, respectively.   If such phospholipids were complexed with proteins in the neuronal membrane, then GABA and glutamate may be able to dissociate these complexes, so freeing charged groups within the membranes which would then be available to mediate fluxes of ions across the membranes and cause the conduct-ance and/or potential changes underlying their neuroactive effects. This idea was born in 1960, but I did not decide to publish it until 1965, and only then because of the considerable amount of audience interest which it seemed to generate in the seminars I gave between these dates.

In parenthesis, it might be of interest to report here, 20 years after conducting the ex-periments, that in an attempt to gain experimental support for this lipoprotein receptor idea, I made some 'liposomes' (Bangham, Standish & Watkins, 1965; Papahadjopoulos & Watkins, 1967) from serum albumin and various phospholipids, filled them with radioactive ions and observed the effect on the efflux of these ions of the presence in the external medium of GABA or L-glutamate. High concentrations of these amino acids did induce an increase in the rate of appearance of $^{42}K^+$ in the medium (Fig 6), but the specificity of the effects was not investigated at the time, nor has been since.   The recent elegant experiments of Michaelis (this volume) emphasize the prematurity of my own endeavours in this area, but also that perhaps they might have been pursued to advantage.

It may also be worth reporting a more recent experiment where phosphatidylserine (PS) was continuously perfused over the isolated hemisected frog spinal cord, leading to decreased glutamate-induced depolarizations and dorsal root-evoked ventral root potentials. This would be predicted by the PS-protein receptor hypothesis, the PS possibly inactivating receptors by combining non-functionally with receptor protein, or simply competing favorably with the transmitter for active sites on the receptor protein.

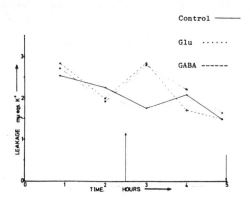

Fig. 6   Increased release of $^{42}K^+$ from phospholipid-protein liposomes by L-glutamate (14 mM) and GABA (28 mM) added to the superfusion medium (J.C. Watkins, unpublished experiments).

POTENT SYNTHETIC AND NATURALLY-OCCURRING EXCITANT AMINO ACIDS.

Whatever the nature of the molecular entities involved, choosing between a "two-point" and "three-point" receptor ultimately hinged to a considerable extent on discovering marked stereoselectivity between enantiomeric pairs of agonists.   While some stereoselectivity might be expected even for a two-point receptor, it would be almost mandatory for a three-point receptor.   Also, it would be expected that stereoselectivity would be more easily demonstrated the bigger the groups substituting the asymmetric carbon atom, thus increasing the likelihood that the binding of one enantiomer would be more hindered (or, in the case of hydrophobic binding by the bulky substituent, perhaps even more enhanced) than the binding of the other.   Therefore we synthesized N-alkyl D- and L-aspartic and glutamic acids, and immediately found the stereoselectivity we were looking for.   In one case, that of N-methyl-D-aspartic acid (NMDA) (Fig 7) we also found a large increase in potency compared with that of the parent substance which, at the time, we put down tentatively to increased affinity for the receptor,

but which, of course we now consider to be due predominantly to a
greatly reduced rate of uptake of NMDA compared with D-aspartate*.
Little did I realize at the time (1961) the importance that NMDA
would ultimately assume, and the six months it took me to devise a
satisfactory synthesis for this substance (Watkins, 1962) certainly
turned out to be a worthwhile investment of time and energy.

Around this time, Krnjevic & Phillis (1963) demonstrated the
excitatory actions of acidic amino acids in higher centres of the
CNS and in so doing, recorded some apparent differences in relative
potencies compared with spinal cord, though this was later refuted
by Crawford & Curtis (1964).   Several other investigators confirmed the 'non-specificity' of the excitatory action in the sense that
all cells tested in the mammalian CNS seemed to be sensitive to the
amino acids.   But, basically, between 1961 and 1965, there were
very few advances in the field (Curtis & Watkins, 1965).   Perhaps
the most novel finding was our discovery that a component of an
extract from the Indian plant *Lathyrus sativus*, which causes the
neurodegenerative disease, lathyrism, was highly potent as a
neuronal excitant (Watkins, Curtis & Biscoe, 1966).   This
substance was β-N-oxalyl-α,β-diaminopropionic acid (ODAP)
(Fig 7).   Other potent naturally occurring excitatory amino acids,

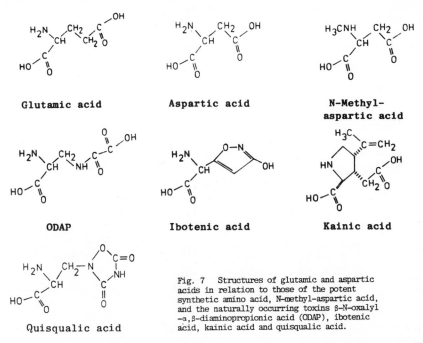

Glutamic acid   Aspartic acid   N-Methyl-
                                               aspartic acid

ODAP   Ibotenic acid   Kainic acid

Quisqualic acid

Fig. 7   Structures of glutamic and aspartic
acids in relation to those of the potent
synthetic amino acid, N-methyl-aspartic acid,
and the naturally occurring toxins β-N-oxalyl
-α,β-diaminopropionic acid (ODAP), ibotenic
acid, kainic acid and quisqualic acid.

* Thus, NMDA is not an inhibitor of the uptake of ^{3}H-L-glutamate (Balcar & Johnston, 1972a); is
not itself accumulated rapidly (Skerritt & Johnston, 1981); and does not have higher affinity for
the ^{3}H-D-AP5 binding site on brain membranes than does D-aspartate(H.J. Olverman & J.C. Watkins,
unpublished observations).

ibotenic acid (Johnston, Curtis, De Groat and Duggan, 1968) and
kainic acid (Shinozaki and Konishi, 1970) (Fig 7) were to be discovered in the near future and to assume important roles in studies
related to neurodegenerative diseases.

IONIC FLUXES

The mid 1960's saw the first use of mammalian CNS tissue in
excitatory amino acid studies *in vitro* with McIlwain introducing the
cerebral cortex slice for this purpose, as well as for related
metabolic studies (Bradford and McIlwain, 1966; Harvey and McIlwain
1968; Ramsay and McIlwain, 1970). Together with earlier and con-
current work on retina by Ames (1956, 1967) these studies
constituted the first demonstration of ionic distribution changes
accompanying excitatory amino acid-induced depolarization, notably
intracellular accummulation of Na^+ and Ca^{2+} and the release of intracellular K^+ into the extracellular medium.

Although the neuronal location of the accumulated Na^+ was not
definitely established, these studies nevertheless provided considerable support for the concept that the mechanism by which glutamate-
related amino acids caused depolarization was by inducing an
increase in neuronal membrane permeability to Na^+. They also
raised the possibility that Ca^{2+} influx may also play a role in such
actions. In addition, the painstaking development of this isolated
brain slice technique in the early 1960's indicated the potential
of the technique, which today is used so widely in electrophysiological, pharmacological and neurochemical studies.

EXCITOTOXICITY

Mention of the late 1960's must feature the introduction by
Olney (see Olney, 1978) of the excitotoxic story as the major event.
Following up a report that subcutaneous glutamate caused rapid
degeneration of retinal neurones in infant mice (Lucas and
Newhouse, 1957), Olney confirmed and extended these findings to
other amino acids including a range of novel substances obtained
from my old stocks (left behind in Canberra when I changed jobs in
1965) and established the connection between toxicity and excitatory
potency (Olney, Ho and Rhee, 1971). This finding can be linked in
importance with that of Shinozaki and Konishi (1970) in showing that
kainic acid, which is the anthelmintic principle present in extracts of the seaweed, *Digenea simplex,* was a very potent
neuroexcitant. Later, of course, kainic acid was shown to be more
potent as an excitotoxic agent than practically all other excitatory
amino acids tested to date, and to share with them the immensely
useful property of being toxic only when coming into contact with
cell bodies or dendrites (Olney, Rhee and Ho, 1974). Terminals
and fibres (axons en-passage) whose cell bodies and associated
dendrites were outside the area of kainic acid administration and/
or diffusion, were resistant to the toxic effect.

The importance of these findings was two-fold. Firstly they

allowed the investigation of biochemical, electrophysiological
and behavioural effects of the selective destruction of discrete
populations of cells, and secondly, they raised the possibility
that the etiology of certain neurodegenerative diseases involved
just such a toxic action of endogenous acidic amino acids.

COMPARTMENTED GLUTAMATE METABOLISM

Despite the fact that glutamate metabolism, especially in brain,
has been intensively studied for more than fifty years, interpreting
metabolic aspects of excitatory amino acid actions has been fraught
with difficulty, particularly in distinguishing those processes
which are associated with amino acid and/or energy metabolism generally from those which are unique to transmitter function.   This
difficulty arises because of the involvement of amino acids in
protein synthesis and breakdown and because of the close association
of glutamate and aspartate with intermediates of the tricarboxylic
acid cycle and thus with energy metabolism. The added complication
that discrete cellular reservoirs, or 'pools', of glutamate are
metabolised differently from each other was recognised in the early
1960's by Waelsch and collaborators and subsequently gave rise to a
whole new field of study under the name of "compartmentation".   A
specific metabolic pool was associated with the uptake of glutamate
from extracellular space and the formation of glutamine (Waelsch,
1961).   Avid uptake of glutamate had been reported by Krebs and
colleagues as early as 1949 (Stern, Eggleston, Hems and Krebs,1949).
However, it was the work of Johnston and colleagues in the early
1970's which focussed attention on uptake as a means of terminating
the action of excitatory amino acids and that differentiated such
uptake processes from the excitatory actions *per se* (Balcar and
Johnston, 1972a,b).   This latter distinction was important since
suggestions had been made that the depolarizations produced by
excitatory amino acids were perhaps a reflection of cellular uptake
processes in operation (Krnjevic and Phillis, 1963; Curtis and
Watkins, 1965).   A clearer (though still immensely complicated)
picture of the processes likely to be involved in the transmitter
function of excitatory amino acids began to emerge (Watkins, 1972).
A particular cycle that was considered (Scheme I) involved release
of glutamate from synaptic terminals following arrival of an action
potential, activation by the released glutamate of excitatory
postsynaptic transmitter receptors, uptake of the amino acid from
extracellular space into glia, conversion of the uptaken glutamate
into glutamine, passage of glutamine back into synaptic endings and
its re-conversion therein to transmitter glutamate.   Direct
evidence supporting such a cycle was later obtained by the groups
of Bradford and Cotman (Bradford, Ward and Thomas, 1978;   Cotman
and Hamberger, 1978).

EMERGING IDEAS OF RECEPTOR DIFFERENTIATION

Entering the 1970's, it would still have been readily admitted
that progress in understanding the role of excitatory amino acids
in mammalian central nervous function had been extremely limited

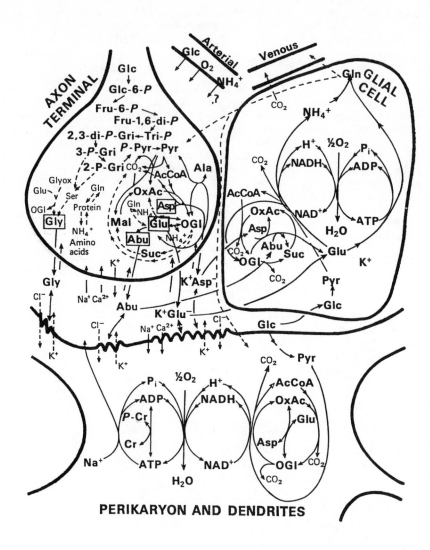

SCHEME I. Metabolic pathways involved in the release, action and uptake of putative amino acid transmitters in the central nervous system. Abbreviations: Glc and Glc-6-P, glucose and glucose 6-phosphate; Fru-6-P and Fru-1,6-di-P, fructose 6-phosphate and fructose 1,6-diphosphate; Tri-P, triose phosphates; 2-P-Gri, 3-P-Gri and 2,3-di-P-Gri, 2-phosphoglycerate, 3-phosphoglycerate and 2,3-diphosphoglycerate; Pyr and P-Pyr, pyruvate and phosphoenolpyruvate; AcCoA, acetyl-coenzyme A; OxAc, oxaloacetate; OGl, α-oxoglutarate; Suc, succinate; Mal, malate; Glyox, glyoxalate; Abu, γ-aminobutyrate; Asp, aspartate; Glu and Gln, glutamate and glutamine; Gly, glycine; Cr and P-Cr, creatine and phosphocreatine. The complicated interrelationships and proposed transcellular movements of metabolites are more clearly apparent in the original (coloured) representation of this scheme (Watkins, 1972).

though it was becoming increasingly well established, emanating from the release studies of Usherwood, that glutamate was probably the transmitter at invertebrate neuromuscular junctions (Usherwood, Machili and Leaf, 1968; Usherwood, 1981). In the previous decade the evidence that inhibitory amino acids were transmitters in both vertebrates and invertebrates had become overwhelming and indeed it had become firmly established that, in the mammalian CNS, at least two different inhibitory amino acids, glycine and GABA, probably functioned in this way (Curtis and Johnston, 1974).

Establishing the transmitter function of inhibitory amino acids and characterizing different receptor systems for these amino acids became possible primarily through the recognition of the specific antagonist properties of strychnine and bicuculline. (These antagonists also, of course, put paid to the old idea that inhibitory and excitatory amino acids might interact with the same receptors!) Similar antagonists had still not been found for excitatory amino acids despite a continuing search throughout the decade; consequently a transmitter function for such amino acids was still in doubt. Signs were, however, beginning to emerge that excitatory amino acids may also act on more than one receptor.

In retrospect, the first indications of this may have been the apparent differences in relative potencies of excitants in different regions of the CNS such as the cerebral cortex (Krnjevic and Phillis, 1963) and spinal cord (Curtis and Watkins, 1960), though the possibility of differential distributions of multiple types of receptors for excitatory amino acids was not mooted at that time. Later, McLennan and colleagues (1968) paid more attention to such regional differences in potency, while in the most thorough study, Duggan (1974) concluded that neurones in the dorsal horn of the spinal cord were significantly more sensitive to L-glutamate than to L-aspartate, while the reverse was true for neurones in the ventral horn. This latter work was followed up by Curtis, Johnston and colleagues (McCulloch *et al*, 1974) who proposed that NMDA and kainic acid acted predominantly on different types of receptors: one type, perhaps associated with a transmitter action of L-aspartate, was more prevalent in the ventral horn of the spinal cord, and the other type, perhaps associated with a transmitter action of L-glutamate, was more prevalent in the dorsal horn.

This work focussed attention on the need to test potential antagonists against different agonists. Indeed, some observations had already been made by this time that were to set the scene for more rapid advances later. Amongst a variety of substances proposed - perhaps not very convincingly - as excitatory amino acid antagonists, including lysergic acid diethylamide (LSD), methionine sulfoximine, α-methyl-DL-glutamic acid and 2-methoxyaporphine (see Curtis and Johnston, 1974) were L-glutamic acid diethyl ester (GDEE) and 3-amino-1-hydroxypyrrolid-2-one (HA-966 or HAP) (Fig 8), both of which later proved to be of considerable value. Thus, McLennan and colleagues (Haldemann and McLennan, 1972) noted that L-glutamic acid diethyl ester differentially depressed responses to

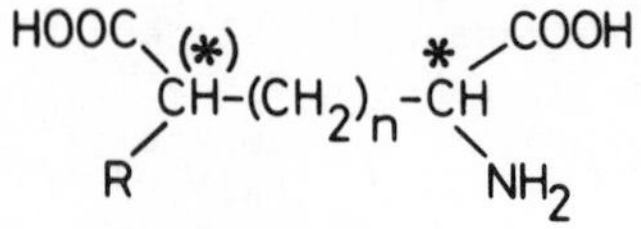

n	R	*	(*)	Compound	Abbreviation
2	H	D	–	D-α-AMINOADIPIC ACID	D-α-AA
3	H	D	–	D-α-AMINOPIMELIC ACID	D-α-AP
4	H	D	–	D-α-AMINOSUBERIC ACID	D-α-AS
3	NH_2	D	D or L	α,ε-DIAMINOPIMELIC ACID	α,ε-DAP

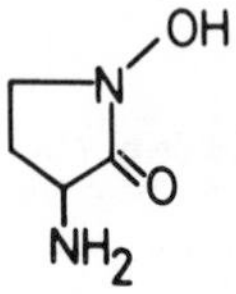

3-AMINO-1-HYDROXY-2-PYRROLIDONE
(HA-966 or HAP)

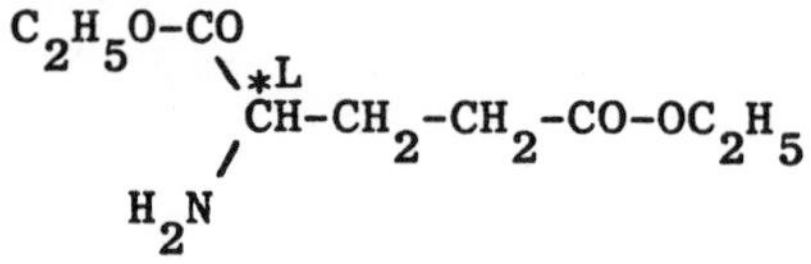

L-Glutamic acid diethyl ester
(GDEE)

Fig. 8   Structures of some amino acid antagonists.

excitatory amino acids, L-glutamate being slightly more sensitive
than L-aspartate and DL-homocysteate (DLH), while John Davies
and I (Davies and Watkins, 1973) formed the impression, without
critically investigating the phenomenon, that HA-966 was more
reliably effective against L-aspartate than against L-glutamate
induced excitations.   Moreover, there were indications in the
results of Curtis et al (1973) that DLH may be more susceptible
to antagonism by HA-966 than was either L-glutamate or L-aspartate.

At this time, in the early to mid-1970's, although we were
continuing to test large numbers of glutamate analogues as potential antagonists, we were also investigating other means whereby
excitatory amino acid-induced responses could be differentiated.
We did, in fact, obtain quite startling results when we varied the
monovalent cation composition of the medium perfusing our isolated
frog cords.   Thus, excitatory amino acids could be classified into
two main groups according to whether responses were (a) increased or
unchanged by decreasing the medium $[K^+]$, and (b) increased or decreased by low $[Na^+]$.   A third group of amino acids showed
intermediate responses (Evans, Francis and Watkins, 1977a).   The
two main groups were best exemplified by L-aspartate and L-
homocysteate, respectively.   L-Aspartate (and L-glutamate-)
induced responses were greatly enhanced in low $[K^+]o$, while

L-homocysteate-induced responses were relatively unaffected.   In
low [Na$^+$], responses to the two agonists were also affected
differentially.   As would be expected for any excitant whose
action was associated with an increase in membrane permeability to
Na$^+$, the actions of L-aspartate (and L-glutamate) were decreased.
However, responses to L-homocysteate were increased (Fig 9).   These
effects were later correlated with differences in ion-dependent
rates of uptake of the amino acids (Watkins, Evans, Headley, Cox,
Francis and Oakes, 1978), and did not, as we had first hoped,
necessarily reflect any fundamental differences in the mechanism of
action of the amino acids (though such differences were not, of
course, precluded).

   Definite, though still only partial, success in our attempts to
differentiate excitatory amino acid receptors came soon after in
related studies on amino acid uptake.   In one study we compared
time courses of recovery from the excitatory effects of amino
acids that were known to be actively taken up by CNS tissue with
those that were considered not to be subject to rapid uptake (Cox,
et al, 1977).   A case in point was constituted by the D and L isomers of homocysteate, where the L-isomer is selectively

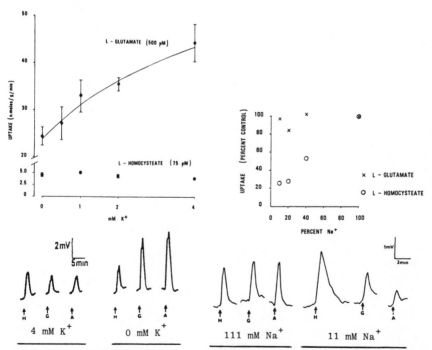

Fig 9. Effects of altered ionic composition of the medium on the uptake into the tissue and on
the responses produced by L-homocysteate, L-glutamate and L-aspartate in the frog spinal cord.
Top, dependence of uptake of [^{3}H]-L-glutamate (500 µM) and [^{35}S]-L-homocysteate (75 µM) on
medium [K$^+$] and [Na$^+$], respectively.   Bottom, motoneuronal depolarizations produced by the
same concentrations of L-glutamate and L-homocysteate (and also by 500 µM L-aspartate) in 4
(control) and 0 mM K$^+$, and in 111 (control) and 11 mM Na$^+$. Watkins et al, 1978 and
unpublished observations.

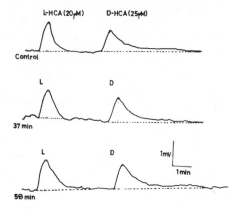

EFFECT OF ⏐-CMPS(10µM) ON FROG VENTRAL
ROOT RESPONSES TO L- AND D- HOMOCYSTEATE

Fig. 10 Prolongation of the time course of recovery of L-homocysteate induced depolarizations by the uptake inhibitor p-chloromercuriphenyl sulphonate (10 µM) in the isolated frog spinal cord.   Over a period of approximately one hour, the recovery time of the response produced by the L isomer was transformed into one resembling that of the response produced by the D isomer.   The latter response was unchanged by the uptake inhibitor (A.A. Francis and J.C. Watkins, unpublished observations).

accumulated by CNS tissue.   We had shown that para-chloromercuri-phenyl sulphonate, a known inhibitor of amino acid uptake (Curtis, Duggan and Johnston 1970), could transform the fast recovery from L-homocysteate-induced responses into the slow offset time characteristic of the D-isomer (Fig 10)(A.A. Francis and J.C. Watkins, unpublished observations).   Since chlorpromazine had also been reported to be an inhibitor of excitatory amino acid uptake (Balcar and Johnston 1972a) we therefore tried this drug in a similar type of experiment (Evans, Francis and Watkins, 1977b). However, we now found that L-homocysteate responses were greatly reduced in magnitude, this effect obscuring any prolongation of recovery.   When other agonists were tested, considerable variation was found in their sensitivity to the drug (Fig 11) and a ranking order of agonist susceptibility to depression by chlor-promazine was obtained which gained major significance only later. Interestingly, high concentrations of diazepam produced a similar range of effects.   On the other hand, pentobarbitone produced a quite different profile of depression which was almost the reverse of that shown by chlorpromazine and diazepam in that quisqualate, the least sensitive to chlorpromazine of the agonists tested, was the most sensitive agonist to the effects of pentobarbitone.   A similar selective effect of barbiturates on quisqualate-induced responses in striatal slices has recently been reported (Teichberg, Tal, Goldberg and Luini, 1984).

But these drugs all possess a variety of other properties

and were unlikely to prove useful in defining sites and/or modes of action of different excitatory amino acids.

NMDA RECEPTORS: Mg^{2+}-SENSITIVITY

Our first unequivocal success in receptor differentiation evolved from an extension of our studies on the effects of variations in the ionic composition of the frog Ringers solution when we investigated the effects of divalent metal ions (Evans, Francis and Watkins, 1977c; Ault, Evans, Francis, Oakes and Watkins, 1980). We found that addition of Mg^{2+} to the normally Mg^{2+}-free frog Ringers solution had a profound depressant effect on the depolarizing responses of some amino acids but not those of others; notably, responses to NMDA were the most markedly depressed while responses to other amino acids, including kainate and quisqualate, were but little, if at all, affected. (Fig 12)   L-Aspartate was more susceptible to this effect of Mg^{2+} than was L-glutamate.   The threshold concentration of Mg^{2+} was in the low micromolar range, and an IC_{50} of approximately 200 µM can be calculated (Fig 13). No effects of Mg^{2+} were observed on rates of amino acid uptake. Significantly, these low concentrations of Mg^{2+} also markedly reduced dorsal root-evoked ventral root potentials in the frog spinal cord (Fig 14) and the effect was clearly different from the well known inhibition of Ca^{2+} influx into pre-synaptic terminals brought about by high Mg^{2+} in a low-Ca^{2+} medium, which depresses synaptic release of transmitter at peripheral neuro-effector junctions

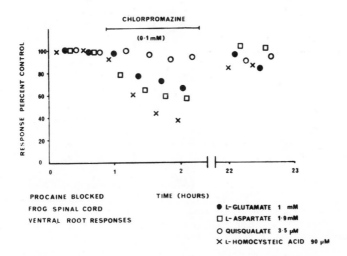

Fig. 11   Differential depression by chlorpromazine (0.1 mM) of frog motoneuronal depolarizations *in vitro* produced by L-glutamate, L-aspartate, quisqualate and L-homocysteate.   Control responses were carefully matched to the same magnitude.   The medium contained tetrodotoxin. (Evans, Francis & Watkins, 1977b).

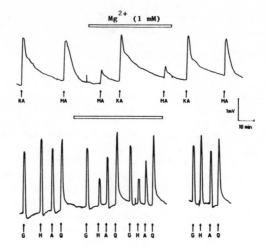

Fig. 12 Differential depression by Mg²⁺ (1 mM) of frog motoneuronal depolarizations *in vitro* produced by (upper sequence) kainate (KA) and N-methyl-D-aspartate (MA) and (lower sequence) L-glutamate (G), L- homocysteate (H), L-aspartate (A) and quisqualate (Q). Control medium was Mg-free and contained tetrodotoxin. (Ault *et al*, 1980)

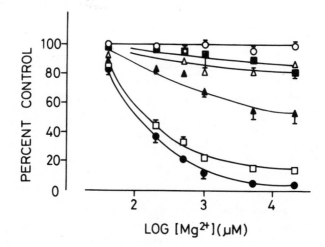

Fig. 13 Relation between Mg²⁺ concentration and extent of depression of frog motoneurone depolarization. Note the plateau level of depression achieved in each case. The IC$_{50}$ for the depressant effect on each agonist is approximately 200 μM. The medium contained tetrodotoxin. O, quisqualate; ■,kainate; △ , L-glutamate; ▲ , L-aspartate; ▢ , L-homocysteate; ●, NMDA. (From Ault *et al*, 1980)

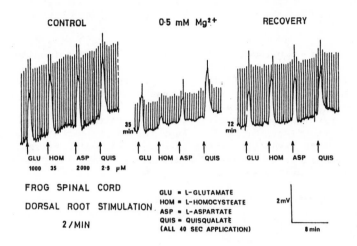

CONTROL   0·5 mM Mg²⁺   RECOVERY

GLU   HOM   ASP   QUIS   GLU   HOM   ASP   QUIS   GLU   HOM   ASP   QUIS
1000   35   2000   2·5 μM

FROG SPINAL CORD

DORSAL ROOT STIMULATION

2 / MIN

GLU = L-GLUTAMATE
HOM = L-HOMOCYSTEATE
ASP = L-ASPARTATE
QUIS = QUISQUALATE
(ALL 40 SEC APPLICATION)

2mV

8 min

Fig. 14   Depression by Mg²⁺(0.5 mM) of dorsal root-evoked ventral root potentials and of
amino acid-induced depolarizations in frog spinal cord (From Evans *et al*, 1977c).

(Del Castillo and Engbaek, 1954).   Indeed Ca^{2+} acted similarly
to Mg^{2+} but was some 20-fold weaker (Ault *et al*, 1980).

The simplest explanation of the Mg^{2+} effect was that there
existed at least two types of excitatory amino acid receptors -
one type activated by NMDA and susceptible to the depressant
action of micromolar concentrations of Mg^{2+}, and at least one
other type, which was activated by such other agonists as kainate
and quisqualate, and which was relatively insensitive to even
quite high (e.g. 20 mM) concentrations of Mg^{2+}.   Many agonists
appeared to act at both Mg-sensitive and Mg-insensitive sites and
a ranking order of agonist susceptibilities to Mg^{2+}, presumably
reflecting the relative contribution of the two types of receptors to the responses produced by each agonist, could be
constructed (Table 2).   This ranking order was similar to one
which had already emerged for chlorpromazine and diazepam and
which was soon to be reproduced by a plethora of much more useful
compounds, many of them newly synthesized in our laboratory.

NMDA-RECEPTORS: ORGANIC ANTAGONISTS

At this time (1976/7) the search for specific antagonists had
been continuing without spectacular success already for 18 years
or so, but, until recently, L-glutamate (especially), L-aspartate
and DL-homocysteate (DLH) had been the only agonists that had
been routinely tested against prospective antagonists. In addition

to these agonists we were now routinely using NMDA, kainate, and
the potent anthelmintic, quisqualate which had just recently been
discovered (Fig. 7) (Shinozaki and Shibuya, 1974; Biscoe *et al*, 1976).

Our first 'acceptable' excitatory amino acid antagonist devoid
of most other complicating actions, was (±)-α,ε-diaminopimelic acid
(DAP), originally suggested by my colleague Dr. Dick Evans as a
"bis-glycine" analogue of the anticholinergic bis-quarternary
ammonium compounds.   Results obtained with this compound, coupled
with those from a re-investigation of HA-966, gave further support
to the concept of two major receptor types first proposed on the
basis of the effects of Mg^{2+}.   When tested against a range of
agonists, both these substances gave an identical pattern to Mg^{2+}
(Fig 15 and Table 2) (Evans, Francis and Watkins, 1978 and unpublished observations).

To determine the minimum structural requirements necessary
for the DAP action (which was more readily susceptible to investigation than was structural modification of the HA-966 molecule),
we needed to test (a) substances of greater and shorter chain length
than DAP, (b) substances without either or both of the amino groups,

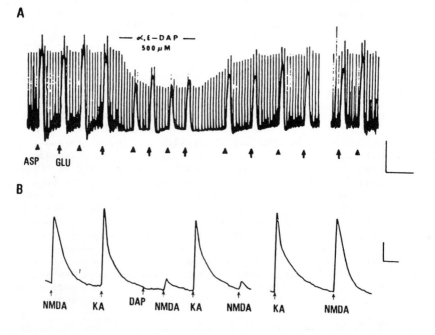

Fig. 15 Depression by α,ε-diaminopimelic acid of dorsal root-evoked ventral root potentials and
of amino acid-induced depolarizations in frog spinal cord. A. Responses measured in absence of
tetrodotoxin. B. Amino acid-induced depolarizations measured in presence of tetrodotoxin.
α,ε-DAP = α,ε-diaminopimelic acid, Asp = L-aspartate, Glu = L-glutamate, NMDA = N-methyl-D-
aspartate, KA = kainate.   Recovery responses recorded *ca* 25 and 40 min (A and B, repectively)
after wash out of α,ε-DAP.   Calibration (both records): 1 mV, 10 min. (From Evans *et al*, 1978).

Table 2 Depressant effects of Mg^{2+} and α,ϵ-diaminopimelic
acid (DAP) on depolarizing responses of frog
motoneurones to a range of excitatory amino acids[1].

Excitant	Percent control reponses in presence of	
	Mg^{2+} (1 mM)	α,ϵ-DAP (0.5 mM)
Quisqualate	101±3 (4)	101±4 (4)
Kainate	96±3 (4)	98±2 (4)
L-Glutamate	83±1 (5)	83±3 (4)
L-Aspartate	69±4 (4)	71±3 (4)
L-Homocysteate	22±1 (5)	34±3 (4)
NMDA	12±3 (3)	14±1 (4)

[1] Evans, Francis and Watkins, 1978, and unpublished observations.

and (c) pure stereoisomers; commercially available DAP, for
example, is a mixture of three isomers, one with the configuration
D (or R)* at both ends (i.e., the DD form); another with the L(S)
configuration at both ends (i.e., the LL form); and one with the
D configuration at one end and the L configuration at the other
(the so-called *meso* form) + (see Fig. 8).

By June, 1977, we were just completing a study of longer and
shorter chain analogues of DAP (all optically inactive mixtures,
i.e., the DL + *meso* forms). This study indicated that the activity
was highest in the case of (±)-DAP but shared to a considerable extent also by both the one-carbon shorter and one-carbon longer chain
compounds (2,4-diaminoadipic and 2,6-diaminosuberic acids, respectively). We were about to embark on phases (b) and (c) above of our
proposed structure-activity study when the effort we would need to
exert in this direction was greatly reduced by a report from
McLennan's group (Hall, McLennan and Wheal 1977), to the effect that
DL-α-aminoadipate was an antagonist of L-glutamate and L-aspartate
induced responses, whereas L-α-aminoadipate had L-glutamate-like
excitatory effects. This latter effect suggested that the weak
depolarizing action of the DL form that Curtis, Phillis and I had
reported 15 years previously (Curtis and Watkins, 1960; Curtis *et
al*, 1961) was due predominantly to the L form. More importantly,

* In the modern system of stereochemical nomenclature the D configuration of α amino acids becomes R, and L configuration becomes S. Clearly R and S are generally preferable to D and L,
but for amino acids, the old system has the advantage of allowing the non-chemical reader to
differentiate 'unnatural' (D) forms from the 'natural' (L) forms.

+ An equimolecular mixture of DD (RR) and LL (SS) forms comprises the DL (RS) form. Both
meso and DL forms are, of course, optically inactive.

the results of McLennan *et al.* implied that the D form possessed
excitatory amino acid <u>antagonist</u> properties, which, of course, they
clearly pointed out.   On the basis of our work with DAP and its
analogues, we considered that the excitatory amino acid antagonist
properties of the DL form of α-aminoadipate and, thus, probably of
the D form of this compound, were highly likely to resemble the NMDA
selective properties of the diamino substances that we had already
defined (but not yet published), and, if so, that this was likely to
identify the D-configuration of an α-amino-α-carboxylic acid terminal as being a key structural feature of such NMDA-selective
antagonists.   We first confirmed the predicted NMDA-selective
nature of the antagonist action of DL-α-aminoadipate (Biscoe, Davies,
Dray, Evans, Francis, Martin and Watkins, 1977).   Then, as
D-α-aminoadipate (DαAA) was not available commercially, we set about
resolving the DL form, as well as the DL forms of higher homologues,
and confirmed the NMDA-selective action of the D enantiomers of this
whole series of compounds, the most potent substance being a higher
homologue of DαAA, D-α-aminosuberate (DαAS) (Fig 8) (Biscoe, Evans,
Francis, Martin, Watkins, Davies and Dray, 1977;   Evans *et al,* 1978;
Evans, Francis, Hunt, Oakes and Watkins, 1979).   In the case of
α,ε-diaminopimelic acid, highest activity was found with the D
configuration at both asymmetric centres, i.e., with the DD form.

The discovery of the antagonists α,ε-diaminopimelate, D-α-
aminoadipate and like substances, was undoubtedly a major turning
point in excitatory amino acid research.   Not only were these
substances highly specific for excitatory amino acid receptors,
being inactive in respect of responses produced by all other depolarizing substances tested (Evans and Watkins, 1978), but, in addition, were selective with respect to different excitatory amino acids
themselves.   Most importantly, they also selectively depressed
synaptic excitation, as exemplified by their depressant action on
the excitation of Renshaw cells following dorsal root stimulation,
without affecting the cholinergic excitation of these cells following
ventral root stimulation (Fig 16) (Biscoe, *et al* 1977).

For the first time - after almost 20 years - it was thus
possible for us to conclude fairly confidently that at least some
excitatory synaptic pathways, particularly in the spinal cord, were
indeed "amino acid-ergic".

NON-NMDA RECEPTORS

In contrast to the previous 20 years, progress since 1978 has
indeed been quite rapid, particularly in relation to the development of yet more potent and specific NMDA antagonists and also in
defining amino acid-ergic pathways and identifying the particular
type of amino acid receptors involved.   Thus, where in early 1977,
only two organic NMDA-antagonists (DAP and HA-966) were known,
now more than 50 such compounds have been recognised (Watkins and
Evans, 1981; Davies *et al,* 1982).   The dicarboxylic amino acids
DαAA and DαAS have been superseded as the most useful NMDA antagonists by substances in which the ω-carboxyl terminal has been

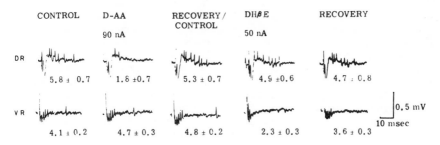

Fig. 16   Selective synaptic depressant actions of D-α-aminoadipate (D-AA) and dihydro-β-erythoidine (DHβE) on dorsal root (DR)- and ventral root (VR)-evoked excitation of a cat Renshaw cell.   The records show representative oscilloscope sweeps of the synaptic responses of the same cell to a constant DR or VR stimulus.   The numbers beneath each trace represent the numbers of spikes (mean±sem) in 15 responses and show, from left: control responses;   responses observed during the iontophoretic administration of D-AA (90 nA for 5 min);   recovery, 2 min after terminating D-AA ejection, these responses serving as controls prior to DHβE ejection;   responses observed during iontophoretic administration of DHβE (50 nA for 2 min);   recovery, 3 min after terminating DHβE ejection.   D-AA and DHβE specifically reduced the DR and VR responses, respectively. (From Biscoe *et al*, 1977).

replaced in each case by the phosphono group, leading to the antagonists D(-)-2-amino-5-phosphonopentanoate (D-AP5) and D(-)-2-amino-7-phosphonoheptanoate (D-AP7) (Davies, Francis, Jones and Watkins, 1981;   Evans, Francis, Jones, Smith and Watkins, 1982; Perkins, Collins and Stone, 1982) and two   dipeptide analogues β-D-aspartyl aminomethyl phosphonate (ASP-AMP) and γ-D-glutamyl-aminomethylphosphonate (GLU-AMP) (Jones, Smith and Watkins, 1984; Davies, Jones, Sheardown, Smith and Watkins, 1984).   The structures of these substances are shown in Fig 17.   Progress has been slower in the development of antagonists effective against those excitatory amino acids that are resistant to NMDA antagonists, such as kainate and quisqualate.   Even here, however, substances are now known which are more potent in antagonizing reponses produced by these two excitants than are DAP and DL-αAA against NMDA-induced responses.   Of substances now known to antagonize kainate-and/or quisqualate-induced responses, the most useful (in terms of potency and/or selectivity) are γ-D-glutamylglycine (γDGG), *cis*-2,3-piperidine dicarboxylic acid (PDA), γ-D-glutamylaminomethyl sulphonate (GAMS), γ-D-glutamyltaurine (GLU-TAU), kynurenic acid (KY-AC), *p*-chlorobenzoylpiperazine-2,3-dicarboxylic acid (*p*CB-PzDA) and *p*-bromobenzoylpiperazine-2,3-dicarboxylic acid (*p*BB-PzDA) (Davies and Watkins, 1983, 1985;   Jones *et al*, 1984;   Davies *et al*, 1984; Perkins and Stone, 1982;   Ganong, Lanthorn and Cotman, 1983;   Erez, Frenk, Golberg, Cohen and Teichberg, 1985;   Herrling, 1985). The structures of these compounds are shown in Fig 18.

It has not been quite so straightforward to classify non-NMDA

types of excitatory amino acid receptors and to assess the likelihood (or otherwise) of their involvement in discrete excitatory
synaptic pathways.   However, considerable progress has been made,
particularly in regard to receptors activated by kainate and quisqualate.   That quisqualate probably acted at receptors that were
different from those mediating the major part of responses to kainate was first recognised by McLennan and Lodge (1979) and by Davies
and Watkins (1979) who found that quisqualate induced responses
were considerably more sensitive to suppression by GDEE than were
responses induced by kainate.   The reverse sensitivity of responses
produced by these two agonists was observed with the antagonist
γDGG (Davies and Watkins, 1981).   As further evidence supporting
the existence of different quisqualate and kainate receptors Evans
identified excitatory amino acid receptors on dorsal root fibres
from baby rats which are considerably more sensitive to kainate
than to quisqualate (Davies, Evans, Francis and Watkins, 1979).

These observations were the foundation of the now fairly
generally accepted concept of at least three types of excitatory
amino acid receptors, termed the NMDA, quisqualate and kainate types
of receptors, though additional receptor types have also been proposed (McLennan, 1983; Luini, Goldberg and Teichberg, 1981).
Further characterization of these multiple receptor sites awaits
the development of more selective antagonists.

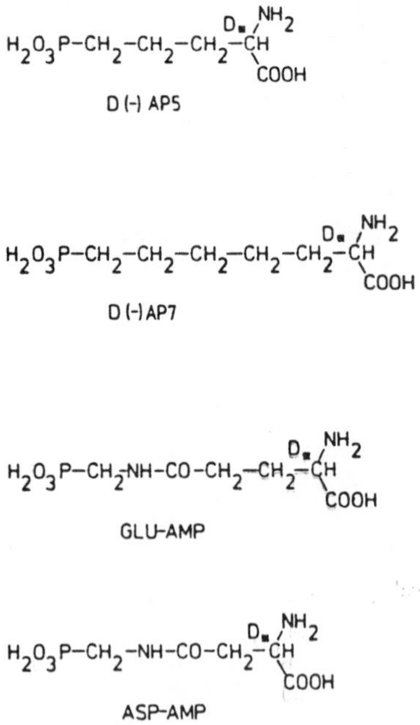

$$H_2O_3P-CH_2-CH_2-CH_2-\overset{D_\blacksquare\diagup NH_2}{\underset{\diagdown COOH}{C}H}$$

D (-) AP5

$$H_2O_3P-CH_2-CH_2-CH_2-CH_2-CH_2-\overset{D_\blacksquare\diagup NH_2}{\underset{\diagdown COOH}{C}H}$$

D (-) AP7

$$H_2O_3P-CH_2-NH-CO-CH_2-CH_2-\overset{D_\blacksquare\diagup NH_2}{\underset{\diagdown COOH}{C}H}$$

GLU-AMP

$$H_2O_3P-CH_2-NH-CO-CH_2-\overset{D_\blacksquare\diagup NH_2}{\underset{\diagdown COOH}{C}H}$$

ASP-AMP

Fig. 17   Structures of the most potent NMDA receptor antagonists.

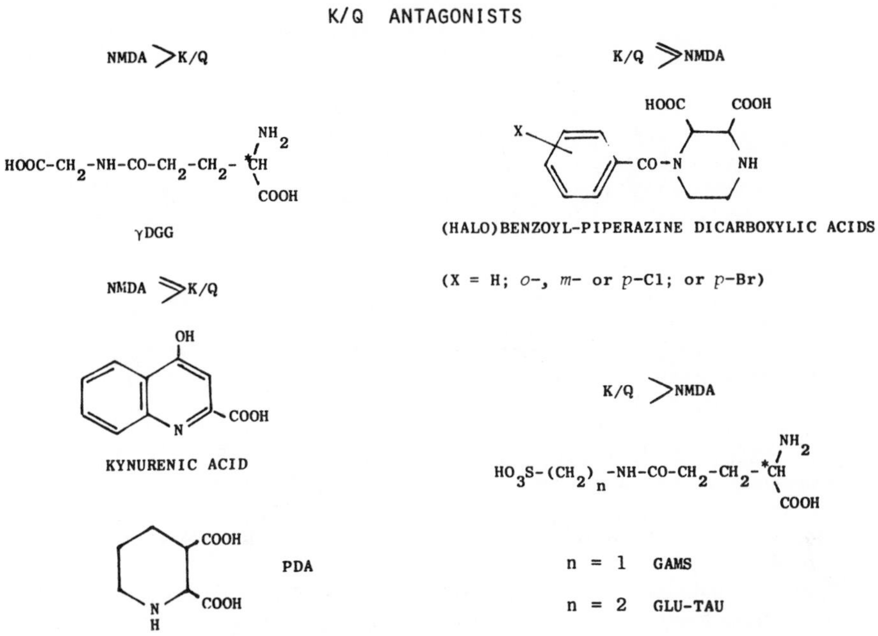

K/Q ANTAGONISTS

NMDA $>$ K/Q

$$HOOC-CH_2-NH-CO-CH_2-CH_2-\overset{\overset{\displaystyle NH_2}{|}}{C}H$$
$$\underset{COOH}{}$$

γDGG

NMDA $>$ K/Q

OH

[quinoline structure]

COOH

KYNURENIC ACID

[piperidine structure]

COOH

COOH

PDA

K/Q $>$ NMDA

HOOC   COOH

X—[benzene ring]—CO—N   NH

(HALO)BENZOYL-PIPERAZINE DICARBOXYLIC ACIDS

(X = H; o-, m- or p-Cl; or p-Br)

K/Q $>$ NMDA

$$HO_3S-(CH_2)_n-NH-CO-CH_2-CH_2-\overset{\overset{\displaystyle NH_2}{|}}{\overset{*}{C}}H$$
$$\underset{COOH}{}$$

n = 1   GAMS

n = 2   GLU-TAU

Fig. 18   Structures of some kainate/quisqualate (K/Q) antagonists.   The substances vary in their relative ability to antagonize NMDA and K/Q receptors, as indicated.   Abbreviations: γDGG, γ-D-glutamylglycine;   PDA, cis-2,3-piperidine dicarboxylic acid;   GAMS, γ-D-glutamyl-aminomethyl sulphonic acid;   GLU-TAU, γ-D-glutamyltaurine.

AMINO ACID-ERGIC SYNAPTIC EXCITATION

Using a combination of antagonists it has been possible to designate a number of synaptic pathways in the mammalian CNS as amino acid-ergic.   Most of the pathways investigated have proved to be much less sensitive to depression by highly selective NMDA antagonists than to depression by substances that antagonize kainate and/or quisqualate, irrespective of whether or not the depressants also antagonize NMDA-induced responses.   A notable exception is the polysynaptic excitation of neurones in the spinal cord following primary afferent stimulation (Biscoe et al, 1977; 1978;   Davies and Watkins, 1979;   Evans et al; 1978, 1979), which is highly sensitive to depression by specific NMDA antagonists and relatively insensitive to those antagonists which do not possess marked NMDA antagonist activity.   Indeed, the relative potencies of substances as depress-ants of polysynaptic excitation in the frog spinal cord closely correlated with their relative potencies as NMDA antagonists (Fig 19) (Evans et al, 1979; 1982).   In contrast monosynaptic excitation of spinal neurones is resistant to highly specific NMDA antagonists, but suppressed by substances that are effective kainate/quisqualate antagonists (Evans, Smith and Watkins, 1981;   Watkins, Davies, Evans, Francis and Jones, 1981;   Davies and Watkins, 1983). A differential role of NMDA and non-NMDA receptors has also been reported in the red

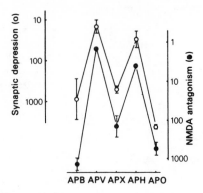

Fig. 19   Parallelism of antagonism of synaptic responses and NMDA-induced depolarizations
by a series of homologous phosphonates in frog spinal cord.   Points show concentrations
(μM) of antagonists necessary to produce a 20% depression in the dorsal root-evoked ventral
root potentials or pA₂ values for antagonism of NMDA-induced motoneuronal depolarizations
(measured in presence of tetrodotoxin).   Abbreviations: APB, (±)-2-amino-4-phosphono-
butyrate;   APV, (±)-2-amino-5-phosphonovalerate;   APX, (±)-2-amino-6-phosphonohexanoate;
APH, (±)-2-amino-7-phosphonoheptanoate;   APO, (±)-2-amino-8-phosphono-octanoate. (From Evans
et al, 1982).

nucleus (Davies,Miller and Sheardown,1985). Other examples have been
recorded in recent reviews (Watkins and Evans,1981; McLennan 1983).

CONCLUSION

The 'twenty five years' I have given prominence to under the constraints of this occasion have been the years between Hayashi's
first reports on the convulsant action of glutamate in the early
1950's and the discovery of truly specific excitatory amino acid
antagonists in the late seventies.   The advent of these antagonists
and the demonstration of their selective depressant effects on
transmission in discrete central synaptic pathways has ushered in a
new era of excitatory amino acid research.   No longer the province
only of a specialized and dedicated few, the field of excitatory
amino acids is now attracting ever-increasing numbers of neuro-
scientists from all disciplines, confident at long last of the
physiological and/or medical relevance of topics within the field.

     Reflecting this explosion of interest and recruitment, advan-
ces during the last few years have already eclipsed those of the
first twenty five.   If I may be excused the bias of one who has
seen the substance known as NMDA progress from a first faint-spot on
a chromatographic paper (Watkins, 1962) to its present prominent
status (Watkins, 1984), I would highlight the following recent
advances as being of particular personal interest:

(a)   the recognition of the high density of NMDA receptors in the
hippocampus (Monaghan, Holets, Toy and Cotman, 1983; Olverman,
Jones and Watkins, 1984; Monaghan, Yao, Olverman, Watkins and
Cotman, 1984), and the demonstration that these receptors play a

role in 'long term potentiation' (Collingridge, Kehl and McLennan, 1983), which is considered likely to be associated with learning and memory.

(b) the possibility that changes in extracellular concentrations of Mg^{2+} (Ault *et al*, 1980) and/or the voltage sensitivity of the Mg^{2+} effect (Nowak *et al*, 1984; Mayer, Westbrook and Guthrie, 1984), plays an important physiological role in regulating the activity of NMDA receptors

(c) the possibility that NMDA receptors are modulating receptors, producing 'plateau responses' similar to those observed in the caudate nucleus (Herrling, 1983), and

(d) the anticonvulsant actions of NMDA antagonists and the potential value of these and other types of excitatory amino acid antagonists as new centrally acting drugs (Croucher, Collins and Meldrum, 1982; Meldrum, 1985).

Beyond these, however, recent advances have been legion in many distinct branches of the field. May I just instance the detailed biochemical, autoradiographic and immunocytochemical studies that have led to the identification of discrete synaptic pathways likely to use L-glutamate (or, in some cases, L-aspartate) as the excitatory transmitter(s), the continuing electrophysiological characterization of excitatory amino acid receptors in these same and other synaptic pathways, the increasing success of efforts to correlate pharmacologically-defined receptors in electrophysiological studies with membrane binding sites of labelled ligands, the new light thrown on the actions of excitatory amino acids by sophisticated membrane conductance measurements, and the increasing evidence of the relation between excitatory amino acid transmitter systems and the etiology and prospective treatment of certain neuropathologies.

It is unnecessary for me to comment further on this wealth of recent progress, In the ensuing pages, many of those responsible for these advances will report to us first hand. One can only feel that, after the slow and laboured beginnings I have described, our field is on the threshold of a truly momentous future.

Acknowledgements. Without first, the far-sightedness of Sir John Eccles in recognizing a place for chemistry in electrophysiological studies (at a time when this was not, unlike now, self-evident) and then the close collaboration over the past 27 years of David Curtis (1958-65 and intermittently thereafter), John Davies (1971-present), Dick Evans (1974-present) and Tim Biscoe (1973-8) in the pharmacological testing programmes conducted, none of this work is likely to have appeared under my name as co-author. The role of Professors Arthur Buller, Tim Biscoe and James Mitchell in inviting me to Bristol in 1973 is also gratefully acknowledged, as is that of Professor Mitchell and Professor B. Matthews in continuing to provide research facilities in the Departments of Pharmacology and Physiology, respectively, following my retirement from the Teaching Staff in 1983. I also wish to acknowledge with gratitude the assistance I have been given in the continuing chemical programme by Keith Hunt, Arwel Jones and Ken Mewett, and by Harry Olverman in initiating and conducting our membrane binding programme. Others who have contributed significantly to the work described include John Phillis (in the early days), Alison Francis and Dave Smith, while Dan Oakes has provided excellent technical assistance of all kinds for the last 12 years. Special thanks are due to Joan Keeling in the preparation of this manu-script. To the many others, too numerous to mention, who have from time to time worked within our group, or conducted collaborative studies with our group, I also express my sincere apprecia-tion. The generous support of the Medical Research Council and the Wellcome Trust, during the past twelve years, is also gratefully acknowledged.

REFERENCES

Ames, A. (1956) Studies on water and electrolytes in nervous tissue.
J. Neurophysiol. *19* 213-223.

Ames, A., Tsukada, Y. and Nesbitt, F.B. (1967) Intracellular Cl^-,
Na^+, K^+, Ca^{2+} and P in nervous tissue: response to glutamate and to
changes in extracellular calcium. J. Neurochem. *14* 145-159.

Ault, B., Evans, R.H., Francis, A.A., Oakes, D.J. and Watkins, J.C.
(1980) Selective depression of excitatory amino acid-induced
depolarization by magnesium ions in isolated spinal cord preparations. J. Physiol. (Lond) *307* 413-428.

Balcar, V.J. and Johnston, G.A.R. (1972a) The structural specificity
of the high affinity uptake of L-glutamate and L-aspartate by rat
brain slices. J. Neurochem. *19* 2657-2666.

Balcar, V.J. and Johnston, G.A.R. (1972b) Glutamate uptake by brain
slices and its relation to the depolarization of neurones by acidic
amino acids. J. Neurobiol. *3* 295-301.

Bangham, A.D. Standish, M.M. and Watkins, J.C. (1965) Diffusion of
univalent ions across the lamellae of swollen phospholipids. J.Mol.
Biol. *13* 238-252.

Barron, D.H. and Matthews, B.H.C. (1938) The interpretation of
potential changes in the spinal cord. J. Physiol. (Lond) *92* 276-321.

Biscoe, T.J., Davies, J., Dray, A., Evans, R.H., Francis, A.A.,
Martin, M.R. and Watkins, J.C. (1977) Depression of synaptic
excitation and of amino acid-induced excitatory responses of spinal
neurones by D-α-aminoadipate, α,ε-diaminopimelic acid and HA-966.
Eur. J. Pharmac. *45* 315-316.

Biscoe, R.J., Davies, J., Dray, A., Evans, R.H., Martin, M.R. and
Watkins, J.C. (1978) D-α-Aminoadipate,α,ε-diaminopimelic acid, and
HA-966 as antagonists of amino acid-induced and synaptic excitation
of mammalian spinal neurones, *in vivo*. Brain Res. *148* 543-548.

Biscoe, T.J., Evans, R.H., Francis, A.A., Martin, M.R., Watkins, J.C
Davies, J. and Dray, A. (1977) D-α-Aminoadipate as a selective
antagonist of amino acid-induced and synaptic excitation of
mammalian spinal neurones. Nature (Lond) *270* 743-745.

Bradford, H.F. and McIlwain, H. (1966) Ionic basis for the
depolarization of cerebral tissues by excitatory acidic amino acids.
J. Neurochem. *13* 1163-1177.

Bradford, H.F., Ward, H.K. and Thomas, A.J. (1978) Glutamine: a
major substrate for nerve endings. J. Neurochem. *30* 1453-1459.

Brooks, V.B., Ransmeier, R.E. and Gerard, R.W. (1949) Action of acetylcholinesterase, drugs, and intermediates on respiration and electrical activity of the isolated frog brain. Am. J. Physiol. *157* 299-316.

Collingridge, G.L. Kehl, S.J. and McLennan, H. (1983) Excitatory amino acids in synaptic transmission in the Schaffer-commissural pathway of the rat hippocampus. J. Physiol. (Lond) *334* 33-46.

Cotman, C.W. and Hamberger, A. (1978) Glutamate as a CNS neuro-transmitter. Properties of release, inactivation and biosynthesis. In: *Amino Acids as Chemical Transmitters*. Ed. by F. Fonnum. Plenum Press. New York. pp 379-412.

Cox, D.W.G., Headley, P.M. and Watkins, J.C. Actions of L- and D-homocysteate in rat CNS: a correlation between low-affinity uptake and the time courses of excitation by microelectrophoretically applied L-glutamate analogues. J. Neurochem. *29* 579-588.

Crawford, J.M. and Curtis, D.R. (1964) The excitation and depression of mammalian cortical neurones by amino acids. Br. J. Pharmacol. *23* 313-329.

Croucher. M.J., Collins, J.F. and Meldrum, B.S. (1982) Anticon-vulsant action of excitatory amino acid antagonists. Science *216* 899-901.

Curtis, D.R., Duggan, A.W. and Johnston, G.A.R. (1970) The inactivation of extracellularly administered amino acids in the feline spinal cord. Exp. Brain Res. *10* 447-462.

Curtis, D.R. and Johnston, G.A.R. (1974) Amino acid transmitters in the mammalian central nervous system. Ergebn. Physiol. *69* 97-188.

Curtis, D.R., Johnston, G.A.R., Game, C.J.A. and McCulloch, R.M. (1973) Antagonism of neuronal excitation by 1-hydroxy-3-amino-pyrrolidone-2. Brain Res. *49* 467-470.

Curtis, D.R., Perrin, D.D. and Watkins, J.C. (1960) The excitation of spinal neurones by the iontophoretic application of agents which chelate calcium. J. Neurochem. *6* 1-20.

Curtis, D.R., Phillis, J.W. and Watkins, J.C. (1959) Chemical excitation of spinal neurones. Nature (Lond) *183* 611.

Curtis, D.R., Phillis, J.W. and Watkins, J.C. (1960) The chemical excitation of spinal neurones by certain acidic amino acids. J. Physiol. *150* 656-682.

Curtis, D.R., Phillis, J.W. and Watkins, J.C. (1961) Actions of amino acids on the isolated hemisected spinal cord of the toad. Br. J. Pharmac. *16* 262-283.

Curtis, D.R. and Watkins,J.C.(1960) The excitation and depression of
spinal neurones by structurally related amino acids. J. Neurochem.
6 117-141.

Curtis, D.R. and Watkins, J.C. (1965) The pharmacology of amino
acids related to γ-aminobutyric acid. Pharm. Rev. *17* 347-392.

Davies, J., Evans, R.H., Francis, A.A. and Watkins, J.C. (1979)
Excitatory amino acid receptors and synaptic excitation in the
mammalian central nervous system. J. Physiol. (Paris) *75* 641-645.

Davies, J., Evans, R.H., Jones, A.W., Smith, D.A.S. and Watkins,
J.C. (1982) Differential activation and blockade of excitatory
amino acid receptors in the mammalian and amphibian central nervous
system. Comp. Biochem. Physiol. *72C* 211-224.

Davies, J., Francis, A.A.. Jones, A.W. and Watkins, J.C. (1981)
2-Amino-5-phosphonovalerate (2APV), a potent and selective
antagonist of amino acid-induced and synaptic excitation. Neurosci.
Lett. *21* 77-81.

Davies, J., Jones, A.W., Sheardown, M.J., Smith, D.A.S. and Watkins,
J.C. (1984) Phosphonodipeptides and piperazine derivatives as
antagonists of amino acid-induced and synaptic excitation in
mammalian and amphibian spinal cord. Neurosci. Lett. *52* 79-84.

Davies, J., Miller, A.J. and Sheardown, M.J. (1985) Excitatory amino
acid receptors and neurotransmittion in the feline red nucleus. J.
Physiol.(Lond) abstract, in press.

Davies, J. and Watkins, J.C. (1973) Microelectrophoretic studies
on the depressant action of HA-966 on chemically and synaptically-
excited neurones in the cat cerebral cortex and cuneate nucleus.
Brain Res. *59* 311-322.

Davies, J. and Watkins, J.C. (1979) Selective antagonism of amino
acid-induced and synaptic excitation in the cat spinal cord. J.
Physiol. (Lond) *297* 621-636.

Davies, J. and Watkins, J.C. (1981) Differentiation of kainate and
quisqualate receptors in the cat spinal cord by selective antagonism
with γ-D(and L)-glutamylglycine. Brain Res. *206* 172-177.

Davies, J. and Watkins, J.C. (1983) Role of excitatory amino acid
receptors in mono and polysynaptic excitation in the cat spinal
cord. Exp. Brain Res. *49* 280-290.

Davies, J. and Watkins, J.C. (1985) Depressant actions of γ-D-
glutamylaminomethyl sulfonate (GAMS) on amino acid-induced and
synaptic excitation in the cat spinal cord. Brain Res. *327* 113-120.

Del Castillo, J. and Engbaek, L. (1954) The nature of the neuro-muscular block produced by magnesium. J. Physiol. (Lond) *124* 370-384.

Del Castillo, J. and Katz, B. (1955) On the localization of acetyl-choline receptors. J. Physiol. (Lond) *128* 157-181.

Duggan, A.W. (1974) The differential sensitivity to L-glutamate and L-aspartate of spinal interneurones and Renshaw cells. Exp. Brain Res. *19* 522-528.

Eccles, J.C. (1946) Synaptic potentials of motoneurones. J. Neurophysiol. *9* 87-120.

Erez, U., Frenk, H., Goldberg, O., Cohen, A. and Teichberg, V.I. (1985) Anticonvulsant properties of 3-hydroxy-2-quinoxalinecarboxy-lic acid, a newly found antagonist of excitatory amino acids. Eur. J. Pharmacol. *110* 31-39.

Evans, R.H., Francis, A.A., Hunt, K., Oakes, D.J. and Watkins, J.C. (1979) Antagonism of excitatory amino acid-induced responses and of synaptic excitation in the isolated spinal cord of the frog. Br.J. Pharmacol. *67* 591-603.

Evans, R.H., Francis, A.A., Jones, A.W., Smith, D.A.S. and Watkins, J.C. (1982) The effects of a series of ω-phosphonic α-carboxylic amino acids on electrically evoked and amino acid induced responses in isolated spinal cord preparations. Br. J. Pharmacol. *75* 65-75.

Evans, R.H., Francis, A.A. and Watkins, J.C. (1977a) Effects of monovalent cations on the responses of motoneurones to different groups of amino acid excitants on frog and rat spinal cord. Experientia (Basel) *33* 246-248.

Evans, R.H., Francis, A.A. and Watkins, J.C. (1977b) Differential antagonism by chlorpromazine and diazepam of frog motoneurone depolarization induced by glutamate-related amino acids. Eur. J. Pharmacol. *44* 325-330.

Evans, R.H., Francis, A.A. and Watkins, J.C. (1977c) Selective antagonism by Mg^{2+} of amino acid-induced depolarization of spinal neurones. Experientia (Basel) *33* 489-491.

Evans, R.H., Francis, A.A. and Watkins, J.C. (1978) Mg^{2+}-like selective antagonism of excitatory amino acid-induced responses by α,ε-diaminopimelic acid, D-α-aminoadipate and HA-966 in isolated spinal cord of frog and immature rat. Brain Res. *148* 536-542.

Evans, R.H., Smith, D.A.S. and Watkins, J.C. (1981) Differential role of excitant amino acid receptors in spinal transmission. J. Physiol. *320* 55P.

Evans, R.H. and Watkins, J.C. (1978) Specific antagonism of
excitant amino acids in the neo-natal rat isolated spinal cord
preparation. Eur. J. Pharmacology. *50* 123-129.

Foster, A.E., Fagg, G.E., Harris, E.W. and Cotman, C.W. (1982)
Regulation of glutamate receptors: a possible role of phosphatidyl
serine. Brain Res. *242* 374-377.

Ganong, A.H., Lanthorn, T.H. and Cotman, C.W. (1983) Kynurenic acid
inhibits synaptic and acidic amino acid-induced responses in the
rat hippocampus and spinal cord. Brain Res. *273* 170-174.

Giambalvo, C. and Rosenberg, P. (1976) The effect of phospholipase
and proteases on the binding of gamma-aminobutyric acid to junctional complexes of rat cerebellum. Biochim. Biophys. Acta.*436* 741-756.

Haldemann, S., Huffman, R.D., Marshall, K.C. and McLennan, H. (1972)
The antagonism of the glutamate-induced and synaptic excitation of
thalamic neurones. Brain Res. *39* 419-425.

Haldemann, S. and McLennan, H. (1972) The antagonistic action of
glutamic acid diethyl ester towards amino acid-induced synaptic
excitation of thalamic neurones. Brain Res. *45* 393-400.

Hall, J.G., McLennan, H. and Wheal, H.V. (1977) The actions of
certain amino acids as neuronal excitants. J. Physiol. (Lond) *272*
52-53P.

Harvey, J.A. and McIlwain, H. (1968) Excitatory acidic amino acids
and the cation content and sodium ion flux of isolated tissues from
the brain. Biochem. J. *108* 269-274.

Hayashi, T. (1954) Effects of sodium glutamate on the nervous
system. Keio J. Med. *3* 183-192.

Herrling, P.L. Morris, R. and Salt, T.E. (1983) Effects of
excitatory amino acids and their antagonists on membrane and action
potentials of cat caudate neurones. J. Physiol. *339* 207-222.

Herrling, P.L. (1985) Pharmacology of the cortico-caudate e.p.s.p.
in the cat: evidence for its mediation by quisqualate- or kainate
receptors. Neuroscience (in press).

Johnston, G.A.R., Curtis, D.R., De Groat, W.C. and Duggan, A.W.
(1968) Central actions of ibotenic acid and muscimol. Biochem.
Pharmacol. *17* 2488-2489.

Johnston, G.A.R. and Kennedy, S.M.E. (1978) GABA receptors and
phospholipids. In: *Amino Acids as Chemical Transmitters,* ed. by
F. Fonnum, Plenum Press, New York, pp 507-516.

Jones, A.W., Smith, D.A.S. and Watkins, J.C. (1984) Structure-
activity relations of dipeptide antagonists of excitatory amino
acids. Neuroscience *13* 573-581.

Krnjevic, K. and Phillis, J.W. (1963) Iontophoretic studies of
neurones in the mammalian cerebral cortex. J.Physiol.(Lond) *165*
274-304.

Lucas, D.R. and Newhouse. J.P. (1957) The toxic effects of sodium L-glutamate on the inner layers of the retina. A.M.A. Arch. Opthalmol. *58* 193.

Luini, A., Goldberg, O. and Teichberg, V.I. (1981) Distinct pharmacological properties of excitatory amino acid receptors in the rat striatum: study by Na^+ efflux assay. Proc. Natn. Acad. Sci. U.S.A. *78* 3250-3254.

Mayer, M.L., Westbrook, G.L. and Guthrie, P.B. (1984) Voltage-dependent block by Mg^{2+} of NMDA responses in spinal cord neurones. Nature (Lond) *309* 261-263.

McCulloch, R.M. Johnston, G.A.R., Game, C.J.A. and Curtis, D.R. (1974) The differential sensitivity of spinal interneurones and Renshaw cells to kainate and N-methyl-D-aspartate. Exp. Brain Res. *21* 515-518.

McLennan, H., Huffman, R.D. and Marshall, K.C. (1968) Patterns of excitation of thalamic neurones by amino acids and by acetylcholine. Nature (Lond) *219* 387-388.

McLennan, H. (1983) Receptors for the excitatory amino acids in the mammalian central nervous system. Progr. Neurobiol. *20* 251-271.

McLennan, H. and Lodge, D. (1979) The antagonism of amino acid-induced excitation of spinal neurones in the cat. Brain Res. *169* 83-90.

Meldrum, B.S. (1985) Possible therapeutic applications of antagonists of excitatory amino acid neurotransmitters. Clinical Science *68* 113-122.

Monaghan, D.R., Holets, V.R., Toy, D.W. and Cotman, C.W. (1983) Anatomical distribution of four pharmacologically distinct ^3H-glutamate binding sites. Nature (Lond) *306* 176-179.

Monaghan, D.T., Yao, D., Olverman, H.J. Watkins, J.C. and Cotman, C.W. (1984) Autoradiography of D-2-[^3H]-amino-5-phosphono-pentanoate binding sites in rat brain. Neurosci. Lett. *52* 253-258.

Nastuk, W.L. (1953) Membrane potential changes at a single muscle end-plate produced by transitory application of acetylcholine with an electrically controlled microjet. Federation Proc. *12* 102.

Nowak, L., Bregestovski, P., Ascher, P., Herbet, A and Prochiantz, A. (1984) Magnesium gates glutamate-activated channels in mouse central neurones. Nature (Lond) *307* 462-465.

Olney, J.W., Ho, O.L. and Rhee, J. (1971) Cytotoxic effects of acidic and sulphur-containing amino acids on the infant mouse central nervous system. Exp. Brain Res. *14* 61-76.

Olney, J.W., Rhee, V. and Ho, O.L. (1974) Kainic acid: a powerful neurotoxic analogue of glutamate. Brain Res. *77* 507-512.

Olney, J.W. (1978) Neurotoxicity of excitatory amino acids. In: *Kainic Acid as a Tool in Neurobiology*, (eds) E.G. McGeer, J.W. Olney and P.L. McGeer, Raven Press, New York, pp 95-121.

Olverman, H.J., Jones, A.W. and Watkins, J.C. (1984) L-Glutamate has higher affinity than other amino acids for [^{3}H]-D-AP5 binding site in rat brain membranes. Nature (Lond) *307* 460-462.

Papahadjopoulos, D. and Watkins, J.C. (1967) Phospholipid model membranes. II. Permeability properties of hydrated lipid crystals. Biochim. Biophys. Acta *135* 639-652.

Perkins, M.N., Collins, J.F. and Stone, T.W. (1982) Isomers of 2-amino-7-phosphonoheptanoic acid as antagonists of neuronal excitants. Neurosci. Lett. *32* 65-68.

Perkins, M.N. and Stone, T.W. (1982) An iontophoretic investigation of the action of convulsant kynurenins and their interaction with the endogenous excitant quinolinic acid, Brain Res. *247* 184-187.

Purpura, D.P., Girado, M., Smith, T.G., Callan, D.A. and Grundfest, H. (1959) Structure-activity determinants of pharmacological effects of amino acids and related compounds on central synapses. J. Neurochem. *3* 238-268.

Ramsey, R.L. and McIlwain, H. (1970) Calcium content and exchange in neocortical tissues during the cation movements induced by glutamates. J. Neurochem. *17* 781-787.

Robbins, J. (1959) The excitation and inhibition of crustacean muscle by amino acids. J. Physiol. (Lond) *148* 39-50.

Shanes, A.M. (1958) Electrochemical aspects of physiological and pharmacological action in excitable cells. Pharmacol. Rev. *10* 59-274.

Shinozaki, H. and Konishi, S. (1970) Actions of several anthelmintics and insecticides on rat cortical neurones. Brain Res. *24* 368-371.

Shinozaki, H. and Shibuya, I. (1974) A new potent excitant, quisqualic acid: effects on crayfish neuromuscular junction. Neuropharmacology *13* 665-672.

Skerritt, J.H. and Johnston, G.A.R. (1981) Uptake and release of N-methyl-D-aspartate by rat brain slices. J. Neurochem. *36* 881-885.

Stern, J.R., Eggleston, L.V. Hems, R. and Krebs, H.A. (1949) Accumulation of glutamic acid in isolated brain tissue. Biochem. J. *44* 410-418.

Teichberg, V.I., Tal, N., Goldberg, O. and Luini, A. (1984) Barbiturates, alcohols and the CNS excitatory neurotransmitters: specific effects on the kainic and quisqualic receptors. Brain Res. *291* 285-292.

Usherwood, P.N.R., Machili, P. and Leaf, G. (1968) L-Glutamate at insect excitatory nerve-muscle synapses. Nature (Lond) *219* 1169-1172.

Usherwood, P.N.R. (1981) Glutamate synapses and receptors on insect muscle, In: *Glutamate as a Neurotransmitter* (eds) G. Di Chiara and G.L Gessa, Raven Press, New York, pp 183-193.

Van Harreveld, A. (1959) Compounds in brain extracts causing spreading depression of cerebral cortical activity and contraction of crustacean muscle. J. Neurochem. *3* 300-315.

Waelsch, H. (1961) Compartmentalized biosynthetic reactions in the central nervous system, In: *Regional Neurochemistry* (eds) S.S. Kety and J. Elkes, Pergamon Press, Oxford, pp 57-64.

Watkins, J.C. (1962) The synthesis of some acidic amino acids possessing neuropharmacological activity. J. Med. Pharm. Chem. *5* 1187-1199.

Watkins, J.C. (1965) Pharmacological receptors and general permeability phenomena of cell membranes. J. Theoret. Biol. *9* 37-50.

Watkins, J.C. (1967) Acidic amino acids and excitation. Biochem. J. *102* 14P.

Watkins, J.C. (1972) Metabolic regulation in the release and action of excitatory amino acids. Biochem. Soc. Symp. *36* 33-47.

Watkins, J.C. (1984) Excitatory amino acids and central synaptic transmission. Trends in Pharmacol. Sci. *5* 373-376.

Watkins, J.C. Curtis, D.R. and Biscoe, T.J. (1966) Central effects of β-N-oxalyl-α,ε-diaminopropionic acid and other lathyrus factors. Nature *211* 637.

Watkins, J.C., Davies, J., Evans, R.H., Francis, A.A. and Jones, A. W. (1981) Pharmacology of receptors for excitatory amino acids, In: *Glutamate as a Neurotransmitter* (eds) G. Di Chiara and G.L. Gessa, Raven Press, New York, pp 263-273.

Watkins, J.C. and Evans, R.H. (1981) Excitatory amino acid transmitters. Ann. Rev. Pharmacol. Toxicol. *21* 165-204.

Watkins, J.C. Evans, R.H., Headley, P.M. Cox, D.W.G., Francis, A.A. and Oakes, D.J. (1978) Role of uptake in excitation of central neurones by glutamate-related amino acids: possible value in transmitter identification, In: *Iontophoresis and transmitter mechanisms in the mammalian central nervous system.* (eds) J.S. Kelly and R.W. Ryall, Elsevier/North Holland Biomedical Press, pp 397-399.

2

Regulatory Aspects of Endogenous Glutamate in Brain

E. Kvamme

GLUTAMATE POOLS IN BRAIN

Glial and Neuronal Pools

Glutamate is a major amino acid in the mammalian brain and together with the other members of the glutamine family, glutamine, GABA and aspartate, it constitutes 70-80% of the free amino acid nitrogen (Timiras et al., 1973). In human cerebral cortex the concentration of glutamate, glutamine and GABA is 10.8 mM, 4.4 mM and 2.1 mM, respectively (Perry et al., 1971). In synaptosomes the dominant amino acid is glutamate, followed by glutamine, aspartate, GABA and taurine (Bradford and Thomas, 1969 and Kontro et al., 1980). Glutamate is taken up by high affinity system in neuronal constituents and astrocytes. There is ample evidence that glutamate is compartmentalized in at least two metabolic pools in brain, a small, probably glial pool, which has a rapid turnover into a large glutamine pool, and a large, probably neuronal pool. This has formed the basis for the hypothetical glutamine cycle, assuming that glutamate is taken up by glial cells, converted to glutamine by the glutamine synthetase reaction, which is predominantly localized in glial cells, and the newly formed glutamine enters neuronal cells to form glutamate and GABA (Balázs et al., 1970, Benjamin and Quastel, 1972 and Van den Berg et al., 1975). Additional evidence for neuronal compartmentalization of glutamate has also been produced. Thus synaptosomes contain sufficient glutamate to inhibit phosphate activated glutaminase, but this glutamate appears not to be available to the enzyme, in contrast to exogenous glutamate (Kvamme and Lenda, 1981). Furthermore, Storm Mathisen et al. (1983) have demonstrated using immunohistochemical methods, that glutamate most likely is stored in neuronal vesicles.

Neurotransmitter Pool of Glutamate

Glutamate is assumed to be an excitatory transmitter and a

41

distinction has been made among a transmitter pool and a metabolic
pool of this amimo acid. Since ^{14}C labelled glutamate can be
released from brain slices by an electric stimulus following incubation with ^{14}C-glucose, such distinction cannot easily be made,
and it is suggested that both pools are regulated by the same processes (Potashner, 1978).

The pool of endogenous glutamate in brain is most likely highly
regulated and possible precursors are 2-oxoglutarate (e.g. derived
from glucose), glutamine, ornithine and arginine. For extensive
discussion of potential precursors of neurotransmitter glutamate
see Shank and Campbell (1983). Experiments supporting the concept
that glutamine is a precursor of transmitter glutamate have been
reported, using rat brain synaptosomes (Bradford et al., 1978),
toad brain hemisections (Shank and Aprison, 1977), and slices from
pigeon optic tectum (Reubi et al., 1978), mouse brain cortex (Tapia
and González, 1978) and dentata gyrus of the hippocampal formation
(Hamberger et al., 1979 a and Hamberger et al., 1979 b). However,
double labelled experiments using [^{14}C]-glucose with [^{3}H]-glutamine
(Bradford et al., 1978), and experiments with single labelled
^{14}C-glucose and glutamine, show that glutamate released by depolarization also to a great extent is derived from glucose
(Hamberger et al., 1979 b).

DEVELOPMENT OF THE ENDOGENOUS GLUTAMATE POOL AS CORRELATED TO THAT
OF GLUTAMATE METABOLIZING ENZYMES

Since experimental evidence favors 2-oxoglutarate and glutamine
as precursors for transmitter glutamate, the most important regulatory enzymes are expected to be aspartate aminotransferase (Asp-T)
(EC 2.6.1.1), glutamate dehydrogenase (GDH) (EC 1.4.1.3), phosphate
activated glutaminase (PAG) (EC 3.5.1.2), ornithine aminotransferase (Orn-T) (EC 2.6.1.13), GABA aminotransferase (GABA-T) (EC
2.6.1.19), and membrane bound enzymes that translocate glutamate
and closely related precursors, such as 2-oxoglutarate and glutamine, across cell membranes. Gamma-glutamyl transferase (Gamma-GT)
(EC 2.3.2.2) has been associated with cellular uptake of glutamine
(Dass and Wu, 1984, McFarlane-Anderson and Alleyne, 1979 and Minn
and Besagni, 1983) and glutamate (Lisý et al., 1983 and Sikka and
Kalra, 1980) and is therefore of interest in this connection.

It is possible that the above mentioned enzymes are all more or
less engaged in maintaining the endogenous pool, including the
neurotransmitter pool of glutamate. In order to throw some light
on this we studied the developmental change of the endogenous glutamate pool as correlated to that of glutamate metabolizing enzymes
(Kvamme et al., 1985, Drejer et al., 1985 and Larsson et al.,
1985), using cultured mouse cerebral cortex interneurons and cerebellar granule cells. In addition in vivo investigations were performed using mouse cerebral cortices and cerebella from birth to 28
days post partum.

We found that the developmental profile of endogenous glutamate and gamma-GT in cultured cerebral cortex interneurons and granule cells are very similar. Endogenous glutamate as well as the activity of gamma-GT decreased about 50% in the cerebral cortex interneurons during the culturing period, but were maintained at almost the same level in the cultured granule cells.

We also found a rather good correspondence in cerebellum in vivo between the developmental profile of endogenous glutamate and gamma-GT for the first 14 days post partum, but in cerebral cortex there is only a good correspondence during the first three days. In cerebral cortex and cerebellum the level of endogenous glutamate was almost doubled 28 days post partum to about 10 mM from about 5 mM at birth, whereas the activity of gamma-GT was little changed. The developmental profile of the other glutamate metabolizing enzymes (PAG, Asp-T, GABA-T, Orn-T and GDH) showed no good correspondence with that of endogenous glutamate neither in the cultured cells nor in vivo, and it is of particular interest that the developmental profiles of GDH and Asp-T were completely dissimilar (Kvamme et al. 1985).

GAMMA-GT AS A POSSIBLE REGULATOR OF ENDOGENOUS GLUTAMATE

As discussed above, endogenous glutamate may be regulated in two fundamentally different ways. 1) by the activity of glutamate metabolizing enzymes, and 2) by cellular or subcellular (e.g. vesicular) transport of glutamate or closely related precursors, such as glutamine and 2-oxoglutarate. The activity of the glutamate metabolizing enzyme reactions PAG, Asp-T, GDH and GABA-T are probably not rate limited by the enzyme content itself, but by available substrate, cofactors, etc. (Kvamme et al., 1984, Drejer et al., 1985 and Larsson et al. 1985). Moreover, Orn-T at least in cerebellar granule cells has a K_m for ornithine which renders the enzyme essentially inactive (Drejer and Schousboe, 1984). Therefore, enzyme reactions catalyzing transport of various substrates are likely to have a regulatory function of the endogenous glutamate pool.

Gamma-GT has been assumed to participate in the cellular transport of amino acids. It is a glycoprotein which is present on the external surface of cellular plasma membranes (Tate and Meister, 1981). Gamma-GT may regulate endogenous glutamate by rate limiting the cellular or subcellular uptake of glutamate, glutamine or closely related precursors. Glutamine has been shown to be a good substrate for gamma-GT (Tate and Meister, 1974, Tate and Meister, 1981) in synaptosomes (Minn and Besagni, 1983), kidney cells (Dass and Wu, 1984), and brush-border membrane vesicles (McFarlane-Anderson and Alleyne, 1979). In addition, lecithin vesicles (Sikka and Kalra, 1980), have active uptake mechanisms for glutamine, as probably mediated by gamma-GT. It has also been shown that gamma-GT catalyzes the synthesis of gamma-glutamyl-

glutamine from glutamine (Tate and Meister, 1974) which may be a
precursor of endogenous glutamate.

The close correspondance between the developmental profiles of
endogenous glutamate and gamma-GT in cultured cells and in vivo in
the early time period post partum, suggest that gamma-GT may have a
regulatory function of glutamate, at least under these conditions.
Whether the point of control is localized to the cellular plasma
membrane or to subcellular membranes is unknown.

CONCLUDING REMARKS

Although glutamine is an important precursor of endogenous glutamate and transmitter glutamate, no conclusive evidence has been
produced to show that glutamine is the main precursor. Other precursors, and in particular 2-oxoglutarate, may contribute as well,
depending on substrate availability and the direction of flow
through the enzyme reactions involved. Thus, GABA-ergic neurons
are able to oxidatively deaminate glutamate (Yu et al., 1984),
whereas in glutamatergic neurons the process occurs to a very
limited extent (Hertz et al., 1983), suggesting that 2-oxoglutarate
is not equally important as precursor of endogenous glutamate in
GABA-ergic and glutamatergic cells.

REFERENCES

Balázs, R., Machiyama, Y., Hammond, B.J., Julian, T. and Richter,
    D. (1970). The operation of the gamma-aminobutyrate bypath of
    the tricarboxylic acid cycle in brain tissue in vitro.
    Biochem. J., 116, 445-467.
Benjamin, A.M. and Quastel, J.H. (1972). Locations of amino acids
    in brain slices from the rat. Biochem. J., 128, 631-646.
Bradford, H.F. and Thomas, A.J. (1969). Metabolism of glucose and
    glutamate by synaptosomes from mammalian cerebral cortex. J.
    Neurochem., 16, 1495-1504.
Bradford, H.F., Ward, H.K. and Thomas, A.J. (1978). Glutamine -
    a major substrate for nerve endings. J. Neurochem. 30,
    1453-1459.
Dass, P O. and Wu, M.-C. (1984). Role of gamma-GT in glutamine
    uptake and metabolism in NRK. Fed. Proc. 43, 1693.
Drejer, J. and Schousboe, A. (1984). Ornithine-aminotransferase
    exhibits different kinetic properties in astrocytes, cerebral
    cortex interneurons, and cerebellar granule cells in primary
    culture. J. Neurochem. 42, 1194-1197.
Drejer, J., Larsson, O.M., Kvamme, E., Svenneby, G., Hertz, L. and
    Schousboe, A. (1985). Ontogenetic development of glutamate
    metabolizing enzymes in cultured cerebellar granule cells and
    in cerebellum. Neurochem. Res. 10, 49-62.
Hamberger, A., Chiang, G.H., Nylén, E.S., Scheff, S.W. and Cotman,
    C.W. (1979 a). Glutamate as a CNS transmitter. I. Evaluation

of glucose and glutamine as precursors for the synthesis of preferentially released glutamate. Brain Res. 168, 513-530.

Hamberger, A., Chiang, G.H., Sandoval, E. and Cotman, C.W. (1979 b). Glutamate as a CNS transmitter. II. Regulation of synthesis in the releasable pool. Brain Res. 168, 531-541.

Hertz, L., Yu, A.C.H., Potter, R.L., Fischer, T.E. and Schousboe, A. (1983). Metabolic fluxes from glutamate and towards glutamate in neurons and astrocytes in primary cultures. In Glutamine, glutamate and GABA in the central nervous system. (eds. L. Hertz, E. Kvamme, E.G. McGeer and A. Schousboe). Alan Liss, New York.

Kontro, P, Marnela, K.-M. and Oja, S.S. (1980). Free amino acids in the synaptosome and synaptic vesicle fractions of different bovine brain areas. Brain Res. 184, 129-141.

Kvamme, E. (1984). Enzymes of cerebral glutamine metabolism. In Glutamine Metabolism in Mammalian Tissues. (eds. D. Häussinger and H. Sies). Springer Verlag, Berlin.

Kvamme, E. and Lenda, K. (1981). Evidence for compartmentalization of glutamate in rat brain synaptosomes using the glutamate sensitivity of phosphate activated glutaminase as a functional test. Neurosci. Lett. 25, 193-198.

Kvamme, E., Schousboe, A., Hertz, L., Torgner, I. Aa. and Svenneby, G. (1985). Developmental change of endogenous glutamate and gamma-glutamyl transferase in cultured cerebral cortical interneurons and cerebellar granule cells, and in mouse cerebral cortex and cerebellum in vivo. Neurochem. Res. in press.

Larsson, O.M., Drejer, J., Kvamme, E., Svenneby, G., Hertz, L. and Schousboe, A. (1985). Ontogenetic development of glutamate and GABA metabolizing enzymes in cultured cerebral cortex interneurons and in cerebral cortex in vivo. Int. J. Devel. Neurosci. 3, 177-185.

Lisý, V., Stastný, F., Murphy, S. and Hájková, B. (1983). Glutamate uptake into cerebral cortex slices is reduced in the presence of a gamma-glutamyl transpeptidase inhibitor. Experientia 39, 111.

McFarlane-Anderson, N. and Alleyne, G.A.O. (1979). Transport of glutamine by rat kidney brush-border membrane vesicles. Biochem. J. 182, 295-300.

Minn, A. and Besagni, D. (1983). Uptake of L-glutamine into synaptosomes. Is the gamma-glutamyl cycle involved? Life Sci. 33, 225-232.

Perry, T.L., Berry, K., Hansen, S., Diamond, S. and Mok, C. (1971). Regional distribution of amino acids in human brain obtained at autopsy. J. Neurochem., 18, 513-519.

Potashner, S.J. (1978). The spontaneous and electrically evoked release, from slices of guinea-pig cerebral cortex, of endogenous amino acids labelled via metabolism of d-[U-^{14}C]glucose. J. Neurochem. 31, 177-186.

Reubi, J.-C., Van den Berg, C. and Cuénod, M. (1978). Glutamine as a precursor for the GABA and glutamate transmitter pools. Neurosci. Lett. 10, 171-174.

Shank, R.P. and Aprison, M.H. (1977). Glutamine uptake and metabo-

lism by the isolated toad brain: Evidence pertaining to its
proposed role as a transmitter precursor. J. Neurochem. 28,
1189-1196.

Shank. R.P. and Campbell, G.LeM. (1983). Metabolic precursors of
glutamate and GABA. In Glutamine, glutamate and GABA in the
central nervous system. (eds. L. Hertz, E. Kvamme, E.G. McGeer
and A. Schousboe). Alan Liss, New York.

Sikka, S C. and Kalra, V K. (1980). Gamma-glutamyl transpeptidase
mediated transport of amino acid in lecithin vesicles. J.
Biol. Chem. 255, 4399-4402.

Storm-Mathisen, J., Leknes, A.K., Bore, A.T., Vaaland, J.L., Edminson, P., Haug, F.-M.S., and Ottersen, O.P. (1983). First
visualization of glutamate and GABA in neurones by immunocytochemistry. Nature 301, 517-520.

Tapia, R. and González, R.M. (1978). Glutamine and glutamate precursors of the releasable pool of GABA in brain cortex slices.
Neurosci. Lett. 10, 165-169.

Tate, S S. and Meister, A. (1974). Interaction of gamma-glutamyl
transpeptidase with amino acids, dipeptides, and derivatives
and analogs of glutathione. J. Biol. Chem. 249, 7593-7602.

Tate, S.S. and Meister, A. (1981). Gamma-glutamyl transpeptidase:
catalytic, structural and functional aspects. Mol. Cell.
Biochem. 39, 357-368.

Timiras, P.S., Hudson, D.B. and Oklund, S. (1973). Changes in
central nervous system. Free amino acids with development and
aging. Prog. Brain Res. 40, 267-275.

Van den Berg, C.F., Matheson, D.F., Ronda, G., Reijnierse, G.L.A.,
Blokhuis, G.G.D., Kroon, M.C., Clarke, D.D. and Garfinkel, D.
(1975). A model of glutamate metabolism in brain: Biochemical
analysis of a heterogeneous structure. In Metabolic Compartmentation and Neurotransmission. (eds. S. Berl, D.D. Clarke
and D. Schneider). Plenum Press, New York.

Yu, A.C.H., Fisher, T.E., Hertz, E., Tildon, J.T., Schousboe, A.
and Hertz, L. (1984). Metabolic fate of [^{14}C]-glutamine in
mouse cerebral neurons in primary cultures. J. Neurosci. Res.
11, 351-357.

3

Metabolic Precursors of the Transmitter Pools of Glutamate and Aspartate

R.P. Shank and G. Le M. Campbell

ABSTRACT

Current evidence indicates that glutamine, α-ketoglutarate
and possibly malate are supplied by astrocytes to neurons to replenish the neurotransmitter pools of glutamate, aspartate and
GABA. Evidence supporting the concept that α-ketoglutarate and
malate are supplied by astrocytes to synaptic terminals include
the immunohistochemical and biochemical demonstration that pyruvate carboxylase, the principal anaplerotic enzyme in CNS tissues,
is located predominantly or exclusively in astrocytes, and that
synaptic terminals possess avid high-affinity transport systems
for these citrate cycle intermediates. These transport systems
appear to be highly regulated by factors that include glutamine,
glutamate, N-acetylaspartate and possibly a protein found in the
post-microsomal supernatant of brain homogenates.

INTRODUCTION

The experimental inquiry into the metabolic precursors of the
neurotransmitter pools of glutamate and aspartate was initiated
in the early 1970's by several groups of neurochemists. To date,
research has been focused primarily on glutamine and glucose even
though at least six or eight compounds have the metabolic potential for replenishing the neurotransmitter pools of these excitatory amino acids (Shank and Campbell, 1983a). The progress made
in elucidating the role of glutamine as a major precursor of the
transmitter pools of glutamate and GABA has been the subject of
several recent reviews (Shank and Aprison, 1978; 1981; Bradford,
1981; Nicklas, 1983; Cotman et al., 1981), and will be touched on
in another chapter in this book (Nicklas, 1985). In this review
we will focus primarily on the contribution of glucose carbon in
replenishing the neurotransmitter pools of glutamate and
aspartate.

Interest in the possibility that glucose might serve a role in replenishing the neurotransmitter pools of glutamate and aspartate initially arose from the knowledge that a large portion of the glucose carbon oxidatively metabolized in CNS tissues passes through glutamate and aspartate prior to the conversion to CO_2. It now appears that this metabolic phenomenon is not due primarily to any role of glucose in replenishing the transmitter pools, but rather to an important role of the malate-aspartate shuttle in oxidative metabolism (Shank and Campbell, 1983b). Investigations into the significance of glucose in replenishing the transmitter pools of glutamate and aspartate is complicated by the metabolic remoteness of these amino acids from the proposed precursor, and the fact that a net synthesis of glutamate and aspartate from glucose requires an anaplerotic (filling up) pathway in order to prevent a depletion of the pool of citrate cycle intermediates.

THE ANAPLEROTIC PATHWAY IN CNS TISSUES.

When pyruvate is metabolized to acetyl CoA the subsequent formation of citrate requires a stoichiometric consumption of oxaloacetate (OAA). Therefore, since OAA must be regenerated for the citrate cycle to continue operating, the acetyl CoA pathway can not serve as a means of restoring pools of citrate cycle intermediates if they are consumed for the net synthesis of glutamate and aspartate. Although several metabolic pathways have the potential for restoring the pool of citrate cycle intermediates, in CNS tissues this process is mediated predominantly if not exclusively, by the carboxylation of pyruvate with the resultant formation of oxaloacetate (Patel, 1974; Figure 1). This "CO_2 fixation" reaction is mediated by pyruvate carboxylase, a biotin containing enzyme that is activated by acetyl CoA, and requires ATP to drive the synthesis of OAA.

Anaplerotic activity mediated by pyruvate carboxylase is potentially very active in CNS tissues (Hawkins and Mans, 1983), and under normal metabolic conditions in CNS tissues it may account for as much as one-tenth of all the pyruvate carbon moieties entering the citrate cycle (Waelsch et al., 1964; Cheng et al., 1967).

EVIDENCE THAT SYNAPTIC TERMINALS ARE NOT THE PRINCIPAL SITE OF ANAPLEROTIC ACTIVITY IN CNS TISSUES

Metabolic studies undertaken by Waelsch and his colleagues in the early 1960's demonstrated that CO_2 fixation is not only prevalent in CNS tissues, but that it also occurs within a small metabolic compartment of unknown cellular origin in which glutamate is metabolized selectively to glutamine (see Berl and Clarke, 1969; 1983). In 1974 when we initiated our investigation

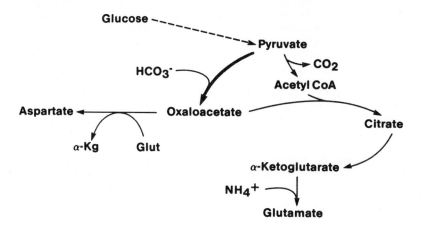

Figure 1:   The anaplerotic pathway in CNS tissues as related to the replenishment of the neurotransmitter pools of glutamate and aspartate. The key reaction in this pathway is the conversion of pyruvate to oxaloacetate.

into the role of glucose as a metabolic precursor of the transmitter pools of glutamate and aspartate we were well acquainted with the observations of Waelsch and his colleagues.   In a preliminary study, we demonstrated that pyruvate carboxylation, as expected, occurred predominantly within the compartment that selectively metabolizes glutamate to glutamine (see Shank and Aprison, 1981).   Based on this knowledge, we postulated that CO_2 fixation in the form of pyruvate carboxylation might occur predominantly within synaptic terminals that utilize glutamate, aspartate or GABA as their neurotransmitter (this assumed that the small metabolic compartment resided in synaptic terminals). We further postulated that a determination of the rate at which these amino acids are synthesized in synaptic terminals via this anaplerotic process should provide an estimate of the contribution of glucose in replenishing the neurotransmitter pools.

In a preliminary study to test our hypothesis we administered (1-^{14}C) pyruvate into the cisterna magna of rats anesthetized lightly with ether, then after a period of five minutes we decapitated the rats and fractionated the brain tissue using the Whittaker procedure (Gray and Whittaker, 1962).   The significance of using pyruvate labeled in only the 1 carbon position is that the label can enter the citrate cycle, and therefore aspartate and glutamate, only via the anaplerotic pathway.   We then determined the amount of ^{14}C in glutamate, aspartate and several other amino acids in extracts from each of the subcellular fractions.   To our surprise the relative amount of label in glutamate and aspartate was appreciably lower in the synaptosomal fraction than in other fractions (Table 1). These results indicated that pyruvate carboxylation occurred in all subcellular fractions, even though metabolic studies in vivo indicated that it should occur selectively in a small compartment. We concluded that this discrepancy was due, at least partly, to a redistribution of metabolites during the subcellular fractionation process, and that such an experimental approach could not provide definitive information regarding the role of glucose as a precursor for replenishing the neurotransmitter pools of glutamate, aspartate and GABA. Accordingly we discontinued this experimental approach.

Table 1. Pyruvate Carboxylation in Subcellular Fractions of Rat Brain: Relative Specific Activity of Aspartate and Glutamate after Intracisternal Administration of $(1-^{14}C)$ Pyruvate.

Subcellular Fraction	Specific Radioactivity Relative to Homogenate	
	Aspartate (3840)*	Glutamate (1130)*
Synaptosomes	0.52 ± 0.04	0.71 ± 0.11
Mitochondria	0.64 ± 0.08	1.17 ± 0.34
Myelin	0.84 ± 0.10	1.22 ± 0.11
Microsomes	1.06 ± 0.13	0.97 ± 0.10
Supernatant	0.83 ± 0.06	1.02 ± 0.22

$(1-^{14}C)$ pyruvate (5 μCi in 10 μL) was injected into the cisternic magna of Wistar rats (250-300 g) anesthetized with ether. Each rat was killed five minutes after the injection and the whole brain was excised and homogenized in ten volumes of 0.32 M sucrose solution. The homogenate was fractionated by the Whittaker procedure. The data represent the mean ±S.D. of four rats.
* The numbers in parentheses represent the actual specific radioactivity in the homogenate (dpm/μmole).

EVIDENCE THAT ASTROCYTES REPRESENT THE PRINCIPAL SITE OF ANAPLEROTIC ACTIVITY AND SERVE AS A SOURCE OF CITRATE CYCLE INTERMEDIATES FOR REPLENISHING GLUTAMATE AND ASPARTATE TRANSMITTER POOLS

In 1979 we began to pursue the precursor role of glucose using a different approach. By this time immunocytochemical studies had conclusively shown that glutamine synthetase in the CNS is localized predominantly to astrocytes (see Norenberg, 1979; 1983); the significance of this finding is that these glial cells must be the physical counterpart of the small metabolic compartment, and therefore, the principal site of CO_2 fixation (anaplerotic activity). Accordingly, if glucose does serve a role in replenishing the neurotransmitter pools, it must first be metabolized within astrocytes to a citrate cycle intermediate, which must then be supplied to nearby synaptic terminals. For such a metabolic shuttle to occur, mechanisms must exist for translocating one or more citrate cycle intermediates across the plasma membrane of astrocytes and neurons (synaptic terminals). One likely mechanism underlying such a translocation process is membrane transport mediated by specific carriers. To test for the existence of such carriers we determined the ability of several preparations enriched in synaptic terminals, or cell bodies derived from either astrocytes or neurons, to transport (accumulate) citrate, α-ketoglutarate and malate (Shank and Campbell; 1981, 1984a;b). Our studies revealed the existence of specific high-affinity carriers for α-ketoglutarate and malate but not for citrate. A Na^+-dependent, high-affinity carrier selective for α-ketoglutarate is prevalent in synaptosomal fractions prepared using Percoll® (Pharmacia, Uppsula) continuous density gradients (Shank and Campbell, 1984b). Another carrier which apparently exhibits no dependency on Na^+ and possesses a high-affinity for malate is also prevalent in these synaptosomal fractions (Fig. 2). The plasma membrane of astrocytes also contains carriers for α-ketoglutarate as indicated by uptake into astrocyte cell bodies (Shank and Campbell, 1984a), and by the formation of glutamine from some of the α-ketoglutarate accumulated by synaptosomal preparations, particularly "high density synaptosomes" (Shank and Campbell, 1984b).

The existence of transport carriers for α-ketoglutarate and malate in the plasma membrane of synaptic terminals and astrocytes supports the concept that a flux of these citrate cycle intermediates can occur between astrocytes and neurons, but it does not establish the direction of that flux. Recently, we and others have provided compelling immunocytochemical and biochemical evidence supporting our assumption that pyruvate carboxylase, and therefore, anaplerotic activity, is restricted predominantly or exclusively to astrocytes (Shank et al. 1981, 1985; Yu et al., 1984). The net synthesis of α-ketoglutarate and malate (via pyruvate carboxylation) in astrocytes, coupled with the concept that synaptic terminals should represent a site of net utilization (to replenish the molecules of citrate cycle intermediates consumed to replenish the neurotransmitter pools of glutamate, aspartate and GABA) indicate that the direction of such a flux must be from astrocytes to neurons.

FACTORS THAT MAY REGULATE THE NET SYNTHESIS AND FLUX OF ALPHA-KETOGLUTARATE AND MALATE FROM ASTROCYTES TO SYNAPTIC TERMINALS

Pyruvate carboxylase has not been characterized extensively

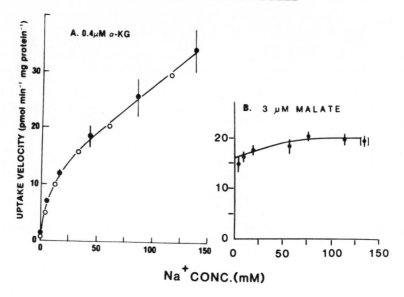

Figure 2: The Na$^+$ dependency of α-ketoglutarate and malate uptake by synaptosomal preparations from rat brain (from Shank and Campbell, 1984b). In A the open and closed circles represent data obtained when NaCl and NaHCO$_3$ were replaced with sucrose and Tris Cl, and LiCl and LiHCO$_3$, respectively. In B NaCl and NaHCO$_3$ were replaced with LiCl and Tris Cl. The data represent the mean ±S.D. of 3 or 4 experiments, except for the open circles which are the mean of 2 experiments.

in CNS tissues; however, studies in other tissues have revealed it to be a highly regulated enzyme. Acetyl CoA is an essential activating factor of this ATP requiring enzyme, and Mg^{++}, K$^+$ and NH$_4{}^+$ all enhance activity at physiologically relevant concentrations. In contrast, glutamate inhibits (K$_i$ ~5 mM) the liver enzyme (Scrutton and White, 1974). We have recently examined the effects of glutamate on pyruvate carboxylase activity in homogenates derived from rat brain tissue and found an inhibitory effect similar to that reported for the rat liver enzyme (unpublished observations). The inhibition of pyruvate carboxylase by glutamate may represent a feedback inhibitory mechanism that reduces anaplerotic activity when excessive amounts of glutamate accumulate within astrocytes.

The transport carriers mediating the translocation of α-ketoglutarate and malate across the membrane of synaptic terminals (and possibly astrocytes) appear to be regulated by several related metabolites, including glutamate, glutamine and aspartate.

Additional factors that may regulate α-ketoglutarate transport include N-acetylaspartate and a heat-labile substance present in the post-microsomal supernatant of brain homogenates (Shank and Campbell, 1984b). This activating substance precipitates in 60% saturated ammonium sulfate and appears to have a molecular weight greater than 25,000.

Glutamate and aspartate inhibit the uptake of both α-ketoglutarate and malate (K_i ~0.02 mM). This inhibitory effect of extracellular glutamate and aspartate could reduce their net synthesis within synaptic terminals by limiting precursor availability. Functionally, this may serve to reduce the amount of glutamate and aspartate released from synaptic terminals during periods of extracellular accumulation. Glutamine reduces the affinity between the substrates and their respective carriers, but greatly increases the Vmax. Functionally glutamine may serve as an activator of α-ketoglutarate and malate uptake into synaptic terminals. In this respect it is noteworthy that the net synthesis of glutamate from α-ketoglutarate and aspartate from malate require a source of an amino moiety, and that glutamine is an ideal substrate to supply this moiety.

N-acetylaspartate inhibits the uptake of α-ketoglutarate, and its potency varies depending on the tissue. It is a comparatively weak inhibitor of α-ketoglutarate uptake into synaptosomes prepared from the cerebrum (whole brain minus cerebellum and brain stem) of 6 to 8 week old rats (Shank and Campbell, 1984b). Furthermore the Hill coefficient was exceptionally low (~0.4). In contrast N-acetylaspartate was a comparatively potent inhibitor of the uptake of α-ketoglutarate by a synaptospmal preparation derived from the cerebellum of 10 to 14-day-old mice (Shank and Campbell, 1981). This discrepency in the potency of N-acetylaspartate may be due to species differences or changes during maturation. Another possibility is that N-acetylaspartate effectively inhibits only one of two or more types of transport carriers. Such a concept is supported by the low Hill coefficient. It may be that N-acetylaspartate inhibits only the carrier in astrocytes or GABAergic terminals; however, metabolic studies do not support this possibility (Shank and Campbell, 1984c).

CONCLUDING REMARKS

Current evidence indicates that glutamine, α-ketoglutarate and possibly malate are supplied by astrocytes to synaptic terminals to serve as metabolic precursors for replenishing the neurotransmitter pools of glutamate and aspartate. It is tempting to speculate that the utilization of α-ketoglutarate and malate is matched to that of glutamine since the latter can provide the amino moiety required by α-ketoglutarate and malate to replenish glutamate and aspartate, respectively. This would minimize the formation of free ammonia from glutamine. However, there is currently no information regarding a possible stoichiometric relationship in the utilization of these transmitter precursors.

ACKNOWLEDGMENTS

We thank Ms. Maria Castagno and Ms. Maria Ciliberto for able assistance in the preparation of this manuscript.

REFERENCES

Berl, S. and Clarke, D.D. (1969). Metabolic Compartmentalization of Glutamate in the CNS. In Handbook of Neurochemistry, Vol. 1. (ed. A. Lajtha). Plenum Press, New York.

Berl, S. and Clarke, D.D. (1983). The Metabolic Compartmentation Concept. In Glutamine, Glutamate, and GABA in the Central Nervous System. (eds. L. Hertz, E. Kvamme, E.G. McGeer, and A. Schousboe. A.R. Liss, Inc., New York.

Bradford, H.F. (1981). Glutamate as a Neurotransmitter: Studies on Its Synthesis and Release. In Chemisms of the Brain. (ed. R.B. Rodnight). Churchill Livingstone, Edinbourgh.

Cotman, C.W., Foster, A. and Lanthorn, T. (1981). An Overview of Glutamate as a Neurotramsmitter. In Glutamate as a Neurotransmitter. (eds. G. DiChiara and G. L. Gessa). Raven Press, New York.

Cheng, S.C., Nakamura, R. and Waelsch (1967). Relative Contribution of Carbon Dioxide Fixation and Acetyl CoA Pathways in Two Nervous Tissues. Nature, 216, 928-929.

Gray, E.G. and Whittaker, V.P. (1962). The Isolation of Nerve Endings from Brain: An Electron Microscopic Study of Cell Fragments Derived by Homogenization and Centrifugation. J. Anat., 96, 79-88.

Hawkins, R.A. and Mans, A.M. (1983). Intermediary Metabolism of Carbohydrates and Other Fuels. In Handbook of Neurochemistry, Vol. 3, 2nd Edition. (ed. A. Lajtha). Plenum Press, New York.

Nicklas, W.J. (1983). Relative Contributions of Neurons and Glia to Metabolism of Glutamate and GABA. In Glutamine, Glutamate, and GABA in the Central Nervous System. (eds. L. Hertz, E. Kvamme, E. G. McGeer, and A. Schousboe). A. R. Liss, Inc., New York.

Norenberg, M.D. (1979). The Distribution of Glutamine Synthetase in the Rat Central Nervous System. J. Histochem. Cytochem., 27, 756-762.

Norenberg, M.D. (1983). Immunohistochemistry of Glutamine Synthetase. In Glutamine, Glutamate, and GABA in the Central Nervous System. (eds. L. Hertz, E. Kvamme, E. G., McGeer, and A. Schousboe). A. R. Liss, Inc., New York.

Patel, M.S. (1974). The Relative Significance of CO_2-Fixing Enzymes in the Metabolism of Rat Brain. J. Neurochem., 22, 717-724.

Scrutton, M.C. and White, M.D. (1974). Purification and Properties of Human Liver Pyruvate Carboxylase. Biochemical Medicine, 9, 271-292.

Shank, R.P. and Aprison, M.H. (1981). Present Status and Significance of the Glutamine Cycle in Neural Tissues. Life Sci., 28, 837-842.

Shank, R.P. and Campbell, G.LeM. (1981). Avid Na^+-Dependent, High-Affinity Uptake of Alpha-Ketoglutarate by Nerve Terminal Enriched Material From Mouse Cerebellum. Life Sci., 28, 843-850.

Shank, R.P. and Campbell, G.LeM. (1983). Glutamate. In Handbook of Neurochemistry, Vol. 3, 2nd Edition. (ed. A. Lajtha). Plenum Press, New York.

Shank, R.P. and Campbell, G.LeM. (1983b). Metabolic Precursors of Glutamate and GABA. In Glutamine, Glutamate and GABA in the Central Nervous System. (eds. L. Hertz, E. Kvamme, E G. McGeer, and A. Schousboe). A. R. Liss, Inc., New York.

Shank, R.P. and Campbell, G.LeM. (1984a). Amino Acid Uptake, Content, and Metabolism by Neuronal and Glial Enriched Cellular Fractions from Mouse Cerebellum. J. Neurosci., 4, 58-69.

Shank, R. P. and Campbell, G.LeM. (1984b). α-Ketoglutarate and Malate Uptake and Metabolism by Synaptosomes: Further Evidence for an Astrocyte-to-Neuron Metabolic Shuttle. J. Neurochem, 42, 1153-1161.

Shank, R.P. and Campbell, G.LeM. (1984c). Glutamine, Glutamate, and Other Possible Regulators of α-Ketoglutarate and Malate Uptake by Synaptic Terminals. J. Neurochem., 42, 1162-1169.

Shank, R.P., Campbell, G.LeM., Freytag, S.O., and Utter, M.F. (1981). Evidence that Pyruvate Carboxylase is an Astrocyte Specific Enzyme. Soc. Neurosci. Abstrs., 7, 936.

Shank, R.P., Bennett, G.S., Freytag, S.O. and Campbell, G.LeM. (1985). Pyruvate Carboxylase: An Astrocyte-Specific Enzyme Implicated in the Replenishment of Amino Acid Neurotransmitter Pools. Brain Res., 329, 364–368.

Waelsch H., Berl, S., Rossi, C.A., Clarke, D.D., and Purpura, D. (1964). Quantitative Aspects of CO_2 Fixation in Mammalian Brain In Vivo. J. Neurochem., 11, 717–728.

Yu, A.C.H., Drejer, J., Hertz, L. and Schousboe, A. (1983). Pyruvate Carboxylase Activity in Primary Cultures of Astrocytes and Neurons. J. Neurochem., 41, 1484–1487.

4

Glia–Neuronal Interrelationships in the Metabolism of Excitatory Amino Acids

W.J. Nicklas

ABSTRACT

Neurotransmitter glutamate metabolism is thought to involve
neuronal-glial interactions. Glutamate is accumulated by
astrocytes by a high affinity, high capacity transporter. In
addition, glutamine synthesis and catabolism are intimately
associated with neurotransmitter glutamate. The enzyme, glutamine
synthetase, is highly enriched in astrocytes. The catabolic
enzyme, glutaminase, may or may not be enriched in glutamergic
nerve endings. Experiments from various laboratories are
described which illustrate these aspects of neurotransmitter
glutamate metabolism. Even complex events such as those
associated with the sequelae of excitotoxic amino acid analogs,
e.g., kainate and quisqualate, can be explained in some measure by
understanding how neurons and neighboring glial cells interact.

INTRODUCTION

A great deal of evidence has been amassed over the last decade
or so which indicates that glia, specifically astrocytes,
contribute significantly to the metabolism and, perhaps, the
functional regulation of the neurotransmitter amino acids,
glutamate/aspartate and GABA (for previous reviews of glial
contributions to neurotransmitter metabolism and function, see
Schoffeniels et al., 1978; Varon and Somjen, 1979; Hertz, 1979;
Bradford, 1982 and Hertz et al., 1983). Earlier studies focussed
on static nutritional and structural interactions between glial
cells and neurons, but recently, more dynamic aspects of the
relationship between neurons and glia have been studied. As Van
Gelder (1983) has rightly pointed out, examination of the close
apposition of glial processes and synapses at the electron
microscopic level graphically reinforces the concept that glial
processes, the extracellular space, and synapses form a unique,
dynamic compartment. Transfer of material between glia and

57

Table 1. Enzyme activities in rat brain striatal homogenates: Effect of kainic acid lesions.

Enzyme	Lesioned striatum	Contralateral striatum	Per Cent difference
	Activity ± S.D.		
Glutamine synthetase (11)	1.46 ± 0.43[a]	0.94 ± 0.23	+55.3
Glutaminase (5)	1.86 ± 0.24	2.09 ± 0.47	-
Glutamate dehydrogenase (6)	0.34 ± 0.08	0.72 ± 0.09	-53.3
Aspartate aminotransferase			
'Cytoplasmic' (5)	53.9 ± 10.4[b]	77.1 ± 13.1	-30
'Mitochondrial' (5)	23.4 ± 2.4	21.6 ± 5.5	-
GABA transaminase (6)	0.11 ± 0.05[c]	0.37 ± 0.10	-72.3
Succinate dehydrogenase (5)	6.23 ± 1.42	6.37 ± 0.27	-
Lactate dehydrogenase (5)	42.9 ± 8.7[a]	74.7 ± 9.6	-42.1

[a]Significantly different from contralateral striatum paired t-test, P < 0.01.
[b]Significantly different from contralateral striatum, P < 0.001.
[c]Significantly different from contralateral striatum, P < 0.02.
Activity is the average, from (n) pairs of striata, of μmol product formed (substrate utilized) $\cdot mg$ $protein^{-1}.h^{-1}$. Lactate dehydrogenase, aspartate aminotransferase and glutamate dehydrogenase were measured at 22°C; the other enzymes were measured at 37°C. Lesions were made by unilateral stereotaxic injection of 2 μg kainic acid 6-8 days prior to killing animals. Data taken from Nicklas et al., 1979.

neurons runs the gambit from the macromolecular (e.g., growth factors) to small ions such as K^+ and Cl^-. With regard to the metabolism of excitatory amino acids, most of the work has involved various aspects of glutamine metabolism. This chapter will primarily concern itself with such studies. Others have studied the possible involvement of anaplerotic mechanisms in the glia as the source of net synthesis of the carbon skeleton of transmitter amino acids (see Shank and Campbell, this volume). This, too, may ultimately be tied to glutamine metabolism in that early studies on the compartmentation of amino acid metabolism in the CNS showed a special association between CO_2 fixation and glutamine synthesis in the CNS (Berl et al., 1962; 1970). It may be that both processes are essential to maintain short and long term homeostasis of amino acid metabolism.

Table 2.   Ratio of labelling of amino acids from D-[2-^{14}C]-glucose
vs. that from [^{3}H]-acetate in striatal slices.

Specific activities (dpm/μmol)	Lesioned striatum Ratio of tritium	Contralateral striatum vs. carbon-14± S.D.
Glutamate	226 ± 29[a]	36.3 ± 2.5
Aspartate	69.6 ± 7.2[a]	9.0 ± 1.2
Glutamine	433 ± 300	359.2 ± 82.9
GABA	85.1 ± 10.4[b]	30.3 ± 4.0
Total dpm/g tissue	48.6 ± 4.4[a]	22.3 ± 1.2

[a]Significantly different from contralateral striatum, P < 0.001.
[b]Significantly different from contralateral striatum, P < 0.005.
Striatal slices prepared from unilateral kainate-lesioned animals
(1 week after 2 μg kainate) were preincubated 60 min at 37°C in
Krebs-Ringer bicarbonate fortified with glucose.   D-[2-^{14}C]-
glucose and [^{3}H]-acetate were added and incubation continued for a
further 15 min.   Data were obtained with four sets of two striata
per incubation.   Taken from Nicklas et al., 1979.

Similarly, most of what is known now of the cellular
localization of various aspects of the metabolism of glutamate
(and GABA) are quite consistent with the earlier studies on the
biochemical compartmentation of this metabolism (for review, see
Van den Berg et al., 1975; Berl and Clarke, 1983).   Such studies
indicated that exogenously supplied glutamate, aspartate and GABA
(as well as non-amino acids such as acetate) appeared to be
metabolized in the CNS largely in a "small" compartment which in
now thought to be the astroglial cell; astrocytes are now known to
have high affinity uptake systems for these substances (see Hertz,
1979).   Central to this "compartmentation" theory was the
experimentally-observed rapid conversion in vivo and in vitro, of
exogenous glutamate and NH$_3$ to glutamine which suggested a
disparate distribution of the synthesizing enzyme, glutamine
synthetase.   This hypothesis was confirmed by Norenberg and
Martinez-Hernandez (1979) who localized glutamine synthetase to
astrocytes in rat brain by immunohistochemistry.   Since then it
has become commonplace to use glutamine synthetase as a glial
marker.   It may be misleading to conclude that this enzyme is
present only in astrocytes but it does seem that it is highly
enriched in this cell type.   An example from previous work
illustrates some of these points (Nicklas et al., 1979).   When the
excitotoxin kainic acid, was stereotaxically injected unilaterally
into rat striatum, it caused destruction of cell bodies and
concomitant astrocytosis.   When we examined the animals one week
later we found that glutamine synthetase activity was increased

50-60% on the lesioned side (Table 1) with some interesting
changes in other enzymes. At the same time, there was a greatly
increased capacity of slices from the lesioned striatum to utilize
^{3}H-acetate relative to D-[2-^{14}C]-glucose (Table 2). This was due
both to a decrease in absolute rate of glucose incorporation and
an increase in acetate utilization. The increased GS activity and
alterations in metabolism are consistent with the hypothesis
concerning these parameters, i.e., GS is enriched in astrocytes
and ^{3}H-acetate is a tracer for this compartment.

In the remainder of this contribution, we shall review how
these concepts have been utilized to answer specific questions
concerning the neuronal-glial components of transmitter amino acid
metabolism in the CNS. Since an understanding of the metabolism
of glutamine is pivotal to elaborating the regulation of these
interactions, the review will focus in that direction.

Studies on Flow of Glutamine from Glia to Nerve Endings

Glutamine has been postulated to be a major precursor of
neuronal, i.e., transmitter, pools of the neuroactive amino acids,
including glutamate and GABA (Van den Berg and Garfinkel, 1971;
Benjamin and Quastel, 1974; Bradford and Ward, 1976; Weiler et
al., 1979) as well as being a "deactivated" form of the
neuroexcitatory, and potentially neurotoxic substance, glutamate
(Krespan et al., 1982). Long before the immunohistochemical
demonstration of GS in astrocytes, Van den Berg and Garfinkel
(1971), on the basis of computer simulation studies of radioactive
labelling data, proposed that GABA levels in nerve endings are
quantitatively maintained by glutamine from the extracellular
space which in turn is replenished by synthesis in a non-nerve
ending compartment via GS. By extrapolation a similar argument is
made for neurotransmitter glutamate metabolism. Phosphate-
stimulated glutaminase is thought to be the enzyme responsible for
glutamine catabolism. It probably does not have a singular
cellular localization (see Table 1, e.g.) although there may be
more or less activity, e.g., in synaptic mitochondria than in
mitochondria from glial cells or neuronal perikarya (Dennis et
al., 1977; Weiler et al., 1979). This topic is discussed in more
detail elsewhere in this volume by Wenthold. It is interesting
that in preliminary studies we have found glutaminase activity to
be significantly decreased in the cerebellum of the Weaver mouse
which lacks glutamergic granule cells; the levels of other related
enzymes, such as glutamate dehydrogenase and GS, are unaffected.
However, the localization of glutaminase to a specific cell type
or a specific carrier-mediated uptake system for glutamine may not
be overriding considerations. What is important is the in situ
functional activity of glutaminase. The brain enzyme is identical
to the kidney isozyme, immunologically and kinetically, and,
therefore, is subject to multiple effectors and inhibitors (see

Kovacevic and McGivan, 1983). For example, a decrease in glutamate levels during neurotransmission has been suggested to increase the in situ activity of this enzyme (Bradford et al., 1978; Benjamin 1981).

A variety of approaches have been taken to investigate the contributions of neurons and glia to the metabolism of glutamate and GABA. These range from in vivo studies on whole animals to tissue culture studies with isolated cell types. In examining the possibility of a functional flow of glutamine into transmitter pools, it is obvious that some degree of intactness of structure is needed in an experimental preparation; this introduces complexities but judicious use of labelling techniques and enzyme inhibitors has proven fruitful. A number of investigators have made use of the "compartmentalization" of amino acid metabolism in the CNS observed with various labelled precursors. As mentioned above, ^{3}H-or ^{14}C-acetate has been shown to more selectively label the "small" compartment in which glutamine is synthesized than do other substrates such as ^{14}C-glucose (see Van den Berg et al., 1975). Glucose, on the other hand, effectively labels depolarization-releasable pools of amino acids (Bradford et al., 1978; Hamberger et al., 1979). One consequence of this more effective incorporation of glucose label into nerve ending pools is that when intact preparations, such as was first shown with spinal cord slices (Minchin, 1977), are labelled with a combination of radioactive acetate/glucose and then depolarized, there is a preferential efflux of glutamate and GABA labelled from glucose. In our own studies we have found that stimulation of voltage-dependent Na^+ channels in brain slices with veratridine causes a similar preferential release of glutamate labelled from glucose, whereas other stimuli such as treatment with the excitotoxin, kainic acid, also causes massive release of glutamate labelled from ^{3}H-acetate (Krespan et al., 1982) (Table 3). Thus, these kinds of experiments allow one to determine, to some degree, the site of action of substances which cause release of amino acids. If one looked only at endogenous release, one might be confused since glutamate would be released in either of the above cases.

A further hypothesis arose from these kinds of experiments. If the glutamine-flow hypothesis were correct, then inhibition of either glutamine synthesis or glutaminase should decrease the flow of acetate-labelled material into synaptic pools of amino acids. Under either of these conditions, subsequent depolarization should show a greatly decreased efflux of amino acid labelled by acetate. This is precisely what occurred (Nicklas, 1982; 1983), and is illustrated in Table 4.

Methionine sulfoximine (MSO) is a potent, irreversible inhibitor of several glutamyl-transferring enzymes including GS (Griffith and Meister, 1979). Gutierrez and Norenberg (1977) found in rats injected with MSO that the preictal, primary effect of MSO is on astrocyte morphology and speculated that astrocytic

Table 3.   Release of labelled glutamate from cerebellar slices.

| Precursor | Fraction released[a] | |
	$[2-^{14}C]$-glucose	$[^3H]$-acetate
Control (4)	0.017 ± 0.002	0.079 ± 0.007
1 mM-Kainic acid (4)	0.126 ± 0.012[b]	0.475 ± 0.034[b]
Control (3)	0.014 ± 0.003	0.046 ± 0.015
10 μM-Veratridine (3)	0.051 ± 0.007[b]	0.049 ± 0.010

Values are the mean ± S.D. for (n) preparations. Cerebellar slices
were preincubated for 10 min, at which time D-$[2-^{14}C]$-glucose and
$[^3H]$-acetate were added. Twenty minutes later, the slices were
transferred to fresh Krebs-Ringer bicarbonate medium. Either
kainic acid or no drug was added immediately for a period of 20
min or veratridine (in ethanol) or ethanol alone (control) was
added 5 min later for a period of 6 min. Taken from Krespan et
al., 1982.

[a]Fraction of tissue glutamate released = $\dfrac{\text{Glutamate in medium}}{\text{Total glutamate}}$

[b]p < 0.005 from corresponding control.

abnormalities may play a role in MSO-induced seizures. In our
experiments, when rat cerebellar slices were incubated with MSO,
GS was inhibited as shown by the large decrease in the radio-
specific activity of glutamine relative to that of glutamate both
from glucose and acetate (Table 4). The relative labelling of
GABA in the tissue from acetate was also greatly decreased but
that from glucose was actually increased. This is especially
interesting in that the absolute specific activity of total tissue
glutamate labelled by glucose was lower in the presence of MSO.
When these slices were depolarized with veratridine, there was a
decreased release of acetate-labelled GABA and glutamate measured
as relative to the specific activity of glutamate in the tissue.
Since the acetate-derived specific activity of tissue glutamate
was not significantly altered by MSO, these changes in medium
amino acids represent genuine decreases in labelling from
acetate. Similar experiments with a glutaminase inhibitor also
showed a decrease in acetate-labelled neuronal stores (Nicklas,
1982; 1983).

These studies with MSO also enabled us to further examine the
problem of the release of glutamate by excitotoxins such as
kainate. Earlier work had indicated that a large part of the
glutamate released was from the acetate-labelled compartment, i.e,
astrocytic (Krespan et al., 1982). In contrast to the veratridine
data given above, kainate releases glutamate from tissue slices in
exact proportion to its increased absolute level and labelling by

Table 4. Effect of MSO on amino acid labelling in rat cerebellar slices from [^{14}C]-glucose and [^{3}H]-acetate.

	Tissue Specific Activity Glutamate [^{14}C] dpm/nmole ± S.D. [^{3}H]	
Control	18.3 ± 3.9	272 ± 19
1 mM MSO	9.9 ± 1.5[a]	278 ± 52

	Tissue Relative Specific Activity*			
	GLN		GABA	
	[^{14}C]	[^{3}H]	[^{14}C]	[^{3}H]
Control	0.33	3.13	1.25	0.23
1 mM MSO	0.12[a]	0.05[a]	1.84[a]	0.08[a]

	Ver-Released Amino Acids Relative Specific Activity*			
	GLUT		GABA	
	[^{14}C]	[^{3}H]	[^{14}C]	[^{3}H]
Control	0.94	1.79	1.33	0.38
1 mM MSO	0.89	0.50[a]	2.05[a]	0.05[a]

*Specific activity of glutamate in tissue taken as one.
[a]$p < 0.01$, different from corresponding control.
Rat cerebellar slices were pretreated with methionine sulfoximine (MSO). After working to remove the MSO, they were then incubated for 15 min with a mixture of D-[2-^{14}C]-glucose and [^{3}H]-acetate. After washing to remove the medium radioactivity, the slices were incubated with 10 μM veratridine for 3 min and medium and tissue rapidly separated.

acetate (Nicklas, 1983; in press). This is an excellent example of how understanding neuronal-glial interactions can help to elucidate quite complex situations. This subject is beyond the scope of this review but we continue to be interested in excitotoxic mechanisms and now believe Cl⁻transport to be intimately associated with such phenomena. It is of interest that Olney and his colleagues have concluded that influx of Cl⁻ but not Ca^{2+} is a necessary factor in the neurotoxic action of glutamate and analogs such as kainate and N-methyl-D-aspartate (J.W. Olney, personal communication; Rothman, 1984; Olney et al., 1984).

These in vitro experiments would indicate that glutamine flow does indeed form a communicative pathway between neurons and glia. In an interesting series of recent studies, Rothstein and Tabakoff (1984) have made use of in vivo MSO administration to examine the glutamergic system in the rat striatum. They found that MSO given intraventricularly decreased striatal GS activity and glutamine for a period of days with no change in total glutamate levels. However, the Ca^{2+}-dependent, K^+-stimulated glutamate release from striatal slices of MSO-treated animals was diminished by 50%. When glutamine was present in the perfusion medium in which release was measured, there was no difference in glutamate efflux from striatal slices of treated and control animals. It appears reasonable to suggest that MSO depleted available neuronal stores of glutamate in their experiments because it inhibited formation of its precursor, glutamine. Similar experiments in vivo may greatly add to our knowledge of the functionality of putative glutamergic pathways.

SUMMARY

The hypothesis of a flow of glutamine from its synthesis in the astrocyte to nerve ending pools of amino acids seems to be well-founded. More studies using labelling techniques coupled with specific enzyme inhibitors should help in elucidating the functionally relevant role for this interaction between neurons and glia both in normal and diseased CNS.

REFERENCES

Benjamin, A. (1981). Control of glutaminase activity in rat brain cortex in vitro: Influence of glutamate, phosphate, ammonium, calcium and hydrogen ions. Brain Res., 208, 363-377.

Benjamin, A. and Quastel, J.H. (1974). Fate of L-glutamate in the brain. J. Neurochem., 23, 457-464.

Berl, S. and Clarke, D.D. (1983). The metabolic compartmentation concept. In Glutamine, Glutamate and GABA in the Central Nervous System. (eds. L. Hertz, E. Kvamme, E.G. McGeer and A. Schousboe). Liss, New York.

Berl, S., Nicklas, W.J. and Clarke, D.D. (1970). Compartmentation of citric acid cycle metabolism in brain: Labeling of glutamate, glutamine, aspartate and GABA by several radioactive tracer metabolites. J. Neurochem., 17, 1009-1015.

Berl, S., Takagaki, G., Clarke, D.D. and Waelsch, H. (1962). Carbon dioxide fixation in the brain. J. Biol. Chem., 237, 2570-2573.

Bradford, H.F. (1982). Neurotransmitter Interactions and Compartmentation. Plenum Press, New York.

Bradford, H.F., deBelleroche, J.S. and Ward, H.K. (1978). On the metabolic and intrasynaptic origin of amino acid transmitters. In Amino Acids as Chemical Neurotransmitters. (ed. F. Fonnum). Plenum Press, New York.

Bradford, H.F. and Ward, H.K. (1976). On glutaminase activity in mammalian synaptosomes. Brain Res., 110, 115-125.

Dennis, S.C., Lai, J.C.K. and Clark, J.B. (1977). Comparative studies on glutamate metabolism in synaptic and non-synaptic rat brain mitochondria. Biochem. J., 164, 727-736.

Griffith, O. and Meister, A. (1979). Potent and specific inhibition of glutathione synthesis by buthionine sulfoximine. J. Biol. Chem., 254, 7558-7566.

Gutierrez, J.A. and Norenberg, M.D. (1977). Ultrastructural study of methionine sulfoximine-induced Alzheimer Type II astrocytosis. Am. J. Pathol., 86, 285-294.

Hamberger, A.C., Chiang, G.H., Nylen, E.S., Scheff, S.W. and Cotman, C.W. (1979). Glutamate as a CNS transmitter: Evaluation of glucose and glutamine as precursors for the synthesis of preferentially released glutamate. Brain Res., 168, 513-530.

Hertz, L. (1979). Functional interaction between neurons and astrocytes. Prog. Neurobiol., 13, 277-323.

Hertz, L., Kvamme, E., McGeer, E.G. and Schousboe, A. (1983). Glutamine, Glutamate, and GABA in the Central Nervous System. Liss, New York.

Kovacevic, Z. and McGivan, J.D. (1983). Mitochondrial metabolism of glutamine and glutamate and its physiological significance. Physiol. Rev., 63, 547-605.

Krespan, B., Berl, S. and Nicklas, W.J. (1982). Alteration in neuronal-glial metabolism of glutamate by the neurotoxin kainic acid. J. Neurochem., 38, 509-518.

Minchin, M.C.W. (1977). The release of amino acids synthesized from various compartmental precursors in rat spinal cord slices. Exp. Brain Res., 29, 515-526.

Nicklas, W.J. (1982). Is glutamine formed in glial cells a precursor for neuronal glutamate and GABA? Trans. Am. Soc. Neurochem., 13, 221.

Nicklas, W.J. (1983). Relative contributions of neurons and glia to metabolism of glutamate and GABA. In Glutamine, Glutamate and GABA. (eds. L. Hertz, E. Kvamme, E.G. McGeer and A. Schousboe). Liss, New York.

Nicklas, W.J., Nunez, R., Berl, S. and Duvoisin, R.C. (1979). Neuronal-glial contributions to transmitter amino acid metabolism: Studies with kainic acid-induced lesions of rat striatum. J. Neurochem., 33, 839-844.

Norenberg, M.D. and Martinez-Hernandez, A. (1979). Fine structural localization of glutamine synthetase in astrocytes of rat brain. Brain Res., 161, 303-310.

Olney, J.W., Price, M.T., Samson, L. and Labruyere, J. (1984). The ionic basis of excitotoxin-induced neuronal necrosis. Soc. Neurosci. Abstr., Vol. 10, p. 24.

Rothman, S.M. (1984). Excitatory amino acid neurotoxicity is produced by passive chloride influx. Soc. Neurosci. Abstr., Vol. 10, p. 24.

Rothstein, J.D. and Tabakoff, B. (1984). Alteration of striatal glutamate release after glutamine synthetase inhibition. J. Neurochem., 43, 1438-1446.

Schoffeniels, E., Franck, G., Hertz, L. and Tower, D.B. (1978). Dynamic Properties of Glial Cells. Pergamon Press, Oxford.

Van den Berg, C.J. and Garfinkel, D. (1971). A simulation study of brain compartments. Biochem. J., 123, 211-218.

Van den Berg, C.J., Reijnierse, A., Blockhuis, G.G.D., Kroon, M.C., Ronda, G., Clarke, D.D. and Garfinkel, D. (1975). A model of glutamate metabolism in brain: A biochemical analysis of a heterogeneous structure. In Metabolic Compartmentation and Neuro-transmission. (eds. S. Berl, D.D. Clarke and D. Schneider). Plenum Press, New York.

Van Gelder, N.M. (1983). Metabolic interaction between neurons and astroglia: Glutamine synthetase, carbonic anhydrase and water balance. In Basic Mechanisms of Neuronal Excitability. Liss, New York.

Varon, S. and Somjen, G. (1979). Neuron-Glia Interactions. MIT Press, Cambridge, Mass.

Weiler, C.T., Nystrom, B. and Hamberger, A. (1979). Glutaminase and glutamine synthetase activity in synaptosomes, bulk isolated glia and neurons. Brain Res., 160, 539-543.

5

The Compartmentation and Turnover of Glutamate and GABA: A Better Understanding by the Use of Drugs

F. Fonnum, V.M. Fosse and R. Paulsen

COMPARTMENTATION OF GLUTAMATE-GLUTAMINE-GABA IN THE BRAIN

Kinetic studies on the metabolism of amino acid precursors in the brain have shown that the synthesis and metabolism of glutamate, aspartate, glutamine and GABA can be explained from a two-compartment model. (Berl et al, 1961; Berl and Clarke, 1978; van den Berg and Garfinkel, 1971; Balazs et al, 1970). Such a simple classification is surprising in view of the heterogenous composition of the brain both in anatomical and biochemical terms. The two compartments are called the large and small glutamate compartment respectively, and each contains its own tricarboxylic acid cycle. The compartmentation was first demonstrated by comparing the specific radioactivity of glutamine to that of glutamate after injection of different labelled precursors. In the case of [^{14}C]bicarbonate, short chain fatty acids (acetate, propionate, butyrate) and amino acids (glutamate, aspartate, GABA, leucine) the ratio exceeds 1.0 a short time after injection. In the case of glucose, pyruvate, lactate, glycerol and hydroxybutyrate the ratio stays well below 1.0. The results can be explained if the first group of precursors are either preferentially taken up or metabolized in a small pool of glutamate which are responsible for the synthesis of glutamine. The other group of precursors are preferentially taken up or metabolized in a large pool of glutamate, where

the synthesis of glutamine is negligible. Detailed studies of the
different precursors have lead to the suggestion of a 5 compartment
model (van den Berg et al, 1978).

The two major compartments have been identified in morphological terms and the small compartment corresponds to glial cells.
Glutamine synthetase (EC 6.3.1.2),the enzyme responsible for glutamine synthesis, has been localized in glial cells by an immunohistochemical technique (Martinez-Hernandez et al, 1977).
Ultrastructural studies have shown the enzyme to be present in
astrocytes (Norenberg and Martinez-Hernandez, 1979). On the other
hand numerous studies support the concept that the large glutamate
compartment is the neuron. Autoradiographs of the dorsal ganglion
after incubation with [^{14}C]glucose show most of the labelling over
neuronal cell bodies (Minchin and Beart, 1974). A decrease in amino
acids from labelling with glucose was found in tissues with dege-
nerated neural structures (Minchin and Fonnum, 1979; Nicklas,
1983).

THE BASAL GANGLIA AS A MODEL FOR AMINO ACID TRANSMITTERS

The neurotransmitter contents of the different connections in
the basal ganglia are well known. In particular the amino acid
transmitters play a dominant role in neostriatum, globus pallidus
and substantia nigra. In neostriatum the heavy cortical input
releases glutamate as its chemical transmitter (Fonnum et al, 1981)
whereas GABA is present as a transmitter of the local inhibitory
neurons which project heavily to globus pallidus and substantia
nigra (Fonnum et al, 1978). In the two latter structures there is
a very small if any glutamate input.

This situation makes it possible to produce several interesting models which allow us to focus in more detail on the composition of the glutamergic and GABAergic terminals. Unilateral decortication allows us to compare neostriatum in the presence and absence of glutamergic terminals. Likewise a unilateral lesion in neostriatum will destroy the GABAergic terminals unilaterally in globus pallidus and substantia nigra. Kainic acid lesion of neostriatum destroys the many interneurons and leaves a region with glial structures, a relative increased proportion of glutamergic terminals and deprived of the many non-glutamergic neurons. We have taken advantage of the different models to study the neurochemical details of the glutamergic and GABAergic terminals.

LOCALIZATION OF ENZYMES LINKED TO GLUTAMATE-GLUTAMINE-GABA

Enzymes have been localized by several different techniques such as histochemistry, immunohistochemistry, lesion studies, subcellular fractionation and inhibitors. As mentioned above glutamine synthetase seems to have an almost exclusive astrocyte localization. The other enzymes involved in glutamate metabolism are usually present in both glial and neuronal structures, but often with a specific cellular preference.

Aspartate aminotransferase (EC 2.6.1.1.AspT) is localized in the brain as two isoenzymes with different kinetic properties and localized in the cytoplasm and mitochondria respectively (Fonnum, 1968). The mitochondrial enzyme activity dominates, but the cytoplasmic enzyme shows product inhibition (Fonnum, 1968). After degeneration of certain neuronal tracts such as the habenula-interpeduncular tract (Sterri and Fonnum, 1980) there was a reduction of mitochondrial AspT activity in the terminal region to the same extent as other synaptic mitochondrial enzymes such as glutaminase and carnitine acetyltransferase, but different from GABA-T and

citrate cleavage enzyme. This should at least indicate a preferential localization of the enzyme activity in nerve terminals. The
finding that amino-oxyacetic acid in high concentration inhibited
glutamate but not glutamine formation in brain slices accords with a
localization of the enzyme mainly to the large compartment (Berl and
Clarke, 1978). Further β-methyleneDL$_1$-aspartate, an inhibitor of
AspT, inhibited oxygen consumption from glucose to the
same extent as AspT inhibition (Fitzpatrick et al, 1983).
Immunohistochemical studies of the cytoplasmic isoenzyme showed that
the enzyme was localized to the auditory nerve, photoreceptors,
ganglion cells in retina and basket and stellate cells in cerebellum
and in cells in neocortical layer II and III (Altschuler et al.,
1981; Wenthold and Altschuler, 1983). These studies should
indicate a preferential localization to aspartate/glutamate neurons.
Histochemical pattern of AspT in hippocampus did not coincide with
the autoradiography of D-Aspartate uptake (Schmidt and Wolf,
1985; Aamodt et al, 1984). In decorticated neostriatum Sandberg et
al (1985) did not detect any significant reduction in AspT activity.
The recent study of Asp-T in fiber tracts in rat brain indicates
that mitochondrial content rather than the transmitter of the fibre
was important for the enzyme level (Godfrey et al, 1984). Further,
there was no correlation between the aspartate level and total AspT
activity in regions of the spinal cord where the aspartate level
varies considerably and aspartate interneurons are localized
(Graham Jr. and Aprison, 1969). We (Fonnum and Iversen,
unpublished) have found only small variations in the ratio of
cytoplasmic AspT and lactate dehydrogenase in synaptosomes from
different brain regions. In conclusion, AspT probably is mainly
localized in neurons. It may have a preference for aspartergic
neurons, but is not a very reliable marker in all brain regions.

The phosphate-stimulated glutaminase (EC3.5.1.2.)has been
localized to mitochondria, particularly the synaptosome

mitochondria, by subcellular fractionation technique (Salganicoff
and De Robertis, 1965; Bradford and Ward, 1976). The results of
lesion studies are unequivocal with regard to the specific localization in different types of neurons. Cortical ablation was accompanied by 20% loss of glutaminase activities in neostriatum (Ward
et al., 1982; Sandberg et al, 1985), but intrastriatal injection of
kainic acid was accompanied by 50% reduction in neostriatum (McGeer
and McGeer, 1978; Nicklas et al, 1979). McGeer and McGeer (1978)
have claimed that the major proportion of glutaminase in
neostriatum originated from GABAergic interneurons. Glutaminase was
shown to be present in higher concentration in enriched Purkinje
and granule cell fractions than in astrocyte fraction from cerebellum (Patel, 1982). We found much higher level of glutaminase,
Asp-T and fumarase in neostriatum than in globus pallidus indicating that mitochondrial content rather than transmitters are of
importance. Immunohistochemical studies showed specific localization of glutaminase activity in certain amacrine cells in retina,
mossy fibres and pyramidal cells in hippocampus, granule cells in
cerebellum and in cells of neocortical layers V and VI (Wenthold
and Altschuler, 1983), all of which probably use glutamate as a
transmitter. In conclusion phosphate stimulated glutaminase seems
to be more dominant in neuronal than glial structures. It is at
present difficult to relate the enzyme to one transmitter type of
neurons only, but it is obviously highly localized to at least some
GABAergic and glutamergic neurons. It may be that it is not the
localization, but perhaps the regulation of the enzyme activity
which is of importance for its function (Kvamme and Olsen, 1980).
The importance of glutamine as a precursor for transmitter glutamate and GABA is well recognized (Hamberger et al, 1979).

Glutamate dehydrogenase (EC1.4.1.2.) has by subcellular fractionation studies been localized to non-synaptic mitochondria
(Reijnierse et al., 1975). The enzyme activity was found to be

higher in astroglial fraction than in Purkinje cell or granule cell fractions from cerebellum (Patel, 1982). After infusion with $[^{15}N]$ ammonium acetate the labelling of the α-amino group of glutamine was 10-fold higher than that of glutamate. This is consistent with a localization of the enzyme in the small compartment. But the enzyme must also be present in the neuronal mitochondria to account for the liberation of ammonia from endogenous glutamate in a glucose saline medium containing amino-oxyacetic acid (Benjamin and Quastel, 1975). It is therefore with some surprise we see that Nicholas et al (1979) observed a 50% loss of glutamate dehydrogenase after kainic acid treatment of neostriatum. Evidence favours a preferential localization of the enzyme in the astroglial cells, but does not exclude its presence in neurons.

Subcellular fractionation and lesion studies indicate that GABA-T is predominantly localized to non-terminal mitochondria (Van Kempen et al, 1965; Kataoka et al, 1974). Histochemical detection of GABA-T after inhibition has shown a rapid recovery of enzyme activity in nerve terminals (Vincent et al, 1980). This has been exploited as a method for detecting GABAergic terminals.

Pyruvate carboxylase catalyzes the formation of oxaloacetate from pyruvate and leads to a net synthesis of a TCA cycle constituent. The enzyme has been shown by immunohistochemical technique to be highly concentrated in the astrocytes (Shank et al, 1981). It was also found to be present in high concentration in a primary culture of astrocytes (Yu et al, 1983).

QUANTIFICATION OF GLUTAMATE IN THE DIFFERENT POOLS IN NEOSTRIATUM AND OTHER BRAIN REGIONS

Glutamate has a central role in the brain metabolism as well as being a transmitter and a precursor for GABA. It should therefore be

localized to 4 pools in brain: (1) transmitter pool present in glutamergic axons and terminals: (2) glutamate precursor pool present
in GABAergic neurons: (3) non-glutamergic neurons called the neuronal metabolic pool: (4) glial pool. The size of these pools will
differ from area to area and from species to species.

The transmitter pool of glutamate may be quantified after
selective degeneration of the glutamate terminals by surgical or
chemical lesions. By this technique the transmitter glutamate pool
was found to be 20-30% of the total glutamate content in
neostriatum, thalamus, lateral septum, nucleus accumbens and fornix
terminal regions (Fonnum et al, 1981; Walaas and Fonnum, 1980; Lund
Karlsen and Fonnum, 1978). Glutamate dominates as the transmitter
of the hippocampal pyramidal cells (Malthe-Sørenssen et al, 1979).
Destruction of the glutamergic pyramidal cells in hippocampus was
accompanied by 50% of the glutamate level in hippocampus (Fonnum
and Walaas, 1981) The transmitter pool in hippocampal neurons could
therefore be 50%.

The pool of glutamate as a precursor for GABA can be estimated
after selective destruction of GABAergic terminals. In substantia
nigra, the region with the highest content of GABAergic terminals,
selective destruction of GABA terminals was accompanied by small
decrease in glutamate levels (Korf and Venema, 1983; Minchin and
Fonnum, 1979). The precursor pool of glutamate is therefore probably
less than 10% of the total glutamate pool in any brain region.

The proportion of glutamate in the small glutamate pool, ie
the glial cells, can be estimated from the relative specific
radioactivity of glutamine compared to glutamate after administration of group 1 precursors (acetate, aspartate, $NaHCO_3$), In
several regions of cat brain this ratio was 5, which should indicate that the glial pool was 20% of the total glutamate pool (Berl

et al, 1961). In agreement we found a 15% reduction in glutamate
after administration of fluoroacetate which presumably inhibits the
tricarboxylic acid cycle in the glial pool. In this respect it
should be born in mind that acetate may not be exclusively localized to the small pool and that the figure is possibly an
overestimation of the glutamate content. Based on kinetic models
Cremer et al (1975) suggested that the glial pool of glutamate was
very small, perhaps in the order of 5%.

The remaining glutamate must be ascribed to other neuronal
structures and coupled to the general metabolism in neurons. This
is by far the largest pool and it probably accounts for about half
of the glutamate content in the brain. In neostriatum, the neuronal cell bodies are destroyed by kainic acid injection. Such
treatment was accompanied by 50% loss of glutamate in the
neostriatum (Nicklas et al, 1979, unpublished) and suggest this
figure for metabolic glutamate pool.

HYPOGLYCEMIA

Glucose is one of the most important precursors for the
transmitter pools of amino acids. It was therefore considered of
interest to study the amino acid changes in the absence of glucose.
Rats were therefore made hypoglycemic by injection of insulin
Actrapid (20 I.E) i.p (Engelsen and Fonnum, 1984). The animals
were denied food, but not water, 20 h prior to the experiment. In
intact neostriatum, which contains both metabolic and transmitter
pool of glutamine, there was first an increase in aspartate and
then a decrease in glutamate. Later when glutamine was severely
reduced, there was also a reduction in glutamate whereas GABA
decreased after isoelectric EEG (Agardh et al, 1978; Engelsen and
Fonnum, 1984). The turnover of GABA was reduced during the last 2

hours of hypoglycemia (Paulsen and Fonnum, unpublished). Since glutamine was reduced prior to glutamate it may be taken as evidence
for its role as precursor for glutamate in vivo.

Other neurotransmitters such as dopamine, noradrenaline and
5HT, are also reduced in neostriatum during hypoglycemia, but
usually at the later stages (Agardh et al, 1979). The extreme sensitivity of acetylCoA to hypoglycemia explains the sensitivity of
acetylcholine synthesis (Gibson and Blass, 1976).

After decortication when neostriatum only contains the metabolic
pool of glutamate, the changes in amino acids were similar but less
pronounced (Engelsen and Fonnum, 1984). We believe that this indicates a slower turnover of glutamate in the operated neostriatum.
This could well mean that the transmitter pool has a higher turnover than the metabolic pool. An alternative explanation is that
the glutamergic input is the driving force of the neostriatum.

When striatal slices were stimulated by a high concentration of
K^+ in the presence of glucose, the Asp/Glu ratio was constant and
about 0.25. When the slices were stimulated in the absence of glucose i.e. hypoglycemic conditions, the released Asp/Glu ratio
increased from 0.25 to 0.85 after 3 stimuli. In the tissue the
final ratio reached 1.7. This indicates that the neostriatal
synaptosome cannot differentiate between the release of Asp and
Glu. This is in accordance with our previous work on the use of
D-Aspartate as a false transmitter (Malthe-Sørenssen et al, 1979).
It is also expected from studies when neostriatal slices were incubated with aspargine and aspartate was released predominantly to
glutamate (Reubi et al, 1980). But it underlines the difficulties
involved in differentiating between the presynaptic release of glutamate and aspartate. Also under these conditions the release of
GABA was more resistant to changes than the release of glutamate.

INHIBITION OF GLUTAMINE SYNTHESIS

Methionine sulphoximine is an inhibitor of glutamine synthetase which is located in the astroglial cells. Injection of this compound is accompanied by a rapid decrease of glutamine and a slow decrease of glutamate in intact neostriatum.

The fall of glutamine prior to glutamate support the concept that glutamine may be a precursor for glutamate. In the decorticated neostriatum the decrease of glutamine was even faster than in the unoperated side. We take this as evidence that the release of transmitter glutamate on the normal side is a major source of glutamine. The other sources of small compartment glutamate appear to be reductive deamination or transamination of small compartment 2-oxoglutarate. It is again interesting that although glutamine was reduced to almost zero there was no effect on GABA. As in hypoglycemia GABA is very resistant to changes. It would in this respect be very interesting to see if there is any effect on GABA turnover.

INHIBITION OF GABA-T

GABA-T can be inhibited by a series of compounds, the most specific being γ-vinyl GABA, GABAculine and ethanolamine-O-sulphate (review Fonnum, 1981). When γ-vinyl GABA was locally injected into neostriatum in concentrations high enough to inhibit GABA-T 80-90%, there was an almost linear increase of GABA for 4 hours. At the same time we observed a decrease in glutamine, and to a lesser extent in glutamate and aspartate. Similar results were obtained by Bernasconi et al (1982) who injected aminooxyacetic acid and GABAculine i.p., but not by Chapman et al (1982) who injected γ-vinyl GABA i.p. The sum of the 4 amino acids (GABA, glutamine, glutamate and aspartate) was only moderately changed. The results demonstrates the tight coupling between GABA and glutamine.

CONCLUDING REMARKS

It is generally agreed that glutamine synthetase and glutamate decarboxylase are speficly localized in astrocytes and GABAergic neurons respectively. But the other enzymes such as cytoplasmic and mitochondrial AspT and glutaminase must have a broad distribution which do not always correspond to a preferential localizaton in GABAergic or Glu/Asp neurons preferentially. Comparisons with other mitochondrial or cytoplasmic enzymes are often neglected before conclusions are drawn.

Studies with different drugs indicate that the turnover of glutamate, glutamine, aspartate and GABA changes with different rate in intact or decorticated neostriatum. In general we see a large loss in glutamine, before we see a change in glutamate and GABA. It is also striking that the changes in GABA level is very resistant to decrease in its precursors glutamate and glutamine. This must indicate that the glutamate precursor pool is more resistant than the other glutamate pool to changes.

REFERENCES

Aamodt, A., Aambø, A., Walaas, I., Søreide, J. and Fonnum, F. (1984). Autoradiographic demonstration of glutamate structues after stereotaxic injection of kainic acid in rat hippocampus. Brain Res., 294, 341-345.

Agardh, C.-D., Folbergrova, J. and Siesjö, B.K. (1973). Cerebral metabolic changes in profound insulin-induced hypoglycemia and in the recovery period following glucose administration. J. Neurochem., 31, 1135-1142.

Agardh, C.D., Carlsson, A., Lundquist, M. and Siesjö, B.K. (1979). The effect of pronounced hypoglycemia on the monoamine metabolism in rat brain. Diabetes 28, 804-809.

Altschuler F.R., Neises G.R., Harmison G.G., Wenthold R.J. and Fex J. (1981). Immunocytochemical localization of aspartate aminotransferase immunoreactivity in cochlear nucleus of the guinea-pig. Proc. natn. Acad. Sci. U.S.A. 78, 6553-6557.

Balazs R., Machiyama Y., Hammond B.J., Julian T. and Richter D. (1970). The operation of the γ-aminobutyrate bypath of the tricarboxylic acid cycle in brain tissue in vitro. Biochem. J., 116, 445-467.

Benjamin, A.M. and Quastel, J.H. (1975). Metabolism of amino acids and ammonio in rat brain cortex slice in vitro: A possible role of ammonia in brain function. J. Neurochem.

Berl S. and Clarke D.D. (1978). Metabolic compartmentation of the glutamate-glutamine system; glial contribution. In: Amino Acids as Chemical Transmitters (Ed. F. Fonnum), pp. 691-708, Plenum Press, New York.

Berl, S., Lajhta, A. and Waelsch, H. (1961). Amino acid and protein metabolism. VI. Cerebral compartments of glutamic acid metabolism. J. Neurochem., 7, 186-197.

Bernasconi R., Maitre L., Martin P. and Raschdorf F. (1982). The use of inhibitors of GABA-transaminase for the determination of GABA turnover in mouse brain regions: An evaluation of aminooxyacetic acid and GABAculine. J. Neurochem., 38, 57-66.

Bradford H.F. and Ward H.K. (1976). On glutaminase in mammalian synaptosomes. Brain. Res., 110, 115-125.

Chapman A.G., Riley K., Evans M.C. and Meldrum B.S. (1982). Acute effects of sodium valproate and γ-vinyl GABA on regional amino acid metabolism 2-[^{14}C] glucose into amino acids. Neurochem. Res., 9, 1089-1105.

Cremer J.E., Heath D.F., Patel A.J., Balazs R. and Cavanagh, J.B. (1975). An experimental model of CNS changes assciated with chronic liver disease: Portocaval anastomosis in the rat. In: Metabolic Compartmentation and Neurotransmission. (Eds. S. Berl, D.D. Clarke and D. Schneider). Plenum Press New York, pp 461-478.

Engelsen B. and Fonnum F. (1983). Effects of hypoglycemia on the transmitter pool and the metabolic pool of glutamate in rat brain. Neuroscience Letts., 42, 317-322.

Fitzpatrick, S.M., Cooper A.J.L. and Duffy T.E. (1983). Use of β-methylene-D,L-aspartate to assess the role of aspartate aminotransferase in cerebral oxidative metabolism. J. Neurochem., 41, 1370-1383.

Fonnum, F. (1968). The distribution of glutamate decarboxylase and aspartate transaminase in subcellular fractions of rat and guinea-pig brain. Biochem. J., 106, 401-412.

Fonnum, F., Gottesfeld, Z. and Grofova, I. (1978). Distribution of glutamate decarboxylase, choline acetyltransferase and aromatic amino acid decarboxylase inb the basal ganglia of normal and operatef rats. Evidence for striatopallidal, striatoentopeduncular and striatonigral GABAergic fibres. Brain Re., 143, 125-138.

Fonnum, F. and Walaas, I. (1978). The effect of intrahippocampal kainic acid injections and surgical lesions on neurotransmitters in hippocampus and septum. J. of Neurochem., 31, 1173-1181.

Fonnum, F. (1981). The turnover of transmitter amino acids with special reference to GABA. In C.J. Pycock and P.V. Taberner (Eds.) Central neurotransmitter turnover. University Park Press, Baltimore, pp. 105-124.

Fonnum, F., Storm-Mathisen, J. and Divac, I. (1981). Biochemical evidence for glutamate as neurotransmitter in the cortico-striatal and cortico-thalamic fibres in rat brain. Neuroscience 6, 863-875.

Gibson, G.E. and Blass, P.J. (1983). Metabolism and neurotransmission. In: Handbook of Neurochemistry, Vol 3 (2nd Ed.), A. Lajhta (Ed.). Plenum Publishing Corporation.

Graham, Jr. L.T. and Aprison M.H. (1969). Distribution of some enzymes associated with the metabolism of aspartate, glutamate γ-aminobutyrate and glutamine in cat spinal cord. J. Neurochem., 16, 559-566.

Hamberger, A., Chiang, G.H., Nylén, E.S., Scheff, S.W. and Cotman, C.W. (1979). Glutamate as a CNS transmitter. I. Evaluation of glucose and glutamine as precursors for the synthesis of preferentially released glutamate. Brain Res., 168, 513-530.

Kataoka, K., Bak, I.J., Hassler, R., Kim J.S. and Wagner, A. (1974). L-glutamate decarboxylase and cholineacetyltransferase activity the substantia nigra and the striatum after surgical interruption of the strio-nigral fibres of the baboon. Exp. Brain. Res., 19, 217-227.

Korf, J. and Venema, K. (1983). Amino acids in the substantia nigra in rats with striatal lesions produced by kainic acid. J. Neurochem., 40, 1171-1173.

Kvamme, E. and Olsen, B.E. (1980). Substrate mediated regulation of phosphate-activated glutaminase in nervous tissue. Brain Res., 181, 228-233.

Lund-Karlsen, R. and Fonnum, F. (1978). Evidence for glutamate as a neurotransmitter in the corticofugal fibres to the dorsal lateral geniculate body and the superior colliculus in rats. Brain Res., 151, 457-467.

Malthe-Sørenssen, D., Skrede, K.K. and Fonnum, F. (1979). Calcium dependent release of D-[^{3}H]aspartate evoked by selective electrical stimulation of excitatory afferent fibres to hippocampal pyramidal cells in vitro. Neurosci., 4 , 1255-1263.

Martinez-Hernandez, A., Bell, K.P. and Norenberg, M.D. (1977). Glutamine synthetase: Glial localization in brain. Science 195, 1356-1358.

McGeer, E.G. and McGeer, P.L. (1978). Localization of glutaminase in the rat neostriatum. J. Neurochem., 32, 1071.

Minchin, M.C.W. and Beart, P.M. (1974). Compartmentation of amino acid metabolism in the rat dorsal ganglia: A metabolic and autoradiographic study. Brain Res., 83, 437-449.

Minchin, M.C.W. and Fonnum, F. (1979). The metabolism of GABA and other amino acids in rat substantia nigra slices following lesions of the striato-nigral pathway. J. Neurochem., 32, 203-210.

Nicklas, W.J. (1983). Relative controbutions of neurons and glia to metabolism of glutamate and GABA. In: Glutamine, Glutamate and GABA in the Central Nervous System. (Eds. L. Hertz, E. Kvamme, E.G. McGeer and A. Schousboe), 219-231. Alan R. Liss, New York.

Nicklas, W.J., Nunez, R., Berl, S. and Duvoisin, R. (1979). Neuronal glial contributions to transmitter amino acid metabolism: studies with kainic acid-induced lesions of rat striatum. J. Neurochem., 33, 839-844.

Norenberg, M.D. and Martinez-Hernandez, A. (1979). Fine structural localization of glutamine synthetase in astrocytes of rat brain. Brain Res., 161, 303-310.

Patel, J.A. (1982). The distribution and regulation in nerve cells and astrocytes of certain enzymes associated with the metabolic compartmentation of glutamate. In: Neurotransmitter Interaction and Compartmentation. (Ed. A.F. Bradford), pp 411-429, Plenum Press, New York.

Reijnierse, G.L.A., Veldstra, H. and van den Berg, C.J. (1975).
Subcellular localization of γ-aminobutyrate transaminase and glutamate dehydrogenase in adult rat brain. Biochem. J., 152, 469-475.

Reubi, J.C., Toggenburger, G. and Cuenod, M. (1980). Aspargine as
precursor for transmitter aspartate in corticostriatal terminals.
J. Neurochem., 35, 1015-1017.

Salganicoff, L. and De Robertis, E. (1965). Subcellular distribution of the enzymes of the glutamic acid, glutamine and γ-
aminobutyric acid cycles in brain. J. Neurochem., 12, 287-309.

Sandberg, M., Ward, H.K. and Bradford, H.F. (1985). Effect of
cortico-striate pathway lesion on the activities of enzymes
involved in synthesis and metabolism of amino acid neurotransmitters in the striatum. J. Neurochem., 44, 42-47.

Shank, R.P., Campbell, G., Le, M., Freitag, S.V. and Utter, M.F.
(1981). Evidence that pyruvate carboxyleserase is an atrocyte specific enzyme. Soc. Neurosci. Abst., 7, 936.

Schmidt, W. and Wolf, G. (1984). Histochemical localization of
aspartate aminotransferase activity in the hippocampal formation
and in peripheral ganglia of the rat with special reference to the
glutamate transmitter metabolism. J. Hirnforsch., 25, 505-510.

Sterri, S.H. and Fonnum, F. (1980). Acetyl-CoA synthesizing enzymes in cholinergic nerve terminals. J. Neurochem., 35, 249-254.

van den Berg, C.J. and Garfinkel, D. (1971). A simulation study of
brain compartments. Biochem. J., 123, 211-218.

van den Berg, C.J., Matheson, D.F., Ronda, G., Reijnierse, G.L.A.,
Blokhuis, G.G.D., Kroon, M.C., Clarke, D.D. and Garfinkel, D.
(1975). A model of glutamate metabolism in brain: a biochemical
analysis of a heterogenous structure. In: Metabolic Compartmentation and Neurotransmission. (Eds. S. Berl., D.D. Clarke and
D. Schneider), NATO Advanced Study Institute Series Vol. 16,
Plenum Press, New York, pp 709-723.

van Kempen, G.M.J., van den Berg, C.J., van der Helm, H.J. and
Veldstra, H. (1965). Intracellular localization of glutamate
decarboxylase, γ-aminobutyrate transaminase and some other enzymes
in brain tissue. J. Neurochem., 12, 581-588.

Vincent, S.R., Kimura, H. and McGeer, E.G. (1980). The phar-
macohistochemical demonstration of GABA transaminase. Neurosci.
Lett., 8, 354-358.

Walaas, I. and Fonnum, F. (1980). Biochemical evidence for glutamate as a transmitter in hippocampal efferents to the basal
forebrain and hypothalamus in rat brain. Neuroscience 5,
1691-1698.

Ward, H.W., Thanki, C.M. and Bradford, H.F. (1983). Glutamine and
glucose as precursors of transmitter amino acids: ex vivo studies. J. Neurochem., 40, 855-860.

Wenthold, R.J. and Altschuler, R.A. (1983). Immunocytochemistry of
aspartate aminotransferase and glutaminase. In: Glutamine,
Glutamate and GABA in the Central Nervous System. (Eds. L. Hertz,
E. Kvamme, E. McGeer and A. Schousboe), 33-50, Alan Liss, New
York.

Yu, A.C.H., Drejer, J., Hertz, L. and Schousboe, A. (1983).
Pyruvate carboxylase activity in primary cultures of astrocytes and
neurons. J. Neurochem., 41, 1484-1487.

6

Immunocytochemical Localization of Enzymes Involved in the Metabolism of Excitatory Amino Acids

R.J. Wenthold and R.A. Altschuler

SUMMARY

The immunocytochemical localizations of several glutamate/aspartate related enzymes are discussed in relation to the neurotransmitter roles of these amino acids. Glutaminase is found to be enriched in several putative excitatory amino acid neurons including the auditory nerve, neurons of the hippocampus and cerebral cortex and granule cells of the cerebellum. Cytoplasmic aspartate aminotransferase is enriched in some excitatory amino acid neurons, but also in many, but not all, GABAergic neurons. Mitochondrial aspartate aminotransferase is found in several populations of neurons, but no pattern related to a particular neurotransmitter is seen. Both glutamate dehydrogenase and glutamine synthetase are enriched in glia. These findings show very different distributions for these enzymes and support other results that suggest glutaminase plays a central role in the production of neurotransmitter glutamate. High levels of glutaminase in neurons may serve as an immunocytochemical marker for excitatory amino acid neurons.

INTRODUCTION

The use of antibodies has had a major impact on our understanding of the molecular organization of the central nervous system by allowing the purification, localization and characterization of many neuronal molecules, some of which are present in minute quantities. This has been especially useful to the study of neurotransmission through the immunocytochemical localization of neurotransmitters, neurotransmitter related enzymes and neurotransmitter receptors. The presence of neurotransmitters in even minor populations of neurons can be determined, and such issues as colocalization of neurotransmitters within the same neuron, which previously could be studied only indirectly, can now be addressed. Such studies depend on the availability of a specific marker for the neurotransmitter. Unlike many of the traditional neurotransmitters, such as acetylcholine, the catecholamines and GABA, the excitatory amino acid neurotransmitters, glutamate (glu) and aspartate (asp), are not known to have distinct synthetic and degradative enzymatic pathways which are used only when these amino acids are functioning as neurotransmitters. Glu and asp themselves are not only neurotransmitters, but serve a wide range of functions in the nervous system. In fact, most of the glu and asp in the brain is probably associated with non-neurotransmitter functions. The multiple roles of glu and asp in the brain, along with the complex state of receptors for these amino acids, has made it exceptionally difficult to study their neurotransmitter properties and even to identify the neurons which use excitatory amino acids as neurotransmitters. Furthermore, it is uncertain if glu, asp, both or some closely-related substance is the actual neurotransmitter for any given population of neurons. One approach to addressing some of these questions is through a characterization of the synthetic and degradative pathways of these amino acids. We have been studying the immunocytochemical localization of several glu/asp related enzymes to determine how the presence of these enzymes relates to the distributions of putative glu/asp neurons. One of our goals is to determine if the localizations of these enzymes might be useful in identifying neurons which use glu/asp as neurotransmitters. Because of the many functions of these amino acids in the CNS, it cannot be expected that the localization of glu/asp pathways will be as straightforward as that of other transmitters. All enzymes known to be associated with the metabolism of glu or asp in the CNS can be expected to be present in all cells; however, the amounts may vary, and this may be related to the neurotransmitter used by the neuron. Production, storage and release of a neurotransmitter is a major function of the presynaptic terminal, and it would appear to be inefficient for all neurons to have an equally high capability for making glu or asp. Not only must a neuron expend energy for the synthesis of the enzyme, but the enzyme must be axonally transported to the synaptic terminal, a process also requiring energy. We have hypothesized that neurons which release glu or asp would have a greater capacity for synthesis of these amino acids than neurons that do not. However, even if this hypothesis is correct, a critical question concerns how much more enzyme these terminals contain. It might also be expected that different glutamergic and aspartergic neurons would have different capabilities for the synthesis of glu or asp.

Our work has concentrated on three enzymes, glutaminase (GLNase), cytoplasmic aspartate aminotransferase (cAAT) and glutamate dehydrogenase (GDH). Two other enzymes, glutamine synthetase (GS) and the mitochondrial form of AAT (mAAT) have also been studied immunocytochemically in the nervous system. This discussion will center on these five enzymes and their relationships to the neurotransmitter roles of glu and asp. One must recall that many other enzymes capable of metabolizing glu or asp are present in brain (See Roberts, 1981 for review).

GENERAL PROPERTIES OF GLU/ASP RELATED ENZYMES

Glutaminase

GLNase is a mitochondrial enzyme that catalyzes the hydrolysis of glutamine to glu and ammonium ions. Its activity is affected by a number of factors including phosphate and glu as well as the state of polymerization of the enzyme (for recent review, see Kvamme, 1983). GLNase obtained from kidney is immunologically identical to the major form of brain GLNase as determined using polyclonal antibodies (Curthoys et al., 1976) and monoclonal antibodies (Cangro et al., 1985). Multiple forms of GLNase have been reported to exist in brain, however, so it remains possible that a minor population of the enzyme immunologically distinct from the kidney enzyme is present.

Many studies have implicated GLNase in the production of neurotransmitter glu. Radioactive glutamine is readily converted into releasable glu in brain slices (Hamberger et al., 1979), in synaptosome preparations (Bradford et al., 1978), and in slice preparations after in vivo labeling (Ward et al., 1983). GLNase is enriched in synaptosomes (Bradford and Ward, 1976) and in neurons relative to astrocytes (Hamberger et al., 1978; Patel et al., 1982). Biochemical studies have given mixed results on whether GLNase is specifically enriched in glutamergic neurons. We found a decrease of 16% of total GLNase activity in the ventral cochlear nucleus two weeks after auditory nerve lesion and found the auditory nerve has 2-5 times more GLNase than several presumably non-glutamergic nerves (Wenthold, 1980). In the hippocampus, however, GLNase does not decrease after lesion of the entorhinal cortex (Nadler et al., 1978). One study reported that lesion of the cortico-striatal tract does not change GLNase activity in the striatum (McGeer and McGeer, 1979). More recently, Sandberg et al. (1985) reported a 15% decrease in striatal GLNase after lesion of the cortico-striatal pathway. A similar decrease was found in the olfactory cortex after lesion of the olfactory bulb (Sandberg et al., 1984).

Aspartate Aminotransferase

AAT is a pyridoxal phosphate dependent enzyme that catalyzes the reversible conversion of asp to oxaloacetate and alpha-ketoglutarate to glutamate. AAT obtained from non-neuronal sources has been purified and biochemically characterized. Two isoenzymes of AAT exist, a mitochondrial form (m-AAT) and a cytoplasmic form (c-AAT). While sequence analysis shows considerable homology between the two enzymes (48% in pig; Graf-Hausner et al., 1983), polyclonal antibodies show no cross-reactivity between the two forms. Antibodies made against both isoenzymes from liver and heart cross-react with their respective form of the enzyme from neuronal tissue.

The biochemical evidence supporting AAT in the metabolism of neurotransmitter glu and asp is much less than for GLNase. AAT is present in synaptosomes, but not enriched (Balazs et al.,1966; Fonnum, 1968), and it is higher in bulk prepared neurons than in bulk prepared astrocytes (Hertz, 1979). Since it appears to be the major neuronal enzyme that can catalyze the synthesis and degradation of asp, AAT might be expected to be involved in the metabolism of neurotransmitter asp, and labeling studies have implicated AAT in such a role (Shank and Campbell, 1982). Biochemical evidence linking AAT with putative aspartergic/glutamergic pathways is limited to the auditory nerve where a decrease of 14% was found in the

ventral cochlear nucleus two weeks after cutting the auditory nerve
(Wenthold, 1980). The auditory nerve is also enriched in AAT relative to
other nerves (Wenthold, 1980; Godfrey et al., 1984). Lesioning the cortico-
striatal pathway does not cause a decrease in striatal AAT (McGeer, 1983;
Sandberg et al., 1985), AAT does not decrease in the olfactory cortex after
destruction of the olfactory bulb (Sandberg et al., 1984) and it does not
decrease in the hippocampus after lesion of the perforant pathway (Nadler
et al., 1978). On the other hand, AAT has been reported to decrease after
cutting cholinergic neurons (Sterri and Fonnum, 1980).

Glutamate Dehydrogenase

GDH catalyzes the reversible conversion of glu to alpha-ketoglutarate
and ammonium ions. It is a mitochondrial enzyme and has been purified
from brain and liver. Both enzymes appear to be identical by peptide
mapping studies (McCarthy and Tipton, 1983). Our results show that
polyclonal antibodies against the liver enzyme cross-react with the brain
enzyme. GDH is low in synaptosomes and it has been suggested that it is
concentrated in glial cells (Berl and Clarke, 1983). Studies on bulk isolated
and cultured neurons and astrocytes have produced mixed results concerning
the distributions of GDH. Patel et al. (1982) found that GDH is distributed
much like GS, being significantly higher in astrocytes than neurons. GDH is
also enriched in C6 glioma cells compared to neuroblastoma cells
(Passonneau et al., 1977). On the other hand, Hertz (1979) found GDH to be
higher in neurons than astrocytes. Lesion studies on putative glutamergic
pathways have generally resulted in no change in GDH activity, although an
increase in activity has been reported in the striatum after cutting the
cortico-striatal pathway (Sandberg et al., 1985). Measuring the enzyme
histochemically, Wolf et al. (1984) report a 19% decrease in GDH in areas of
the hippocampus after lesion of the Schaffer's collaterals.

Glutamine Synthetase

GS catalyzes the conversion of glu and ammonium ion to glutamine.
Studies on cultured astrocytes and neurons have shown that GS is enriched in
glia relative to neurons. GS has been purified from several sources and
mono-specific antibodies have been produced. The immunocytochemistry of
glutamine synthetase has been thoroughly studied in the CNS, and these
studies show the enzyme to be enriched in glial cells in all areas studied (For
review, see Norenberg, 1983).

IMMUNOCYTOCHEMISTRY

Immunocytochemistry provides a method for the precise localization
of antigens at both the light and electron microscopic levels in complex
tissue such as the central nervous system. While this technique has been
invaluable in the identification of neurotransmitters, it has several potential
limitations. Foremost is the fact that the antibodies used must be specific

for the intended antigens. While an indication of this specificity can be obtained with techniques such as immunoblotting or immunoprecipitation, it is usually not possible to show that an antibody is completely specific, and this must be considered when analyzing immunocytochemical data. Furthermore, an antibody may not recognize related forms of a particular antigen, which, for example, may occur when several isoenzymes are present. Therefore, separate biochemical data corroborating the immunocytochemical results are often desirable. For example, our immunocytochemical results (Altschuler et al., 1981; 1984) showing that AAT and GLNase are enriched in terminals of the auditory nerve are supported by an independent study showing that these enzymes are enriched in the auditory nerve based on enzymatic activity (Wenthold, 1980).

As it is usually applied, immunocytochemistry is not quantitative. In fact, results are often obtained by titrating the several reactions involved to give a high sample to blank ratio. The background labeling that one sees can be due to a general nonspecific binding of one of the antibodies or can be due to a specific binding of the primary antibody, but at a much lower level than found in structures considered positive. The presence of low level specific binding may be important to the correct interpretation of results, for example, in identifying an immunocytochemical marker for a particular type of neuron.

A final point, which is pertinent to enzymes, is that immunocytochemical labeling reflects the number of antigen molecules and does not give information on the activity of the enzymes. Localized enzymes could in fact be inactive molecules or fragments of the original molecule; however, a greater limitation is that the activity of the enzymes, as regulated by availability of substrates, products and cofactors, is not determined.

The fact that the glu/asp related enzymes in brain appear to be immunologically identical to the enzymes in other tissues has facilitated the immunocytochemical localizations of the enzymes. With the exception of GS, all enzymes studied in brain have used antibodies produced against non-neuronal enzymes. We have used antibodies against pig heart cAAT and rat kidney GLNase and their characterizations have been previously reported (Altschuler et al., 1981; Curthoys et al., 1976). We have recently produced antibodies in rabbits and guinea pigs against GDH from bovine liver. These antibodies inhibit the enzymatic activities of GDH from both liver and brain. Immunoblot analysis shows a major immunoreactive band with a molecular weight of 55,000, although minor reactive species were also present. To eliminate possible contaminating antibodies, antibodies specific for the 55,000 molecular weight subunit were affinity purified using the method of Olmsted (1981). The affinity purified antibodies showed a single immunoreactive band corresponding to the subunit of GDH. No cross-reaction was seen with GS as determined by immunoblot analysis.

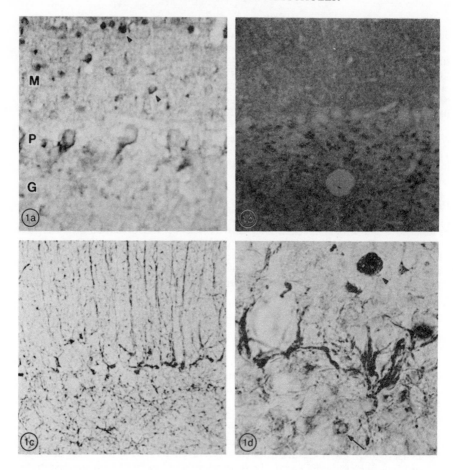

Figure 1. Immunocytochemical localization of AAT (a,d), GLNase (b) and GDH (c) in the cerebellum of the rat. Granule cell bodies are labeled with antibodies against GLNase and AAT (arrows) while stellate and basket cells are labeled only with antibodies against AAT (arrowheads). In the granule cell layer, both antibodies also give labeling not confined to granule cell bodies. Antibodies against GDH label predominately glia as shown by the intense labeling of Bergmann glial cells. M, molecular layer; P, Purkinje cell layer; G, granule cell layer.

IMMUNOCYTOCHEMISTRY OF GLU/ASP RELATED ENZYMES

Aspartate Aminotransferase

cAAT has now been immunocytochemically localized in   CNS

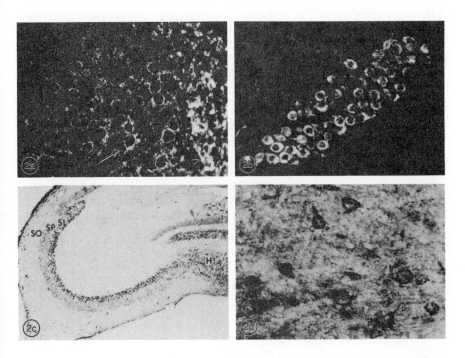

Figure 2.   Immunocytochemical localization of GLNase in putative glutamergic/aspartergic neurons including auditory nerve terminals (a) and cell bodies (b) of the guinea pig and neurons of the hippocampus (c) and cerebral cortex (d) of the rat.   In cochlear nucleus (a) labeling is seen in large axo-somatic terminals (arrows) and granule cell bodies (arrowheads). In hippocampus intense labeling is present in mossy fiber terminals in SL and lighter labeling in cell bodies in SP. Pyramidal cells of layer VI in cerebral cortex are labeled.   SL, stratum lucidum; SP, stratum pyramidale; SO, stratum oriens; H, hilus.

structures by several independent groups using different antibodies, one being a monoclonal (Altschuler et al., 1981; 1982; Recasens and Delaunoy, 1981; Lin et al., 1983a; Kamisaki et al., 1984). In general, these studies are in agreement. cAAT is enriched in several putative glutamergic/aspartergic neurons, including photoreceptors, auditory nerve and cerebellar granule cells (Figure 1). However, cAAT is also enriched in other neurons which are not believed to use glu or asp as a neurotransmitter. These include stellate and basket cells of the cerebellum which are believed to be GABAergic and cells in layers 2 and 3 of the neocortex which have the same distribution as GABAergic neurons (Wenthold and Altschuler, 1983; Kamisaki et al., 1984; Donoghue et al., 1985). Staining of glial cells has also been seen with cAAT antibodies (Altschuler et al., 1981). The immunocytochemical results with AAT in the cerebellum are supported by earlier histochemical studies. These studies show intense labeling of stellate and basket cell bodies, terminal baskets beneath Purkinje cell bodies and Purkinje cell bodies

(Martinez-Rodriguez et al., 1974; 1976). Synaptic labeling in the granule cell layer is also reported. Since the histochemical stain does not differentiate between the two forms of the enzyme, this closely fits the expected pattern of the sum of the cAAT and mAAT labelings seen immunocytochemically.

Glutamate Dehydrogenase

We have studied the immunocytochemical distribution of GDH in cerebellum, cochlear nucleus and hippocampus. In all three areas GDH appears concentrated in glial cells. Most apparent is the heavy labeling of Bergmann glial cells and their processes (Figure 1). Neuronal cell bodies were not labeled. Since electron microscopic localization has not yet been done, it cannot be ruled out that some presynaptic terminals are enriched in GDH. However, labeling is not seen in synaptic terminals that can be recognized at the light level, including the auditory nerve terminals, mossy fiber terminals of the hippocampus and synaptic baskets of the cerebellum. The predominant glial localization of GDH is supported by histochemical studies in the cerebellum showing intense labeling of Bergmann glial cells (Martinez-Rodriguez et al., 1974; 1976). In the cerebral cortex in cases of experimental portal-systemic encephalopathy, GDH was also enriched in astrocytes (Norenberg, 1976).

Glutaminase

GLNase has been immunocytochemically localized in several putative glutamergic/aspartergic neurons (Figure 2). In the auditory nerve intense immunoreactive staining is present in cell bodies, fibers and terminals (Altschuler et al., 1984). In the hippocampus, mossy fiber terminals are the heaviest labeled structures (Altschuler et al., 1985). The cell bodies giving rise to the mossy fiber pathway, the granule cells of the dentate gyrus, are also labeled, as are the pyramidal cells and cells in the entorhinal cortex which give rise to the perforant pathway. Puncta of immunoreactivity which would correspond to terminals of Schaffer collaterals and commissural fibers and to terminals of the perforant pathway were not seen. Pyramidal cells in layers V and VI of the cerebral cortex, which are believed to be glutamergic, contain GLNase-like immunoreactivity (Donoghue et al., 1985). In the cerebellum granule cell bodies contain GLNase-like immunoreactivity. Reaction product is also present which is not confined to granule cell bodies which may represent labeled granule cell axons or dendrites or terminals of mossy fibers. Stellate, basket, Purkinje and Golgi cell bodies are not labeled. Svenneby and Storm-Mathisen (1983) have also studied the immunocytochemical distribution of GLNase in the hippocampus using an antibody prepared in their laboratory. They report a general light labeling of all structures, while our results show heavy labeling of mossy fiber terminals and granule cells. The fact that they used a fixative of 5% glutaraldehyde could very well explain these differences. Most large molecule antigens cannot tolerate such a high concentration of glutaraldehyde and we find that even low concentrations of glutaraldehyde (0.1%) appreciably diminishes GLNase labeling.

DISCUSSION

The basic patterns of distributions of the enzymes involved in glu and asp metabolism are summarized in Table 1. Of the five enzymes GDH and GS are enriched in glial cells, while the others are enriched in neurons. Only GLNase appears to be specifically enriched in neurons thought to use glu or asp as a neurotransmitter. However, since there is detectable labeling in most structures, it is apparent that the enrichment of GLNase in these neurons is not nearly as great as, for example, the enrichment of glutamate decarboxylase in GABAergic neurons. Furthermore, the lesion studies done on putative glutamergic/aspartergic neurons demonstrate only modest decreases in GLNase after degeneration of the presynaptic terminals. An indication of the enrichment of GLNase can be obtained from biochemical analysis of the auditory nerve. Measurement of GLNase enzymatic activity in the auditory nerve shows it to have nearly 3 times the specific activity of that in the trigeminal nerve which contained the next highest specific activity and more than 5 times that of the facial nerve. Therefore, if this relationship is consistent throughout the nervous system, it suggests that the neurons which are immunocytochemically positive for GLNase contain 3 to 5 times more enzyme than surrounding structures.

In the hippocampus we have not seen labeling in terminals of Schaffer's collaterals or the perforant pathway, although the cell bodies are labeled. Since electron microscopic analysis has not been done, the lack of labeling could simply reflect the limit of resolution for these small terminals at the light microscopic level. Other factors such as loss of antigenicity during fixation could also lead to these results. Alternatively, it suggests that GLNase is not enriched in these terminals. The lesion study done on the perforant pathway also suggested that GLNase was not enriched in this population of terminals (Nadler et al., 1978). Since the synthesis of neurotransmitter takes place primarily at the presynaptic terminal, the enzyme would be expected to be concentrated at this site rather than the cell body. This may indicate that the amount of GLNase in the presynaptic terminal varies among different populations of putative excitatory amino acid neurons. Whether this means that different metabolic routes are used in the production of neurotransmitter in these cases or that GLNase is regulated differently, remains to be determined.

In many parts of the CNS, cAAT is seen to be rather uniformly distributed. Only in the cochlear nucleus, cerebellum and retina does AAT appear concentrated in putative glu/asp neurons. In the cochlear nucleus and cerebellum, AAT is present with GLNase, while in the retina, only AAT appears elevated in photoreceptors. AAT is found to be elevated in some GABAergic neurons in the cerebellum and cerebral cortex, suggesting it may play a role in the synthesis of GABA. However, it is not elevated in all GABAergic neurons such as Purkinje and Golgi cells of the cerebellum and GABAergic neurons in the hippocampus and retina (Altschuler et al., 1981; 1985). In the retina the immunocytochemical distributions of AAT and GAD are reported to differ significantly (Lin et al.,1983b). In the cerebellum it has been suggested that taurine may be a neurotransmitter of basket and stellate cells (Chan-Palay et al., 1982; Okamoto et al., 1983). Since AAT and cysteine-sulfinic acid transaminase may be the same protein (Recasens

Table 1.   Summary of Immunocytochemical Distributions of Enzymes Related to Glutamate and Aspartate Metabolism

Glutaminase	Enriched in putative excitatory amino acid neurons in cochlear nucleus, hippocampus, cerebellum and cerebral cortex.
c-Aspartate aminotransferase	Enriched in putative excitatory amino acid neurons in cochlear nucleus, retina and cerebellum. Found in GABAergic neurons in cerebellum. Present in glia in cochlear nucleus.
m-Aspartate aminotransferase	Enriched in some neurons with no direct relationship to any neurotransmitter. Does not correspond to distribution of cAAT.
Glutamate dehydrogenase	Enriched in glia in cerebellum, cochlear nucleus and hippocampus.
Glutamine synthetase	Enriched in glia throughout nervous system.

et al., 1980), it might be expected that this enzyme is enriched in these cells. However, studies on the retina show that AAT and cysteine-sulfinic acid decarboxylase, a marker for taurine-containing neurons (Wu, 1982), have different immunocytochemical distributions (Lin et al., 1983b). We do not find GLNase enriched in any GABAergic neurons which is somewhat unexpected since labeling studies have implicated GLNase in the production of GABA.

mAAT and cAAT have been immunocytochemically studied in the cerebellum, cerebral cortex, olfactory bulb and retina, and in all cases, the distributions of the two enzymes are different (Kamisaki et al., 1984; Inagaki et al., 1985). It has been suggested that the primary function of AAT in the brain is its participation in the malate-aspartate shuttle, which would require identical distributions of both forms of the enzyme (Shank and Campbell, 1982). Since this distribution is not seen, it suggests that much of the AAT has other functions, some of which may be neurotransmitter related. This, however, does not rule out the likelihood that the malate-aspartate shuttle is widespread in the brain.

Our immunocytochemical results suggest that GDH, like GS, is concentrated in glia. They do not support the results which indirectly implicate GDH in the synthesis of glu in the presynaptic terminal. Glucose can serve as a precursor for releasable glu (Hamberger et al., 1978), and alpha-ketoglutarate is taken up by synaptosomes through a high affinity process and converted to glu (Shank and Campbell, 1983). The conversion of alpha-ketoglutarate to glu could also be mediated by AAT, but as pointed out by Shank and Campbell (1983), this would require that the pool of asp be stoichiometrically replenished. Alternatively, GDH may be enriched in glia, but sufficient enzyme may be present in the presynaptic terminal to produce neurotransmitter.

It has recently been shown that antibodies can be made against small molecule neurotransmitters conjugated to larger molecules and used to immunocytochemically localize the neurotransmitter. With this method the immunocytochemical localization of glu has been reported (Storm-Mathisen et al., 1983). These studies show that glu is not specifically concentrated in putative excitatory amino acid neurons, but enriched in many, but not all neurons, apparently unrelated to the neurotransmitter. It was suggested that this glu represents the metabolic pool and is larger than the transmitter pool.   However, putative excitatory amino acid terminals were immunoreactive for glu after conditions of high affinity uptake. With our antibodies against glu and asp, we do not see a correlation between the distributions of these amino acids and GLNase or AAT. An exception is the mossy fiber terminals in the hippocampus where both glu and GLNase are immunocytochemically localized. These findings suggest that presynaptic terminals that release glu and asp usually have relatively low concentrations of the amino acids, but a high capacity to produce them either through uptake or synthesis.   This notion is consistent with the function of a neurotransmitter requiring that it be rapidly replaced during periods of sustained release. It also suggests that the larger neuronal pool of glu which does not have high levels of enzymes cannot be rapidly replaced.

In conclusion, the studies outlined above show that the glu/asp related enzymes have very different distributions in the brain. One question that we wished to address with these studies is whether or not any of these enzymes is enriched in all excitatory amino acid neurons such that its presence would serve as a marker to identify these neurons. GLNase appears to fulfill this requirement in that it is consistently enriched in putative excitatory amino acid neurons.   However, GLNase is not a straightforward immunocytochemical marker as is available for other neurotransmitters.   The fact that GLNase is ubiquitous but enriched in putative excitatory amino acid neurons requires that results on its immunocytochemical distribution be cautiously interpreted. Other questions such as the possible low levels of GLNase in some putative excitatory amino neurons also need to be resolved. However, our results, together with the lesion studies, show a correlation between neurons enriched in GLNase and neurons believed to use an excitatory amino acid as a neurotransmitter. Therefore immunocytochemical analysis of GLNase coupled with other techniques, such as high affinity uptake, amino acid levels and release, will give additional support in neurotransmitter identification.

REFERENCES

Altschuler, R.A., Monaghan, D.T., Haser, W.G., Wenthold, R.J., Curthoys, N.P. and Cotman, C.W. (1985). Immunocytochemical localization of glutaminase-like and aspartate aminotransferase-like immunoreactivities in the rat and guinea pig hippocampus. Brain Res., 330, 225-233.

Altschuler, R.A., Mosinger, J.L., Harmison, G.G., Parakkal, M.H. and Wenthold, R.J. (1982). Aspartate aminotransferase-like immunoreactivity as a marker for aspartate/glutamate in guinea pig photoreceptors. Nature, 298, 657-659.

Altschuler, R.A., Neises, G.R., Harmison, G.G., Wenthold, R.J. and Fex, J. (1981). Immunocytochemical localization of aspartate aminotransferase immunoreactivity in cochlear nucleus of the guinea pig. Proc. Natl. Acad. Sci., 78, 6553-6557.

Altschuler, R.A., Wenthold, R.J., Schwartz, A.M., Haser, W.G., Curthoys, N.P., Parakkal, M. and Fex, J. (1984). Immunocytochemical localization of glutaminase-like immunoreactivity in the auditory nerve. Brain Res., 291, 173-178.

Balazs, R., Dahl, D. and Harwood, J.R. (1966). Subcellular distribution of enzymes of glutamate metabolism in rat brain. J. Neurochem., 13, 897-905.

Berl, S. and Clarke, D.D. (1983). The metabolic compartmentation concept. In Glutamine, Glutamate and Gaba in the Central Nervous System. (eds. L. Hertz, E. Kvamme, E.G. McGeer and A. Schousboe). Alan R. Liss, Inc., New York, pp. 205-217.

Bradford, H.F. and Ward, H.K. (1976). On glutaminase activity in mammalian synaptosomes. Brain Res., 110, 115-125.

Bradford, H.F., Ward, H.K. and Thomas, A.J. (1978). Glutamine - a major substrate for nerve endings. J. Neurochem., 30, 1452-1459.

Cangro, C.B., Sweetnam, P.M., Wrathall, J.R., Haser, W.G., Curthoys, N.P. and Neale, J.H. (1985). Glutaminase-immunoreactivity in a subpopulation of DRG neurons. Amer. Soc. Neurochem. Absts., 16, 117, 1985.

Chan-Palay, V., Lin, C.T., Palay, S., Yamamoto, M. and Wu, J.Y. (1982). Taurine in the mammalian cerebellum: Demonstration by autoradiography with (^{3}H) taurine and immunocytochemistry with antibodies against the taurine-synthesizing enzyme, cysteine - sulfinic acid decarboxylase. Proc. Natl. Acad. Sci. U.S.A., 79, 2695-2699.

Curthoys, N.P., Kuhlenschmidt, T., Godfrey, S.S., and Weiss, R.F. (1976). Phosphate-dependent glutaminase from rat kidney. Arch. Biochem. Biophys., 172, 162-167.

Donoghue, J.P., Wenthold, R.J. and Altschuler, R.A. (1985). Localization of glutaminase-like and aspartate aminotransferase-like immunoreactivity in neurons of cerebral neocortex. J. Neurosci., In press.

Fonnum, F. (1968).   The distribution of glutamate decarboxylase and aspartate transaminase in subcellular fractions of rat and guinea-pig brain. Biochem. J., 106, 401-411.

Godfrey, D.A., Bowers, M., Johnson, B.A. and Ross, C.D. (1984). Aspartate aminotransferase activity in fiber tracts of the rat brain.   J. Neurochem., 42, 1450-1455.

Graf-Hausner, U., Wilson, K. J. and Christen, P.   (1983).   The covalent structure of mitochondrial aspartate aminotransferase from chicken.   J. Biol. Chem., 258, 8813-8826.

Hamberger, A.C., Chiang, C.H., Nylen, E.S., Scheff, S.W. and Cotman, C.W. (1979).   Glutamate as a CNS transmitter.   I. Evaluation of glucose and glutamine as precursors for the synthesis of preferentially released glutamate. Brain Res., 168, 513-530.

Hamberger, A., Cotman, C.W., Sellstrom, A. and Weiler,C.T. (1978). Glutamine, glial cells and their relationship to transmitter glutamate.   In Dynamic Properties of Glial Cells.   (eds. G. Franck, L. Hertz and D. B. Tower). Plenum Press, New York, pp. 163-172.

Hertz, L. (1979).   Functional interactions between neurons and astrocytes. I. Turnover and metabolism of putative amino acid transmitters.   Prog. Neurobiol. 13, 277-323.

Inagaki, N., Kamisaki, Y., Kiyama, H., Horio, Y., Tohyama, M. and Wada, H. (1985).   Immunocytochemical localizations of cytosolic and mitochondrial glutamic oxaloacetic transaminase isozymes in rat retina   as markers for the glutamate-aspartate neuronal system. Brain Res., 325, 336-339.

Kamisaki, Y., Inagaki, S., Tohyama, M., Horio, Y. and Wada, H. (1984). Immunocytochemical localization of cytosolic and mitochondrial glutamic oxaloacetic transaminase isoenzymes in rat brain. Brain Res., 297, 363-368.

Kvamme, E. (1983).   Glutaminase (PAG).   In Glutamine, Glutamate and GABA in the Central Nervous System.   (eds. L. Hertz, E. Kvamme, E. G. McGeer and A. Schousboe). Alan R. Liss, Inc., New York, pp. 51-67.

Lin, C. T. and Chen, L. H. (1983a). Production and characterization of an antibody to cytosolic aspartate aminotransferase and immunolocalization at the enzyme in rat organs. Laboratory Invest., 48, 718-725.

Lin, C.T., Li, H.Z. and Wu, J.Y. (1983b). Immunocytochemical localization of L-glutamate decarboxylase, gamma-aminobutyric acid transaminase, cysteine sulfinic acid decarboxylase, aspartate aminotransferase and somatostatin in rat retina. Brain Res., 270, 273-283.

Martinez-Rodriguez, R., Fernandez, B., Cevallos, C. and Gonzalez, M. (1974).   Histochemical location of glutamate dehydrogenase and aspartate aminotransferase in chicken cerebellum. Brain Res., 69, 31-40.

Martinez-Rodriguez, R., Garcia-Segura, L.M., Toledano, A. and Martinez-Marillo, R. (1976). Aspartate aminotransferase activity and glutamic dehydrogenase in the cerebellar cortex in several species of animals. A histochemical study. J. Hirnforsch., 17, 387-397.

McCarthy, A.D. and Tipton, K.F. (1983). Glutamate dehydrogenase. In Glutamine, Glutamate and GABA in the Central Nervous System. (eds. L. Hertz, E. Kvamme, E.G. McGeer and A. Schousboe). Alan R. Liss, Inc., New York, pp. 19-32.

McGeer, E.G. and McGeer, P.L. (1979). Localization of glutaminase in the rat neostriatum. J. Neurochem., 32, 1071-1075.

McGeer, E.G., McGeer, P.L. and Thompson, S. (1983). GABA and glutamate enzymes. In Glutamine, Glutamate and GABA in the Central Nervous System. (eds. L. Hertz, E. Kvamme, E.G. McGeer and A. Schousboe). Alan R. Liss, Inc., New York, pp. 3-17.

Nadler, J.V., White, W.F., Vaca, K.W., Perry, B.W. and Cotman, C.W. (1978). Biochemical correlates of transmission mediated by glutamate and aspartate. J. Neurochem., 31, 147-155.

Norenberg, M.D. (1976). Histochemical studies in experimental portal-systemic encephalopathy. Arch. Neurol., 33, 265-269.

Norenberg, M.D. (1983). Immunohistochemistry of glutamine synthetase. In Glutamine, Glutamate and GABA in the Central Nervous System. (eds. L. Hertz, E. Kvamme, E.G. McGeer and A. Schousboe). Alan R. Liss, Inc., New York, pp. 95-111.

Okamoto, K., Kimura, H. and Sakai, Y. (1983). Evidence for taurine as an inhibitory neurotransmitter in cerebellar stellate interneurons: Selective antagonism by TAG (6-aminomethyl-3-methyl-4H, 1, 2,4-benzothiadiazine-1,1-dioxide). Brain Res., 265, 163-168.

Olmsted, J.B. (1981). Affinity purification of antibodies from diazotized paper blots of heterogenous protein samples. J. Biol. Chem., 256, 11955-11957.

Passonneau, J.V., Lust, W.D. and Crites, S.K. (1977). Studies on the GABAergic system in astrocytoma and neuroblastoma cells in culture. Neurochem. Res., 2, 605-617.

Patel, A.J., Hunt, A., Gordon, R.D. and Balazs, R. (1982). The activities in different neural cell types of certain enzymes associated with the metabolic compartmentation of glutamate. Dev. Brain Res., 4, 3-11.

Recasens, M., Benzra, R., Basset, P. and Mandel, P. (1980). Cysteine sulfinate aminotransferase and aspartate aminotransferase isoenzymes of rat brain. Purification, characterization, and further evidence for identity. Biochemistry, 19, 4583-4589.

Recasens, M. and Delaunoy, J.P. (1981). Immunological properties and immunohistochemical localization of cysteine sulfinate or aspartate aminotransferase-isoenzymes in rat CNS. Brain Res., 205, 351-361.

Roberts, E. (1981). Strategies for identifying sources and sites of formation of GABA-precursor or transmitter glutamate in brain. In Glutamate as a Neurotransmitter. (eds. G. DiChiara and G. L. Gessa). Raven Press, New York, pp. 91-102.

Sandberg, M., Bradford, H.F. and Richards, C.D. (1984). Effect of lesions of the olfactory bulb on the levels of amino acids and related enzymes in the olfactory cortex of the guinea pig. J. Neurochem., 43, 276-279.

Sandberg, M., Ward, H.K. and Bradford, H.F. (1985). Effect of cortico-striate pathway lesion on the activities of enzymes involved in synthesis and metabolism of amino acid neurotransmitters in the striatum.   J. Neurochem., 44, 42-47.

Shank, R.P. and Campbell, G.L. (1982). Glutamine and alpha-ketoglutarate and metabolism by nerve terminal enriched material from mouse cerebellum., Neurochem. Res., 7, 595-610.

Shank, R.P.. and Campbell, G.L. (1983). Glutamate. In Handbook of Neurochemistry, Vol. 3, (ed. A. Lajtha). Plenum Publishing Corp., New York, pp. 381-404.

Shank, R.P. and Campbell, G.L. (1983). Metabolic precursors of glutamate and GABA. In Glutamine, Glutamate and GABA in the Central Nervous System. (eds. L. Hertz, E. Kvamme, E.G. McGeer and A. Schousboe). Alan R. Liss, Inc., New York, pp. 355-369.

Sterri, S.H. and Fonnum, F. (1980). Acetyl-CoA synthesizing enzymes in cholinergic nerve terminals. J. Neurochem., 35, 249-254.

Storm-Mathisen, J., Leknes, A.K., Bore, A.J., Vaaland, J.L., Edminson, P., Haug, F.M.S. and Ottersen, O.P. (1983). First visualization of glutamate and GABA in neurones by immunocytochemistry. Nature, 301, 517-520.

Svenneby, G. and Storm-Mathisen, J. (1983). Immunological studies of phosphate activated glutaminase. In Glutamine, Glutamate and GABA in the Central Nervous Systems. (eds. L. Hertz, E. Kvamme, E.G. McGeer and A. Schousboe). Alan R. Liss, Inc., New York. pp. 69-76.

Ward, H.K., Thanki, C.M. and Bradford, H.F. (1983). Glutamine and glucose as precursors of transmitter amino acids: Ex vivo studies. J. Neurochem., 40, 855-860.

Wenthold, R.J. (1980). Glutaminase and aspartate aminotransferase decrease in the cochlear nucleus after lesion of the auditory nerve. Brain Res., 190, 293-297.

Wenthold, R.J. and Altschuler, R.A. (1983).   Immunocytochemistry of aspartate aminotransferase and glutaminase. In Glutamine, Glutamate and GABA in the Central Nervous System. (eds. L. Hertz, E. Kvamme, E.G. McGeer and A. Schousboe). Alan R. Liss, Inc., New York, pp. 33-50.

Wolf, G., Schunzel, G. and Storm-Mathisen, J. (1984). Lesions of Schaffer's collaterals in the rat hippocampus affecting glutamate dehydrogenase and succinate dehydrogenase activity in the stratum radiatum of CA1.   J. Hirnforsch., 25, 249-253.

Wu, J.Y. (1982).   Purification and characterization of cysteic acid and cysteine-sulfinic acid decarboxylase and L-glutamate decarboxylase from bovine retina. Proc. Natl. Acad. Sci., 79, 4270-4275.

7

Antibodies for the Localization of Excitatory Amino Acids

J. Storm-Mathisen, O.P. Ottersen and T. Fu-long*

SUMMARY

Antisera were raised against glutamate (Glu) or aspartate (Asp) coupled to bovine serum albumin by glutaraldehyde. After immunosorbent purification the antisera reacted selectively with brain protein-glutaraldehyde conjugates of Glu or Asp, as assessed in a model system that permitted testing under conditions closely similar to those of the immunocytochemical procedure. Specific staining was abolished after solid phase adsorption with fixation complexes of the respective amino acids, and inhibited by free Glu or Asp at concentrations above 10-50 mM. These antisera were utilized to demonstrate soluble Glu and Asp separately. To retain the amino acids the tissue was fixed with glutaraldehyde. In perfusion-fixed tissue, labelling of perikarya and dendrites predominated over nerve terminal labelling, apparently reflecting the existence of large metabolic pools of Glu and Asp. In contrast, a nerve terminal pool was selectively labelled in hippocampal slices fixed by immersion after incubation in Krebs' medium. This pool could be depleted Ca^{2+}-dependently under depolarizing conditions, with a concomitant accumulation of immunoreactivity in glia. The findings give proof for the existence of a Glu/glutamine shuttle between neurons and glia and suggest that it is possible to selectively demonstrate transmitter Glu and Asp in brain tissue fixed in vitro.

INTRODUCTION

For anatomical tracing of neurons assumed to use glutamate (Glu) or aspartate (Asp) as transmitters, methods based on high affinity uptake of radiolabelled Glu or Asp seem to be the best available (see refs. in Taxt and Storm-Mathisen, 1984). One such method takes advantage of the fact that radiolabelled D-Asp, a

---

*On leave of absence from the Brain Research Institute, Academia Sinica, Shanghai, China. Study supported by NAVF and NORAD.

metabolically inert uptake substrate, is taken up in nerve endings
in vivo and transported to the parent cell bodies where it can be
detected by autoradiography (Streit, 1980; Storm-Mathisen and Wold,
1981; this volume). Nonetheless, these methods are rather indirect
and do not distinguish between Glu and Asp. Specific enzyme markers
have not yet been found. Aspartate aminotransferase and glutaminase
may be concentrated in certain Glu-ergic or Asp-ergic neurons
(Wenthold and Altschuler, 1983). However, a high content of
cytoplasmic aspartate aminotransferase also occurs in some
GABA-ergic neurons (Wenthold and Altschuler, this volume), and it
has been suggested that the level of this enzyme reflects the metabolic activity of the cell rather than the transmitter identity
(Bolz et al., 1985). Further, the actual rate of strongly regulated
enzymes, such as phosphate activated glutaminase (Kvamme, this
volume), is not necessarily reflected by the amount of enzyme
detected immunocytochemically. Useful additional information could
be provided by methods for visualization of Glu and Asp in tissue
sections.

An important criterion for establishing the identity of the
transmitter at a synapse is the demonstration of the putative
transmitter in the presynaptic neuron, and, most critically, in the
terminals. Whereas the presence of gamma-aminobutyric acid (GABA)
in the presynaptic element is considered as strong evidence for
GABA-ergic transmission, the presence of Glu or Asp is more difficult to interpret, since the latter two amino acids are heavily
involved in intermediary metabolism and in protein synthesis and
breakdown (Roberts, 1981). Thus estimates based on the effects of
lesions and other lines of evidence suggest that the "metabolic
pool" of Glu may be considerably larger than the "transmitter pool"
(Fonnum, 1984). The availability of antibodies for the localization of transmitter amino acids (Storm-Mathisen et al., 1983) has
enabled us to demonstrate Glu-like and Asp-like immunoreactivities
(Glu-LI and Asp-LI) microscopically. In brains of animals fixed by
rapid perfusion with glutaraldehyde Glu-LI and Asp-LI occur predominantly in neuronal perikarya and dendrites. However, as the
"transmitter pool" may be selectively visualized in certain in
vitro conditions (Fu-long et al., 1984, 1986; Ottersen and
Storm-Mathisen, 1985; present report), the stage is now set for the
use of Glu/Asp immunocytochemistry as an aid to transmitter identification.

CHARACTERIZATION OF ANTISERA

The antisera employed in the present study were raised in
rabbits against L-Glu or L-Asp conjugated to bovine serum albumin
(BSA) by glutaraldehyde (G) (Storm-Mathisen et al., 1983; Ottersen
and Storm-Mathisen, 1984a,b, 1985). Antibodies directed against
glutaraldehyde-treated protein were removed by immunosorption with
Sepharose beads (S) on which BSA had been immobilized and reacted
with glutaraldehyde. Both antisera (undiluted) were adsorbed with

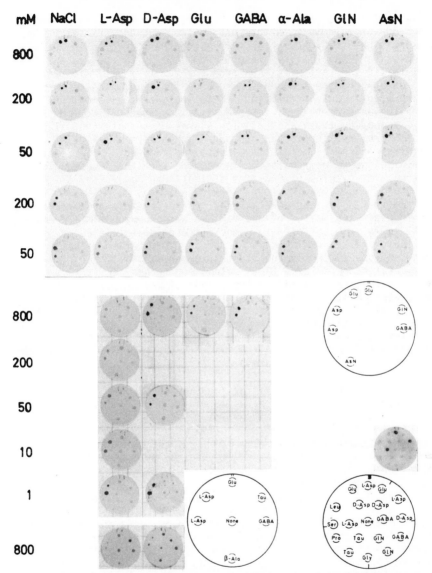

Fig. 1. Characteristics of antisera raised against BSA-conjugated
Glu (serum 13; upper 3 rows) and Asp (serum 18; rest). The sera
were purified by immunosorption with S-BSA-G (all), and then
S-BSA-G-GABA (rows 1-3, dilution 1:2000; rows 4 and 5, dilution
1:1000; lower right corner, dilution 1:300), or S-BSA-G-GABA
followed by absorption to S-ovalbumin-G-Asp (bottom left, dilution
1:75) and elution at pH 2.2 (rows 6-10, nominal dilution 1:150).
Amino acids fixed to rat brain macromolecules (Fig. 2) were spotted
on 13 mm Millipore filters and processed with the sera. Free amino
acids or NaCl were added to the diluted sera (except bottom right).

S-BSA-G-GABA; some batches of the Glu antiserum were also adsorbed with S-BSA-G-glutamine. After purification the Asp antiserum did not show significant crossreactivity with the stereoisomer D-Asp, or with L-Glu or GABA (Fig. 1; inset, Fig. 3A), while the Glu antiserum failed to react with GABA or Asp (Fig. 1; inset, Fig. 4A). A range of other amino compounds were also screened, none of them showing significant interaction. These included those mentioned in Figure 1 plus N-acetylaspartate, N-methylaspartate (fixation probably impeded by the N-substitution), phosphoethanolamine, alanine, valine, tyrosine, tryptophan, threonine, histidine, lysine, ornithine, arginine, methionine, cysteine, cadaverine, putrescine, spermidine, spermine, carnosine, homocarnosine, and reduced glutathione. Amino acids of proteins did not interfere (Fig. 1).

The testing of antisera was carried out against glutaraldehyde-brain protein conjugates of the various amino compounds (2.5 umol/mg protein). The conjugates were immobilized on cellulose ester (Millipore) filters which could be processed together with the brain sections (Ottersen and Storm-Mathisen, 1984b), using the peroxidase-antiperoxidase (PAP) method of Sternberger (1979). The present procedure secured identical conditions for testing and immunocytochemistry. This is particularly important as the staining intensity depends on the amount of tissue pr. incubation volume. To obtain comparable intensity of staining it was necessary to use sera at higher concentrations when tissue sections were processed together with the test filters (Figs. 2-5) than when the filters were incubated alone (Fig. 1).

The sera were also evaluated in inhibition experiments with free amino acids. Free L-Asp in the low mM range added to the

Fig. 2. Photomicrographs showing closely spaced horizontal sections through area dentata of a guinea pig stained with GABA antiserum 26 diluted 1:250 (A), Asp antiserum 18 diluted 1:100 (B), Glu antiserum 13 diluted 1:500 (C) or thionin (D), or processed according to the Timm/Haug procedure (large insets in A-C). Arrows indicate dentate basket cells; these are usually devoid of Glu-LI (C) but contain GABA-LI (A) and/or Asp-LI (B). Small inset in A shows test filter treated as the section. The spots contained 0.1 ul of suspensions of brain macromolecules (8 ug protein/ul) reacted with glutaraldehyde and an amino acid: 1 Asp, 2 Glu, 3 GABA, 4 taurine, 5 none. Arrowheads in A denote GABA-LI negative cells in the second reflected blade. Arrowhead in smaller inset in C shows Glu-LI negative neuron in the outer plexiform layer (frame indicates area enlarged in inset). Anatomical abbreviations in this and the following figures: CA4, field of Ammon's horn (first reflected blade); F, layer of fusiform cells or second reflected blade; G, granular cell layer; IP, inner plexiform layer; LM, stratum lacunosum moleculare; LU, stratum lucidum; M, Mi, molecular layer of fascia dentata, inner zone; O, stratum oriens; OP, outer plexiform layer; P, pyramidal cell layer; Ra, stratum radiatum. Bar, 200 um.

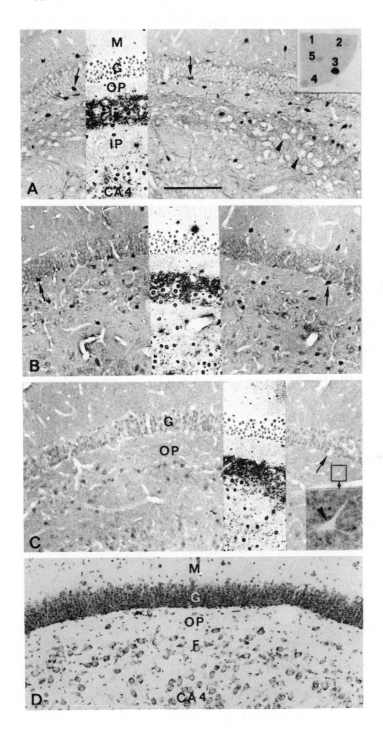

affinity purified Asp antiserum 18 inhibited staining of the L-Asp conjugates, whereas D-Asp and other amino acids had no effect even at a concentration of 800 mM (Fig. 1). Free Glu or GABA at concentrations above 50 mM likewise inhibited staining by the Glu antiserum 13 (Fig. 1) or by the GABA antiserum 26 (not shown). Higher amino acid concentrations seemed to be required when the sera were not affinity purified (Fig. 1).

These findings show that the free amino acids, at concentrations comparable to those present in the tissue and in the reaction mixture when preparing the conjugates (20 mM), are able to interact selectively with the specific antibodies. Previously we have mentioned (discussion in Ottersen and Storm-Mathisen, 1984b) that 200 mM of Glu or GABA added to the diluted serum did not inhibit staining by anti Glu and anti GABA sera, respectively; those results were based on few data with different serum preparations.

In keeping with the specific nature of our antisera we found that adsorption of the Glu antiserum with S-BSA-G-Glu or the Asp antiserum with S-ovalbumin-G-Asp virtually abolished the specific staining, but that the latter could be reproduced by the antibodies subsequently eluted from the affinity column at low pH (Fig. 1). As an additional control measure, portions of the Glu antiserum and Asp antiserum were adsorbed with, respectively, S-BSA-G-Asp and S-BSA-G-Glu. This adsorption step did not lead to changes in the staining pattern, further confirming that the sera contained insignificant amounts of crossreacting antibodies (Ottersen and Storm-Mathisen, 1985).

IMMUNOREACTIVITIES IN PERFUSION-FIXED MATERIAL

The hippocampus was chosen as a model for immunocytochemical testing of our antisera because here the roles of Glu and Asp have been extensively studied. In a first series of experiments we studied Glu-LI and Asp-LI in 20 um Vibratome sections obtained from hippocampi of mice, rats, guinea pigs, cats, and Senegalese baboons (Papio papio) that had been perfusion fixed under pentobarbital anesthesia with 5% glutaraldehyde (diluted from 25% in 0.1 M sodium phosphate buffer pH 7.4). The fixative was preceded by a brief wash with 2% dextran (MW 70,000) in the same buffer. One guinea pig was initially perfused with a sulphide solution instead of dextran (Ottersen and Storm-Mathisen, 1985) to permit the preparation of parallel Timm stained sections (Haug, 1973). (Such pretreatment did not affect the immunostaining.) Figure 2 shows results obtained in the latter animal, the photomicrographs representing closely spaced horizontal sections through area dentata. In contrast to rats, guinea pigs exhibit a distinct lamination of the hilar region (Blackstad, 1985). The various laminae are depicted in a thionin stained (Fig. 2D) and a Timm stained section. The photomicrograph of the latter section was divided into three stripes and mounted to cover matching regions in the immunostained sections (Fig. 2A-C).

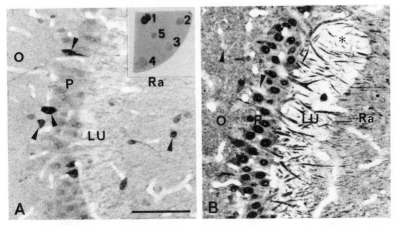

Fig. 3. Photomicrographs showing Asp-LI (A, rat, serum dilution
1:200) and Glu-LI (B, guinea pig, serum dilution 1:500) in CA3.
(Rat and guinea pig exhibited similar patterns of immunoreactivity
in this part of the hippocampus.) Arrowheads indicate cells
(presumably interneurons) that are strongly Asp-LI positive (A) but
Glu-LI negative (B). The pyramidal cells are rich in Glu-LI but
low in Asp-LI. Asterisk indicates end bulb. Inset in A shows test
filter incubated as the section (key, see Fig. 2). Bar, 100 um.

The zinc poor outer plexiform layer (OP) borders on the granular cell layer and contains few neurons (Fig. 2D). Deep to it is
a cell rich zone which is intensely stained in Timm preparations;
this is the layer of fusiform cells (F) or the second reflected
blade. The inner plexiform layer (IP) contains, like the outer
plexiform layer, few cells and little zinc, and separates the
second from the first reflected blade or CA4, which merges with the
pyramidal cell layer of the hippocampus proper.

Figure 2A shows, for reference, a section stained with GABA
antiserum 26. (This serum was characterized in Ottersen and
Storm-Mathisen, 1984b,c). GABA-LI occurs in more than 80% of the
neurons in the outer plexiform layer and also in a sizable proportion of the neurons in the molecular layer and the second reflected
blade. The inner plexiform layer and first reflected blade contain
only scattered GABA-LI positive neurons. The immunoreactive
neurons that were associated with the granular cell layer were
typically situated at its basal aspect, and their location and
morphological features suggested that they were dentate basket
cells. The granular cells were unstained, but were outlined by
bouton-like immunoreactive dots. The zinc-rich zone corresponding
to the second reflected blade overlapped precisely with a zone of
increased GABA-LI in the neuropil.

The distribution of neurons intensely stained for Asp-LI
(Fig. 2B) matched that of neurons stained for GABA-LI. In the
outer plexiform layer the two cell populations must overlap

extensively since the proportion of strongly Asp-LI positive cells
in this layer was about 80%, corresponding to the proportion of
GABA-LI positive cells. In contiguous sections incubated respectively for GABA-LI and Asp-LI the two immunoreactivities were also
found to coexist in neurons of the dentate molecular layer (not
illustrated). Similar to GABA-LI, intense Asp-LI occurred in dentate basket cells at the hilar aspect of the granular cell layer,
in a considerable proportion of the neurons in the second reflected
blade, and in scattered neurons in CA4. The remaining cells (i.e.,
the cells without intense immunoreactivity) were not as pale as the
negative cells in the GABA preparations (cf. Fig. 2A), but contained modest levels of Asp-LI. In fact, all cells of the hippocampal formation, including the granular and pyramidal cells,
appeared to contain some Asp-LI. The only exceptions were the
granular cells of baboon which were devoid of immunoreactivity (not
illustrated).

Glu-LI occurred in most of the neurons of the area dentata,
i.e., in the granular cells and in the majority of the neurons in
the first (CA4) and second reflected blade (Fig. 2C). Interestingly,
the distribution of the Glu-LI negative cells matched that of the
neurons strongly positive for Asp-LI or GABA-LI. Thus, the outer
plexiform and dentate molecular layers, which showed the highest
proportions of intensely stained Asp-LI and GABA-LI positive
neurons, were poor in neurons stained for Glu-LI. Glu-LI negative
perikarya also occurred at the site of the dentate basket cells.

Figure 3 compares the distribution of Asp-LI (A, from rat) and
Glu-LI (B, from guinea pig) in the CA3. Asp-LI is high in scattered neurons, probably corresponding to interneurons, in the pyramidal cell layer, and is also concentrated in neurons in the
neuropil layers. The pyramidal cells are modestly labelled. As in
the dentate area, Glu-LI shows a distribution strikingly different
from that of Asp-LI, being high in pyramidal cells but low in interneurons (Fig. 3B, 4B). The stratum lucidum is weakly labelled with
both antisera in this type of material.

To sum up, we have found that the distribution of Asp-LI in
the perfusion-fixed hippocampus is in some respects similar to that
of GABA-LI, and distinctly different from that of Glu-LI. These
results raise the question of whether our Asp antiserum could have
crossreacted with fixed GABA. This is very unlikely for several
reasons. First, the Asp antiserum did not produce significant
staining of GABA conjugates prepared at a GABA concentration
(2.5 umol/mg protein) about 100 times higher than the average concentration in the hippocampus (inset, Fig. 3A). Second, several
other classes of strongly GABA-LI positive neurons, such as interneurons in the neocortex and dorsal subdivision of the lateral
geniculate body, were not particularly enriched in Asp-LI. Third,
pretreatment of the animal with gammavinyl GABA (4-amino-hex-5-enoic
acid), a GABA aminotransferase inhibitor (Jung et al., 1977),
altered the distribution of GABA-LI in the hippocampus without

inducing similar changes in the pattern of Asp-LI (Ottersen and Storm-Mathisen, 1985).

The most likely explanation of the high Asp-LI in interneurons is that Asp is concentrated in these cells, playing an as yet unknown role. Although Asp has been proposed as a transmitter of interneurons in other parts of the central nervous system, most notably the spinal cord (Davidoff et al., 1967), the case for Asp subserving a similar role in the hippocampus is rather weak. Of significance in this context is the apparent failure of hippocampal interneurons to accumulate D-[^{3}H]Asp after local injections in vivo (Storm-Mathisen and Wold, 1981; Fischer et al., 1982, 1985). In contrast, an accumulation of this compound was found in hilar neurons following injections in the contralateral area dentata (Fischer et al., 1982, 1985). This observation is consistent with the finding of Nadler et al. (1976, 1978) of a decreased Ca^{2+}-dependent, K^+-stimulated efflux of endogenous Asp in the area dentata after commissurotomy, and raises the possibility that some of the strongly Asp-LI positive hilar neurons may use Asp as transmitter. In the guinea pig, the site of origin of the commissural fibers is probably the second reflected blade (Sørensen, 1980).

The observation that the distribution of Glu-LI negative cells matched that of GABA-LI positive neurons is consistent with data indicating a low Glu content in GABAergic neurons (Minchin and Fonnum, 1979; Korf and Venema, 1983). In these neurons, Glu is consumed by Glu decarboxylase (GAD) to form GABA.

Our findings suggest that the total contents of Glu and Asp, i.e., the "metabolic pool" plus the "transmitter pool", are revealed in immunocytochemical preparations of perfusion fixed tissue. Of these pools, the "metabolic pool" is probably the larger (see above) and would therefore tend to mask the "transmitter pool". Thus, the data from perfusion fixed tissue are not easily interpretable in terms of transmitter identity.

GLUTAMATE-LIKE AND ASPARTATE-LIKE IMMUNOREACTIVITIES IN IMMERSION-FIXED MATERIAL

As part of our strategy aimed at distinguishing "metabolic" Glu/Asp from "transmitter" Glu/Asp we decided to exploit the in vitro hippocampal slice, in which the different pools of Glu/Asp should be accessible to experimental manipulation. Transverse slices (200 um) of hippocampi, rapidly dissected out from rats killed by decapitation, were cut in the cold on a tissue chopper and quickly transferred to oxygenated Krebs' solution on ice. The slices were then immersion fixed (in 5% glutaraldehyde in sodium phosphate buffer, as for perfusion), immediately, or after incubation for up to 1 h in the Krebs' medium at 30°C (medium replaced every 10 min to avoid accumulation of metabolites and

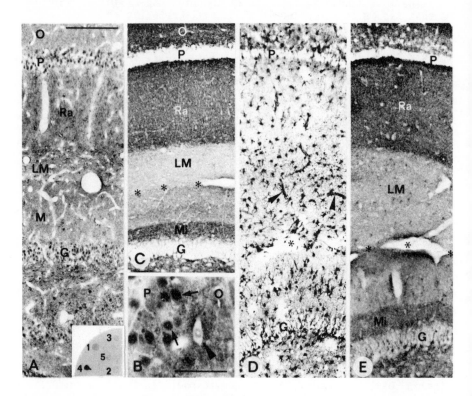

Fig. 4. Photomicrographs showing Glu-LI in sections of immersion
fixed hippocampal slices resectioned at 20 um (C-E), and, for comparison, in the hippocampus of a perfusion fixed rat brain (A,B).
Inset in A shows test filter incubated as the sections.
Key: 1 GABA, 2 Asp, 3 taurine, 4 Glu, 5 none. B: Arrowhead, Glu-LI
poor nonpyramidal neuron at the border between stratum oriens and
pyramidale of CA3; arrows, pyramidal cells enriched in Glu-LI.
C-E: Sections from slices incubated in Krebs' medium for 1 h in the
presence of 5 mM K+ (C), 55 mM K+ (D), and 55 mM K+ + 0.5 mM glutamine (E). Note laminar pattern of nerve terminal staining in C and
E, and intense Glu-LI in glial cells and around vessels (arrowheads) in D. Asterisks (C-E) indicate obliterated hippocampal
fissure. Transverse sections, lateral to the right. Antiserum 13
diluted 1:250 (A,B), or 1:400 (C-E). Bar in A (valid also for
C-E), 200 um; in B, 70 um.

released substances). After fixation the slices were soaked in 30%
sucrose, mounted flat, and resectioned at 20 um on a freezing
microtome. Figure 4C shows the pattern of Glu-LI obtained after

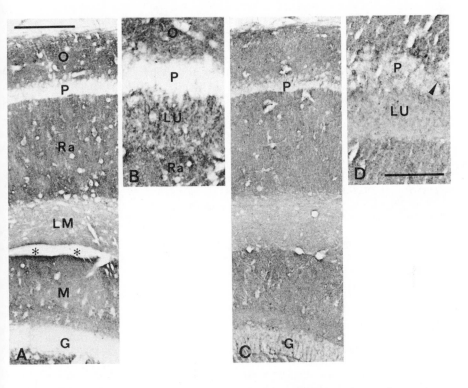

Fig. 5. Photomicrographs showing closely spaced sections of a rat hippocampal slice. The sections were treated with Glu antiserum 13 diluted 1:400 (A,B) or Asp antiserum 18 diluted 1:60 (C,D). The slice was cut into ice cold Krebs' medium and immediately fixed. Glu-LI and Asp-LI are similarly distributed in the neuropil layers of CA1 and fascia dentata (A,C), but only Glu-LI is concentrated at the site of the mossy fiber boutons (B). Scattered Asp-LI positive bouton-like dots (arrowhead in D) occur among the unlabelled peri-karya of granular and pyramidal cells. Asterisks in A, obliterated hippocampal fissure. Bars: A,C, 200 um; B,D, 100 um.

1 h incubation at physiological K$^+$ concentration (5 mM). Glu-LI has been lost from cell bodies and dendrites (compare with Fig. 4A,B from perfusion fixed material) and occurs selectively in nerve terminal-like dots forming a distinct laminar pattern. Thus, the density of Glu-LI positive boutons is high in strata oriens and radiatum and the inner zone of the dentate molecular layer, and shows a stepwise decrease on moving from the inner zone through the intermediate and outer zones of the dentate molecular layer to the hippocampal stratum lacunosum moleculare. This laminar pattern is similar to that observed in autoradiograms of D-[^{3}H]Asp uptake (Taxt and Storm-Mathisen, 1984). Glu-LI was also concentrated in mossy fiber boutons (not illustrated). The presence of Glu-LI in

such boutons, after immersion fixation, has been verified by electron microscopy (Storm-Mathisen et al., 1983).

When K^+ (55 mM) or veratrine (50 uM) were added to the medium to cause transmitter release, Glu-LI disappeared from boutons and accumulated in glial cells (Fu-long et al., 1984, 1986; Storm-Mathisen and Ottersen, 1985) (Fig. 4D). (Glia are normally not discernible.) These changes were abolished in media containing 0.1 mM Ca^{2+} and 10 mM Mg^{2+}. Supplementing the medium with 2 mM D,L-methionine sulfoximine to inhibit glutamine synthetase (Cooper et al., 1983) reduced the minimum K^+ concentration required for inducing glial labelling from about 30 to about 15 mM. The K^+-induced glial labelling was reduced and the nerve terminal staining reappeared when glutamine was added to the medium at physiological concentration (0.5 mM) (Fig. 4E). The addition of 0.1 mM NH_4Cl in the high K^+ medium removed most of the glial staining, but did not influence the labelling of nerve terminals.

These findings represent direct histological evidence for the existence of a Glu-glutamine shuttle between neurons and glia, originally proposed on the basis of biochemical data (see Hamberger et al., 1979). This shuttle involves uptake of Glu in glia and conversion of Glu to glutamine (by glutamine synthetase), and, subsequently, transfer of the synthesized glutamine to neurons for conversion to Glu (by glutaminase). The reason for the appearance of Glu-LI in glia under prolonged depolarization-induced release of Glu and other transmitters from nerve endings might be depletion of sources of NH_4^+ and perhaps ATP for the formation of glutamine by glutamine synthetase. Under these conditions Glu would accumulate in glia instead of being fed back to nerve endings as glutamine. Glutamine and NH_4Cl are likely to serve as nitrogen donors in these experiments. Treatment with NH_4Cl was not sufficient to recover the nerve terminal staining pattern, probably because the glutamine formed would be diluted too much in the incubation medium.

The distribution of Asp-LI in the immersion-fixed hippocampal slice was similar to that of Glu-LI, with two exceptions: 1. Asp-LI was lower than Glu-LI in stratum lucidum, the mossy fiber layer (cf. Fig. 5D with Fig. 5B). 2. The granular and pyramidal layers were devoid of Glu-LI, but contained distinct punctate Asp-LI positive structures between the cell bodies (Fig. 5).* These results suggest that both Asp and Glu are concentrated in CA3 Schaffer collateral/commissural axon terminals, in the hilar projections to the inner zone of the dentate molecular layer, and depending on the exact conditions, in perforant path terminals in the intermediate and outer zones of the dentate molecular layer (see Swanson, 1979, for a review on hippocampal pathways).

---

*The differences between this experiment and that represented in Fig. 4C, with respect to the distribution of Glu-LI in the dentate molecular layer, are probably due to the different treatment of the slices before fixation.

We have recently found (Gundersen et al., 1986) that under depolarization Asp-LI accumulates in glia and is lost from nerve endings in the same way as Glu-LI. Both direct uptake of Asp and conversion of Glu to Asp by Asp aminotransferase may contribute to the glial accumulation of Asp-LI.

The data presented above indicate that the in vitro conditions used here favor the visualization of the transmitter pools of the excitatory amino acids. Thus, the antisera label Glu or Asp pools that 1) appear to be restricted to nerve terminals thought to be excitatory on other grounds (see Storm-Mathisen and Ottersen, 1984), 2) are releasable by high K^+ or veratrine in a Ca^{2+}-dependent manner, 3) participate in the Glu-glutamine shuttle, which is thought to be partly responsible for the replenishment of transmitter Glu (Hamberger et al., 1979). (Asp may enter this shuttle after conversion to Glu in glia.)

If the selective retention of Glu and Asp in excitatory terminals is due exclusively to the ability of the latter to take up these amino acids, the in vitro immunocytochemical approach would not add much to the $D-[^3H]Asp$ or $L-[^3H]Glu$ uptake data and would not be expected to distinguish between Glu and Asp. However, the differences between the patterns of Glu-LI and Asp-LI, most notably in the mossy fiber zone, indicate that other factors than uptake, e.g., synthetic capacity, also play a role. Thus, the present in vitro results can be interpreted to suggest that the mossy fiber terminals synthesize Glu rather than Asp and that Glu is the more likely transmitter in this fiber system. It must be remembered, however, that the in vitro conditions used here differ in many ways from the in vivo situation.

An intriguing question to be addressed in further studies is whether Glu and Asp occur in the same or in separate terminals in the fiber systems where the two amino acids coexist. The presence of Glu and Asp in the same terminals would not be surprising, since Glu and Asp are interconvertible (Roberts, 1981) and may act as precursor for each other. Demonstration of immunoreactivity within synaptic vesicles would be strong evidence for transmitter identity (cf. Naito and Ueda, 1983), although some observations indicate that release may occur directly from cytosol (De Belleroche and Bradford, 1977). Electronmicroscopy of slices fixed in vitro suggested a vesicular location of Glu-LI in the mossy fiber boutons (Storm-Mathisen et al., 1983; Storm-Mathisen and Ottersen, 1983), but the use of the PAP technique involves the danger of diffusion of the reaction product and limited resolution because of the large size of the PAP complex (Sternberger, 1979). Studies with high resolution methods such as those based on colloidal gold will be required to establish whether the immunoreactivities are present within synaptic vesicles.

CONCLUSION

The present immunocytochemical approach is not useful for transmitter studies in perfusion-fixed brains where "transmitter" Glu and Asp and "metabolic" Glu and Asp seem to be indiscriminately retained. In immersion-fixed brain slices, however, it is possible to selectively visualize Glu and Asp pools that show localizations and responses to experimental manipulations as expected of the "transmitter pools". It seems that the application of specific Glu and Asp antisera represents a novel and potent approach for research on excitatory amino acid transmitters.

REFERENCES

Blackstad, T.W. (1985). Laminar specificity of dendritic morphology: Examples from the guinea pig hippocampal region. In Quantitative Neuroanatomy in Transmitter Research. (eds. L.F. Agnati, K. Fuxe and T. Hökfelt). Macmillan Press, London, in press.

Bolz, J., Thier, P. and Brecha, N. (1985). Localization of aspartate aminotransferase and cytochrome oxidase in the cat retina. Neurosci. Lett., 53, 315-320.

Cooper, A.J.L., Vergara, F. and Duffy, T.E. (1983). Cerebral glutamine synthetase. In Glutamine, Glutamate and GABA in the Central Nervous System. (eds. L. Hertz, E. Kvamme, E.G. McGeer and A. Schousboe). Alan R. Liss, New York, pp. 77-93.

Davidoff, R.A., Graham, L.T. jr., Shank, R.P., Werman, R., and Aprison, M.H. (1967). Changes in amino acid concentrations associated with loss of spinal interneurons. J. Neurochem., 14, 1025-1031.

De Belleroche, J.S. and Bradford, H.F. (1977). On the site of origin of transmitter amino acids released by depolarization of nerve terminals in vitro. J. Neurochem., 29, 335-343.

Fischer, B.O., Ottersen, O.P. and Storm-Mathisen, J. (1982). Anterograde and retrograde axonal transport of D-[^{3}H]aspartate (D-Asp) in hippocampal excitatory neurones. Neuroscience 7 (Suppl.), S68.

Fischer, B.O., Storm-Mathisen, J. and Ottersen, O.P. (1985). Hippocampal excitatory neurons: Anterograde and retrograde axonal transport of D-[^{3}H]aspartate. This volume.

Fonnum, F. (1984). Glutamate: a neurotransmitter in mammalian brain. J. Neurochem., 42, 1-11.

Fu-long, T., Ottersen, O.P. and Storm-Mathisen, J. (1984). Immunocytochemical visualization of glutamate compartmentation in hippocampal slices during K$^+$-induced release: effect of glutamine. Neurosci. Lett., Suppl. 18, S196.

Fu-long, T., Ottersen, O.P. and Storm-Mathisen, J. (1986). Compartmentation of glutamate between nerve endings and glia studied by immunocytochemistry. In preparation.

Gundersen, V., Nordbø, G., Laake, J.H., Ottersen, O.P. and
    Storm-Mathisen, J. (1986). Experimental manipulations of the
    cellular localization of aspartate: an immunocytochemical
    study in brain slices. In preparation.
Hamberger, A.C., Chiang, G.H., Nylén, E.S., Scheff, S.W. and
    Cotman, C.W. (1979). Glutamate as a CNS transmitter.
    I. Evaluation of glucose and glutamine as precursors for the
    synthesis of preferentially released glutamate.
    Brain Res., 168, 513-530.
Haug, F.-M.S. (1973). Heavy metals in the brain. A light
    microscopic study of the rat with Timm's sulphide silver
    method. Methodological considerations and cytological and
    regional staining patterns. Adv. Anat. Embryol. Cell. Biol.,
    47(4), 1-71.
Jung, M.J., Lippert, B., Metcalf, B.W., Böhlen, P. and Schechter,
    P.J. (1977). y-Vinyl GABA (4-amino-hex-5-enoic acid), a new
    selective irreversible inhibitor of GABA-T: effects on brain GABA
    metabolism in mice. J. Neurochem., 29, 797-802.
Korf, J. and Venema, K. (1983). Amino acids in the substantia nigra
    of rats with striatal lesions produced by kainic acid.
    J. Neurochem., 40, 1171-1173.
Minchin, M.C.W. and Fonnum, F. (1979). The metabolism of GABA and
    other amino acids in rat substantia nigra slices following
    lesions of the striato-nigral pathway. J. Neurochem.,
    32, 203-209.
Nadler, J.V., Vaca, K.W., White, W.F., Lynch, G.S. and Cotman, C.W.
    (1976). Aspartate and glutamate as possible transmitters of
    excitatory hippocampal afferents. Nature (Lond.), 260,
    538-540.
Nadler, J.V., White, W.F., Vaca, K.W., Perry, B.W. and Cotman, C.W.
    (1978). Biochemical correlates of transmission mediated by
    glutamate and aspartate. J. Neurochem., 31, 147-155.
Naito, S. and Ueda, T. (1983). Adenosine triphosphate-dependent
    uptake of glutamate into protein I-associated synaptic
    vesicles. J. Biol. Chem., 258, 696-699.
Ottersen, O.P. and Storm-Mathisen, J. (1984a). Neurons containing
    or accumulating transmitter amino acids. In Handbook of
    Chemical Neuroanatomy. Vol. 3 (eds. A. Björklund, T. Hökfelt
    and M.J. Kuhar). Elsevier/North Holland, Amsterdam.,
    pp. 141-246.
Ottersen, O.P. and Storm-Mathisen, J. (1984b). Glutamate- and
    GABA-containing neurons in the mouse and rat brain, as
    demonstrated with a new immunocytochemical technique.
    J. Comp. Neurol., 229, 374-392.
Ottersen, O.P. and Storm-Mathisen, J. (1984c). GABA-containing
    neurons in the thalamus and pretectum of the rodent: An immu-
    nocytochemical study. Anat. Embryol., 170, 197-207.
Ottersen, O.P. and Storm-Mathisen, J. (1985). Different neuronal
    localization of aspartate-like and glutamate-like
    immunoreactivities in the hippocampus of rat, guinea pig, and
    Senegalese baboon (Papio papio), with a note on the distribu-
    tion of GABA. Neuroscience, in press.

Roberts, E. (1981). Strategies for identifying sources and sites of formation of GABA-precursor or transmitter glutamate in brain. In Glutamate as a Neurotransmitter. (eds. G. Di Chiara and G.L. Gessa). Raven Press, New York, pp. 91-102.

Sørensen, K.E. (1980). Distribution of an ipsilateral-commissural projection from the second reflected blade of the dentate hilus to the fascia dentata in guinea pig. Neurosci. Lett., Suppl. 5, S166.

Sternberger, L.A. (1979). Immunocytochemistry. Wiley, Chichester.

Storm-Mathisen, J. and Ottersen, O.P. (1983). Immunohistochemistry of glutamate and GABA. In Glutamine, Glutamate and GABA in the Central Nervous System. (eds. L. Hertz, E. Kvamme, E.G. McGeer and A. Schousboe). Alan R. Liss, New York, pp. 185-201.

Storm-Mathisen, J. and Ottersen, O.P. (1984). Neurotransmitters in the hippocampal formation. In Cortical Integration. (eds. F. Reinoso-Suárez and C. Ajmone-Marsan). Raven Press, New York, pp. 105-130.

Storm-Mathisen, J. and Ottersen, O.P. (1985). Antibodies against amino acid neurotransmitters. In Neurohistochemistry Today. (eds. P. Panula, H. Päivärinta and S. Soinila). Alan R. Liss, New York, in press.

Storm-Mathisen, J. and Wold, J.E. (1981). In vivo high-affinity uptake and axonal transport of D-[2,3-^{3}H]aspartate in excitatory neurons. Brain Res., 230, 427-433.

Storm-Mathisen, J., Leknes, A.K., Bore, A.T., Vaaland, J.L., Edminson, P., Haug, F.M.Š. and Ottersen, O.P. (1983). First visualization of glutamate and GABA in neurones by immunocytochemistry. Nature (Lond.), 301, 517-520.

Streit, P. (1980). Selective retrograde labeling indicating the transmitter of neuronal pathways. J. Comp. Neurol., 191, 429-463.

Swanson, L.W. (1979). The hippocampus - new anatomical insights. Trends Neurosci. 2, 9-12.

Taxt, T. and Storm-Mathisen, J. (1984). Uptake of D-aspartate and L-glutamate in excitatory axon terminals in hippocampus: autoradiographic and biochemical comparison with y-aminobutyrate and other amino acids in normal rats and in rats with lesions. Neuroscience, 11, 79-100.

Wenthold, R.J. and Altschuler, R.A. (1983). Immunocytochemistry of aspartate aminotransferase and glutaminase. In Glutamine, Glutamate and GABA in the Central Nervous System. (eds. L. Hertz, E. Kvamme, E.G. McGeer and A.Schousboe). Alan R. Liss, New York, pp. 33-50.

8

Identification of Pathways for Acidic Amino Acid Transmitters and Search for New Candidates: Sulphur-containing Amino Acids

M. Cuénod, K.Q. Do, C. Matute and P. Streit

In this paper, the emphasis will be given on two topics: (I) The information which can be drawn from selective retrograde labeling of CNS pathways while using D-[^{3}H]-Asp as a marker. (II) The search for new neuroactive substances, possibly involved in neurotransmission.

I. D-[^{3}H]-ASP SELECTIVE RETROGRADE LABELING

In the past 20 years, great efforts have been made by neurobiologists in order to develop a biochemical neuroanatomy, and one aspect has been the characterization of CNS pathways according to their neurotransmitter(s). Various histochemical and immunohistochemical methods have been developed, which allowed to characterize catecholaminergic, cholinergic, GABAergic, and peptidergic neurons. The CNS pathways using excitatory amino acids such as glutamate (Glu) and/or aspartate (Asp) have been more difficult to be defined morphologically. Immunohistochemistry of Glu bound to bovine serum albumine has been relatively successful in the hands of Storm-Mathisen and his colleagues (see this volume) to define neurons rich in Glu. We have developed a method based on selective uptake of labeled transmitters or their metabolites, followed by retrograde transport of the label which can then be analysed by autoradiography (for review see Cuénod et al., 1982; Cuénod and Streit, 1983).

This approach has been successful in cholinergic and aminergic systems, as well as with glycine, GABA and D-aspartate. For pathways using excitatory amino acids, presumably L-Glu and/or L-Asp, as transmitters, D-[^{3}H]-Asp which uses the same uptake carriers but is much less metabolized has turned out to be a good marker candidate, as demonstrated in the following examples. In the pigeon optic nerve (Beaudet et al., 1981), the rat primary afferents (Barbaresi et al., in prep.), and olfactory afferents (Fischer et al., 1982), a subpopulation of neurons are retrogradely labeled. D-[^{3}H]-Asp injection in the cerebellar cortex labeled selectively the olivary climbing fibers afferents in the rat (Wiklund et al., 1982,1984; Toggenburger et al., 1983) an in the monkey (Matute et al., in preparation). D-[^{3}H]-Asp retrograde labeling has been very effective in the corticofugal systems, in which there is accumulating evidence that Glu and/or Asp play an essential role. The cortical projections to the striatum (Streit, 1980), amygdala (Fischer et al., 1982), substantia nigra (Streit, 1980), thalamus: lateral geniculate nucleus (Baughman and Gilbert, 1980, 1981; LeVay and Sherk, 1981), and ventrobasal complex (Rustioni et al., 1983), dorsal column nuclei (Rustioni and Cuénod, 1982) and red nucleus (Bernays and Streit, in preparation) have all been labeled after D-[^{3}H]-Asp injections (see fig. 1).

Wiklund and Cuénod (1984), after D-[^{3}H]-Asp injections in centromedian-parafascicular thalamic nuclei, observed retrograde labeling of large numbers of cortical neurons in layers V and VI, mainly in the ipsilateral insular and precentral areas, but also in the anterior cingulate, prelimbic, infralimbic and sensorimotor areas.

Wiklund and collaborators (this symposium) also investigated the afferents to the dorsal raphe nucleus (DR) and to the raphe magnus nucleus (RM) by retrograde labeling with D-[^{3}H]-Asp. D-[^{3}H]-Asp injection in RM led to selective labeling of the following structures: frontal cortex, medial preoptic area, dorsomedial hypothalamus, and periaqueductal gray (PAG). Devoid of D-[^{3}H]-Asp labeled cells, which were however labeled by [^{125}I]-wheat germ agglutinin (WGA), were RM afferents originating in bed nucleus of stria terminalis, basal magnocellular nucleus, zona incerta, nucleus parafascicularis prerubralis, superior colliculus, cuneiform nucleus, cerebellar and vestibular nuclei.

Following D-[^{3}H]-Asp injections in DR retrograde labeling was found in lateral habenular nucleus, substantia nigra, PAG, while the non selective marker, WGA, retrogradely labeled many more structures, including the ventral tegmental area, parabrachial nuclei, RM, diagonal Band of Broca, cuneiform nucleus, superior vestibular nucleus, and some hypothalamic and reticular areas. The projection of the lateral habenular nucleus to the

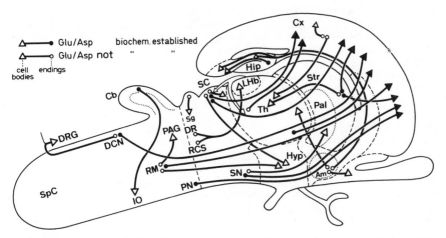

Figure 1: Schematic representation of the main pathways shown to be retrogradely labeled with D-[³H]-Asp
in the rat brain.
Abbreviations:

Am	Amygdala	Pal	Pallidum
Cb	Cerebellum	PN	Pontine nuclei
Cx	Cortex	RCS	Raphe Centralis Superior
DCN	Dorsal Column Nuclei	RM	Raphe magnus
DR	Dorsal Raphe	SC	Superior Colliculus
DRG	Dorsal Root Ganglion	Sg	Sagulum
Hip	Hippocampus	SN	Substantia Nigra
Hyp	Hypothalamus	SpC	Spinal cord
IO	Inferior Olive	Str	Striatum
LHb	Lateral Habenula	Th	Thalamus
PAG	Periaqueductal gray		

raphe centralis superior nucleus was also labeled by
D-[³H]-Asp (Pritzel, Kalén, and Wiklund, personal communication).

Matute and Streit (submitted) undertook a study of
the projections to the superior colliculus (SC). This
investigation originated from a discrepancy in the literature on the cortico-collicular pathways; neurochemical
investigations indicated glutamate and/or aspartate as
transmitter(s) in the rat (Lund-Karlsen and Fonnum,
1978), however no retrograde labeling was observed in
the cat visual cortex following collicular injection of
tritiated D-aspartate (Baughman and Gilbert, 1981).

In rats, selectivity of retrograde perikaryal labeling was assessed by comparing patterns observed in

D-[^{3}H]-aspartate experiments with those found after administration of a non-selective tracer, horseradish peroxidase labeled wheat germ agglutinin (HRP-WGA), and of the tritiated neurotransmitter gamma-aminobutyric acid (GABA). Following D-[^{3}H]-aspartate injection into the superior colliculus, labeling was intense in a large number of cortical and hypothalamic neurons. Other afferents to SC, however, such as those originating from the ventrolateral geniculate nucleus, the pars reticulata of the substantia nigra, the locus coeruleus, the pontine nuclei or the retinal ganglion cells, were not labeled. Similar results were obtained in rabbits.

In cats, the analysis was focused on the cerebral cortex. Retrogradely labeled neurons were observed in various cortical areas including a few in the visual area 18 but none in area 17. The paucity of this visuo-cortical labeling is more consistent with the negative findings obtained with this tracing method as mentioned previously (Baughman and Gilbert, 1981) than with the very recent neurochemical observations in the cat (Fosse et al., 1984). - Thus, this work demonstrates that there is no discrepancy between results obtained with the retrograde tracing method using tritiated D-aspartate and with neurochemical methods in the rat, whereas in the cat such a discrepancy seems to exist.

On the other hand, the fact that retrograde labeling with D-[^{3}H]-Asp was observed only in a subset of the afferents to SC emphasizes its selectivity. This selectivity does not display major differences among the mammalian species studied. Moreover, according to the information available about the distribution of neurotransmitters in the brain, the findings favor the idea that tritiated D-aspartate is a retrograde tracer selective for glutamatergic and/or aspartatergic pathways.

In conclusion, D-[^{3}H]-Asp is an interesting selective marker for pathways which seem to use an excitatory amino acid as transmitter, although the conclusion that it is Glu or Asp must be taken with caution. It is important, in any application of this method, to include systematically controls with non-selective retrograde tracers as well as with other transmitter amino acids or transmitter related substances.

II. ENDOGENOUS RELEASE OF SULFUR CONTAINING AMINO ACIDS

The evidence for Glu and/or Asp as the real transmitter(s) is not definitively established in the pathways mentioned above and the respective role of Glu and

Asp is unclear in most of them. Beside that, many important CNS connections are still undefined, as far as
their transmitters are concerned. To those, for instance, belong many afferent systems and the massive
thalamo-cortical projections. Furthermore, the presence
of 3 or 4 pharmacologically distinct classes of receptors activated by excitatory substances suggests the
possible existence of other endogenous agonists in addition to Glu and Asp.

Thus, Do et al. (in preparation) undertook an investigation aimed at screening and discovering endogenous
neuroactive substances probably involved in neurotransmission. In order to select them, we took advantage of
the fact that compounds involved in intercellular communication are released from neurons upon membrane depolarization in a Ca^{++}-dependant manner. We analysed the superfusates of nerve tissue for trace amount of amines,
amino acids, oligopeptides and related molecules, comparing the outflow under resting and stimulated conditions. The analysis was performed with high pressure
liquid chromatography (HPLC) and gas chromatography combined with mass spectrometry. Although the evidence that
a given compound is released does by no means establish
its neuroactive properties, release allows a selection,
among the innumerable substances present in tissue, of
compounds worth to be further investigated for their
physiological effects.

Rat brain slices of neocortex, hippocampus, striatum, mesodiencephalon, cerebellum, pons-medulla and spinal cord were superfused by Earl's bicarbonate-buffered
salt solution. The depolarisation was induced by raising
the $[K^+]$ to 50mM or by adding 33µg/ml of veratrine. In
parallel experiments, the $[Ca^{++}]$ was decreased from 2mM
to 0.1mM and the $[Mg^{++}]$ increased from 1mM to 12mM. The
superfusion solutions, which contained a high salt concentration, were derivatized with 4-N,N-dimethylamino-
azobenzene-4'-isothiocyanate (DABITC) according to a modification of the method of Chang (1981). The amino
acids and peptides were then analysed by reversed phase
HPLC. Using various improvements, this precolumn derivatization reversed phase HPLC allowed to resolve all the
natural amino acids including taurine, GABA and β-alanine as well as the quantitation of proline and for all
at the picomole level. Furthermore the very polar sulfur
containing amino acids (S-aa) derivatives, such as cysteine sulfinic acid (CSA), cysteic acid (CA), homocysteine sulfinic acid (HCSA) and homocysteic acid (HCA) mi-
grated early in the chromatogram. These very polar compounds, which usually coeluted with the void volume in
the standard amino acid analytical methods, have been

detected simultaneously and well resolved for the first time, and that at the picomole level in biological material. All four are present in TCA-extracts of rat neocortex.

The analysis of the superfusates of cortical slices, either under resting or K^+-depolarization conditions, gave the following results (see fig. 2): a massive release of GABA, Glu and Asp was confirmed, the ratio of stimulated over resting fractions being respectively 6.6, 14.1 and 15.3. In addition, a discrete release of CSA (ratio 1.5), HCSA (1.9) as well as a rather massive release of HCA (7.3) were observed, while CA was unaffected. A compound eluting between CSA, CA and their homo analogues, named P18, appeared to be increased by a factor of 2.1 during stimulation (see table 1). With low $[Ca^{++}]$ superfusion, the S-aa could not be detected any more, either under resting or stimulation conditions.

The depolarization induced release of the S-aa was different in the various rat brain regions investigated,

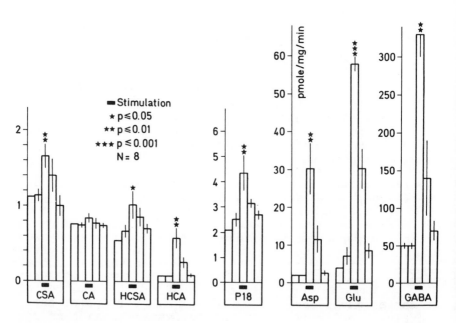

Figure 2: Time course of the K^+ depolarization induced release of some amino acids from cortex slices. Concentration expressed in pmole/mg protein/min in superfusates of five consecutive fractions, the third one corresponding to the stimulation. Abbreviations in text.

as shown in table 1: CSA and HCSA were released moderately but significantly in neocortex, hippocampus and mesodiencephalon, and, for HCSA only, in striatum. A striking release of HCA was observed in all the regions, the neocortex and the hippocampus showing the highest values. In contrast CA was practically unaffected by the depolarization except in the hippocampus and the mesodiencephalon where a very weak release was detected. When slices of all regions were superfused with 33 µg/ml of veratrine, the concentration of the four S-aa was similarly increased as described for K^+ depolarization.

It should be pointed out that the depolarization induced efflux of Glu varies in the brain regions determined, between 40 and 90 pmole/mg protein/min, while that of the S-aa ranges between 0.5 and 2.5 pmole/mg protein/min, so that the efflux of Glu (stimulation condition) is about 100 times larger than that of HCA for

Table 1: Relative increase of the K^+ depolarization induced release of CSA, CA, HCSA, HCA, P18, Asp and Glu in various rat brain regions.
* p<0.05; ** p<0.01; *** p<0.001 (Student's t-test).

	CSA	CA	HCSA	HCA	P18	ASP	GLU
	**		*	**	**	**	***
Cortex	1.5	1.1	1.9	7.3	2.1	15.3	14.1
	*	*	*	*	*	**	**
Hippocampus	2.4	1.3	1.6	6.1	1.5	20.8	40.6
			*	*	***	**	***
Basalganglia	1.0	0.9	1.3	3.6	1.6	10.4	30.8
	*	*	*	*	***	*	**
Meso-diencephalon	1.5	1.4	1.4	4.5	2.0	13.2	13.5
Cerebellum	1.0	0.8	0.7	2.7	1.1	3.0	4.0
			*	**	*	*	
Medulla-pons	1.1	1.2	1.5	2.5	2.3	1.3	3.0
			**	*	***	***	
Spinal cord	1.0	1.1	1.2	2.3	2.4	2.3	3.1

instance. However, the spontaneous efflux of HCA (resting condition) is also about 100 times smaller than that of Glu, so that their relative release is of the same order of magnitude. Whether these quantitative differences reflect differences in the size of the releasing compartment or other facts such as reuptake or metabolism, cannot be answered at the moment.

The four S-aa have been reported by Watkins and his collaborators to exert an excitatory effect on spinal cord neurons (Curtis and Watkins, 1960; Wu and Dowling, 1978; Watkins and Evans, 1981; Mewett et al., 1983). CSA has been shown to be endogenous in rat brain where it was unevenly distributed (Baba et al., 1980, Ida and Kuriyama, 1983). Its synthesizing and degradative enzymes were associated with synaptosomes (Agrawal et al., 1971; Passantes-Morales et al., 1977, Musra and Olney, 1975, Recasens et al., 1978). The degradative enzymes of CSA were located in nerve endings by immunohistochemistry (Recasens and Delaunoy, 1981; Chan Palay et al., 1982). A sodium dependent high affinity uptake has been reported in rat brain synaptosomes and located by quantitative autoradiography (Recasens et al., 1982; Iwata et al., 1982; Parsons and Rainbow, 1984). CSA also binded in a sodium independent manner to synaptic membranes (Recasens et al., 1982, 1983) and exogenous radioactive CSA was released from various rat brain regions (Iwata et al., 1982; Baba et al., 1983; Recasens et al., 1984). Recasens et al. (1984) observed a similar regional distribution of veratrine stimulated release of L-[^{3}H] CSA, except for the striatum, where it was high while the endogenous one in the present data was low. Finally, CSA elicited an increase in cortical levels of cAMP and cGMP (Ferrendelli et al., 1974; Shimizu et al., 1974; Baba et al., 1982). In hippocampal slices, the CSA induced an increase of cAMP which could be selectively blocked by taurine (Baba et al., 1982).

L-CSA has been shown to induce a Na^+-efflux in rat striatal slices (Luini et al., 1984) and to interact predominantly with non-NMDA receptors in the cat caudate (Herrling and Turski, 1985 a). It is thus surprising that no endogenous release was detected for CSA in the rat striatum in the present study.

To our knowledge, presence and endogenous release of HCA are reported here for the first time. For this S-aa, much less information is available, on the enzymatic aspects, as well as on uptake, binding, and localization. Curtis and Watkins (1960) and Mewett et al. (1983) pointed out the excitatory action of HCA which was sensitive to antagonism by 2-amino-5-phosphono-valeric acid (APV) as well as its high affinity for the Glu binding site. Both Luini et al. (1984) and Baudry et al. (1983) observed that D,L-HCA was one of the most potent activator of the Na^+ efflux, and suggested that it had a specificity for the NMDA receptor, thus confirming the interpretation of Mewett et al. (1983). This was recently confirmed and extended by Herrling and Turski (1985 b, this volume) who, by microiontophoretic application of HCA, induced in cat caudate neurons a discharge of the NMDA type, blocked by 2-amino-7-phosphono-heptanoic acid (AP-7).

Thus, with the demonstration of its endogenous release, HCA may become a candidate as a physiological agonist with a relatively high selectivity for the NMDA receptor. Its localization in specific CNS pathways will be an important step in the elucidation of its putative functional role.

The screening for possible neuroactive endogenous substances involved in intercellular communication by analysing with sensitive methods the material released by depolarization has proven to be a useful approach. In the case of the S-aa presented here, their neuroactivity is already established to a large extent.

However, the DABITC precolumn derivatisation HPLC method has also revealed other polar compounds, such as P18, which are increased in perfusate of depolarized slices. They must be first chemically identified, and then tested for their possible function.

Acknowledgements

The authors wish to thank Dr. J. C. Watkins for many very helpful discussions and the gift of several reference amino acids, as well as Dr. L. Wiklund for many useful suggestions. This work has been supported by grants 3.455.83 and 3.228.82 of the Swiss National Science Foundation and by the Dr. Eric Slack-Gyr-Foundation. The support of the Sandoz-Foundation, the "Jubiläumsstiftung der Schweiz. Lebensversicherung- und Rentenanstalt", the "Geigy-Jubiläums-Stiftung" and the "Hartmann-Müller-Stiftung" is gratefully acknowledged.

References

Agrawal, H.C., Davison, A.N., and Kaczmarek, L.K. (1971) Subcellular distribution of taurine and cysteine sulfinate decarboxylase in developing rat brain. Biochem. J., 122, 759-763.

Baba, A., Yamagami, S., Mizuo, H., and Iwata, H. (1980) Microassay of cysteine sulfinic acid by an enzymatic cycling method. Anal. Biochem., 101, 288-293.

Baba, A., Lee, E., Tatsuno, T., and Iwata, H. (1982) Cysteine sulfinic acid in the central nervous system: Antagonistic effect of taurine on cysteine sulfinic acid-stimulated formation of cyclic AMP in guinea pig hippocampal slices. J. Neurochem., 38, 1280-1285.

Baba, A., Okumura, S., Mizuo, H., and Iwata, H. (1983) Inhibition by diazepam and γ-aminobutyric acid of depolarisation-induced release of $[^{14}C]$cysteine sulfinate and $[^{3}H]$glutamate in rat hippocampal slices. J. Neurochem., 40, 280-284.

Barbaresi, P., Rustioni, A., and Cuénod, M.: Retrograde labeling of dorsal root ganglion neurons after injection of tritiated amino acids in the spinal cord of rats and cats. (in preparation)

Baudry, M., Kramer, K., Fagni, L., Recasens, M., and Lynch, G. (1983) Classification and properties of acidic amino acid receptors in hippocampus. Mol. Pharmacol., 24, 222-228.

Baughman, R.W. and Gilbert, C.D. (1980) Aspartate and glutamate as possible neurotransmitters of cells in Layer VI of the visual cortex. Nature, 287, 848-849.

Baughman, R.W., and Gilbert, C.D. (1981) Aspartate and glutamate as possible neurotransmitters in the visual cortex. J. Neurosci., 1, 427-439.

Beaudet, A., Burkhalter, A., Reubi, J.C., and Cuénod, M. (1981) Selective bidirectional transport of $[^{3}H]$D-aspartate in the pigeon retinotectal pathway. Neurosci., 6, 2021-2034.

Chan-Palay, V., Lin, C.T., Palay, S.L., Yamamoto, M., and Wu, J.Y. (1982) Taurine in the mammalian cerebellum: demonstration by autoradiography with $[^{3}H]$ taurine and immunocytochemistry with antibodies against the taurine-synthesizing enzyme, cysteine sulfinic acid decarboxylase. Proc. Natl. Acad. Sci. USA, 79, 2695-2699.

Chang, J.-Y. (1981) Isolation and characterization of polypeptide at the picomole level. Biochem. J., 199, 537-545.

Cuénod, M., Bagnoli, P., Beaudet, A., Rustioni, A., Wiklund, L., and Streit, P. (1982) Transmitter-specific retrograde labeling of neurons. In Cytochemical Methods in Neuroanatomy (eds. V. Chan-Palay and S.L. Palay), pp. 17-44. Alan R. Liss, New York.

Cuénod, M. and Streit, P. (1983) Neuronal tracing using retrograde migration of labeled transmitter-related compounds. In Methods in Chemical Neuroanatomy, Handbook of Chemical Neuroanatomy Vol. 1 (eds. A. Björklund and T. Hökfelt), pp. 365-397. Elsevier, Amsterdam.

Curtis, D.R. and Watkins, J.C. (1960) The excitation and depression of spinal neurones by structurally related amino acids. J. Neurochem., 6, 117-141.

Do, K.Q., Mattenberger, M., Streit, P., and Cuénod, M.: Endogenous in vitro release of sulphur containing amino acids from various rat brain regions. (in preparation)

Ferrendelli, J.A., Chang, M.M., and Kinscherf, D.A. (1974) Elevation of cyclic GMP levels in central nervous system by excitatory and inhibitory amino acids. J. Neurochem., 22, 535-540.

Fischer, B.O., Ottersen, O.P., and Storm-Mathisen, J. (1982) Labelling of amygdalopetal and amygdalofugal projections after intra-amygdaloid injectios of tritiated D-aspartate. Neurosci., Suppl. 7, S69.

Fosse, V.M., Heggelund, P., Iversen, E., and Fonnum, F. (1984) Effects of Area 17 ablation on neurotransmitter parameters in efferents to area 18, the lateral geniculate body, pulvinar and superior colliculus in the cat. Neurosci. Lett., 52, 323-328.

Herrling, P.L. and Turski, W.A. (1985a) The pharmacological specificity of cysteine sulfinic acid, an endogenous excitatory amino acid, in the cat caudate. Experientia (in press).

Herrling, P.L. and Turski, W.A. (1985b) Interactions of sulphur-containing excitatory amino acids with membrane and synaptic potentials of cat caudate neurons. This volume.

Ida, S. and Kuriyama, K. (1983) Simultanous determination of cysteine sulfinic acid and cysteic acid in rat brain by high-performance liquid chromatography. Anal. Biochem., 130, 95-101.

Iwata, H., Yamagami, S., Mizuo, H., and Baba, H. (1982) Cysteine sulfinic acid in the central nervous system: uptake and release of cysteine sulfinic acid by a rat brain preparation. J. Neurochem., 38, 1268-1274.

LeVay, S. and Sherk, H. (1981) The visual claustrum of the cat. I. Structure and connections. J. Neurosci., 1, 956-980.

Luini, A., Goldberg, O., and Teichberg, V.I. (1984) An evaluation of selected brain constituants as putative excitatory neurotransmitters. Brain Res., 324, 271-277.

Lund Karlsen, R. and Fonnum, F. (1978) Evidence for glutamate as a neurotransmitter in the corticofugal fibres to the dorsal lateral geniculate body and the superior colliculus in rats. Brain Res., 151, 457-467.

Matute, C. and Streit, P.: Selective retrograde labeling using D-[^{3}H]-aspartate in afferents to mammalian superior colliculus. J. Comp. Neurol. (submitted).

Mewett, K.N., Oakes, D.J., Olverman, H.J., Smith, D.A.S., and Watkins, J.C. (1983) Pharmacology of the excitatory actions of sulphonic and sulphinic amino acids. In CNS Receptors-From Molecular Pharmacology to Behavior (eds. P. Mandel and F.V. DeFeudis), pp. 163-174. Raven Press, New York.

Misra, C.H. and Olney, J.N. (1975) Cysteine oxidase in brain. Brain Res., 97, 117-126.

Parsons, B. and Rainbow, T. (1984) Localisation of cysteine sulfinic acid uptake sites in rat brain by quantitative autoradiography. Brain Res., 294, 193-197.

Pasantes-Morales, H., Loriette, C., and Chatagner, F. (1977) Regional and subcellular distribution of taurine synthesizing enzymes in the rat central nervous system. Neurochem. Res., 2, 671-680.

Recasens, M., Gabellec, M.N., Austin, L., and Mandel, P. (1978) Regional and subcellular distribution of cysteine sulfinate transminase in rat nervous system. Biochem. Biophys. Res. Commun., 83, 449-456.

Recasens, M. and Delaunoy, J.P. (1981) Immunological properties and immunohistochemical localistion in CNS of cysteine sulfinate transaminase. Brain Res., 205, 351-361.

Recasens, M., Varga, V., Nanopoulos, D., Saadoun, F., Vincendon, G., and Benavides, J. (1982) Evidence for cysteine sulfinate as a neurotransmitter. Brain Res., 239, 153-173.

Recasens, M., Saadoun, F., Varga, V., DeFeudis, F.V., Mandel, P., Lynch, G., and Vincendon, G. (1983) Separate binding sites in rat brain synaptic membranes for L-cysteine sulfinate and L-glutamate. Neurochem. Int., 5, 89-94.

Recasens, M., Fagni, L., Baudry, M., and Lynch, G. (1984) Potassium and veratrine-stimulated L-[^{3}H]cysteine sulfinate and L-[^{3}H]glutamate release from rat brain slices. Neurochem. Int., 6, 325-332.

Rustioni, A. and Cuénod, M. (1982) Selective retrograde transport of D-aspartate in spinal interneurons and cortical neurons of rats. Brai Res., 236, 143-155.

Rustioni, A., Schmechel, D.E., Spreafico, R., Cheema, S., and Cuénod, M. (1983) Excitatory and inhibitory amino acid putative neurotransmitters in the ventralis posterior complex: An autoradiographic and immunocytochemical study in rats and cats. In Somatosensory Integration in the Thalamus (eds. G. Macchi, A. Rustioni, and R. Spreafico), pp. 365-383. Elsevier, Amsterdam.

Shimizu, H., Ichishita, H., and Odagiri, H. (1974) Stimulated formation of cyclic adenosine 3',5'-monophosphate by aspartate and glutamate in cerebral cortical slices of guinea pig. J. Biol. Chem., 249, 5955-5962.

Streit, P. (1980) Selective retrograde labeling indicating the transmitter of neuronal pathways. J. Comp. Neurol., 191, 429-463.

Toggenburger, G., Wiklund, L., Henke, H., and Cuénod, M. (1983) Release of endogenous and accumulated exogenous amino acids from slices of normal and climbing fibre-deprived rat cerebellar slices. J. Neurochem., 41, 1606-1613.

Watkins, J.C. and Evans, R.H. (1981) Excitatory amino acid transmitters. Ann. Rev. Pharmacol. Toxicol., 21, 165-204.

Wiklund, L., Toggenburger, G., and Cuénod, M. (1982) Aspartate: Possible neurotransmitter in cerebellar climbing fibers. Sience, 216, 78-79.

Wiklund, L. and Cuénod, M. (1984) Differential labeling of afferents to thalamic centromedian-parafascicular nuclei with [³H]-choline and D-[³H]aspartate: further evidence for transmitter specific retrograde labelling. Neurosci. Lett., 46, 275-281.

Wiklund, L., Toggenburger, G., and Cuénod, M. (1984) Selective retrograde labelling of the rat olivocerebellar climbing fiber system with D-[³H]aspartate. Neurosci., 13, 441-468.

Wu, S.M. and Dowling, J.E. (1978) L-aspartate: Evidence for a role in cone photoreceptor synptic transmission in the carp retina. Proc. Natl. Acad. Sci. USA, 75, 5205-5209.

9

Excitatory Amino Acids as Transmitters in the Olfactory System

G.G.S. Collins

SUMMARY

Evidence that L-aspartate (Asp) and/or L-glutamate (Glu) may
be neurotransmitters of the mitral (and tufted?) cells of the
olfactory bulb and also of some olfactory cortical pyramidal cells
is reviewed. Several reports describing the synaptically-evoked
release of endogenous Asp and/or Glu from olfactory cortex slices
support such a role. Surgical and chemical lesions cause loss
of Asp and/or Glu at sites corresponding to the location of the
candidate neurones. In the rat, there is a selective loss of
evoked Asp release from olfactory cortex slices following
bulbectomy. Excitatory amino acid receptors have been identified
in the bulb and also in the olfactory cortex where they have presynaptic, synaptic and extrasynaptic distributions. Mitral cell
self-excitation is mediated by N-methyl-D-aspartate (NMDA) receptors whilst excitatory transmission in the olfactory cortex is
mediated by quisqualate and NMDA receptors. Some receptor agonists
exhibit a 'response desensitization' and there is the possibility
that an as yet unknown category of receptors may contribute to
responses evoked by exogenous Asp and Glu.

INTRODUCTION

The mammalian olfactory system consists of the olfactory
nerves, olfactory bulbs and primary olfactory cortices. The
complex is rich in transmitter candidates (see Halasz and Shepherd,
1983), which, in addition to Asp and Glu, include acetylcholine,
the monoamines dopamine, noradrenaline and 5-hydroxytryptamine, a
number of peptides including angiotensin, carnitine, cholecysto-
kinin, enkephalin, neurotensin, somatostatin and substance P and
the inhibitory amino acids GABA and possibly taurine. The primary
connections of the system are relatively simple. The olfactory
receptors project their axons to the ipsilateral olfactory bulb
where they synapse with the dendrites of mitral, tufted and peri-

131

glomerular cells arranged in complexes known as glomeruli (Shepherd, 1972). The major output from each bulb is carried by the mitral (and tufted) cell axons which together constitute the lateral olfactory tract (LOT). In turn, the tract terminals form excitatory synapses with a population of pyramidal cells located superficially in the olfactory cortex (Stevens, 1969). The aim of this article is to summarize the evidence that the mitral cells of the bulb and some of the cortical pyramidal cells utilize either Asp and/or Glu as their transmitters.

DISTRIBUTION STUDIES

Measuring Asp and Glu distribution gives only limited evidence of a possible neurotransmitter role, largely because of their involvement in metabolic processes. Compared with other regions, relatively high levels of Asp and Glu are found in the olfactory bulb (Popov et al., 1967). The laminar distribution of Glu is rather uniform with somewhat higher levels in the glomerular layer whereas Asp is concentrated in the layer containing the mitral cell bodies and dendrites (Godfrey et al., 1980; Nadi et al., 1980). Immunocytochemical studies confirm an association between Glu-like immunoreactivity and the mitral cells (Ottersen and Storm-Mathisen, 1984) although the authors questioned the significance of this finding. In the olfactory cortex the highest concentrations of both Asp and Glu are located at depths containing the pyramidal cell perikarya (Collins, 1979b). The enzyme aspartate aminotransferase has been used as a marker of aspartergic and glutamatergic fibres and although the LOT possesses high activity, this also may be related to metabolic rather than neurotransmitter functions (Godfrey et al., 1984).

RELEASE STUDIES

The measurement of amino acid release has proved to be one of the more fruitful approaches in providing evidence of a neurotransmitter role for Asp and Glu in the olfactory system. Using slices of guinea-pig olfactory cortex, a series of studies (Matsui and Yamamoto, 1975; Yamamoto and Matsui, 1974, 1976) reported that both LOT stimulation and high K^+ concentrations released pre-loaded $[^{14}C]$-Glu in a Ca^{2+}-dependent manner although the magnitude of release was only significant when measured in the presence of cysteate. These studies suffer from two limitations. First, as Asp and Glu probably share the same uptake system,(Balcar and Johnston, 1972), the $[^{14}C]$-Glu would be accumulated by and thus released from both aspartergic and glutamatergic neurones. Second, although cysteate was used to block $[^{14}C]$-Glu reuptake, it is likely to be a substrate for the uptake system (Balcar and Johnston, 1972) and is also a powerful excitant (Mewett et al., 1983).

A specific, Ca^{2+}-dependent release of endogenous Glu on LOT stimulation of guinea-pig olfactory cortex slices was first reported by Bradford and Richards (1976). However, when similar experiments were carried out using slices from rats, LOT stimulation caused a Ca^{2+}-dependent release of endogenous Asp (+ GABA) but not Glu (Collins, 1979a). These inconsistencies were later investigated (Collins et al., 1981) by comparing the K^+-, protoveratrine A- and synaptically-evoked release of endogenous amino acids from slices of rat and guinea-pig olfactory cortex. The pattern of release was similar whichever species was used; all the depolarizing stimuli released Asp (+ GABA) from the pial surfaces whereas both Asp and Glu (+ GABA) were released from the cut under surfaces. These results were interpreted as evidence that in both species, Asp rather than Glu is more likely to be the LOT transmitter but that both amino acids might be transmitters of some olfactory cortical pyramidal cells.

There are some neuropharmacological studies which indirectly support these conclusions. For example, pentobarbitone, which depresses transmission at the LOT- pyramidal cell synapse (Richards, 1972) reduces synaptically-evoked Asp release (Collins, 1980, 1981a). Similarly, activation of GABA receptors located on the LOT terminals either directly by exogenous muscimol (Collins, 1980) or by synaptically-released GABA (Collins, 1981b) also reduces endogenous Asp release. The GABA analogue baclofen, which in the olfactory cortex blocks neurotransmission from the pyramidal cell collaterals, simultaneously inhibits the synaptically-evoked release of both Asp and Glu (Collins et al., 1982). Finally, the excitatory effects of noradrenaline in the olfactory cortex are probably the result of increased release of transmitter Asp and Glu mediated by populations of adrenoceptors located on the LOT terminals and possibly elsewhere (Collins et al., 1984).

All studies in which the synaptically-evoked release of Asp and Glu have been monitored suffer from the major drawback that the site of release is always in doubt. It is now clear that neuro-anatomically, the olfactory cortex has a much more complex structure than hitherto appreciated with at least 9 distinct cell types recently having been described (Haberly, 1983). It is only when release studies are linked with lesioning experiments that it is possible to identify the origin of an amino acid with any certainty.

LESIONING EXPERIMENTS

Both surgical and chemical lesioning techniques have been used to study the neurotransmitter systems of the olfactory complex. In the guinea-pig, bulbectomy reduces both the levels and synthesis of Asp and Glu in the olfactory cortex (Harvey et al., 1975; Scholfield et al., 1983) and similar changes have been reported for the rat (Godfrey et al., 1980; Graham, 1977). However, Collins

(1979a) reported that bulbectomy in the rat caused a selective loss
of tissue Asp with an accompanying failure of LOT stimulation to
evoke Asp release. The loss of Asp was confined to the superficial
layers of the olfactory cortex (Collins, 1979b): this is of some
significance for the LOT fibres also terminate superficially
(Heimer and Kalil, 1978; Stevens, 1969). When olfactory cortex
slices were challenged with protoveratrine A, evoked Asp release
from bulbectomized slices was 50% less than from control
preparations whereas Glu and GABA release was unaffected (Collins
and Probett, 1981).

     All these reports are consistent with Asp and/or Glu being
transmitters of the mitral cells although the results of the
present author would tend to favour Asp. One problem associated
with the interpretation of bulbectomy experiments is that surgery
is accompanied by a rapid transneuronal degeneration of cortical
pyramidal cells (Heimer and Kalil, 1978): loss of Asp and Glu from
deeper cortical layers (Godfrey et al., 1980; Graham, 1977) may
well reflect this phenomenon. One paradoxical finding is that
bulbectomy in the rat fails to reduce high affinity D-[^{3}H]-Asp
uptake in pyriform cortex (Walker and Fonnum, 1983). A recent
study of amino acid levels in various regions of the olfactory
cortex following bulbectomy suggests transmitter heterogeneity of
the LOT fibres (Collins, 1984), a conclusion consistent with neuroanatomical findings (Price and Sprich, 1975). Thus it is possible
that only a fraction of the LOT fibres utilise Asp and/or Glu as
transmitters.

     Chemically-induced lesions have also provided useful data.
Destruction of the olfactory receptors by $ZnSO_4$ has no effect on
Asp or Glu levels or uptake in the bulb suggesting that neither
amino acid is a transmitter for the olfactory neurones (Margolis
et al., 1974). Both local and systemic administration of kainic
acid evokes a massive cell loss in the pyriform cortex (Olney and
Gubareff, 1978) which is paralleled by a significant reduction in
high affinity Glu uptake (Heggli et al., 1981). Similarly,
ibotenic acid, a selective NMDA receptor agonist (Watkins and
Evans, 1981), also causes marked cell loss in the pyriform cortex
(Schwarcz et al., 1979). Kainate also damages cells in the
olfactory bulb although cell loss seems to be confined to the
mitral and tufted cells (Krammer et al., 1980). These results
imply a widespread distribution of excitatory amino acid receptors
in the olfactory system and hence provide indirect support of a
neurotransmitter role for Asp and Glu.

ELECTROPHYSIOLOGICAL STUDIES

     Application of exogenous Asp or Glu to single neurones of the
olfactory bulb (Frosch and Dichter, 1984) or cortex (Constanti et
al., 1980; Hori et al., 1982; Legge et al., 1966; Richards and
Smaje, 1976) characteristically causes an increase in the frequency

of EPSPs and/or cell firing. The threshold concentration of Glu
for pyriform cortical neurones has been calculated to be between
0.2 and 0.4mM (Misell and Richards, 1979). When population
responses are recorded, Asp and Glu are approximately equipotent
in depolarizing olfactory cortical neurones (Surtees and Collins,
1985). In the single study where the actions of Asp and Glu were
recorded intracellularly (Constanti et al., 1980), the excitants
evoked complex responses which included depolarizations both with
and without increases in input conductance and hyperpolarizations
which were invariably associated with increased membrane conductance. The complexity of the responses was no doubt in part due to
the presence of more than one receptor category (see below). It is
significant that many of the actions of Asp and Glu in this study
were mimicked by LOT stimulation. Similarly complex responses have
been reported to occur on application of Glu to cultured olfactory
bulb neurones (Frosch and Dichter, 1984). It also seems that Glu
depolarizes olfactory cortex glial cells (Constanti and Galvan,
1978) although the mechanism involved remains to be explored.

EXCITATORY AMINO ACID RECEPTORS

In the mammalian CNS there are probably at least 4 categories
of excitatory amino acid receptor known as NMDA, quisqualate,
kainate (Watkins and Evans, 1981) and 2-amino-4-phosphonobutyrate
(APB) receptors (see Foster and Fagg, 1984). It is likely that Asp
and Glu interact with all 4 types. The contribution made by each
receptor category to the composite responses evoked by Asp and Glu
will depend upon; (1) the relative affinities and efficacies of the
agonists for the receptors; (2) the relative proportions of
receptor categories at the relevant location; (3) the agonist
concentration (Mayer and Westbrook, 1984).

The presence of excitatory amino acid receptors in the
olfactory system has been demonstrated in a number of ways. The
neurotoxic actions of kainate and ibotenate indirectly establish
the occurrence of kainate and NMDA receptors respectively (see
above). Autoradiographic studies have confirmed the localization
of moderately high concentrations of kainate binding sites both in
the olfactory bulb and cortex (Monaghan and Cotman, 1982) and of
very high densities of [^{3}H]-AMPA (α-amino-3-hydroxy-5-methyl isoxa-
zolepropionic acid, a quisqualate receptor ligand) binding sites in
the outer layers of the pyriform cortex with rather lower densities
in the olfactory tubercle (Monaghan et al., 1984). The binding of
[^{3}H]-Glu is high in all areas of the rat primary olfactory cortex
and bulb (Halpain et al., 1984).

Electrophysiological and neurochemical techniques have
revealed NMDA, kainate and quisqualate receptors in the rat
olfactory cortex (Auker et al., 1982; Collins et al., 1983; Hori
et al., 1982; Surtees and Collins, 1985). A proportion of the
NMDA receptors are located presynaptically where they selectively

inhibit the K^+-evoked release of Asp; in contrast, a population
of presynaptic kainate receptors mediate increased release of both
Asp and Glu.   These results strongly suggest that Asp and Glu are
located in different neuronal populations (Collins et al., 1983).
It is unclear whether APB receptors are present for although APB
blocks transmission, relatively high concentrations are required
(Collins, 1982; Hori et al., 1982) (see below).

Selective receptor antagonists have proved useful both in
confirming the identity of receptors and also in providing indirect
evidence of a neurotransmitter role for Asp and Glu.   In the
olfactory bulb, the selective NMDA receptor antagonist α-amino
adipate (Watkins and Evans, 1981) blocks self-excitation of mitral
cells, a process reflecting direct feedback of dendritically-
released transmitter onto the same and adjacent neurones (Nicholl
and Jahr, 1982).   By comparison of the relative antagonist
potencies of 2-amino-5-phosphonopentanoate, γ-D-glutamylglycine
and cis-2,3-piperidine dicarboxylate on synaptic transmission in
the rat olfactory cortex, Collins (1982) concluded that the
effects of the LOT transmitter are mediated by quisqualate
receptors (note that the autoradiographic distribution of $[^3H]$-
AMPA binding sites (Monaghan et al., 1984) supports this conclusion)
whereas NMDA receptors are involved in pyramidal cell transmission.
This raises the interesting possibility that the mitral cell
transmitter released dendritically acts on NMDA receptors (Nicholl
and Jahr, 1982) whereas that released from the axon (LOT) terminals
acts on quisqualate receptors.   One anomolous finding is that at
concentrations which block transmission at the LOT-pyramidal cell
synapse(Collins, 1982; Hori et al., 1982) APB does not antagonize
Asp and Glu responses (Hori et al., 1982; Surtees and Collins,
1985).   A presynaptic site of action of APB might explain this
inconsistency (Collins, 1982; Neal et al., 1981).

An attempt has also been made to estimate the contributions
made by the receptor subtypes to the composite depolarizations
evoked by exogenous Asp and Glu in the olfactory cortex (Surtees
and Collins, 1985).   By use of selective antagonists, it was
calculated that 36% of the response evoked by Asp and 21% of that
by Glu is mediated by NMDA receptors with quisqualate and kainate
receptors making a significant but quantitatively unknown
contribution to the remainder.

One interesting characteristic of excitatory amino acid
responses in the olfactory cortex is that they 'desensitize';
that is, responses become progressively smaller on repeated
agonist application.   This phenomenon was first described for Glu-
evoked excitation of pyriform cortex neurones (Drostrovsky and
Richards, 1973) and has since been extended to include NMDA
(Surtees and Collins, 1985).   Similar response 'desensitizations'
occur in hippocampal slice preparations (Fagni et al., 1983).   At
low agonist concentrations, the 'desensitization' shows some
selectivity (Surtees and Collins, 1985) but as agonist concentra-

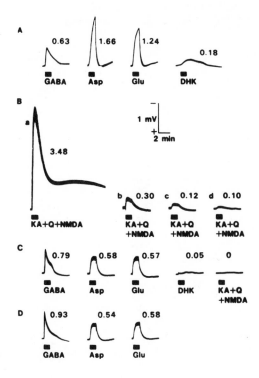

Fig. 1. Residual responses to GABA (10mM), Asp (10mM), Glu (10mM)
and dihydrokainate (DHK, 5mM) following excitatory amino acid
response 'desensitization' and receptor blockade, measured using
the technique described by Brown and Galvan (1979). Each panel
represents the depolarization accompanying drug application to the
pial surface of a slice for 1 min (filled bars). The number
adjacent to each response is the peak depolarization in mV. A.
Control responses to standard agonist doses. B. 1st (a), 4th (b),
6th (c) and 7th (d) response to a mixture containing 1mM each of
kainate (KA), quisqualate (Q) and NMDA. The 'desensitization' in
response was irreversible. C. Agonist responses following
'desensitization' and during perfusion of the non-selective
excitatory amino acid antagonist cis-2,3-piperidine dicarboxylate.
Note that although responses mediated by kainate, quisqualate and
NMDA receptors have been abolished, responses to Asp and Glu are
only reduced by some 60% and there is no reduction in the GABA
response. D. Agonist responses in the presence of cis-2,3-piperi-
dine dicarboxylate plus DHK (5mM). Note that DHK, an inhibitor of
excitatory amino acid uptake, does not reduce the residual
responses to Asp and Glu suggesting that they are not a reflection
of electrogenic amino acid uptake.

tions are increased, 'desensitization' of responses to Asp, quisqualate and kainate can also be detected (Collins, 1985). It is possible that different mechanisms operate at low and high agonist concentrations. It has been suggested that 'desensitizations' recorded at high agonist concentrations reflect an increase in intracellular Na^+ concentrations which occurs to such a degree that the Na^+/K^+ pump capacity is exceeded with a consequent loss of neuronal excitability (Collins, 1985).

It has also been noted that responses evoked by the selective agonists NMDA, quisqualate and kainate are markedly more sensitive to excitatory amino receptor antagonists than are those evoked by Asp and Glu (Surtees and Collins, 1985). Recently, (Collins, 1985), by the combined use of agonist 'desensitization' (see above) plus receptor blockade, responses evoked by the selective agonists have been abolished (Fig. 1). Even under these extreme circumstances, responses evoked by Asp and Glu are reduced by only some 60%. The resistance of the residual Asp and Glu responses to dihydrokainate and APB suggest that they reflect neither electrogenic uptake nor are mediated by APB receptors. One possibility is that the residual responses reflect an interaction of Asp and Glu with an as yet unknown receptor category.

CONCLUSIONS

When viewed as a whole, there is strong experimental support for the likelihood that the mitral cells utilise either Asp and/or Glu as neurotransmitters; evidence that olfactory cortical pyramidal cells are aspartergic and/or glutamatergic is less convincing. At present, the critical evidence supporting such conclusions stems from the results of release and lesioning experiments and even here, results are not entirely consistent. Studies concerning the excitatory amino acid receptor populations of the olfactory cortex must be treated with some caution. First, when exogenous agonists have been investigated it is likely that a significant and perhaps major proportion of the evoked responses originates from extrasynaptic receptors. Second, inhibition of synaptic transmission by receptor antagonists at best provides only indirect evidence of a neurotransmitter role for Asp and Glu and has also given rise to some experimental inconsistencies. In spite of the difficulties involved, application of sophisticated electrophysiological techniques would appear to be the next essential step in clarifying the roles of Asp and Glu in the mammalian olfactory system.

REFERENCES

Auker, C.R., Braitman, D.J. and Rubinstein, S.L. (1982). Electrophysiological action of kainic acid and folates in the in vitro olfactory cortex slice. Nature, 297, 583-584.

Balcar, V.J. and Johnston, G.A.R. (1972). The structural specifi-
    city of the high affinity uptake of L-glutamate and L-aspartate
    by rat brain slices. J. Neurochem., 19, 2657-2666.
Bradford, H.F. and Richards, C.D. (1976). Specific release of
    endogenous glutamate from piriform cortex stimulated in vitro.
    Brain Res., 105, 168-172.
Brown, D.A. and Galvan, M. (1979). Responses of the guinea-pig
    isolated olfactory slice to γ-aminobutyric acid recorded with
    extracellular electrodes. Br. J. Pharmac., 65, 347-353.
Collins, G.G.S. (1979a). Evidence of a neurotransmitter role for
    aspartate and γ-aminobutyric acid in the rat olfactory cortex.
    J. Physiol. (Lond.), 291, 51-60.
Collins, G.G.S. (1979b). Effect of chronic bulbectomy on the depth
    distribution of amino acid transmitter candidates in rat
    olfactory cortex. Brain Res., 171, 552-555.
Collins, G.G.S. (1980). Release of endogenous amino acid neuro-
    transmitter candidates from rat olfactory cortex slices:
    possible regulatory mechanisms and the effects of pento-
    barbitone. Brain Res., 190, 517-528.
Collins, G.G.S. (1981a). Effects of pentobarbitone on the synap-
    tically evoked release of the endogenous amino acid neurotrans-
    mitter candidates aspartate and GABA from rat olfactory cortex.
    Adv. Biochem. Psychopharmacol., 27, 147-156.
Collins, G.G.S. (1981b). The effects of chlordiazepoxide on
    synaptic transmission and amino acid neurotransmitter release
    in slices of rat olfactory cortex. Brain Res., 224, 389-404.
Collins, G.G.S. (1982). Some effects of excitatory amino acid
    receptor antagonists on synaptic transmission in the rat
    olfactory cortex slice. Brain Res., 244, 311-318.
Collins, G.G.S. (1984). Amino acid transmitter candidates in
    various regions of the primary olfactory cortex following
    bulbectomy. Brain Res., 296, 145-147.
Collins, G.G.S. (1985). 'Desensitization' of excitatory amino acid
    responses in the rat olfactory cortex. Neuropharmacology,
    submitted.
Collins, G.G.S., Anson, J. and Kelly, E.P. (1982). Baclofen:
    effects on evoked field potentials and amino acid neurotrans-
    mitter release in the rat olfactory cortex slice. Brain Res.,
    238, 371-383.
Collins, G.G.S., Anson, J. and Probett, G.A. (1981). Patterns of
    endogenous amino acid release from slices of rat and guinea-
    pig olfactory cortex. Brain Res., 204, 103-120.
Collins, G.G.S., Anson, J. and Surtees, L. (1983). Presynaptic
    kainate and N-methyl-D-aspartate receptors regulate excitatory
    amino acid release in the olfactory cortex. Brain Res., 265,
    157-159.
Collins, G.G.S. and Probett, G.A. (1981). Aspartate and not
    glutamate is the likely transmitter of the rat lateral
    olfactory tract fibres. Brain Res., 209, 231-234.
Collins, G.G.S., Probett, G.A., Anson, J. and McLaughlin, N.J.
    (1984). Excitatory and inhibitory effects of noradrenaline on
    synaptic transmission in the rat olfactory cortex slice.

Brain Res., 294, 211-223.

Constanti, A., Connor, J.D., Galvan, M. and Nistri, A. (1980).
Intracellularly-recorded effects of glutamate and aspartate on
neurones in the guinea-pig olfactory cortex slice. Brain Res.,
195, 403-420.

Constanti, A. and Galvan, M. (1978). Amino acid-evoked depolarization of electrically inexcitable (neuroglial?) cells in the
guinea-pig olfactory cortex slice. Brain Res., 153, 183-187.

Dostrovsky, J.O. and Richards, C.D. (1979). Apparent desensitization of guinea-pig olfactory neurones to L-glutamate.
J. Physiol. (Lond.), 300, 73P.

Fagni, L., Baudry, M. and Lynch, G. (1983). Classification and
properties of acidic amino acid receptors in hippocampus I.
Electrophysiological studies of an apparent desensitization and
interactions with drugs which block transmission. J. Neurosci.,
3, 1538-1546.

Foster, A.C. and Fagg, G.E. (1984). Acidic amino acid binding
sites in mammalian neuronal membranes: their characteristics
and relationships to synaptic receptors. Brain Res. Rev., 7,
103-164.

Frosch, M.P. and Dichter, M.A. (1984). Physiology and pharmacology
of olfactory bulb neurons in dissociated cell culture. Brain
Res., 290, 321-332.

Godfrey, D.A., Bowers, M., Johnson, B.A. and Ross, C.D. (1984).
Aspartate aminotransferase activity in fiber tracts of the rat
brain. J. Neurochem., 42, 1450-1456.

Godfrey, D.A., Ross, C.D., Carter, J.A., Lowry, O.H. and
Matschinsky, F.M. (1980). Effect of intervening lesions on
amino acid distributions in rat olfactory cortex and olfactory
bulb. J. Histochem. Cytochem., 28, 1157-1169.

Graham, L.T. (1977). Glutamate and aspartate associated with
lateral olfactory tract fibres. Trans. Am. Soc. Neurochem., 8,
209.

Haberly, L.B. (1983). Structure of the piriform cortex of the
opossum. I. Description of neuron types with Golgi methods.
J. comp. Neurol. 213, 163-187.

Halasz, N. and Shepherd, G.M. (1983). Neurochemistry of the
vertebrate olfactory bulb. Neurosci., 10, 579-619.

Halpain, S., Wieczorek, C.M. and Rainbow, T.C. (1984). Localization of L-glutamate receptors in rat brain by quantitative
autoradiography. J. Neurosci., 4, 2247-2258.

Harvey, J.A., Scholfield, C.N., Graham, L.T. and Aprison, M.H.
(1975). Putative transmitters in denervated olfactory cortex.
J. Neurochem., 24, 445-449.

Heggli, D.E., Aamodt, A. and Malthe-Sørenssen, D. (1981). Kainic
acid neurotoxicity; effect of systemic injection on neurotransmitter markers in different brain regions. Brain Res.,
230, 253-262.

Heimer, L. and Kalil, R. (1978). Rapid transneuronal degeneration
and death of cortical neurones following removal of the
olfactory bulb in adult rats. J. comp. Neurol., 178, 559-610.

Hori, N., Auker, C.R., Braitman, D.J. and Carpenter, D.O. (1982).

Pharmacologic sensitivity of amino acid responses and synaptic activation of in vitro prepyriform neurons. J. Neurophysiol., 48, 1289-1301.

Krammer, E.B., Lischka, M.F. and Sigmund, R. (1980). Neurotoxicity of kainic acid; evidence against an interaction with excitatory glutamate receptors in rat olfactory bulbs. Neurosci. Lett., 16, 329-334.

Legge, K.F., Randic, M. and Straughan, D.W. (1966). The pharmacology of neurones in the pyriform cortex. Br. J. Pharmac., 26, 87-107.

Margolis, F.L., Roberts, N., Ferriero, D. and Feldman, J. (1974). Denervation in the primary olfactory pathway of mice : biochemical and morphological effects. Brain Res., 81, 469-483.

Matsui, S. and Yamamoto, C. (1975). Release of radioactive glutamic acid from thin sections of guinea-pig olfactory cortex in vitro. J. Neurochem., 24, 245-250.

Mayer, M.L. and Westbrook, G.L. (1984). Mixed-agonist action of excitatory amino acids on mouse spinal cord neurones under voltage clamp. J. Physiol. (Lond.), 354, 29-53.

Mewett, K.N., Oakes, D.J., Olverman, H.J., Smith, D.A.S. and Watkins, J.C. (1983). Pharmacology of the excitatory actions of sulphonic and sulphinic amino acids. Adv. Biochem. Psychopharmacol., 37, 163-174.

Misell, D.L. and Richards, C.D. (1979). Estimates of the threshold concentrations of glutamate required to excite nerve cells. J. Physiol. (Lond.), 287, 37-38P.

Monaghan, D.T. and Cotman, C.W. (1982). The distribution of [^{3}H]Kainic acid binding sites in rat CNS as determined by autoradiography. Brain Res., 252, 91-100.

Monaghan, D.T., Yao, D. and Cotman, C.W. (1984). Distribution of [^{3}H]AMPA binding sites in rat brain as determined by quantitative autoradiography. Brain Res., 324, 160-164.

Nadi, N.S., Hirsch, J.D. and Margolis, F.L. (1980). Laminar distribution of putative neurotransmitter amino acids and ligand binding sites in the dog olfactory bulb. J. Neurochem., 34, 138-146.

Neal, M.J., Cunningham, J.R., James, T.A., Joseph, M. and Collins, J.F. (1982). The effect of 2-amino-4-phosphonobutyrate (APB) on acetylcholine release from the rabbit retina : evidence for ON - channel input to cholinergic amacrine cells. Neurosci. Lett., 26, 301-305.

Nicholl, R.A. and Jahr, C.E. (1982). Self-excitation of olfactory bulb neurones. Nature, 296, 441-444.

Olney, J.W. and Gubareff, T. (1978). Extreme sensitivity of olfactory cortical neurons to kainic acid toxicity. In Kainic acid as a tool in neurobiology. (eds. E. G. McGeer, J. W. Olney and P. L. McGeer). Raven Press, N.Y. pp 201-217.

Ottersen, O.P. and Storm-Mathisen, J. (1984). Glutamate- and GABA-containing neurons in the mouse and rat brain, as demonstrated with a new immunocytochemical technique. J. comp. Neurol., 229, 374-392.

Popov, N., Pohle, W., Rösler, V. and Matthies, H. (1967). Regionale Verteilung von γ-Aminobuttersäure, Glutaminsäure, Asparginsäure, Dopamin, Noradrenalin und Serotonin im Rattenhirn. Acta biol. med. ger., 18, 695-702.

Price, J.L. and Sprich, W.W. (1975). Observations on the lateral olfactory tract of the rat. J. comp. Neurol., 162, 321-336.

Richards, C.D. (1972). On the mechanism of barbiturate anaesthesia. J. Physiol. (Lond.), 227, 749-767.

Richards, C.D. and Smaje, J.C. (1976). Anaesthetics depress the sensitivity of cortical neurones to L-glutamate. Br. J. Pharmac., 58, 347-357.

Scholfield, C.N., Moroni, F., Corradetti, R. and Pepeu, G. (1983). Levels and synthesis of glutamate and aspartate in the olfactory cortex following bulbectomy. J. Neurochem., 41, 135-138.

Schwarcz, R., Hökfelt, T., Fuxe, K., Jonsson, G., Goldstein, M. and Terenius, L. (1979). Ibotenic acid-induced neuronal degeneration: a morphological and neurochemical study. Exp. Brain Res., 37, 199-216.

Shepherd, G.M. (1972). Synaptic organization of the mammalian olfactory bulb. Physiol. Rev., 52, 864-917.

Stevens, C.F. (1969). Structure of cat frontal olfactory cortex. J. Neurophysiol., 32, 184-192.

Surtees, L. and Collins, G.G.S. (1985). Receptor types mediating the excitatory actions of exogenous L-aspartate and L-glutamate in rat olfactory cortex. Brain Res., in press.

Walker, J.E. and Fonnum, F. (1983). Effect of regional cortical ablations on high-affinity D-aspartate uptake in striatum, olfactory tubercle, and pyriform cortex of the rat. Brain Res., 278, 283-286.

Watkins, J.C. and Evans, R.H. (1981). Excitatory amino acid transmitters. Ann. Rev. Pharmacol. Toxicol., 21, 165-204.

Yamamoto, C. and Matsui, S. (1974). Facilitated release of glutamic acid from olfactory cortex slices by stimulation of excitatory input. Proc. Jap. Acad., 50, 653-657.

Yamamoto, C. and Matsui, S. (1976). Effect of stimulation of excitatory nerve tract on release of glutamic acid from olfactory cortex slices in vitro. J. Neurochem., 26, 487-491.

10

Amino Acids as Excitatory Transmitters in the Retina

A.M. López Colomé

INTRODUCTION

The retina is a part of the CNS which only recently has been taken into consideration as a model for studying synaptic mechanisms and their regulation. This organ exhibits characteristics which are experimentally convenient, for in addition to being readily accessible, the retina can be considered as a natural undamaged slice of 300 µ thickness through which substances can easily diffuse and which can be maintained in vitro for lasting periods of time due to its limited dependence on vascular support. Besides, specific cell types can be impaled and the whole retinal activity can be registered in the ERG. Physiologically, retinal response to its natural stimulus, light, can be followed, and biochemically, synaptic endings from different cell populations can be easily isolated as a consequence of its well defined layered organization (Neal and Atterwill, 1974).

General Organization of the Retina

Basically, the vertebrate retina possesses five types of neurons: photoreceptors, horizontal cells, bipolar cells, amacrine cells, and ganglion cells, plus the glial or Müller cells. Neuronal elements are arranged in three layers; the outer nuclear layer (ONL) which consists of the cell bodies of the photoreceptor cells; the inner nuclear layer (INL) which contains the somas of the horizontal and bipolar cells in its outer half and those of the amacrine cells in its inner half, and the ganglion cell layer (GCL) which contains ganglion cell bodies. Outer segments of photoreceptors form a special layer external to the ONL. Inbetween these three layers of cells, two synaptic layers exist in which contacts among them occur. At the outer plexiform layer (OPL) the photoreceptor terminals contact

*Partially supported by grant PCCBBNA-000800 from CONACyT.

horizontal and bipolar cell dendritic terminals. In addition, horizontal-horizontal and horizontal-bipolar synapses are present at this level. The whole OPL is $\cong 10\,\mu$ thick, and presents some anatomical variation according to the species and the type of photo-receptor (rod or cone). At the inner plexiform layer (IPL) bipolar cells contact amacrine and ganglion cells simultaneously; amacrine-amacrine, amacrine-bipolar and amacrine-ganglion cell synapses are also included in this layer. The IPL is thicker than the OPL and its degree of complexity varies according to the degree of information processing carried out by the retina. The glial Müller cells extend throughout the whole thickness of the retina and are regarded as astrocytic in nature (Dowling, 1979).

It has been proposed that photoreceptors release tonically in the dark an excitatory compound which maintains horizontal and a subclass of bipolar cells (OFF) depolarized while hyperpolarizes ON bipolar cells (Trifonov, 1968). This effect seems to be due to a continuous sodium entry in the dark; upon light-stimulation the "dark sodium current" would be blocked inducing the hyperpolarization of photoreceptors and the arrest of transmitter release, with the concomitant effect on the membrane of second order neurons (Hagins, 1979).

The excitatory amino acids glutamate (L-glu) and aspartate (L-asp) are the most suitable candidates for performing excitatory transmission between photoreceptors and second order neurons as well as between bipolar and ganglion cells (Neal, 1976). Regarding the photoreceptor transmitter, the problem exists that it should depolarize horizontal and OFF-bipolar cells and hyperpolarize ON-bipolar cells, which would involve two different ionic mechanisms. This problem has been addressed by either proposing different transmitters from the photoreceptors to ON and OFF bipolar cells, or alternatively, a single transmitter which could induce a reciprocal modulation between the two classes of bipolars (Sterling, 1983). However, the identification of these compounds as excitatory amino acids has proved difficult, in part because of the lack of specific antagonists of their synaptic actions (Watkins and Evans, 1981).

EXCITATORY AMINO ACIDS AS RETINAL TRANSMITTERS

Results supporting L-glu and L-asp as excitatory transmitters in the retina come mainly from indirect electrophysiological studies, although biochemical data regarding uptake, release and autoradio-graphy are also available. Recently, evidence has been achieved through selective lesions of the retina as well as receptor binding studies using L-glu, L-asp, kainate (KA) and N-methyl-D-aspartate (NMDA) as ligands.

Biochemical Evidence

One of the main criteria for establishing transmitter functions for a compound are, their concentration in synaptic endings, the presence of a synthesizing mechanism, a high-affinity, Na^+-dependent uptake system and the release of the compound following physiological stimulation (Werman, 1966).

Concentration and distribution. Regarding concentration, glutamate represents 10-15% of the free amino acid pool in the retina, while aspartate represents 2-10% (Pasantes-Morales et al., 1972). The absolute concentration of these compounds is rather constant for most species, in which glutamate concentration is $\cong$ 4 μmoles/g and that of aspartate ranges from 0.2 to 1.5 μmoles/g. In most cases the two amino acids are evenly distributed throughout the retina, however some data indicate a relative concentration in the outer layers (frog) or the ganglion cell layer (monkey); the levels of these compounds are not affected by light (Pasantes-Morales et al., 1973).

Regarding the presence of synthesizing enzymes for glutamate and aspartate, the finding of malic dehydrogenase together with aspartate-glutamate transaminase activity concentrated in the inner segments of photoreceptors, could indicate the participation of these compounds as photoreceptor transmitters (Neal, 1976). More recently, aspartate aminotransferase immunoreactivity has also been localized in photoreceptors and proposed as a marker for glutamatergic/asparta tergic neurons in the retina (Altshuler et al., 1982; Lin et al., 1983).

Uptake. Glutamate and aspartate are accumulated by a common high-affinity uptake mechanism which has been characterized in rat retina (White and Neal, 1976) and in rabbit retinal synaptosomes from both plexiform layers (Thomas and Redburn, 1978).

Autoradiographic studies indicate that glutamate is accumulated preferentially by glial cells in the rat and rabbit retina (White and Neal, 1976; Ehinger and Flack, 1971). Although glia takes-up these compounds extensively, there is also an incorporation into some neuronal elements in the retina (Lund-Karlsen and Fonnum, 1976; Lund-Karlsen, 1978; Hampton and Redburn, 1983). The accumulation of glutamate and aspartate in photoreceptors of several species has also been reported (Brunn and Ehinger, 1974; Ehinger, 1981; Schwartz, 1982; Marc and Lam, 1981). Combining uptake of L-glu and L-asp with immunochemical localization of aspartate aminotransferase, Brandon and Lam, (1983) suggest L-glu as the candidate for cone neurotransmitter in human and rat retina.

Release. Glutamate and aspartate are released by physiological stimulation (high potassium) from toad isolated photoreceptors in a Ca^{++}-dependent fashion (Miller and Schwartz, 1983). A consistent K^+-stimulated, Ca^{++}-dependent release of glutamate and aspartate from chick retina has also been reported (López-Colomé and Somohano, submitted), although other authors (Tapia and Arias, 1982) using the

same preparation have been unable to detect release. Light stimulation has no effect on L-glu or L-asp release from the chick retina
(Pasantes-Morales et al., 1973); however in the rabbit retina, low
frequency light-stimulation has been shown to arrest the spontaneous
efflux of aspartate in the dark which would suggest L-asp as the
photoreceptor transmitter (Neal et al., 1979).

According to the accepted scheme, the neurotransmitter released
from photoreceptors should depolarize horizontal cells. In agreement
with this postulation, L-glu and L-asp (the former more effectively)
induce the release of GABA from toad (Schwartz, 1982) and carp (Ayoub
and Lam, 1984) horizontal cells. Release of GABA from these cells is
also increased by the uptake inhibitor D-asp which is in support of
L-glu as the photoreceptor transmitter (Yazulla, 1985).

Physiological evidence

A main criterion for a substance to be regarded as a neurotransmitter is identity of action. Since excitatory amino acids are
supposed to mediate transmission in the vertical pathway of the
retina, physiological experiments have been aimed to demonstrate the
glutamate and/or aspartate are capable of mimicking the effects of
the natural transmitter from the photoreceptor and bipolar cells,
and show similar pharmacological properties.

Photoreceptor transmitter. Several studies have demonstrated that
glutamate and aspartate (glutamate being more potent) are able to
mimic the effects of the natural transmitter (Murakami et al., 1975;
Kaneko and Shimasaki, 1976; Gerschenfeld and Piccolino, 1979). L-Asp
has been proposed as the transmitter from cone photoreceptors in the
carp retina, since it depolarizes horizontal cells and blocks the
hyperpolarization induced by light (Murakami et al., 1972; Cervetto
and McNichol, 1972; Dowling and Ripps, 1973). This effect as well as
that of the natural transmitter were antagonized by DL-α-amino
adipate (Wu and Dowling, 1978). In contrast, Ishida and Fain (1981)
suggested L-glu as the possible transmitter, due to the potentiation
by D-asp of L-glu effect and of the light responses in horizontal
cells. Since these cells are sensitive to L-glu, KA and quiscualate
(QA) and to NMDA as well, it was proposed that while L-glu could act
as the transmitter, L-Asp could play a modulatory role through NMDA
receptors (Yazulla, 1983). These findings are in agreement with those
obtained in cultured horizontal cells from the carp and the skate
retinas (cone and rod-dominated respectively), in which L-glu, KA
and QA showed a depolarizing effect (Lasater and Dowling, 1982) while
D- and L-asp were uneffective (Ishida et al., 1984). QA was the most
potent agonist in the carp and KA in the skate; NMDA had no effect
on either type of retina.

The ionic basis of depolarizing and hyperpolarizing effects of the
photoreceptor transmitter on the horizontal and bipolar cells is
still not clear. The permeability changes induced by the ionto-

phoretic application of L-glu to carp horizontal cells (Tachibana, 1985) correspond to those expected from the photoreceptor-horizontal cell transmitter. Shiells et al. (1981) have demonstrated in the rod-retina of the dogfish that although L-glu and L-asp induce the same effects as the natural transmitter on the ON-bipolar and horizontal cells, KA was more effective. This finding together with the fact that APB mimicked the transmitter effect only on the ON-bipolars but not the horizontal cells, raised some doubt on the proposed photo-receptor-transmitter role of L-glu. Regarding the ionic mechanism, KA, the natural transmitter and APB close sodium channels on the ON-bipolars, which corresponds with data obtained for the rod-transmit-ter (Saito et al., 1978) and also with the effect of exogenously applied glutamate and aspartate on the response of the ON-bipolar cells (Kondo and Toyoda, 1980).

Using structural analogues of excitatory amino acids, Rowe and Ruddock (1982a) demonstrated that roach-fish horizontal cells are de-polarized by the L-glu agonists KA and QA whereas the antagonists glutamate diethyl-ester (GDEE) and γ-D-glutamyl-glycine failed to block the effect of exogenously applied QA and KA respectively (Rowe and Ruddock, 1982b).

As far as bipolar cells are concerned, Slaughter and Miller (1981) have demonstrated that 2APB selectively blocks responses at the ON-channel of the mudpuppy retina thus mimicking the endogenous photoreceptor transmitter. Using 2,3-piperidine dicarboxylic acid (PDA), a specific excitatory amino acid antagonist, it has been shown that OFF-bipolars and horizontal cells bare synaptic receptors of the L-glu/KA type which are also present in the inner retina, and are different from ON-bipolar cells receptors (Slaughter and Miller, 1983a).

In sum, although some studies have suggested that cones and rods could use different transmitters (Marc and Lam, 1981), studies in cone-dominated (Slaughter and Miller, 1981) and rod-dominated (Shiells et al., 1981) fish retinas using 2APB and 2-amino-3-phospho nopropionate indicate that more likely rods and cones use the same transmitter, which should be either L-glu or a related compound.

Bipolar cell transmitter. Both ON and OFF bipolar cells are sup posed to release an excitatory transmitter (Werblin, 1979). L-glu and L-asp depolarize ganglion cells in the frog (Kishida and Naka, 1967), cat (Straschill and Perwein, 1973) and carp (Negishi et al., 1978).

In the carp retina, PDA not only affects the OFF responses at the proximal retina, but supresses both ON and OFF-responses at the inner retina suggesting both that ON and OFF-bipolars release the same transmitter and that this compound should be an excitatory amino acid (Slaughter and Miller, 1983). In rabbit retina, APB abolishes ON responses (Cunningham et al., 1983), and similar results were observed in the cat retina where APB abolishes ON responses while increases the maintained firing of OFF-ganglion cells (Bolz et al., 1984) supporting the theory that OFF ganglion cells receive a direct inhibitory input from ON bipolars (Sterling, 1983). On the other hand, L-asp has been shown to enhance visually driven excitation of "sustained"but not of "transient" ganglion cells in the cat. Since this effect was reduced by 2APV, the possibility of L-asp being the transmitter at this levels has been proposed (Ikeda and Sherdown, 1981).

Concluding, the only consistent evidence seems to indicate that bipolar cells (ON and OFF) release an excitatory amino acid to amacrines and ganglion cells.

Evidence from Lesion Experiments

The neurotoxic effects of some excitatory amino acids has been used as a tool for selectively degenerating neurons in areas of the CNS, including the retina (Olney, 1969). On these basis, many lesion studies using MSG and KA have been performed in the retina in order to localize cells which are sensitive to excitotoxic amino acids and hence could bare receptors for these compounds..

In the goldfish retina, KA caused reversible edema to rod- and cone-horizontal cells, while pure cone-bipolars and the majority of amacrines were distroyed (Yazulla and Kleinschmidt, 1980; Hampton et al., 1981). Since the lesion was located in the proximal terminals of the cells, a receptor mediated mechanism could be involved. In all cases, cone systems showed a higher sensitivity to KA. In the chick retina, some neurochemical changes in cholinergic, GABAergic and dopaminergic markers (concentration, synthesizing enzymes and uptake) have been detected after the injection of 120 nmoles of KA (Schwarcz and Coyle, 1977).

Recently, a graded sensitivity of chick retinal neurons to intravitreally injected KA has been described: OFF-bipolars>amacrines horizontals>ON-bipolars, photoreceptors and ganglion cells being fairly resistant to the higher dose (200 nmoles). This lesion has been followed by qualitative (Morgan and Ingham, 1981) and quantitative (Dvorak and Morgan, 1983) light microscopy. Pharmacological antagonism of the lesion suggests the existence of two different kinds of KA receptors at the OPL (Morgan, 1983); however the low specificity of the existing antagonists (Rowe and Ruddock, 1982b) throws some doubts on this assumption.

Evidence from Binding Studies

Information regarding biochemical and pharmacological characteri zation of excitatory amino acid receptors in the retina is scarce. Biziere and Coyle (1979) first identified KA-binding receptors in the inner nuclear layer of the chicken retina, which exhibited characteristics similar to those in the brain.

In the bovine retina, Mitchell et al. (1981) reported binding of ^{3}H-L-glu to three sites (K_B=12, 72 and 800 nM) displaced preferentially by L-glu; later studies (Mitchell and Redburn, 1982), showed only two binding sites (K_B=10 and 800 nM). L-glu was displa - ced by 2APB but not by KA or GDEE. In contrast, Höckel and Müller (1982) found only one glutamate-binding site (K_B=3 uM) in the same species; binding was effectively inhibited by DL-$\measuredangle$-amino adipate (DL-$\measuredangle$-AA), L-glu and $\measuredangle$-methyl-glutamate, and moderately by GDEE (44%). In both studies, binding was confined to the inner plexiform layer.

^{3}H-NMDA binding in the bovine retina (K_B = 8.4 and 82 nM) was also located in the inner retina (Mitchell et al., 1981) as well as ^{3}H-L-asp binding (K_B = 6.3 nM); binding of both compounds was preferentially antagonized by NMDA (Mitchell and Redburn, 1982).

In the chicken retina, ^{3}H-L-glu binding sites were also described (López-Colomé, 1981). L-^{3}H-asp binding sites were also described (López-Colomé and Somohano, 1982) which are concentrated in the OPL (K_B = 40 nM but are also present at the IPL (K_B = 11.8 nM). While L-glu binding was antagonized by GDEE, L-asp binding was displaced more effectively by NMDA; KA binding was not displaced by either compound. In contrast to the bovine retina, although NMDA was more effective in inhibiting L-asp than L-glu binding, 2APB had no effect on the binding of either compound.

Studies aimed to localizing L-glu and L-asp receptive neurons have been performed in the chick retina using selective KA-induced degeneration and the subsequent measuring of L-glu and L-asp binding. In order to avoid short term effects described in previous paragraphs, the lesion was allowed to stabilize for two weeks. Lesion was followed by measuring receptors every hour for 12 hours and then every day up to 16 days after intraocular injection of 60 nmoles KA. (López-Colomé, unpublished). Receptors for both amino acids increased significantly during the first three days after lesioning, and came to a 20% reduction after 15 days. Although receptors for both compounds were equally reduced, the time course followed by the lesion (Fig 1) suggested two different types of receptors present on two cell populations which show different sensitivity to KA.

Localization of Excitatory Pathways

Postsynaptic receptors. Localization of L-glu and L-asp receptors in specific cell types was investigated by lesioning the retina with increasing concentrations of KA as to eliminate OFF-bipolars (6 nmoles), amacrines (60 nmoles) horizontal cells (120 nmoles) and ON-bipolars (200 nmoles), and measuring synaptic receptors (López-Colomé and Somohano, 1984). As shown in Fig. 2, the

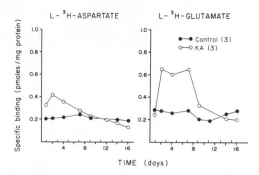

TIME (days)

Fig. 1 Time course of L-glu and L-asp receptor binding following
intraocular injection of 60 n moles KA. Data are the mean of 3
independent experiments performed in triplicate.

decrease of glutamate and aspartate binding differed, indicating the
existence of two populations of receptors. While L-glu receptors con
centrate mainly on OFF-bipolars and horizontal cells, receptors for
L-asp appear to be equally distributed on amacrines, horizontals and
bipolars. Ganglion cells, which resist the higher dosis of KA could
be responsible for the residual 30% binding of both amino acids
(Fig.2). Altogether these data agree with preexisting ones and suggest that OFF- bipolars bare L-glu receptors which are more sensitive
to KA than ON-bipolar ones, while horizontal cells apparently have
only glutamate-like receptors. Amacrines and ganglion cells seem to
possess both types.

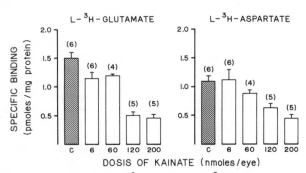

DOSIS OF KAINATE (nmoles/eye)

Fig. 2 Effect of kainate on L-^{3}H-glu and L-^{3}H-asp specific binding.
The doses of KA preferentially eliminated OFF-bipolars (6 nmoles),
amacrines (60 nmoles), horizontals (120 nmoles) and some ON-bipolars
(200 nmoles). Data are the mean ± SEM of the number of experiments
indicated in parentheses. c=control.

Release. Presynaptic release of a neurotransmitter due to
physiological stimulation should parallel the existence of postsynaptic receptors, hence a study of K$^+$-stimulated Ca^{++}-dependent

release of ^{3}H-L-glu and ^{3}H-L-asp from chick retina has been
performed after lesioning with increasing doses of KA in order to
eliminate specific neuronal types (López-Colomé and Somohano, submitted). As can be seen in Table 1, glutamate seems to be released
mainly from the OFF-and aspartate from the ON-bipolars. GABA and
glycine release are in agreement with data which indicate (Neal,1976)
that the former is released from horizontals plus a population of
amacrines, whereas glycine would be released by another group of
amacrines. Release and receptor binding up together suggest that OFF-
bipolars could release L-glu onto glycinergic amacrines while ON-
bipolars may be releasing L-asp onto GABAergic amacrines.

Table 1. Effect of KA selective lesion on the stimulated-release of
glutamate, aspartate, GABA and glycine from chick retina.

Dosis of KA (n moles/eye)	Affected Cell-Type	Decrease in K$^+$-stimulated Ca^{++}-dependent release (%)			
		^{3}H-GABA	^{14}C-Gly	^{3}H-L-Glu	^{3}H-L-Asp
6	OFF-Bipolars	–	40	50	20
60	Amacrines	50	75	70	20
120	Horizontals	80	75	70	50
200	ON-Bipolars	80	95	100	100

Data are referred to control unlesioned retinas from eyes contralateral to those injected. Results are the mean of 5 experiments
in which values varied less than 10%.

Presynaptic Receptors. Although some studies have suggested the
possibility of presynaptic regulation of excitatory neurotransmission in the retina (Shiells et al, Bolz et al) no direct study has
been aimed to clarify this point. We have looked at the effect of
analogues of excitatory amino acids on the K$^+$-stimulated Ca^{++}-de-
pendent release of ^{3}H-D-asp from chick retina (López-Colomé and
Roberts, submitted). L-Glu (since 50 μm) as well as NMDA (200 μM) and
KA (since 100 μM) showed 70-90% inhibition of release.

The phosphonic analogues APB, APV and APH (2-amino 7-phosphono-
heptanoic acid) also inhibited release to different extent. The
effect of L-glu, KA and NMDA seemed to be at different receptors,
since it was not additive. However, AMPA reversed KA-induced inhibition and APV reversed NMDA-induced inhibitory effect (Table 2). Since
this effect was not abolished in the presence of 1 μM TTX, data
indicate it could be mediated by presynaptic autoreceptors for L-glu
and NMDA. In the effect of the photoreceptor transmitter as well as
in the input of ON and OFF bipolar to ganglion cells a dichotomy
exists; the existence of this type of receptors and their localization would be of great help in the tracing of excitatory pathways in

the vertebrate retina. These data would also be of help in clarifying the discrepancies on the type and location of excitatory amino
acid receptors in this organ.

Table 2. Effect of some analogues of excitatory amino acids on the
K^+-stimulated release of ^{3}H-D-Asp.

Compound	Pre-stimulation	Peak-stimulation
Control	0.020	0.137
L-Glu (50 μM)	0.027	0.047
AMPA (1 mM)	0.022	0.075
KA (1 mM)	0.025	0.040
GDEE (1 mM)	0.025	0.160
L-Glu (50 μM) + AMPA (1 mM)	0.026	0.093
L-Glu (50 μM) + KA (1 mM)	0.028	0.050
KA + AMPA (1 mM)	0.027	0.152
NMDA (200 μM)	0.020	0.063
APV (200 μM)	0.020	0.073
NMDA (200 μM) + KA (1 mM)	0.035	0.071
NMDA + APV (200 μM)	0.021	0.073

Values are expressed as % of total radioactivity released per minute.
Results are the mean of 4 independent experiments.

SUMMARY

    Evidence supporting excitatory amino acids as retinal transmit-
ters has been accumulating during the last 10 years. Electrophysiological and biochemical data from several vertebrates point to a
glutamate/kainate type of transmitter mainly at the outer plexiform
layer. Fewer but consistent data indicate the presence of NMDA-like
receptors localized at the inner plexiform layer, which could
indicated the participation of L-aspartate in transmission at this
level. The selective lesion of retina with neurotoxic compounds (i.
e. kainate, veratridine) followed by either release or receptor-
binding experiments have provided useful data as to the localization
of cell types which might use or receive L-glutamate or L-aspartate
as transmitters. The identification of presynaptic receptors for
these compounds with pharmacological properties different from those
of postsynaptic receptors opens the possibility of glutamate and
aspartate exerting a cross-regulation of release. Since most of the
mechanisms related to excitatory amino acid transmission are present
in the retina, this organ offers a suitable model in which studies
can be performed and extrapolated to other less accessible areas of
the CNS.

References

Altschuter, R.A., Monsinger, J.L., Harmison, G.G., Parakkai, M.H. and
    Wenthold, R.J. (1982) Aspartate aminotransferase-like immunoreacti
    vity as a marker for aspartate/glutamate in guinea-pig photorecep-

tors. Nature, 298, 657-659.

Ayoub, G.S., and Lam, D.M.K. (1984). The release of γ-aminobutyric acid from horizontal cells of the goldfish (Carassius auratus) retina. J. Physiol., 355, 191-214.

Biziere K. and Coyle J.T. (1979) Localization of receptors for kainic acid on neurons in the innernuclear layer of retina. Neuropharmacology, 18, 409-413.

Bolz, J., Wässle, H. and Thier, P. (1984). Pharmacological modulation of ON and OFF ganglion cells in the cat retina. Neuroscience, 12, 875-885.

Brandon, C. and Lam, D.M.K. (1983). L-Glutamic acid: A neurotransmit ter candidate for cone photoreceptors in human and rat retinas. Proc. nat. Acad. Sci. USA, 80, 5117-5121.

Bruun, A. and Ehinger B. (1974) Uptake of certain possible neurotransmitters into retinal neurones of some mammals. Expl Eye Res. 19, 435-447.

Cervetto, L. and MacNichol, E.E. (1972) Inactivation of horizontal cells in turtle retina by glutamate and aspartate. Science, 178, 767-768.

Cunningham, J.R., Dawson, C. and Neal, M.J. (1983) Evidence for a cholinergic inhibitory feed-back mechanism in the rabbit retina. J. Physiol., 340, 455-468.

Dowling, J.E. (1979) Information processing by local circuits: The vertebrate retina as a model system. In The Neurosciences; Fourth Study Program (eds. P.O. Schmitt and F.G. Worden). MIT Press, Cambridge, MA.

Dowling, J.E. and Ripps H. (1973) Effect of magnesium on horizontal cell activity in the skate retina. Nature, Lond., 242, 101-103.

Dvorak, D.R. and Morgan, I.G. (1983) Intravitreal kainic acid permanently by eliminates off-pathways from chick retina. Neurosci. Lett, 36, 249-254.

Ehinger, B. (1981) (³H)-D-aspartate accumulation in the retina of pigeon, guinea pig and rabbit. Expl. Eye Res., 33, 381-391.

Ehinger, B. and Falck B. (1971) Autoradiography of some suspected neurotransmitter substances: GABA, glycine, glutamic acid, histamine, dopamine and L-DOPA. Brain Res., 33, 157-172.

Gerschenfeld, H.M. and Piccolino, M. (1979). Pharmacology of the connections of cones and L-horizontal cells in the vertebrate retina. In The Neurosciences: Fourth Study Program (eds. P.O. Schmit and F.G. Worden). MIT Press, Cambridge, MA.

Hagins, W.A. (1979). Excitation in vertebrate photoreceptos. In The Neurosciences: Fourth Study Program (eds. P.O. Schmitt and F.G. Worden). MIT Press, Cambridge, MA.

Hampton, C.K. and Redburn, D.A. (1983). Autoradiographic analysis of ³H-glutamate, ³H-dopamine, and ³H-GABA accumulation in rabbit retina after kainic acid treatment. J. Neurosci. Res. 9, 239-251.

Hampton, C.K., García, C. and Redburn, D.A. (1981) Localization of kainic acid-sensitive cells in mammalian retina. J. Neurosci. Res., 6, 99-111.

Höckel, S.H.J. and Müller, W.E. (1982). L-Glutamate receptor binding in bovine retina. Exp. Eye Res., 35, 55-60

Ikeda, H. and Sheardown, M.J. (1982) Aspartate may be an excitatory transmitter mediating visual excitation of "sustained" but not "transient" cells in the cat retina: Iontophoretic studies in vivo. Neuroscience, 7, 25-36.

Ishida, A.T. and Fain, G.L. (1981) D-aspartate potentiates the effects of L-glutamate on horizontal cells in goldfish. Proc. natn. Acad. Sci. U.S.A. 78, 5890-5894.

Ishida, A.T., Kaneko, A. and Tachibana, M. (1984) Responses of solitary retinal horizontal cells from Carassius auratus to L-glutamate and related amino acids. J. Physiol. 348, 255-270.

Kaneko, A. and Shimazaki, H. (1976) Synaptic transmission from photo receptors to the second-order neurons in the carp retina. In Neural Principles in vision (eds. F. Zettler and R. Weiler). Springer-Verlag, Berlin.

Kishida, K. and Naka, K.E. (1967) Amino acids and the spikes from the retinal ganglion cells. Science 156, 648-650.

Kondo, H. and Toyoda, J.E. (1980). Dual effect of glutamate and aspartate on the on-center bipolar cell in the carp retina. Brain Res., 199 240-243.

Lasater, E.M. and Dowling, J.E. (1982) Carp horizontal cells in culture respond selectively to L-glutamate and its agonists. Proc. natn. Acad. Sci. U.S.A. 79: 936-940.

Lasater, E.M., Dowling, J.E. and Ripps, H. (1984) Pharmacological properties of isolated horizontal and bipolar cells from the skate retina. J. Neurosci., 4, 1966-1975.

Lin, C.T., Li, H.Z. and Wu, J.Y. (1983). Immunocytochemical localization of L-glutamate decarboxylase, gamma aminobutyric acid transaminase, cysteine sulfinic acid decarboxylase, aspartate aminotransferase and somatostatin in rat retina. Brain Res., 270, 273-283.

López-Colomé, A.M. (1981) High-affinity binding of L-glutamate to chick retinal membranes. Neurochem. Res. 6, 1019-1033.

López-Colomé, A.M. and Somohano, F. (1982). Characterization of L-^{3}H-aspartate binding to chick retinal subcellular fractions. Vision Res., 22, 1495-1501.

López-Colomé, A.M. and Somohano, F. (1984). Localization of L-glutamate and L-aspartate synaptic receptors in chick retinal neurons. Brain Res., 298, 159-162.

Lund-Karlsen, R. (1978) The toxic effect of sodium glutamate and DL-⋖-amino adipic acid on rat retina: changes in high affinity uptake of putative transmitters. J. Neurochem. 31, 1055-1061.

Lund-Karlsen, R. and Fonnum, F. (1976) The toxic effect of sodium glutamate on rat retina: Changes in putative transmitters and their corresponding enzymes. J. Neurochem. 27: 1437-1441.

Marc, R.E. and Lam, D.M.K. (1981) Uptake of aspartic and glutamic acid by photoreceptors in goldfish retina. Proc. natn. Acad. Sci. U.S.A. 78, 7185-7189.

Miller, A.M. and Schwartz, E.A. (1983). Evidence for the identification of synaptic transmitters released by photoreceptors of the toad retina. J. Physiol., 334, 325-349.

Mitchell, C.K. and Redburn, D.A. (1982). 2-Amino-4-phosphonobutyric acid and N-methyl-D-aspartate differentiate between ^{3}H-glutamate and ^{3}H-aspartate binding sites in bovine retina. Neurosci. Lett.,

28, 241-246.
Mitchell, C., Hampton, C. and Redburn D. (1981) Localization of
  receptor and transport sites for glutamate and aspartate in inner
  plexiform layer of bovine retina. Invest. Ophthal. visual Sci. 20,
  215.
Morgan, I.G. and Dvorak, D.R. (1983) Physiologically active kainic
  acid preferring receptors in vertebrate retina. In Glutamine,
  Glutamate and GABA in the Central Nervous System. (eds. L. Hertz,
  E. Kvamme, E.G. McGeer and A. Schousboe). Alan R. Liss, New York.
Morgan, I.G. and Ingham, C.A. (1981) Kainic acid affects both
  plexiform layers of chicken retina. Neurosci. Lett. 21, 275-280.
Murakami, M., Ohtsu K. and Ohtsuka, T. (1972) Effects of chemicals
  on receptors and horizontal cells in the retina. J. Physiol.
  Lond. 227, 899-913.
Murakami, M. Ohtsuka T. and Shimasaki H. (1975) Effects of aspartate
  and glutamate on the bipolar cells of the carp retina. Vision Res.
  15, 456-458.
Neal, M.J. (1976) Amino acid transmitter substances in the
  vertebrate retina. Gen. Pharmac. 7, 321-332.
Neal, M.J. and Atterwill C.K. (1974) Isolation of photoreceptor and
  conventional nerve terminals by subcellular fractionation of
  rabbit retina. Nature 251, 331-333.
Neal, M.J., Collins, G.G. and Massey S.C. (1979) Inhibition of
  aspartate release from the retina of the anaesthetised rabbit by
  stimulation with light flashes. Neurosci. Lett. 14, 241-245.
Negishi, K., Kato, S., Teranishi, T. and Laufer, M. (1978) Dual
  actions of some amino acids on spike discharges in the carp retina.
  Brain Res. 148, 67-84
Olney, J.W. (1969) Glutamate-induced retinal degeneration in neo-
  natal mice. Electron microscopy of acutely evolving lesion. J.
  Neuropathol. Exp. Neurol., 28, 455-474.
Pasantes-Morales, H., Klethi, J. Ledig M. and Mandel P. (1972) Free
  amino acids of chicken and rat retina. Brain Res., 41, 494-497.
Pasantes-Morales, H., Klethi, J., Ledig, M., and Mandel, P. (1973).
  Influence of light and dark on the free amino acid pattern of the
  developing chick retina. Brain Res., 57, 59-65.
Rowe, J.S. and Ruddock, K.H. (1982a) Depolarization of retinal
  horizontal cells by excitatory amino acid neurotransmitter agonists
  Neurosci. Lett. 30, 257-262.
Rowe, J.S. and Ruddock, K.H. (1982b) Hyperpolarization of retinal
  horizontal cells by excitatory amino acid neurotransmitter
  antagonists. Neurosci. Lett. 30, 251-256.
Saito, T., Kondo, H. and Toyoda, J. (1978) Rod and cone signals in
  the on-center bipolar cell: their different ionic mechanisms.
  Vision Res. 18, 591-595.
Schwarcz, R. and Coyle, J.T. (1977) Kainic acid: Neurotoxic effects
  after intraocular injection. Invest. Ophthalmol. Vis. Sci., 16,
  141-148.
Schwartz, E.A. (1982) Calcium-independent release of GABA from
  isolated horizontal cells of the toad retina. J. Physiol., 323,
  211-227.

Schwartz, E.A. (1982). Identification and function of synaptic transmitters used in the outer synaptic layer of the toad retina. In Neurotransmitters in the Retina and the Visual Centers (eds A. Kaneko, N. Tsukahara and K. Uchizono). Biomed. Res. Foundation. Tokyo.

Shiells, R.A., Falk, G. and Naghshineh, S. (1981) Action of glutamate and aspartate analogues on rod horizontal and bipolar cells. Nature 294, 592-594.

Slaughter M.M. and Miller, R.F. (1981) 2-amino-4-phosphonobutyric acid. A new pharmacological tool for retina research. Science 211, 182-185.

Slaughter, M.M. and Miller, R.F. (1983a). An excitatory amino acid antagonist blocks cone input to sign-conserving second-order retinal neurons. Science, 219, 1230-1232.

Slaughter, M.M. and Miller, R.F. (1983b). Bipolar cells in the mudpuppy retina use an excitatory amino acid neurotransmitter. Nature, 303, 537-538.

Sterling, P. (1983). Microcircuitry of the cat retina. A rev. Neurosci., 6, 149-185.

Straschill, M. and Perwein, J. (1973) The effect of iontophoretically applied acetylcholine upon the cat's retinal ganglion cells. Pflugers Arch. ges. Physiol 339, 289-298.

Tachibana, M. (1985). Permeability changes induced by L-glutamate in solitary retinal horizontal cells isolated from Carassius auratus. J. Physiol., 358, 153-167.

Tapia, R. and Arias, C. (1982) Selective stimulation of neurotransmitter release from chick retina by kainic and glutamic acids. J. Neurochem. 39, 1169-1178.

Thomas, N.T. and Redburn, D.A. (1978) Uptake of ^{14}C-aspartic acid and ^{14}C-glutamic acid by retinal synaptosomal fractions. J. Neurochem. 31, 63-68.

Trifonov, Y.A. (1968) Study of synaptic transmission between photoreceptors and horizontal cells by means of electrical stimulation of the retina. Biofizika 13, 809-817.

Watkins, J.C. and Evans, R.H. (1981) Excitatory amino acid transmitters. Annu. Rev. Pharmacol. Toxicol 21, 165-204.

Werblin, F.S. (1979). Integrative pathways in local circuits between slow-potential cells in the retina. In The Neurosciences. Fourth Study Program (eds. P.O. Schmitt and F.G. Worden) MIT Press, Cambridge, MA.

Werman, R. (1966). Criteria for identification of a central nervous system transmitter. Comp. Biochem. Physiol. 18, 745-766.

White, R.D. and Neal M.J. (1976) The uptake of L-glutamate by the retina. Brain Res., 111, 79-83.

Wu, S.M. and Dowling, J.E. (1978). L-Aspartate: evidence for a role in cone photoreceptor synaptic transmission in the carp retina. Proc. nat. Acad. Sci., U.S.A., 75, 5205-5209.

Yazulla, S. (1983) Stimulation of GABA release from retinal horizontal cells by potassium and acidic amino acid agonists. Brain Res., 275, 61-74.

Yazulla, S. (1985). Evoked efflux of ^{3}H-GABA from goldfish retina in the dark. Brain Res., 325, 171-180.

Yazulla, S. and Kleinschmidt J. (1980) The effects of intraocular injection of kainic acid on the synaptic organization of the goldfish retina. Brain Res. *182*, 287-301.

11

Mechanisms of Reuptake of Neurotransmitters from the Synaptic Cleft

B.I. Kanner and R. Radian

ABSTRACT

The function of sodium dependent neurotransmitter transport is probably to terminate the overall process of synaptic transmission. Recent studies have shown that in these transport systems, such as those for γ-aminobutyric acid (GABA), glutamate and biogenic amines, solute fluxes are not only coupled to sodium but also to additional ions. The GABA transporter, for instance, appears to catalyze electrogenic cotransport of GABA, sodium and chloride. The glutamate transporter catalyses electrogenic cotransport of sodium and glutamate while potassium is antiported. The role of potassium in the translocation cycle is to allow the return of the "unloaded transporter". The glutamate as well as the GABA transporter have been solubilised from plasma membranes deriveed from rat brain and reconstituted into liposomes. The resulting proteoliposomes exhibited glutamate and GABA transport with the same characteristics as in the native vesicles. The GABA transporter has been purified to near homogeneity using transport activity upon reconstitution as an assay. Purification steps included fractionation of the cholate extract with ammonium sulfate, DEAE cellulose and lectin-chromatography. It appears that the GABA transporter is a glycoprotein with a molecular weight of 82-85 KDa. (Supported by NIH grant NS 16708).

Introduction :

Sodium-coupled neurotransmitter transport systems catalyse simultaneous transport of sodium and neurotransmitter. They are located in the synaptic plasma membrane of nerves (and sometimes in the plasma membrane of glial cells). This neurotransmitter transport plays an important role in the overall process of synaptic transmission, by terminating the process (Fig. 1).

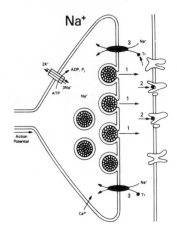

FIG. 1.
An overview of synaptic transmission. Following the arrival of an
action potential at the nerve terminal and Ca^{2+} entry, release of
the transmitter occurs (1). Upon diffusion of transmitter in the
cleft, it interacts at the post-synaptic site with its receptors,
resulting in signal transmission (2).The last step is re-uptake of
the transmitter via sodium-coupled transport systems (3),here
depicted in the membrane of the terminal. Reprinted with permission
from Ref.1; Copyright 1983, Biochim.Biophys. Acta.

   From a bionergetic point of view, the sodium-coupled transport
systems are very interesting, since it has been recently discovered that many systems require other ions in addition to sodium.
It has been demonstrated in several cases that, besides sodium
ions, these additional ions, such as chloride and potassium,
serve as additional coupling ions (reviewed in (1))

   In the following sections we will describe two distinct systems.
One is the system γ-aminobutyric acid (GABA) transport which
appears to be sodium and chloride coupled. The other is the
transport system for L-glutamate which transports sodium and
L-glutamate in one direction and potassium in the other.
Recently, for the first time, a neurotransmitter transporter has
been solubilised, purified and reconstituted. This is the GABA
transporter. In the second part of this chapter, we shall report
on these studies.

Properties of GABA Transport

After early studies with relatively intact preparations such as brain slices and synaptosomes, much progress has been made after studying GABA transport in synaptic plasma membrane vesicles (2-6). The vesicles obtained after osmotic lysis of synaptosomes allow manipulation of the external and internal medium and enable study of the transport in isolation from other processes. The properties of GABA transport are summarized in Table I :

TABLE I : PROPERTIES OF GABA TRANSPORT
1. Influx requires both external Na^+ and Cl^- and is electrogenic
2. To achieve maximal steady state levels a Na^+ as well as a Cl^- gradient (both out in) are required.
3. The requirement for Cl^- is specific.
4. Efflux requires both internal Na^+ and Cl^-.
5. Stoichiometry measurements consistent with $2Na^+ : Cl^-$: 1 GABA.
6. Binding order is possibly assymetric.

The mechanism of transport is chemiosmotic. It is of interest to note that not only sodium but also chloride appears to serve as coupling ion. This latter conclusion deserves some clarification. Three possibilities for the action of chloride come to mind (Fig. 2).

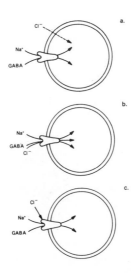

FIG. 2
Possible explanations for the absolute chloride dependency of γ-aminobutyric acid (GABA) transport. As described the following roles for chloride have been considered: (a) Cl^- following sodium-coupled

Υ-aminobutyric acid influx to maintain electroneutrality ; (b)
(Na$^+$ + Cl$^-$)-coupled Υ-aminobutyric acid transport; (c) Cl$^-$activates
the Na$^+$-coupled γ-aminobutyric acid transporter from the outside.
As discussed, only model (b) is compatible with the data. Reprinted
with permission from Ref. 1; Copyright 1983, Biochim. Biophys. Acta.

Since the process is electrogenic, the explanation for the
chloride requirement could be a trivial one. Chloride is a rather
permeant anion and therefore it could maintain electroneutrality
of the sodium-γ-aminobutyric acid co-transport (Fig. 2A) or create
the internal negative membrane potential needed as driving force.
This possibility appears unlikely, since more permeant anions like
thiocyanate and nitrate are only poor substitutes for chloride (2).
Moreover, this chloride requirement persists in the presence of
internal potassium, which is a permeant ion in nerves (7), even
in the presence of the potassium ionophere valinomycin (2,5). Thus,
potassium should be able to move out to charge compensate the
positive charge moving inward via the transporter at least as
efficiently as the inward movement of chloride. The explanation
that the ability of chloride to create a membrane potential is the
reason for its being required can be discarded on the basis of the
following experiment (Fig. 3).

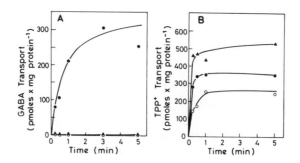

FIG. 3.
The effect of external chloride on γ-aminobutyric acid (GABA)
accumulation and on the membrane potential. Transport of Υ-aminobutyric acid (panel A) or tetraphenylphosphonium cation (TTP$^+$)
(panel B) was performed with KP$_i$ loaded vesicles. In addition to

the isotope and 1 mM $MgSO_4$ the composition of the external media was : (●–●) 0.1 M NaCl; (○–○) 0.1 M NaP_i; (▲–▲) 0.1 M NaP_i + 2.5 µM valinomycin. The external concentrations of the solutes were γ-aminobutyric acid 0.16 µM (29.3 Ci/mmol); TPP^+ 5.4 µM (888 Ci/mmol). Reprinted with permission from Ref. 5 ; copyright 1983, American Chemical Society).

Membrane vesicles are loaded with potassium phosphate and upon creation of an electrochemical sodium ion gradient (by dilution of the vesicles into sodium-containing media) those vesicles are tested for their ability to catalyze sodium-coupled ɣ-aminobutyric acid transport (Fig. 3A) and to maintain a negative-interior membrane potential (Fig. 3B), assayed by the accumulation of the tritiated lipophilic cation tetraphenylphosphonium (TPP^+). When the vesicles are diluted into sodium phosphate, an appreciable membrane potential (-51 mV) is generated which is not much less than when sodium chloride is employed as dilution medium (-59 mV) (Fig. 3B). However, absolutely no γ-aminobutyric acid accumulation is observed in the chloride-free medium (Fig. 3A). In the presence of valinomycin, dilution of the vesicles in the sodium phosphate-containing medium results actually in a more negative membrane potential (-70 mV) than dilution into the sodium chloride-containing medium (Fig. 3B), but γ-aminobutyric acid transport occurs solely in the chloride-containing media (Fig. 3A). Thus the role of chloride in the process must be due to a specific interaction of chloride with the transporter.

Two obvious possibilities come to mind regarding this interaction. First, the transporter may actually function by catalyzing co-transport of three species; sodium, chloride, and γ-aminobutyric acid (Fig. 2B). The alternative to transport would be that chloride binds to the transporter and this causes a conformational change such that now the transporter can accept the sodium and γ-amino-butyric acid and subsequently transport those two species (Fig. 2C). In order to distinguish further between these two possibilities, we have examined the effect of internal ions on the efflux of γ-aminobutyric acid. If co-transport with chloride occurs, then efflux of internal γ-aminobutyric acid by the transport should require not only internal sodium but also internal chloride (Fig. 2B), whereas for the option depicted in Fig. 2C this is not expected to be the case. Indeed, it appears that both internal sodium and chloride are required for efflux of γ-aminobutyric acid (4) and therefore it appears that the model of Fig. 2B is probably the correct one. Consistent with this is that a chloride gradient (out > in) can serve as a driving force for ɣ-aminobutyric acid accumulation (2). The present preparation of membrane vesicles is too leaky to sodium and chloride (5) to test the ultimate prediction of the model, namely that one should be able, in a tight enough system, to observe sodium- and ɣ-aminobutyric acid-dependent chloride fluxes and also chloride- and ɣ-aminobutyric acid-dependent sodium fluxes. However, there are some indications of vesicle heterogeneity. Thus, the γ-aminobutyric acid gradient holds for a

relatively long time in relation to flux of (Na$^+$) (3) and therefore
it seems that at least some of the γ-aminobutyric acid-transporting
vesicles are tight. As a consequence, it might be possible to
isolate the tight population and to perform the above experiments.
The alternative approach is to isolate a highly purified γ-aminobutyric acid transporter preparation and to reconstitute this into
single-walled liposomes and use this for the experiments. As described below, it has been possible to solubilized the transporter and
to reconstitute it (8)and very recently to obtain a very pure
preparation.

The above mechanism for γ-aminobutyric acid transport (Fig.2B),
and the fact that the process is electrogenic, impose restrictions
on the possibilities for the stoichiometry of the process. Assuming
that γ-aminobutyric acid is transported in its predominant form, the
zwitterion (although there is no direct evidence for this), the
stoichiometry obviously cannot be 1 Na$^+$:Cl$^-$:1 γ-aminobutyric acid.
Thus, more than one sodium should be moving per chloride and γ-amino
butyric acid; in other words, the stoichiometry should be n Na$^+$:
m Cl$^-$: γ-aminobutyric acid with n>m. However, the preparation is too
leaky to sodium and chloride to obtain precise stoichiometric data.
Still, it is possible to show that the stoichiometry for sodium is
larger than 1. At the steady state the γ-aminobutyric acid gradient
can be expressed as a function of the other ion gradients as follows

$$\text{Ln} \frac{[\gamma\text{-aminobutyric acid}]_{in}}{[\gamma\text{-aminobutyric acid}]_{out}} = n \text{ Ln} \frac{[Na^+]_{out}}{[Na^+]_{in}}$$

$$+ m\text{Ln} \frac{[Cl^-]_{out}}{[Cl^-]_{in}} - \frac{(n-m)\ F}{RT} \Delta\psi$$

Thus, when the chloride gradient and the membrane potential are kept
constant, a plot of the log of the γ-aminobutyric acid gradient
versus the log of the sodium ion gradient should give a straight
line with a slope equal to the stoichiometry. By the same approach,
the stoichiometry for chloride and the number of charges can be
determined. However, while it is trivial to determine the γ-aminobutyric acid gradient at the steady state, we do not know the sodium
gradient across the very vesicles which transport γ-aminobutyric
acid. The only time at which this parameter is known for the γ-amino
butyric acid-ergic vesicles is at time zero (the time of the impo-
sition of the gradient). It is clear that when that value is used
in the log/log plot, an understimate of the stoiciometry will be
found (at the steady state the actual sodium gradient will be
smaller than the plotted gradient). Yet, when the data are plotted
in this fashion (Fig. 4) a stoichiometry of 1.5 ± 0.2 (20 experiments) is found and thus one can conclud·e that more than one sodium
is transported per translocation cycle (5). The values for chloride
and the number of charges are found to be 0.47 ± 0.02 and 0.9 ±
0.08, respectively (5). Thus, the simplest stoichiometry consistent
with these data is 2Na$^+$:Cl$^-$: γ-aminobutyric acid, but more complex
stoichiometries cannot be excluded.

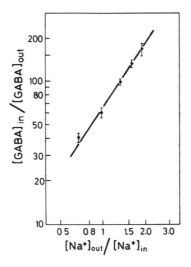

FIG 4. Relationship between sodium gradient and γ-aminobutyric acid (GABA) gradient. Membrane vesicles (15 µl, 76.5µg of protein)loaded with 50 mM NaP_i + 50 mM KP_i, pH 6.8); were diluted with 10 vol. of influx solutions containing mixtures (sum 150 mM) of NaCl and choline chloride + 0.153 µM γ-amino [2,3-^{3}H]butyric acid (29.3 Ci/ mmol). The final ratio of external to internal sodium was that shown on the abscissa. Reactions were stopped at 3 min and the γ-amino-butyric acid concentration ratio was determined as described under "Experimental Procedures".Reprinted with permission from Ref.5, copyright 1983, American Chemical Society.

The ion dependencies of net flux and exchange of γ-aminobutyric acid in membrane vesicles have recently been compared. It has been observed that external γ-aminobutyric acid enhances efflux (4) and the usual interpretation for this is that a step distinct from γ-aminobutyric acid translocation (its binding from one side of the membrane, translocation and release from the other side) is rate-limiting for the efflux. Interestingly, although the stimulation by external γ-aminobutyric acid requires external sodium, it does not require external chloride (6). The interpretation for this is that when labelled γ-aminobutyric acid can rebind there before the chloride (translocated from the inside) is released. Thus unlabelled γ-aminobutyric acid is carried inward where it exchanges with label-led γ-aminobutyric acid and so forth. This is consistent with an ordered mechanism in which chloride binds first to the outside, followed by γ-aminobutyric acid and finally by sodium (Fig. 5). Although kinetic experiments with intact synaptosomes appeared to fit best with another model (9) (which, however, allowed for binding of γ-aminobutyric acid to the transporter in the absence of sodium), a recent study with synaptosomes combining kinetic and exchange experiments indicates that γ-aminobutyric acid has to bind to the

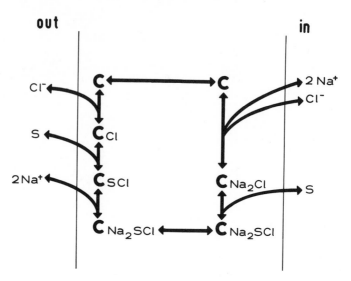

FIG 5.Diagram illustrating one of the possible binding sequences for
chloride, γ-aminobutyric acid (GABA) and sodium. Influx from left to
right.Efflux from right to left.Reprinted with permission from
Ref.6; Copyright 1983, Biochim.Biophys. Acta.

transporter prior to the sodium (10). The studies with the membrane
vesicles are consistent with the idea that on the inside the binding
order to the transporter is different from that on the outside, so
that it is possible that the transporter is asymmetric (Fig. 5).

Properties of L-glutamate transport

Using the membrane vesicle preparation, similar studies to
those on the GABA system have been performed with the one for L-
glutamate. The properties of the system are summarised in Table II.

Table II : Properties of L-glutamate Transport

I. Influx requires both external sodium and internal
   potassium and is electrogenic.
2. To achieve maximal steady state levels gradients of
   Na^+(out > in) and K^+ (in>out) are required.
3. The requirement for K^+ is specific.
4. Efflux requires both internal sodium and external potassium.
5. Binding order is assymetric.
6. The role of potassium is to allow the return of the
   unloaded transporter.

As for GABA, the mechanism of transport is chemiosmotic,sodium
being cotransported with L-glutamate. Moreover also in this system
an additional coupling ion exists, but this is potassium rather
than chloride which is used in the GABA system. For influx of
L-Glutamate internal potassium is required, while for efflux

potassium is required on the outside. Thus potassium is counter-transported with L-glutamate. The electrogenecity of the L-glutamate transporter was more difficult to establish since the addition of valinomycin to vesicles with a potassium ion gradient (in>out) imposed across their membrane leads to two phenomena with opposing effects on L-glutamate transport : (1) Enhancement of the interior negative membrane potential and (2) partial dissipation of the outward potassium gradient and inward sodium gradient, provided there is a "sodium leak". The first effect will stimulate and the second will inhibit L-glutamate transport. Liposomes into which the L-glutamate transporter has been reconstituted have a much lower leakage to ions and reconstituted L-glutamate transport is drama-tically enhanced by valinomycin. This provides strong evidence that the system in fact catalyses electrogenic transport.

As mentioned before, efflux of glutamate requires external potassium (14,15) and internal sodium (14). However, there is also an additional mode of exit for glutamate. This is by exchange which occurs in the presence of external sodium and glutamate (15). This suggests that potassium activates a step which is required for efflux but not for the translocation step (15) (Fig 6).

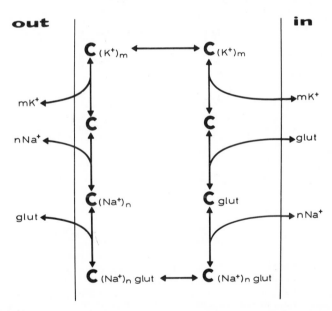

FIG 6. One of the possible models for the mechanism of L-glutamic acid (glut) translocation. The binding/debinding order of sodium and L-glutamic acid as indicated here for the inside and the outside may explain the data, but other possibilities exist. Reprinted with permission from Ref. 15; copyright 1982. American Chemical Society.

The most obvious candidate for such a step is the return of the
unloaded transporter. After sodium and glutamate have been released,
potassium will bind to the transporter. This facilitates reorienta-
tion of the transporter so that upon translocation of potassium and
its release, the binding sites for sodium and glutamate are available again at the original side of the membrane. Now a new translocation cycle can be initiated. This is true for influx (going
counterclockwise)as well as for efflux (going clockwise). Interestingly, it has been observed that the stimulation of efflux by
external glutamate also occurs in the absence of sodium (15). It is
of interest to note also that L-aspartic acid-induced efflux of
L-glutamic acid does not require external sodium. Thus, L-aspartate
is also transported by the L-glutamate transporter via the same
mechanism. The easiest way to explain the independence of exchange
on external sodium is that on the outside sodium binds first,
followed by glutamate (Fig. 6). Thus, during efflux, upon translocation, labelled glutamate is released on the outside and unlabelled
glutamate binds before the sodium is released. Furthermore, it can
be shown that exchange requires internal sodium (15). Thus, the
transporter seems to be asymmetric in its interaction with sodium.
The observation can best be explained by postulating that on the
inside sodium is released prior to L-glutamate (Fig.6)

A Multitude of Sodium-Coupled Neurotransmitter Transport Systems
Requiring Addition Ions.
          Table III summarizes sodium-coupled neurotransmitter transport
systems requiring additional ions.

Table III. Sodium-coupled neurotransmitter transport systems
           requiring additional ions.

Transmitter	Source	Additional ion	Reference
GABA	Rat Brain	Cl^-	2-6
	Insect Nervous System		11
Glutamate	Rat Brain	K^+	12-15
	Kidney		16-19
	Liver		20
	Crab nerve		21
Serotonin	Platelets	$Cl^- + K^+$	22-33
	RBL cells		34
	Rat Brain		35
glycine	Rat Brain	Cl^-	35,36
	Erythrocytes		37,38
Norepinephrine	Rat Brain	Cl^-	35
	Rat Heart Nerve		39
Dopamine	Rat Brain	Cl^-	35

Although it is not yet clear that in all cases the additional ions
are also co-transported, there are several where it is quite obvious.
In addition to the GABA transporter these include the glutamate,
serotonin and glycine transporters from several tissues. Those
cases which still need experimental verification are mainly those
where transport has been measured in relatively intact systems,
such as synaptosomes. The use of membrane vesicles should permit

clear-cut experimental verification.

Solubilization, Reconstitution and Purification of the GABA Transporter

The GABA transporter has been solubilized and reconstituted into liposomes. The transport characteristics, such as sodium and chloride dependence, in the reconstituted system were the same as in the native membrane vesicles. It is of interest to note that in the potassium-loaded proteoliposomes the stimulation of transport by valinomycin is much larger than in the "native" membrane vesicles, suggesting strongly that the former are much tighter to potassium. In view of the higher phospholipid/protein ratio in the proteoliposomes, this is to be expected. The reconstitution assay enables one to assay the transporter functionally (ability to transport upon reconstitution).

Using this criterion we have highly purified this transporter. The development of a very fast reconstitution procedure using "minicolumns" has permitted the fast and simultaneous reconstitution of many samples and this has enabled us to perform the purification of the reconstitutively active transporter. The procedure is based on rapidity, using small fast flowing columns and is completed within 8 hours. This rapidity is absolutely essential since, although the transporter is relatively stable, in the last stages the transporter is very diluted and undergoes partial inactivation. Briefly, the transporter is extracted with cholate in the presence of high salt (0.4M). The soluble extract is fractionated with ammonium sulfate. Most proteins precipitate at 50% saturation but the transporter remains in the supernatant (40). The activity is precipitated at a saturation of 70% and is recovered as a floating pellet. Further purification requires the removal of both the ammonium sulfate and the cholate. This is achieved by gel filtration on Sephadex G-50, which is equilibrated with the detergents of choice for the next column, namely a mixture of Triton X-100 and octyl-glucoside. The void volume is collected and run on a DEAE column equilibrated with the above-mentioned detergents. The activity is eluted with a potassium phosphate gradient.

Until the end of the ammonium sulfate fractionation, little if any inactivation occurs. The change of the detergents is sometimes accompanied with an apparent loss of activity although it is possible that this is due to the presence of the non-ionic detergents (by carry-over) during the reconstitution. At the stage of the DEAE column there is an inactivation of 30-60%, yielding an increase in specific activity of 30-50 fold over the starting material and an actual purification of 100-150 fold. On the last column there is an increase of another 2-2.5 fold in specific activity but 60-75% inactivation. Thus the total actual purification is between 800-1500 fold. The properties of GABA transport by the purified and reconstituted protein are identical to that of the starting material. Thus influx is absolutely dependent on external sodium and chloride is abolished by nigericin which will collapse the sodium gradient by exchanging the external sodium with the internal potassium. Valinomycin stimulates the initial rate of influx up to 20-fold. This is expected since GABA transport is electrogenic and the big

magnititude of the stimulation indicates that the vesicles are
probably very tight. It is of interest to note that the valinomycin
stimulation in the case of the proteoliposomes obtained with the cru
de extract is transient and changes to inhibition at longer times.
This is possibly due to the presence of proteins which may represent
"sodium leaks" and which are purified away along the procedure. The
Km for GABA of the purified transporter is 4 μM which is very similar to the value of 2.5 μM found in native vesicles.

The gel pattern of the purified preparation - as visualized
by silver staining -exhibits a major polypeptide with an apparent
molecular weight of 24 K Daltons. There are a few other polypepti-
des present in the preparation but only one of those with an
apparent molecular weight of 83-85 k Daltons, as well as the 24 k
Dalton polypeptide, correlate very well with the activity. Very
recently another purification step was discovered. It was found
that the GABA transporter was retained on wheat germ agglutinium
coupled to sepharose. The gel pattern disclosed the exclusive
presence of the 83-85 k Dalton polypeptide. Thus it appears that
this polypeptide - a glycoprotein - represents the GABA transporter.

REFERENCES

1. KANNER, B.I. 1983. Biochim. Biophys. Acta 726, 293–316.

2. KANNER, B.I. 1978. Biochemistry 17, 1207–1211.

3. KANNER, B.I. 1980. Biochemistry 19, 692–697.

4. KANNER, B.I., and KIEFER, L. 1981. Biochemistry 20, 3354–3358.

5. RADIAN, R., and KANNER, B.I. 1983. Biochemistry 22, 1236–1241.

6. KANNER, B.I., BENDAHAN, A., and RADIAN, R. 1983. Biochim. Biophys. Acta 731, 54–62.

7. BAKER, P.F., HODGKIN, AL., and SHAW, T.I. 1962. J. Physiol (London) 164, 355–374.

8. KANNER, B.I. 1978. FEBS Lett. 89, 47–50.

9. WHEELER, D.D., and HOLLINGSWORTH, R.G. 1979. J. Neurosci. Res. 4, 265–289.

10. NELSON, M.T., and BLAUSTEIN, M.P. 1982. J. Membrane Biol. 69, 212–223.

11. GORDON, D., ZLOTKIN, E., and KANNER, B.I. 1982. Biochim. Biophys. Acta 688, 229–237.

12. KANNER, B.I., and SHARON, I. 1978. Biochemistry 17, 3949–3953.

13. KANNER, B.I., and SHARON, I. 1978. FEBS Lett. 94, 245–248.

14. KANNER, B.I., and MARVA, E. 1982. Biochemistry 21, 3143–3147.

15. KANNER, B.I., and BENDAHAN, A. 1982. Biochemistry 21, 6327–6330.

16. SCHNEIDER, E.G., and SACKTOR, B. 1980. J. Biol. Chem. 255, 7645–7649.

17. SCHNEIDER, E.G., HAMMERMAN, M.R., and SACKTOR, B. 1980. J. Biol. Chem. 255, 7650–7656.

18. BURSKHARDT, G., KINNE, R., STANGE, G., and MURER, H. 1980. Biochim. Biophys. Acta 599, 181–201.

19. SACKTOR, B., LEPOR, N., and SCHNEIDER, E.G., 1981. Biol. Sci. Rep. 1, 709–713.

20. SIPS, H., DE GROOT, P.A., and VAN DAM, K. 1982. Eur. J. Biochem. 122, 259–264.

21. BAKER, P.F., and POTASHNER, S.J. 1971. Biochim. Biophys. Acta 249, 616–622.

22. RUDNICK, G. 1977. J. Biol. Chem. 252, 2170–2174.

23.   KEYES, S.R., and RUDNICK, G. 1982. J. Biol. Chem. 257, 1172–1176.

24.   PAASONEN, M.K. 1968. Ann. Med. Exp. Biol. Fen-. 46, 416–422.

25.   SNEDDON, J.M. 1973. Prog. Neurobiol. 1, 151–198.

26.   SNEDDON, J.M. 1969. Br. J. Pharmacol. 37, 680–688.

27.   LINGJAERDE, O. JR. 1971. Acta Physiol. Scand. 81, 75–83.

28.   NELSON, P.J., and RUDNICK, G. 1982. J. Biol. Chem. 257, 6151–6155.

29.   NELSON, P.J., and RUDNICK, G. 1979. J. Biol. Chem. 254, 10084–10089.

30.   RUDNICK, G., and NELSON, P.J. 1978. Biochemistry 17, 4739–4742.

31.   TALVENHEIMO, J., NELSON, P.J., and RUDNICK, G. 1979. J. Biol. Chem. 254, 4631–4635.

32.   RUDNICK, G., and NELSON, P.J. 1978. Biochemistry 17, 5300–5303.

33.   TALVENHEIMO, J., and RUDNICK, G. 1980. J. Biol. Chem. 255, 8606–8611.

34.   KANNER, B.I., and BENDAHAN, A., submitted for publication.

35.   KUHAR, M.J., and ZARBIN, M.A. 1978. J. Neurochem. 31, 251–256.

36.   MAYOR, F. JR., MARVIZON, J.G., ARAGON, M.C., GIMENEZ, C. and VALDIVIESO, F. 1981. Biochem. J. 198, 535–541.

37.   VIDAVER, G.A. 1964. Biochemistry 3, 799–803.

38.   IMLER, J.R., and VIDAVER, G.A. 1972. Biophys. Biochim. Acta 288, 153–165.

39.   SANCHEZ-ARMASS, S., and ORREGO, F. 1977. Life Sci. 20, 1829–1838.

40.   AGMON, V., and KANNER, B.I. 1980. Abstr. FEBS, S5-P74.

12

Glutamate Transport in the Synaptic Vesicle

T. Ueda

ABSTRACT

L-glutamate, a strong candidate for the major excitatory neurotransmitter, is now clearly demonstrated to be taken up into virtually pure, isolated synaptic vesicles in an ATP-dependent but Na^+-independent manner. Pharmacological evidence suggests that the vesicular glutamate uptake is driven by an electrochemical proton gradient generated by a proton-pump ATPase. The ATP-dependent vesicular uptake is highly specific for glutamate, and the gluta-mate translocator has properties distinct from the Na^+-dependent glutamate transporter in the plasma membrane as well as from post-synaptic receptors. These lines of evidence provide additional support for the proposed neurotransmitter role of glutamate. It is proposed that this unique vesicular translocator may play a crucial role in selecting glutamate for synaptic transmission.

INTRODUCTION

The neurotransmitter role of L-glutamate in the central ner-vous system has been the subject of intense investigation for the last quarter of a century. The pioneering work of Curtis, Watkins, Krnjević and Phillis in the early 1960's demonstrated (Curtis et al., 1960; Krnjević and Phillis, 1963) that this acidic amino acid exerts a rapid, reversible, excitatory action, with high potency, on the majority of the central neurons, raising the possibility of a neurotransmitter role for glutamate (for reviews, see Curtis and Johnston, 1974; Krnjević, 1974). It has proved difficult, however, to firmly establish the notion that glutamate serves as a neuro-transmitter. Besides the apparent lack of specificity of neurons affected, glutamate is present in all types of cells, neuronal or nonneuronal, eucaryotic or procaryotic, and, as such, it is involved in diverse biological processes such as energy and nitro-gen metabolism, the synthesis of protein, peptides and nucleotides, the formation of the inhibitory neurotransmitter γ-aminobutyric acid, and the regulation of osmotic balance (for reviews, see Johnson, 1972; Shank and Campbell, 1983). Nonetheless, in recent years, a variety of evidence has accumulated to support the proposed transmitter role of glutamate (for reviews, see Cotman et al., 1981; Watkins and Evans, 1981; Fonnum, 1984). In particular, it is now apparent that there exist multiple, but pharmacologically distinct, specific glutamate receptors and binding sites (Johnston, 1979; McLennan, 1981; Roberts, 1981; Watkins and Evans, 1981; Fagg et al., 1982; Olverman et al., 1984; Crunelli et al., 1985; Slaughter and Miller, 1985), which regulate the flux of Na^+, K^+, and Ca^{2+} (Curtis et al., 1960; Krnjević and Phillis, 1963; Bradford and McIlwain, 1966; Nowak et al., 1984; Puil, 1981; Sonnhof and Bührle, 1981). The pharmacologically distinct populations of glutamate binding sites have been shown to have different anatomical distributions in the central nervous system (Monaghan et al., 1983; Greenamyre et al., 1984). Electrophysiological experiments have demonstrated that the reversal potential of glutamate-induced response is similar to that of evoked EPSP (Hablitz and Langmoen, 1982; Crunelli et al., 1984). Moreover, Michaelis et al. have recently isolated a glutamate binding protein from the synaptic membrane, and showed that, when reconstituted into liposomes, it is capable of causing a glutamate-induced Na^+ flux (Michaelis et al., 1983; Stormann et al., 1984). Another important line of evidence, which provided strong support for the proposed neurotransmitter role of glutamate, is that glutamate is released from nerve terminals in a calcium-dependent manner upon their polarization (for reviews, see Cotman et al., 1981; Abdul-Ghani et al., 1981).

It is not known, however, whether glutamate in the nerve terminal is released into the synaptic cleft directly from the synaptic vesicle by an exocytotic or other mechanism. If the calcium-dependent release is mediated by a process which involves synaptic vesicles, loading the vesicles with glutamate would be an important step in glutamate synaptic transmission. Contrary to this premise, present evidence provides no indication that glutamate is substantially concentrated in the isolated synaptic vesicles compared to other subcellular fractions (Mangan and Whittaker, 1966; Rassin, 1972; Kontro et al., 1980). Likewise, there has been until recently no clear demonstration of glutamate uptake into isolated synaptic vesicles. Based upon this and other evidence, De Belleroche and Bradford proposed that the cytoplasmic glutamate, rather than the vesicular, is released from the nerve terminal (De Belleroche and Bradford, 1977). In contrast, Naito and Ueda (1983) have recently provided evidence that highly purified synaptic vesicles are capable of accumulating L-glutamate selectively in an energy-dependent manner. This observation argues for the possibility that synaptic vesicles may play an active role in glutamate synaptic transmission. In this article, evidence for the vesicular uptake of glutamate is reviewed, and this vesicular transport system is compared with the other vesicular transport systems, for neurotransmitters and hormones as well as with the plasma membrane glutamate transport system, with respect to the nature of the driving

force involved and the substrate specificity. It will be argued
that these synaptic vesicles endowed with the unique, glutamate-
specific translocator may play a crucial role in selecting glutamate as a neurotransmitter.

IMMUNOCHEMICAL PURIFICATION, PURITY AND PROMINENT PROTEINS OF THE
SYNAPTIC VESICLE FROM BOVINE BRAIN

In establishing the association of a biological process,
enzyme activity, content, or constituent with the synaptic vesicle,
one cannot overemphasize the importance of the purity of that synaptic vesicle preparation. This point is particularly relevant to
many of the brain synaptic vesicle preparations, which are often
contaminated with plasma membranes. In the studies of Naito and
Ueda (1983), particular efforts were made to prepare synaptic vesicles with a high degree of purity. In these studies, a specific
immunochemical approach was developed, and combined with the conventional, biophysical method for the purification of synaptic
vesicles (Ueda et al., 1979), which was originally developed by
Whittaker et al. (1964) and De Robertis et al. (1962). The synaptic vesicles were immunoprecipitated with affinity-purified antibodies to Synapsin I (Naito and Ueda, 1981; Ueda and Naito, 1982),
a well-characterized, neuron-specific phosphoprotein (Ueda and
Greengard, 1977; Greengard, 1981; Nestler and Greengard, 1983),
which is highly concentrated on the surface of synaptic vesicles
(Bloom et al., 1979; Ueda et al., 1979; De Camilli et al., 1983).
The immunoprecipitated synaptic vesicles are essentially free of
contamination from plasma membranes and other non-vesicular subcellular organelles and have a mean diameter of 400 Å. This is
consistent with the biochemical analysis of non-vesicular marker
enzymes. As expected, the immunoprecipitated synaptic vesicles are
substantially enriched with Synapsin I (Naito and Ueda, 1983). A
number of additional proteins are associated with these vesicles,
among which appears to be actin (Mr=42,000); actin is reported to
be present in the cholinergic synaptic vesicles (Stadler and Tashiro, 1979; Wagner and Kelly, 1979) and chromaffin granules (Winkler
and Westhead, 1980), as well as in another brain vesicle preparation (Zisapel and Zurgil, 1979). These observations altogether
suggest that actin may be widely distributed among various types of
storage vesicles, and may have a function common to all of these
storage vesicles and granules. Further analysis indicates that the
70,000- and 34,000-dalton proteins, prominent components of the
immunoprecipitated vesicles, are also present in the cholinergic
vesicles (Wagner and Kelly, 1979; Stadler and Tashiro, 1979). The
34,000-dalton protein has been suggested to be a nucleotide carrier
protein (Luqmani, 1981), but the 70,000-dalton protein has not been
identified.

EVIDENCE FOR ACTIVE TRANSPORT OF GLUTAMATE IN THE SYNAPTIC VESICLE

The ATP-dependent accumulation of catecholamine in adrenal medullary chromaffin granules has been well established since its first demonstration by Kirshner and Carlsson et al. in 1962. This catecholamine uptake system is the most extensively studied and best understood of all the neurotransmitter/hormone transport systems in the intracellular organelle membrane described thus far (for reviews see Bashford et al., 1976; Njus et al., 1981; Johnson et al., 1982), and has served as a model for studying the uptake of glutamate into the isolated synaptic vesicles. It is only recently that clear evidence (Naito and Ueda, 1982; 1983) was provided that L-glutamate is taken up in an ATP-dependent manner into the identified synaptic vesicles (Fig. 1). This vesicular uptake is dependent upon temperature and requires the hydrolysis of ATP and, interestingly, and perhaps importantly, is remarkably specific for L-glutamate; thus, aspartate, GABA, glycine, and glutamine do not serve as effective substrates (Table I). Product analysis has indicated that this uptake occurs without a significant metabolic conversion of glutamate. These experiments represent the first clear demonstration of an energy-dependent, but Na^+-independent, specific uptake of glutamate into isolated synaptic vesicles. The same basic characteristics of the ATP-dependent glutamate uptake are observed in a less pure but substantially purified synaptic vesicle fraction (Naito and Ueda, 1985). In addition to ATP, Mg^{2+} is essential for the glutamate uptake, whereas neither Ca^{2+} nor Na^+ is required for the uptake process, indicating that Ca^{2+}-ATPase, Ca^{2+}/Mg^{2+}-ATPase,

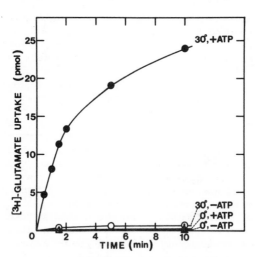

Fig. 1. Time course of glutamate uptake into isolated synaptic vesicles. Synaptic vesicles (20 μg protein) immunoprecipitated with anti-Synapsin I IgG were incubated with 50 μM [^{3}H] glutamate at 0°C (Δ,▲) and 30°C (0,●) in the presence (▲,●) and absence (Δ,0) of ATP. Taken from Naito and Ueda (1983).

TABLE I

ATP-DEPENDENT UPTAKE OF PUTATIVE AMINO ACID NEUROTRANSMITTERS INTO
IMMUNOPRECIPITATED SYNAPTIC VESICLES

Neurotransmitter	Relative Activity (%)
L-Glutamate	100
L-Aspartate	0.89 ± 0.22 (n=6)
GABA	1.43 ± 0.45 (n=4)
Glycine	0.06 ± 0.06 (n=3)

Immunoprecipitated vesicles (20 μg) were incubated for neurotransmitter uptake at 30° for 1.5 min in the absence or presence of ATP,
in a mixture containing 50 μM tritiated neurotransmitter, 4 mM
$MGSO_4$, 5 mM Tris-HCl (pH 7.4), and 0.32 M sucrose. The uptake
activity relative to that of glutamate was expressed as percentage
in each experiment. Values given are mean $\pm$ S.E.M. from n experiments.   Modified from Naito and Ueda (1983).

or Na^+/K^+-ATPase is not involved in the vesicular glutamate uptake.
GTP and ITP are much less effective in providing energy for the
glutamate uptake. The uptake is saturable with glutamate, with Km
of 1.6 mM and Vmax of at least 13 nmol/min/mg. The vesicle's low
affinity for glutamate may relate to the rather high intracellular
concentration of glutamate, which is considered to be in the low
millimolar range. Other evidence indicates that at least two free
sulfhydryl groups are important in the ATP-driven uptake of glutamate. All of these characteristics suggest that the transport of
glutamate into the synaptic vesicle is carried out by a mechanism
involving ATP-derived energy utilization and a specific carrier
protein(s).

An interesting feature of the vesicular glutamate uptake is
the pronounced stimulation produced by low millimolar concentrations of chloride. Chloride ions alone cause very little, if any,
glutamate uptake, and the stimulatory effect of chloride is absolutely dependent upon the presence of ATP. Thus, the synergistic
actions of ATP and chloride are necessary to achieve a maximal uptake of glutamate into synaptic vesicles; this is particularly prominent when the ATP concentration is low. The stimulatory effect
is specific to chloride and bromide, and other anions and cations
are without effect. Although the mechanism of this chloride-
induced stimulation is not known, the effect is reflected in an
increase in Vmax with no significant changes in Km for glutamate.
Since the effective concentration is in the physiologically relevant range (Woodbury, 1965), it is possible that chloride ions in
the nerve terminal may have an important role in in vivo accumula-

tion of glutamate in synaptic vesicles, especially when ATP in the nerve terminal is transiently and locally reduced, perhaps as a result of intense neuronal firing. In this regard, it is of interest to note the evidence obtained by Gallo et al. (1983), suggesting that the GABA agonist-induced chloride influx into the nerve endings may lead to a potentiation of the evoked release of glutamate. Whether this potentiated release is related to a chloride-induced increase in the accumulation of glutamate in the synaptic vesicles is an open question. There is good evidence now that chloride enhances the binding of glutamate and the agonist/antagonist α-phosphonobutyrate to a specific population of binding sites without changing Km for the ligand (Fagg et al., 1982; Monaghan et al., 1983; Butcher et al., 1983; Butcher et al., 1984; Greenamyre et al., 1985). It is also reported that certain glioma cells have a sodium-independent, chloride-dependent high affinity mechanism for cellular uptake of glutamate (Waniewski & Martin (1984). These lines of evidence, including the chloride stimulation of the vesicular glutamate uptake, all suggest an interaction between chloride ions and certain classes of glutamate binding proteins, whether they are the receptor type or the transmembrane translocator type.

The studies by Naito and Ueda (1982, 1983, 1985) clearly show the importance of Mg^{2+}/ATP and chloride in the in vitro uptake of glutamate into synaptic vesicles. Disbrow et al. (1982) also observed an ATP- and temperature-dependent uptake of glutamate into crude synaptic vesicle fractions. However, it is difficult to conclude with certainty that the glutamate uptake they reported is a process that occurs in the synaptic vesicle, because (1) the vesicle fractions they used are known to be contaminated with significant amounts of plasma membranes and (2) the ATP-dependency is rather small. It is likely that this small degree of ATP-dependency is partly due to the absence of chloride from their incubation medium, and partly due to the plasma membrane contamination. In contrast to these studies, Kuriyama et al. (1968) reported that glutamate is taken up into the synaptic vesicle fractions in a Na^+-dependent manner. However, as pointed out by De Belleroche and Bradford (1973), their vesicle preparations may be significantly contaminated with plasma membranes, and it is quite possible that the Na^+-dependent "vesicular" uptake described by these investigators may be largely due to the Na^+-dependent glutamate uptake system present in the contaminating plasma membrane (synaptosomal or glial). Moreover, the high concentrations of Na^+ required for the glutamate uptake into their "vesicle" fractions could not be physiologically relevant; the synaptic vesicles in vivo are in the intracellular milieu containing low concentrations of Na^+. The same arguments may apply to the studies by Lähdesmäki et al. (1977), in which glutamate binding to synaptic vesicle fractions was measured in the presence of extracellular concentrations of Na^+.

Despite the evidence for the energy-dependent vesicular glutamate uptake, glutamate has never been found to be particularly abundant in the synaptic vesicle fractions, as compared to its level in the other subcellular fractions (Mangan and Whittaker, 1966, Rassin, 1972; Kontro et al., 1980). However, such a lack of evidence for the abundance of glutamate in the isolated vesicles may well be attributed to the possible leakage of the amino acid during the vesicle preparation, as suggested by Mangan and Whittaker (1966) and Shank and Campbell (1983). Indeed, when the isolated vesicles were loaded with glutamate in the presence of ATP and chloride, and subsequently subjected to osmotic shock, a substantial amount of the amino acid was released. These observations suggest that the contents of glutamate and perhaps other acidic amino acids determined in isolated synaptic vesicles may not represent those in situ, and these results should be interpreted with caution, particularly in the physiological context. In contrast, recent immunocytochemical evidence indicates that glutamate is concentrated in the synaptic vesicles in those nerve terminals considered to be glutamatergic, but not in the GABAergic nerve endings (Storm-Mathisen et al., 1983). This observation is consistent with the energy-dependent, chloride-stimulated specific vesicular accumulation of glutamate demonstrated in vitro.

THE NATURE OF THE DRIVING FORCE

In recent years, it has become apparent that the chemiosmotic concept, originally formulated by Mitchell (1961, 1966) provides the basis for understanding the mechanism not only of energy transduction, but also of active transport and various other biological processes (for review, see Kaback, 1982, 1983). It states, in its most general form, that the immediate driving force for many processes in energy-coupling membranes is an electrochemical proton gradient ($\Delta\mu_H+$) composed of the membrane potential ($\Delta\Psi$) and the transmembrane pH gradient (ΔpH). The electrochemical proton gradient is generated either via the respiratory chain or by a Mg^{2+}-dependent H^+-ATPase. Bashford et al. (1975, 1976) provided evidence that the proton translocation catalyzed by the H^+-ATPase is coupled to the catecholamine uptake into chromaffin granules. This represents the first demonstration of the link between a H^+-ATPase and the uptake of neurotransmitter/hormone into intracellular storage vesicle/granules. It is now generally accepted (1) that the H^+-ATPase, upon the hydrolysis of ATP, initially generates a membrane potential, positive inside, which leads to the development of a pH gradient in the presence of permeant anions in the extragranular medium; and (2) that either the membrane potential or the pH gradient alone can drive the catecholamine uptake, but both components are required for optimal accumulation to occur (Johnson et al., 1978; Schuldiner et al., 1978; Holz, 1978; Johnson and Scarpa, 1979; Johnson et al., 1979; Apps et al., 1980; Kanner et al., 1980; Sherman and Henry, 1980).

As mentioned earlier, ATP hydrolysis is required for the uptake of glutamate into synaptic vesicles. Mg^{2+} is an essential divalent cation, but Ca^{2+}, Na^+, and K^+ appear to play little or no

role in the vesicular glutamate uptake.   These characteristics
suggest the involvement of a H+-ATPase in the glutamate uptake.
This notion is supported by the observation (Naito and Ueda, 1985)
that the glutamate uptake is diminished by DCCD and trimethyltin,
agents known to inhibit the H+-ATPase and thereby block the catecholamine uptake into chromaffin granules (Bashford et al., 1976;
Johnson et al., 1979; Johnson et al., 1982; Schuldiner et al., 1978;
Apps et al., 1980).   Moreover, the protonophore FCCP and the lipophilic anion thiocyanate, which are known to dissipate the membrane
potential generated by H+-ATPase and thereby inhibit the catecholamine into chromaffin granules (Holz, 1978; Johnson and Scarpa,
1979; Johnson et al., 1979), reduce the ATP-dependent glutamate
uptake.   The ATP-dependent glutamate uptake is also inhibited by
nigericin plus K+ or ammonium, agents known to dissipate the pH
gradient and thereby block the catecholamine uptake into chromaffin
granules (Johnson and Scarpa, 1979; Johnson et al., 1979).   Recently, a good correlation between the ATP-induced development of the
membrane potential across the synaptic vesicle and the glutamate
uptake has been observed (Shioi and Ueda, 1985).   These observations lead to the proposal that the glutamate uptake into the synaptic vesicle is also driven by an electrochemical proton gradient,
a mechanism analogous to that of catecholamine uptake into storage
granules (Fig. 2).   According to this model (Ueda, 1984), a H+-ATPase
in the synaptic vesicle membrane, upon hydrolysis of ATP, translo-
cates protons from the outside to the inside of the vesicle membrane
and thereby generates a membrane potential, positive inside.   This
membrane potential could provide a driving force for the transport
of glutamate into the synaptic vesicle.   When permeant anions such
as chloride are present at relatively high concentrations in the

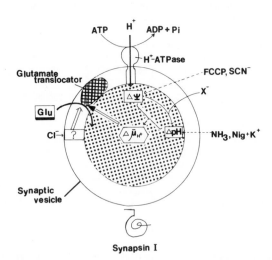

Fig. 2. Proposed mechanism for glutamate uptake into synaptic
vesicles.

extravesicular medium, the membrane potential would induce an in-flux of anions, which would then reduce the membrane potential and result in an increase in the pH gradient across the membrane. The pH gradient thus formed could also cause the glutamate uptake. The relative contribution of these two components to inducing the maxi-mal rate of glutamate uptake is not known. As noted earlier, low millimolar concentrations of chloride stimulate substantially the glutamate uptake largely driven by the membrane potential. Recent evidence indicates that the chloride-induced stimulation is brought about without a detectable increase in ΔpH or H^+-ATPase activity (Naito, Shioi and Ueda, unpublished observations). The mechanism of this chloride stimulation remains to be explored.

The involvement of the electrochemical proton gradient in neurotransmitter/hormone transport is not limited to the catechola-mine uptake into chromaffin granules and the glutamate uptake into synaptic vesicles (Table II). It has been demonstrated, in addi-tion, in (1) serotonin uptake into platelet granules and partially purified brain synaptic vesicles, (2) catecholamine uptake into partially purified brain synaptic vesicles and heart synaptic vesi-cles, and (3) acetylcholine uptake into electric organ synaptic vesicles. It may be noted that the catecholamine uptake system is not specific for catecholamines; thus, it transports serotonin as well. There is also good evidence that the electrochemical proton gradient plays a vital role in the cellular uptake of various sub-stances in procaryotic cells (for review, see Kaback, 1982; 1983). It appears from these studies that ATP-dependent transport proces-ses involve at least two functionally distinct components, a proton-pump ATPase and a substrate-specific translocator. It should be pointed out that Synapsin I, in particular its phosphorylation, is unlikely to play a role in the ATP-dependent vesicular uptake of glutamate or its regulation. It has been observed that the anti-Synapsin I antibodies, which block the phosphorylation of Synapsin I (Naito and Ueda, 1981; Ueda and Naito, 1982) have no effect on the vesicular glutamate uptake, nor is the specific activity of uptake increased in parallel to the marked increase in the specific content of Synapsin I, which occurs accompanying the immunoprecipi-tation (Naito and Ueda, 1983). This suggests that a population of those synaptic vesicles which take up glutamate contain little or no Synapsin I, and hence are not immunoprecipitated. The function of Synapsin I remains to be defined, although these experiments do not rule out the possibility that Synapsin I may be involved in the regulation of synaptic release of glutamate.

VESICULAR GLUTAMATE TRANSLOCATOR

It is evident (Kontro et al., 1980) that glutamate is the major amino acid in the nerve endings in various brain regions; however, the nerve terminals also contain significant amounts of other amino acids, particularly aspartate, another excitatory amino acid neurotransmitter candidate. It is also well known that the

TABLE II

VESICULAR UPTAKE OF NEUROTRANSMITTERS/HORMONES DRIVEN BY
ELECTROCHEMICAL PROTON GRADIENT

Neurotransmitter/hormone	Subcellular organelle	References
a. Catecholamines (dopamine, norepinephrine, epinephrine)	Chromaffin granules (bovine adrenal medulla)	Holz, 1978; Johnson et al., 1982; Schuldiner et al., 1978; Njus & Radda, 1979; Scherman & Henry, 1980; Kanner et al., 1980; Apps et al., 1980.
	Synaptic vesicle (rat brain)	Toll & Howard, 1978.
	Synaptic vesicle (rat heart)	Angelides, 1980.
	PC-12	Rebois et al., 1980.
b. Serotonin	Platelet granule	Rudnick et al., 1980; Carty et al., 1981; Wilkin & Salganicoff, 1981.
	Chromaffin granule	Johnson et al., 1979; Apps et al., 1980; Kanner et al., 1980.
	Synaptic vesicle (rat brain)	Maron et al., 1979.
c. Acetylcholine	Synaptic vesicles (Torpedo electric organ)	Anderson et al., 1982.
	PC-12	Toll & Howard, 1980.
d. Glutamate	Synaptic vesicle (bovine brain)	Naito & Ueda, 1982; 1983; 1985.
	Synaptic vesicle (rat brain)	Disbrow et al., 1982.

sodium-dependent cellular uptake systems for glutamate, including one in the nerve terminal plasma membrane, transport both acidic amino acids without much discrimination (Logan and Snyder, 1972; Roberts and Watkins, 1975; Schousboe, 1981; Gordon and Balázs, 1983; Christensen and Makowske, 1983). The question then arises how glutamate in the nerve terminal is specifically selected to serve the neurotransmitter function. Evidence summarized above suggests that this selection may occur at the level of synaptic vesicles (Ueda, 1984). These isolated synaptic vesicles do take up L-glutamate in a remarkably selective manner; thus, aspartate and other related amino acids such as GABA, glutamine and glycine are hardly accumulated by these vesicles under identical incubation conditions. Not only is aspartate ineffective as a substrate, but also it is ineffective as an inhibitor of the glutamate uptake (Table III), indicating that aspartate is hardly recognized by the glutamate translocator. This property distinguishes the vesicular

TABLE III

EFFECTS OF GLUTAMATE ANALOGS ON THE VESICULAR UPTAKE OF GLUTAMATE

Test agents	L-[^{3}H]Glutamate uptake (%)
None (control)	100
L-Glutamate	27 + 0
DL-γ-Methylene glutamate	39 + 3
DL-α-Methyl glutamate	51 + 2
D-Glutamate	68 + 1
L-Glutamine	117 + 1
Glutarate	97 + 4
α-Ketoglutarate	101 + 5
N-Methyl-L-glutamate	88 + 1
γ-Methyl-L-glutamate	107 + 2
γ-Ethyl-L-glutamate	115 + 3
L-Glutamic acid dimethylester	83 + 1
L-Glutamic acid diethylester	94 + 9
L-Aspartate	104 + 2
D-Aspartate	111 + 1
N-Methyl-D-aspartate	102 + 1
Kainate	97 + 1
2-Amino-4-phosphonobutyrate	95 + 3
L-Homocysteate	86 + 1
D-α-Aminoadipate	81 + 3
L-α-Aminoadipate	80 + 3
DL-α-Aminopimelate	106 + 10
Quisqualate	99 + 2
Ibotenate	102 + 1

The ATP-dependent glutamate uptake was determined, using 20 μg of synaptic vesicles and 50 μM L-[^{3}H]glutamate in the presence of various test agents (each 5 mM). From Naito & Ueda (1985).

translocator from the Na^+-dependent plasma membrane transporter of
acidic amino acids (Table IV). The vesicular translocator also
differs in affinity for glutamate from the Na^+-dependent cellular
transport system; the Km value for the vesicular uptake is in the
neighborhood of 1.6 mM (Naito and Ueda, 1985), whereas that for
the cellular uptake ranges generally from 2 μM to 66 μM (Logan and
Snyder, 1972; Campbell and Shank, 1978; Kanner and Sharon, 1978;
Schousboe, 1981; Schousboe and Hertz, 1981; Gordon and Balázs,
1983; Christensen and Makowske, 1983). Perhaps the substantially
higher concentration of glutamate in the cell than in the extracellular fluid may have made it unnecessary to provide the intracellular organelle synaptic vesicle with a high affinity uptake
system for glutamate. Another difference between the vesicular
and cellular uptake systems lies in the nature of the driving force
involved; while the vesicular glutamate translocator is sensitive
to the electrochemical proton gradient, the plasma membrane transporter responds to sodium and potassium gradients (Bennett et al.,
1974; Kanner and Sharon, 1978; Gordon and Balázs, 1983; Christensen & Makowske, 1983). Finally, the ATP-dependent vesicular uptake
system is stimulated by low concentrations of chloride, whereas the
Na^+-dependent plasma membrane uptake system is potentiated by high

TABLE IV

DIFFERENT PROPERTIES OF VESICULAR AND Na^+-DEPENDENT PLASMA
MEMBRANE UPTAKE SYSTEMS FOR GLUTAMATE TRANSPORT

Parameters	Glutamate Uptake System	
	Synaptic vesicle membrane (bovine brain)	Plasma membrane (Synaptosomes, neurons, glia cells, and hepatocytes)
Driving force	$\Delta\bar{\mu}_{H^+}$	$Na^+_o > Na^+_i$ $K^+_o < K^+_i$
Stimulating factor	Low Cl^- (1-4 mM)	High Cl^- (100 mM)
Affinity for Glu	Low (1.6 mM)	High (2-66 μM)
Interaction with Asp	No	Yes

concentrations of chloride (Kuhar and Zarbin, 1978). These different properties suggest that the vesicular glutamate translocator and the Na^+-dependent high affinity transporter are non-identical proteins, although they may share a partially homologous amino acid sequence.

The vesicular glutamate translocator is also distinct from the glutamate receptors. It does not interact with most of the excitatory amino acids and related compounds or their antagonists, which are known to act on glutamate receptors, except glutamate itself, of course, and its close analogs such as α-methyl glutamate and γ-methylene glutamate (Table III). Thus, not only aspartate, but also quisqualate, kainate, N-methyl-D-aspartate, ibotenate, homocysteate, glutamic acid diethyl ester, and α-amino-adipate are all unable to compete with glutamate for the vesicular uptake of glutamate, nor can they inhibit it. Although affinity for glutamate is rather low, the vesicular glutamate translocator has extremely restrictive structural requirements for a ligand to interact with the active site. The primary (but not secondary) amino group at the α-carbon, the γ-carboxyl group, the five carbon skeleton, and the L-configuration of the α-carbon are all necessary for recognition by the vesicular translocator. It is this unique property, namely the narrow substrate/inhibitor specificity, besides the sensitivity to the electrochemical proton gradient and low concentrations of chloride ions, that could enable the vesicular translocator to select glutamate for synaptic transmission (Fig. 3).

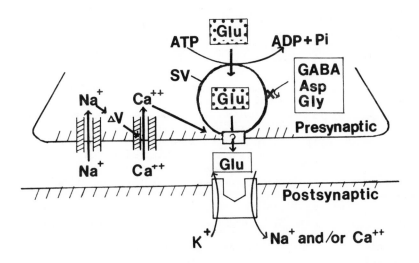

Fig. 3. Proposed model of glutamatergic synaptic transmission.

Thus, unless the synaptic vesicles in a given nerve terminal are endowed with this glutamate-specific translocator, the terminal glutamate may not serve the transmitter function. However, evidence for the direct release of glutamate in a Ca^{2+}-dependent manner from the synaptic vesicle into the synaptic cleft remains to be found.

CONCLUSION

Evidence reviewed in this article suggests that L-glutamate, a major CNS excitatory neurotransmitter candidate, could be taken up in vivo into the synaptic vesicles in certain nerve terminals, utilizing the energy derived from ATP hydrolysis. This evidence provides additional support for the neurotransmitter role of glutamate, and raises the possible involvement of synaptic vesicles in glutamate synaptic transmission. It appears that the vesicular glutamate uptake is driven by an electrochemical proton gradient generated by H^+-ATPase in the vesicle membrane, analogous to the active transport of other neurotransmitter substances into storage vesicles, which is now known to be a common form of energy, central not only to the biological energy transduction but also to the active transport of various substances and other biological processes. In addition to ATP, chloride ions in the nerve terminal may play an important role in the in vivo accumulation of glutamate in the synaptic vesicles. The mechanism of the intriguing chloride stimulation, as well as a detailed picture of the translocation of glutamate, remains to be elucidated.

The vesicular glutamate translocator is highly specific for L-glutamate, and, hence, proposed to play a crucial role in selecting the acidic amino acid for synaptic transmission. However, it is not known whether the synaptic vesicles with those properties described here are present exclusively in those nerve terminals considered to be glutamatergic. It is anticipated that the ATP-dependent vesicular uptake of glutamate will be demonstrated in various parts of the central nervous system, particularly those areas rich in glutamatergic projection, such as the hippocampus, caudate nucleus, and the cerebellar molecular layer. Whether glutamate taken up into the synaptic vesicle in an ATP-dependent manner is directly released into the synaptic cleft in a Ca^{2+}-dependent fashion upon depolarization also remains to be demonstrated. Whether the vesicular glutamate translocator described here represents a marker for the glutamatergic terminal is an interesting and important problem for future investigation. Should it turn out that the vesicular pool is not the immediate source of glutamate for synaptic release, the vesicular glutamate uptake system still could play a role in the storage function of the synaptic vesicle, protecting glutamate from further metabolism by cytosolic enzymes.

ACKNOWLEDGEMENTS

   Our original work described herein was supported in part by
NIH Grant NS15113, NSF Grant BNS8207999 and a Biomedical Research
Support Grant from the University of Michigan. I am thankful to
Mary Roth for excellent assistance in the preparation of the manu-
script.

REFERENCES

Abdul-Ghani, A.-S., Coutinho-Netto, J. and Bradford, H.F. (1981).
In Vivo Superfusion Methods and the Release of Glutamate. In Gluta-
mate: Transmitter in the Central Nervous System. (eds. P.J. Roberts,
J. Storm-Mathisen and G.A.R. Johnston). Wiley, Chichester.

Anderson, D.C., King, S.C. and Parsons, S.M. (1982). Proton Gradient Linkage to Active Uptake of [^{3}H]Acetylcholine by Torpedo Electric Organ Synaptic Vesicles. Biochemistry, 21, 3037-3043.

Angelides, K.J. (1980). Transport of Catecholamines by Native and
Reconstituted Rat Heart Synaptic Vesicles. J. Neurochem., 35, 949-
962.

Apps, D.K., Pryde, J.G. and Phillips, J.H. (1980). Both the Trans-
membrane pH Gradient and the Membrane Potential are Important in the
Accumulation of Amines by Resealed Chromaffin-Granule 'Ghosts'.
FEBS Lett., 111, 386-390.

Apps, D.K., Pryde, J.G., Sutton, R. and Phillips, J.H. (1980). Inhibition of adenosine triphosphatase, 5-hydroxytryptamine transport
and proton-translocation activities of resealed chromaffin-granule
'ghosts'. Biochem. J., 190, 273-282.

Bashford, C.L., Radda, G.K. and Ritchie, G.A. (1975). Energy-linked
Activities of the Chromaffin Granule Membrane. FEBS Lett., 50, 21-
24.

Bashford, C.L., Casey, R.P., Radda, G.K. and Ritchie, G.A. (1976).
Energy-coupling in Adrenal Chromaffin Granules. Neurosci., 1, 399-
412.

Bennett, J.P., Jr., Mulder, A.H. and Snyder, S.H. (1974). Neurochemical Correlates of Synaptically Active Amino Acids. Life Sci.,
15, 1045-1056.

Bloom, F.E., Ueda, T., Battenberg, E. and Greengard, P. (1979).
Immunocytochemical localization, in synapses, of Protein I, an endo-
genous substrate for protein kinases in mammalian brain. PNAS,
USA, 76, 5982-5986.

Bradford, H.F. and McIlwain, H. (1966). Ionic Basis for Depolarization of Cerebral Tissue by Excitatory Amino Acids. J. Neurochem., 13, 41-51.

Butcher, S.P., Collins, J.F. and Roberts, P.J. (1983). Characterization of the binding of DL-[^{3}H]-2-amino-4-phosphonobutyrate to L-glutamate-sensitive sites on rat brain synaptic membranes. Br. J. Pharmac., 80, 355-364.

Butcher, S.P., Roberts, P.J. and Collins, J.F. (1984). Ionic Regulation of the Binding of DL-[^{3}H]2-Amino-4-Phosphonobutyrate to L-Glutamate-Sensitive Sites on Rat Brain Membranes. J. Neurochem., 43, 1039-1045.

Campbell, G.LeM. and Shank, R.P. (1978). Glutamate and GABA uptake by cerebellar granule and glial cell enriched populations. Brain Res., 153, 618-622.

Carlsson, A., Hillarp, N.-Å. and Waldeck, B. (1962). A Mg^{++}-ATP Dependent Storage Mechanism in the Amine Granules of the Adrenal Medulla. Med. Exp., 6, 47-53.

Carty, S.E., Johnson, R.G. and Scarpa, A. (1981). Serotonin Transport in Isolated Platelet Granules. J. Biol. Chem., 256, 11244-11250.

Christensen, H.N. and Makowske, M. (1983). Recognition Chemistry of Anionic Amino Acids for Hepatocyte Transport and for Neurotransmittory Action Compared. Life Sci., 33, 2255-2267.

Cotman, C.W., Foster, A.C. and Lanthorn, T.H. (1981). An overview of glutamate as a neurotransmitter. In Glutamate as a Neurotransmitter. (eds. G. Di Chiara and G.L. Gessa). Raven Press, New York.

Crunelli, V., Forda, S. and Kelly, J.S. (1984). The Reversal Potential of Excitatory Amino Acid Action on Granule Cells of the Rat Dentate Gyrus. J. Physiol., 351, 327-342.

Crunelli, V., Forda, S. and Kelly, J.S. (1985). Excitatory amino acids in the hippocampus: synaptic physiology and pharmacology. TINS, 8, 26-30.

Curtis, D.R., Phillis, J.W. and Watkins, J.C. (1960). The Chemical Excitation of Spinal Neurones by Certain Acidic Amino Acids. J. Physiol., 150, 656-682.

Curtis, D.R. and Johnston, G.A.R. (1974). Amino Acid Transmitters in the Mammalian Central Nervous System. Ergeb. der Physiol., 69, 97-188.

De Belleroche, J.S. and Bradford, H.F. (1973). Amino Acids in Synaptic Vesicles from Mammalian Cerebral Cortex: A Reappraisal. J. Neurochem., 21, 441-451.

De Belleroche, J.S. and Bradford, H.F. (1977). On the Site of
Origin of Transmitter Amino Acids Released by Depolarization of
Nerve Terminals in Vitro. J. Neurochem., 29, 335-343.

De Camilli, P., Harris, S.M., Jr., Huttner, W.B. and Greengard, P.
(1983). Synapsin (Protein I), a nerve terminal-specific phosphoprotein. II. Its specific association with synaptic vesicles demonstrated by immunocytochemistry in agarose-embedded synaptosomes. J.
Cell Biol., 96, 1355-1373.

De Robertis, E., De Lores Arnaiz, G.R. and De Iraldi, A.P. (1962).
Isolation of Synaptic Vesicles from Nerve Endings of the Rat Brain.
Nature, 194, 794-795.

Disbrow, J.K., Gershten, M.J. and Ruth, J.A. (1982). Uptake of L-
[3H] Glutamic Acid by Crude and Purified Synaptic Vesicles from Rat
Brain. Biochem. Biophys. Res. Comm., 108, 1221-1227.

Fagg, G.E., Foster, A.C., Mena, E.E. and Cotman, C.W. (1982). Chloride and Calcium Ions Reveal a Pharmacologically Distinct Population
of L-Glutamate Binding Sites in Synaptic Membranes: Correspondence
between Biochemical and Electrophysiological Data. J. Neurosci., 7,
958-965.

Fonnum, F. (1984). Glutamate: a Neurotransmitter in Mammalian
Brain. J. Neurochem., 42, 1-11.

Gallo, V., Aloisi, F. and Levi, G. (1983). Muscimol Potentiation of
Acidic Amino Acid Release from Cerebellar Synaptosomes is Chloride
Dependent. J. Neurochem., 40, 939-945.

Gordon, R.D. and Balázs, R. (1983). Characterization of Separated
Cell Types from the Developing Rat Cerebellum: Transport of Glutamate and Aspartate by Preparations Enriched in Purkinje Cells, Granule Neurones, and Astrocytes. J. Neurochem., 40, 1090-1099.

Greenamyre, J.T., Young, A.B. and Penney, J.B. (1984). Quantitative
Autoradiographic Distribution of L-[3H]Glutamate Binding Sites in
Rat Central Nervous System. J. Neurosci., 4, 2133-2144.

Greenamyre, J.T., Olson, J.M.M., Penney, J.B., Jr., and Young, A.B.
(1985). Autoradiographic Characterization of N-Methyl-D-Aspartate-,
Quisqualate- and Kainate-Sensitive Glutamate Binding Sites. J.
Pharmacol. & Exp. Ther., in press.

Greengard, P. (1981). Intracellular Signals in the Brain. Harvey
Lect., 75, 277-331.

Hablitz, J.J. and Langmoen, I.A. (1982). Excitation of Hippocampal
Pyramidal Cells by Glutamate in the Guinea-Pig and Rat. J. Physiol.,
325, 317-331.

Holz, R.W. (1978). Evidence that Catecholamine Transport into Chromaffin Vesicles is Coupled to Vesicle Membrane Potential. PNAS, USA, 75, 5190-5194.

Johnson, J.L. (1972). Glutamic Acid as a Synaptic Transmitter in the Nervous Sytem. A Review. Brain Res., 37, 1-19.

Johnson, R.G., Carlson, N.J. and Scarpa, A. (1978). ΔpH and Catecholamine Distribution in Isolated Chromaffin Granules. J. Biol. Chem., 253, 1512-1521.

Johnson, R.G. and Scarpa, A. (1979). Protonmotive Force and Catecholamine Transport in Isolated Chromaffin Granules. J. Biol. Chem., 254, 3750-3760.

Johnson, R.G., Pfister, D., Carty, S.E. and Scarpa, A. (1979). Biological Amine Transport in Chromaffin Ghosts. J. Biol. Chem., 254, 10963-10972.

Johnson, R.G., Beers, M.F. and Scarpa, A. (1982). H^+ ATPase of Chromaffin Granules. J. Biol. Chem., 257, 10701-10707.

Johnson, R.G., Carty, S. and Scarpa, A. (1982). A Model of Biogenic Amine Accumulation into Chromaffin Granules and Ghosts Based on Coupling to the Electrochemical Proton Gradient. Fed. Proc., 41, 2746-2754.

Johnston, G.A.R. (1979). Central Nervous System Receptor for Glutamic Acid. In Glutamic Acid: Advances in Biochemistry and Physiology. (eds. L.J. Filer, Jr., S. Garattini, M.R. Kare, W.A. Reynolds and R.J. Wurtman). Raven Press, New York.

Kaback, H.R. (1982). Membrane Vesicles, Electrochemical Ion Gradients, and Active Transport. Curr. Topics in Membranes & Transport, 16, 393-404.

Kaback, H.R. (1983). The Lac Carrier Protein in Escherichia coli. J. of Membrane Biol., 76, 95-112.

Kanner, B.I. and Sharon, I. (1978). Active Transport of L-Glutamate by Membrane Vesicles Isolated from Rat Brain. Biochem., 17, 3949-3953.

Kanner, B.I., Sharon, I., Maron, R. and Schuldiner, S. (1980). Electrogenic Transport of Biogenic Amines in Chromaffin Granule Membrane Vesicles. FEBS Lett., 111, 83-86.

Kirshner, N. (1962). Uptake of Catecholamines by a Particulate Fraction of the Adrenal Medulla. J. Biol. Chem., 237, 2311-2317.

Kontro, P., Marnela, K.-M. and Oja, S.S. (1980). Free Amino Acids in the Synaptosome and Synaptic Vesicle Fractions of Different

Bovine Brain Areas. Brain Res., 184, 129–141.

Krnjević, K. and Phillis, J.W. (1963). Iontophoretic Studies of Neurones in the Mammalian Cerebral Cortex. J. Physiol., 165, 274–304.

Krnjević, K. (1974). Chemical Nature of Synaptic Transmission in Vertebrates. Physiol. Rev., 54, 418–540.

Kuhar, M.J. and Zarbin, M.A. (1978). Synaptosomal Transport: a Chloride Dependence for Choline, GABA, Glycine and Several Other Compounds. J. Neurochem., 31, 251–256.

Kuriyama, K., Roberts, E. and Kakefuda, T. (1968). Association of the γ-Aminobutyric Acid System with a Synaptic Vesicle Fraction from Mouse Brain. Brain Res., 8, 132–152.

Lähdesmäki, P., Karppinen, A., Saarni, H. and Winter, R. (1977). Amino Acids in the Synaptic Vesicle Fraction from Calf Brain: Content, Uptake and Metabolism. Brain Res., 138, 295–308.

Logan, W.J. and Snyder, S.H. (1972). High Affinity Uptake Systems for Glycine, Glutamic and Aspartic Acids in Synaptosomes of Rat Central Nervous Tissues. Brain Res., 42, 413–431.

Luqmani, Y.A. (1981). Nucleotide Uptake by Isolated Cholinergic Synaptic Vesicles: Evidence for a Carrier of Adenosine 5'-Triphosphate. Neurosci., 6, 1011–1021.

Mangan, J.L. and Whittaker, V.P. (1966). The Distribution of Free Amino Acids in Subcellular Fractions of Guinea-Pig Brain. Biochem. J., 98, 128–137.

Maron, R., Kanner, B.I. and Schuldiner, S. (1979). The Role of a Transmembrane pH Gradient in 5-Hydroxy Tryptamine Uptake by Synaptic Vesicles from Rat Brain. FEBS Lett., 98, 237–240.

Michaelis, E.K., Michaelis, M.L., Stormann, T.M., Chittenden, W.L. and Grubbs, R.D. (1983). Purification and Molecular Characterization of the Brain Synaptic Membrane Glutamate-Binding Protein. J. Neurochem., 40, 1742–1753.

Mitchell, P. (1961). Coupling of Phosphorylation to Electron and Hydrogen Transfer by a Chemi-osmotic Type of Mechanism. Nature, 191, 144–148.

Mitchell, P. (1966). Chemiosmotic Coupling in Oxidative and Photosynthetic Phosphorylation. Biol. Rev., 41, 445–502.

Monaghan, D.T., Holets, V.R., Toy, D.W. and Cotman, C.W. (1983). Anatomical Distributions of Four Pharmacologically Distinct ^{3}H-L-Glutamate Binding Sites. Nature, 306, 176–179.

Naito, S. and Ueda, T. (1981). Affinity-purified Anti-Protein I Antibody. J. Biol. Chem., 256, 10657-10663.

Naito, S. and Ueda, T. (1982). ATP-dependent Glutamate Uptake into Protein I-associated Synaptic Vesicles. Soc. for Neurosci. Abst., 8, 878.

Naito, S. and Ueda, T. (1983). Adenosine Triphosphate-dependent Uptake of Glutamate into Protein I-associated Synaptic Vesicles. J. Biol. Chem., 258, 696-699.

Naito, S. and Ueda, T. (1985). Characterization of Glutamate Uptake into Synaptic Vesicles. J. Neurochem., 44, 99-109.

Nestler, E.J. and Greengard, P. (1983). Protein Phosphorylation in the Brain. Nature, 305, 583-588.

Nicoll, R.A. and Alger, B.E. (1981). Synaptic Excitation May Activate a Calcium-Dependent Potassium Conductance in Hippocampal Pyramidal Cells. Science, 212, 957-959.

Njus, D. and Radda, G.K. (1979). A Potassium Ion Diffusion Potential Causes Adrenaline Uptake in Chromaffin-Granule 'Ghosts'. Biochem. J., 180, 579-585.

Njus, D., Knoth, J. and Zallakian, M. (1981). Proton-Linked Transport in Chromaffin Granules. Curr. Topics in Bioenerg., 11, 107-147.

Nowak, L., Bregestovski, P., Ascher, P., Herbet, A., and Prochiantz, A. (1984). Magnesium Gates Glutamate-Activated Channels in Mouse Central Neurones. Nature, 307, 462-465.

Olverman, H.J., Jones, A.W. and Watkins, J.C. (1984). L-Glutamate Has Higher Affinity Than Other Amino Acids for [^{3}H]-D-AP5 Binding Sites in Rat Brain Membranes. Nature, 307, 460-462.

Puil, E. (1981). S-Glutamate: Its Interactions with Spinal Neurons. Brain Res. Rev., 3, 229-322.

Rassin, D.K. (1972). Amino Acids as Putative Transmitters: Failure to Bind to Synaptic Vesicles of Guinea Pig Cerebral Cortex. J. Neurochem., 19, 139-148.

Rebois, R.V., Reynolds, E.E., Toll, L. and Howard, B.D. (1980). Storage of Dopamine and Acetylcholine in Granules of PC12, a Clonal Pheochromocytoma Cell Line. Biochem., 19, 1240-1248.

Roberts, P.J. and Watkins, J.C. (1975). Sturctural Requirements for the Inhibition for L-Glutamate Uptake by Glia and Nerve Endings. Brain Res., 85, 120-125.

Roberts, P.J. (1981). Binding Studies for the Investigation of Receptors for L-Glutamate and Other Excitatory Amino Acids. In Glutamate: Transmitter in the Central Nervous System. (eds. P.J. Roberts, J. Storm-Mathisen and G.A.R. Johnston). Wiley, Chichester.

Rudnick, G., Fishkes, H., Nelson, P.J. and Schuldiner, S. (1980). Evidence for Two Distinct Serotonin Transport Systems in Platelets. J. Biol. Chem., 255, 3638-3641.

Scherman, D. and Henry, J.-P. (1980). Role of the Proton Electrochemical Gradient in Monoamine Transport by Bovine Chromaffin Granules. Biochim. et Biophys. Acta, 601, 664-677.

Schousboe, A. (1981). Transport and Metabolism of Glutamate and GABA in Neurons and Glial Cells. Int. Rev. of Neurobiol., 22, 1-45.

Schousboe, A. and Hertz, L. (1981). Role of Astroglial Cells in Glutamate Homeostasis. In Glutamate as a Neurotransmitter. (eds. G. Di Chiara and G.L. Gessa). Raven Press, New York.

Schuldiner, S., Fishkes, H. and Kanner, B.I. (1978). Role of a Transmembrane pH Gradient in Epinephrine Transport by Chromaffin Granule Membrane Vesicles. PNAS, USA, 75, 3713-3716.

Shank, R.P. and Campbell, G.LeM. (1983). Glutamate. In Handbook of Neurochemistry, Vol. 3 (2nd Ed.). (ed. A. Lajtha). Plenum Press, New York.

Shioi, J. and Ueda, T. (1985). Involvement of Protonmotive Force in ATP-Dependent Uptake of L-Glutamate into Synaptic Vesicles. Fed. Proc., 44, 1388.

Slaughter, M.M. and Miller, R.F. (1981). 2-Amino-4-Phosphonobutyric Acid: A New Pharmacological Tool for Retina Research. Science, 211, 182-185.

Slaughter, M.M. and Miller, R.F. (1985). Identification of a Distinct Synaptic Glutamate Receptor on Horizontal Cells in Mudpuppy Retina. Nature, 314, 96-97.

Stadler, H. and Tashiro, T. (1979). Isolation of Synaptosomal Plasma Membranes from Cholinergic Nerve Terminals and a Comparison of Their Proteins with Those of Synaptic Vesicles. Europ. J. Biochem., 101, 171-178.

Stormann, T.M., Chang, H.H., Johe, K. and Michaelis, E.K. (1984). Functional Reconstitution of the Synaptic Membrane Glutamate-Binding Protein: Glutamate Receptor-like Activity in Liposomes. Soc. for Neurosci. Abst., 10, 958.

Storm-Mathisen, J., Leknes, A.K., Bore, A.T., Vaaland, J.L., Edminson, P., Haug, F.-M.S. and Ottersen, O.P. (1983). First Visualiza-

tion of Glutamate and GABA in Neurones by Immunocytochemistry.
Nature, 301, 517–520.

Toll, L. and Howard, B.D. (1978). Role of Mg^{2+}-ATPase and a pH
Gradient in the Storage of Catecholamines in Synaptic Vesicles.
Biochemistry, 17, 2517–2523.

Toll, L. and Howard, B.D. (1980). Evidence that an ATPase and a
Protonmotive Force Function in the Transport of Acetylcholine into
Storage Vesicles. J. Biol. Chem., 255, 1787–1789.

Ueda, T. and Greengard, P. (1977). Adenosine 3':5'-Monophosphate-
regulated Phosphoprotein System of Neuronal Membranes. I. Solubilization, Purification, and Some Properties of an Endogenous Phosphoprotein. J. Biol. Chem., 252, 5155–5163.

Ueda, T., Greengard, P., Berzins, K., Cohen, R.S., Blomberg, F.,
Grab, D.J. and Siekevitz, P. (1979). Subcellular Distribution in
Cerebral Cortex of Two Proteins Phosphorylated by a cAMP-dependent
Protein Kinase. J. Cell Biol., 83, 308–319.

Ueda, T. and Naito, S. (1982). Specific Inhibition of the Phosphorylation of Protein I, a Synaptic Protein, by Affinity-purified Anti-
Protein I Antibody. Prog. Brain Res., 56, 87–103.

Ueda, T. (1984). ATP-dependent Uptake of Glutamate into Synaptic
Vesicles. In Transmembrane Signaling and Sensation. (eds. F. Oosa-
wa, T. Yoshioka and H. Hayashi). Japan Scientific Soc. Press,
Tokyo.

Wagner, J.A. and Kelly, R.B. (1979). Topological Organization of
Proteins in an Intracellular Secretory Organelle: The Synaptic
Vesicle. PNAS, USA, 76, 4126–4130.

Waniewski, R.A. and Martin, D.L.(1984). Characterization of L–Glutamic Acid Transport by Glioma Cells in Culture: Evidence for Sodium-
independent, Chloride-dependent High Affinity Influx. J. Neurosci.,
4, 2237–2246.

Watkins, J.C. and Evans, R.H. (1981). Excitatory Amino Acid Transmitters. Annu. Rev. Pharmacol. Toxicol., 21, 165–204.

Whittaker, V.P., Michaelson, I.A. and Kirkland, R.J.A. (1964). The
Separation of Synaptic Vesicles from Nerve-Ending Particles ('Synaptosomes'). Biochem. J., 90, 293–303.

Wilkins, J.A. and Salganicoff, L. (1981). Participation of a Transmembrane Proton Gradient in 5'-hydroxytryptamine Transport by Platelet Dense Granules and Dense-Granule Ghosts. Biochem. J., 198, 113–
123.

Winkler, H. and Westhead, E. (1980). The Molecular Organization of

Adrenal Chromaffin Granules. Neurosci., 5, 1803–1823.

Woodbury, J.W. (1965). The Cell Membrane: Ionic and Potential Gradients and Active Transport. In Physiology and Biophysics. (eds. T.C. Ruch and H.D. Patton). Saunders, Philadelphia.

Zisapel, N. and Zurgil, N. (1979). Studies on Synaptic Vesicles in Mammalian Brain Characterization of Highly Purified Synaptic Vesicles from Bovine Cerebral Cortex. Brain Res., 178, 297–310.

13

Presynaptic Receptors for Excitatory Amino Acid Transmitters

P.J. Roberts

INTRODUCTION

Within recent years, considerable attention has
been paid to the role of postsynaptic receptors for
excitatory amino acids within the mammalian central
nervous system, and this has led to the characterisation
of at least three subtypes (viz. those activated by
N-methyl-D-aspartate (NMDA), quisqualate and kainate)
on the basis of electrophysiological studies (see
Watkins, this volume). In contrast to the postsynaptic
receptors, relatively little attention has been paid to
either the existence (although of course their possible
presence might contribute substantially to the binding
seen on synaptic plasma membranes) or characteristics
of presynaptic receptors for excitatory amino acids.

Broadly speaking, presynaptic receptors may be
considered to be of one of two types: (i) autoreceptors,
which are localised on the presynaptic terminal and
which may be activated by the transmitter, eg glutamate,
released from that terminal, and which results in a
feed-back inhibition of transmitter release. (ii)
heteroreceptors, which, as the name suggests, are
localised on neuronal terminals of a different class;
for example a glutamate neurone might impinge upon an
axon terminal which releases dopamine as its
transmitter. Depending upon the ionic processes
involved, the activation of heteroreceptors might
produce either inhibition or enhancement of transmitter
release. Having said this however, it is worth bearing
in mind that the question of a functional role for the
experimental observations on effects on transmitter
release, is far from settled. Indeed one eminent
researcher has stated that 'neurotransmitter release
does not seem to be regulated by neuronal receptors

mediating feedback and that the mechanism of action of
presynaptically active agents is still uncertain'
(Kalsner, 1985). Supplementary to this, it has been
proposed that the demonstration of axonal transport of
presynaptic heteroreceptors (but apparently not of
autoreceptors) indicates that interneuronal modulatory
processes may indeed have a physiological role (Laduron,
1985).

AUTOREGULATION OF EXCITATORY AMINO ACID RELEASE

In our initial studies we investigated the release
of pre-accumulated D-[^{3}H] aspartate in response to K^+, or
protoveratrine-depolarisation of slices prepared from a
number of regions of rat brain. In cerebral cortex,
striatum and cerebellum, the simultaneous inclusion of
L-glutamate (0.1 mM) in the superfusion medium failed to
influence the release of D-aspartate. In contrast
however, release of label from hippocampal slices was
markedly inhibited in the presence of L-glutamate (Fig
1), (McBean and Roberts, 1981). Tetrodotoxin resistance
indicated that voltage-dependent Na^+ channels were not
involved and that the effect was most likely occurring
through direct interaction with presynaptic receptors,
rather than via an interneuronal mechanism.

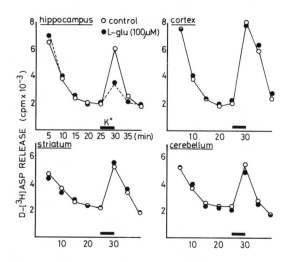

Figure 1.   Release of the false transmitter, D-[^{3}H] –
aspartate from superfused rat brain slices. 0.2 x 0.2 mm
prisms were loaded with D-[^{3}H]aspartate and superfused
at 1 ml min^{-1}. Data represent Ca^{2+}-dependent release in
presence of K^+ and effects of L-glu (0.1mM).

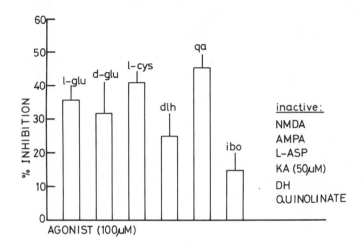

Figure 2. Effects of excitatory amino acid agonists on
the release of D-[³H]aspartate from hippocampal slices.
Preloaded, superfused slices were depolarised with 50mM
K⁺ in the presence or absence of agonist (0.1 mM).
Results are means of at least 6 independent observations.
dlh = DL-homocysteate; qa = quisqualate; ibo = ibotenate;
AMPA = RS-amino-3-hydroxy-5-methyl-4-isoxazolepropionate;
KA = kainate.

    In the presence of a moderately high (0.1 mM)
concentration of L-glutamate, the K⁺-evoked release of
D-aspartate was reduced by approximately 35%. The
D-isomer was also active, as were the compounds
quisqualate, the sulphur-containing amino acid
L-cysteate, and to a lesser extent, DL-homocysteate. The
mixed (though NMDA-preferring) agonist ibotenate, was
only weakly active in mimicking the inhibition observed
by L-glutamate, while kainate, and NMDA-selective
agonists such as quinolinate and NMDA itself, were devoid
of inhibitory activity (Fig. 2). None of the substances
tested at this concentration were able to influence the
spontaneous (in the absence of K⁺) release of D-[³H]-
aspartate. Enhancement of release by substrates for the
transporter was not observed, indicating that
homoexchange did not occur under these conditions. The
data therefore suggest that at least pharmacologically,
the release of D-aspartate (which is basically a
non-metabolizable analogue of glu/asp) from hippocamapal
slices, may be inhibited by excitatory amino acid
agonists able to interact with the quisqualate type of

receptor. However, it would seem that this site differs
from that occurring post-synaptically, since AMPA, which
is a selective quisqualate agonist when studied
electrophysiologically (Krogsgaard-Larsen et al., 1980)
was inactive in inhibiting D-aspartate release. When
comparing other studies, it is interesting to note that
kainate has been demonstrated to produce a marked
stimulation of excitatory amino acid transmitter release
(Ferkany et al., 1982); however, this occurs at high
micromolar to millimolar concentrations.

There are now available a number of reasonably
potent and in some cases, selective inhibitors of
excitatory amino acid receptors (Evans and Watkins,
1981). In one series of experiments, the susceptibility
of the effects of L-glutamate to pharmacological
antagonism was investigated (Fig. 3). In the presence of
the (albeit weak) quisqualate antagonist L-glutamate
diethylester, the inhibitory effect of L-glutamate on
D-[^{3}H]aspartate was reversed. ($\pm$)2-amino-4-phosphono-
butyrate was weakly active, while 2-amino-7-phosphono-
heptanoate (AP7; APH) and D-alpha-aminosuberate, which
are selective for the postsynaptic NMDA receptor were
inactive.

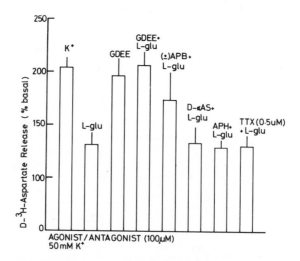

Figure 3. Effects of amino acid antagonists on the
K$^+$-stimulated release of D-[^{3}H]aspartate in the presence
of 0.1 mM L-glutamate. Release is expressed as a
percentage of basal (unstimulated) efflux. Results means
$\pm$ S.E.M. of at least 5 independent observations. GDEE,
L-glutamate diethylester; APB, 2-amino-4-phosphono-
butyrate; D-AS, D-alpha-aminosuberate; APH, 2-amino-7-
phosphonoheptanoate; TTX, tetrodotoxin.

The antagonist 2-amino-5-phosphonovalerate (AP5; APV) exhibited somewhat anomalous properties (not shown). Alone, it had a marked glutamate-like action, suggesting that it may act as a partial agonist at these receptors.

Release of endogenous glutamate. Although D-aspartate is a valuable tool for studying excitatory amino acid uptake and release because in contrast to L-glutamate or L-aspartate, it is essentially not metabolised in nervous tissue (Johnston, 1981),its use suffers from several disadvantages. Firstly, it will not discriminate between presumed glutamatergic and aspartergic terminals, and secondly, it is far from certain that D-aspartate mixes homogeneously with the transmitter pool of endogenous amino acid. Thus its release may be only a rough reflection of the actual processes occurring. It is therefore important when one is considering regulation of transmitter release, to attempt to study the release of the endogenous transmitter. For the hippocampus, there is ample evidence that L-glutamate is the major excitatory transmitter (Schwartzkroin, 1975; Segal, 1981; Storm-Mathisen and Wold, 1981), with little evidence for a role for L-aspartate. Interestingly, in the olfactory cortex, aspartate appears to be involved, and here it has been demonstrated that its release is decreased rather selectively by NMDA (Collins et al., 1983).

The release of endogenous glutamate from hippo-campal slices was studied under conditions identical to those employed for D-[^{3}H]aspartate.

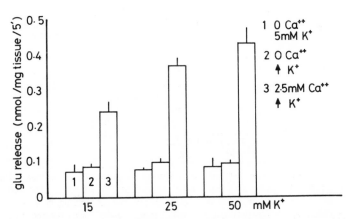

Figure 4. Release of endogenous glutamate from hippo-campal slices: calcium dependence, and effects of increasing concentrations of potassium ions.

In the absence of Ca^{2+}, 15-50 mM K^+ was unable to increase the release of L-glutamate into the superfusion medium above that observed in normal K^+ medium (5 mM) in the absence of calcium ions. The addition of 2.5 mM Ca^{2+} however, resulted in an immediate and robust release. It was clearly and reproducibly observed at the lowest potassium concentration employed, and exhibited a dose dependency up to 50 mM.

Effects of excitatory amino acid agonists. As in the experiments with D-[^{3}H]aspartate, slices were depolarised by the addition of 50 mM K^+; effects of excitatory amino acid agonists were investigated at a single concentration of 0.1 mM, on both the spontaneous and evoked release of glutamate. Since endogenously-released glutamate was being assayed, it was clearly not readily feasible to examine the action of L-glutamate itself on release.

None of the agonists influenced the control release of L-glutamate (elevated K^+, zero calcium). However, upon depolarisation of the slices a similar pattern to that observed for the release of D-aspartate emerged (Figs. 5 and 6)

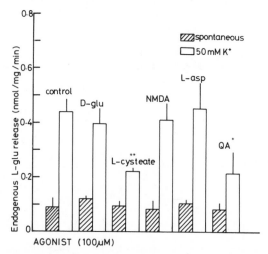

Figure 5.   Spontaneous and evoked release of L-glutamate from hippocampal slices (determined by the method of Graham and Aprison, 1966). Superfusion samples were collected over 5 min periods, and were exposed to both elevated K^+ and excitatory amino acid analogue during that period. Results are means ± S.E.M. of at least 6 independent determinations. * P<0.05; **P<0.01.

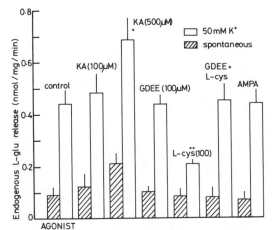

Figure 6.   Effects of agonists and antagonists on the
release of L-glutamate. Experimental details as in
legend to Fig. 5.

        Of the selective postsynaptic receptor agonists,
only quisqualate produced a significant (approx. 50%)
inhibition of the calcium-dependent evoked release of
endogenous glutamate. This was duplicated by L-cysteate,
which is generally considered to be a mixed agonist
which exhibits rather complex binding characteristics
(Recasens et al., 1983). Once again, neither NMDA nor
kainate produced an inhibitory effect; indeed, kainate
at 0.5 mM or above, produced an enhancement of release,
which is in accord with the findings of Ferkany et al.
(1982) where enhanced release was observed in slices of
mouse cerebellum, hippocampus and striatum through a
tetrodotoxin-resistant mechanism. L-glutamate
diethylester (GDEE) reversed the effect of L-cysteate
and quisqualate (Fig. 6). AMPA, apparently was not able
to mimic the effects of quisqualate.

DISCUSSION

        Although L-glutamate exhibits well-described pre-
synaptic regulatory actions, particularly glutamate-
dopamine (Giorguieff et al., 1977; Roberts and Anderson,
1979; Marien et al., 1983; Godukhin et al., 1984), where
glutamate receptors appear to be localised on
nigro-striatal afferent dopaminergic terminals (Roberts
et al., 1982), and excitatory amino acid - acetylcholine
(Lehmann and Scatton, 1982) interactions in the nervous
system in vitro, little attention appears to have been
directed towards the possible existence and physiolog-

ical function of presynaptic autoreceptors which might
control transmitter release via a negative feed-back
process. Because of the complexity of the organisation
of the central nervous system, the existence of such
physiological control mechanisms is not easy to
demonstrate. However, for noradrenaline at least, there
is strong suggestive evidence for such a role (Starke,
1979), although the mechanisms by which presynaptic
regulation occurs are poorly understood.

The data that we have obtained for presynaptic
excitatory amino acid autoreceptors must still be
considered as "suggestive" in light of current
controversy concerning the concept of autoreceptors.
Kalsner (1985) has stated five distinct criteria that
should be met in order to establish feed-back
regulation. Probably none of these have been fully
satisfied for any central transmitter. One of the chief
experimental difficulties is that of establishing
appropriate stimulation parameters to reflect the in
situ situation. However, in hippocampus, in vitro, it is
clearly possible to demonstrate an inhibition of
calcium-dependent evoked release of both the artificial
substrate D-aspartate, and endogenous L-glutamate.
Despite its ubiquity as an excitatory transmitter in
brain, the occurrence of "autoreceptors" appears to be
remarkably restricted, possibly to the hippocampus,
olfactory cortex and retina (Roberts and Lopez-Colome,
in preparation). Skerritt et al (1981) similarly have
reported failure to detect a feed-back inhibitory
mechanism in rat cortical minislices.

The receptors involved in regulation of
L-glutamate release from hippocampal slices are clearly
not identical to those characterised postsynaptically.
They are wholly resistant to activation by NMDA and its
analogues, or by kainate. Quisqualate, L-cysteate (and
cysteine sulphinate) and L-glutamate itself (when
examined versus D-aspartate release) were able to
inhibit release to an extent of around 50% maximum.

The existence of autoreceptors for excitatory
amino acids clearly has potential implications for
therapeutics. A number of disorders including epilepsy,
ischaemia and neurodegenerative diseases may be
associated with increased excitation, possibly related
directly to elevated extracellular concentrations of
these substances. Selective agents which would reduce
excitatory amino acid transmitter release from
presynaptic terminals through a pharmacological action
on these regulatory receptors, might be of considerable
potential value in reducing levels of central

excitation.

This work was supported by grants from The Wellcome Trust and the SERC.

REFERENCES

Collins, G.G.S., Anson, J. and Surtees, L. (1983). Pre-synaptic kainate and N-methyl-D-aspartate receptors regulate excitatory amino acid release in the olfactory cortex. Brain Res., 265, 157-159.

Evans, R.H. and Watkins, J.C. (1981). Pharmacological antagonists of excitatory amino acids. Life Sci., 28, 1303-1308.

Ferkany, J.W., Zaczek, R. and Coyle, J.T. (1982). Kainic acid stimulates excitatory amino acid transmitter release at presynaptic receptors. Nature, 298, 757-759.

Giorguieff, M.F., Kemel, M.L. and Glowinski, J. (1977). Presynaptic effect of L-glutamic acid on the release of dopamine in rat striatal slices. Neurosci. Lett. 6, 73-77.

Godukhin, O.V., Zharikova, A.D. and Budantsev, A.Yu. (1984). Role of presynaptic dopamine receptors in regulation of the glutamatergic neurotransmission in rat neostriatum. Neuroscience, 12, 377-383.

Graham, L.T. Jr., and Aprison, M.H. (1966). Fluorimetric determination of aspartate, glutamate and GABA in nerve tisues using enzymatic methods. Analyt. Biochem., 15, 487-497.

Johnston, G.A.R. (1981). Glutamate uptake and its possible role in neurotransmitter inactivation. In Glutamate: Transmitter in the Central Nervous System. (eds. P.J. Roberts, J. Storm-Mathisen and G.A.R. Johnston). Wiley, Chichester.

Kalsner, S. (1985). Is there feedback regulation of neurotransmitter release by autoreceptors. Biochem. Pharmacol., 34, 4085-4097.

Krogsgaard-Larsen, P., Honore, T., Hansen, J.J., Curtis, D.R. and Lodge, D. (1980). New class of glutamate agonists structurally related to ibotenic acid. Nature, 284, 64-66.

Laduron, P.M. (1985). Presynaptic heteroreceptors in

regulation of neuronal transmission. Biochem.
Pharmacol., 34, 467-470.

Lehmann, J. and Scatton, B. (1982). Characterisation of
the excitatory amino acid-receptor mediated release of
[³H]acetylcholine from rat striatal slices. Brain Res.,
252, 77-89.

Marien, M., Brien, J. and Jhamandas, K. (1983).
Regional release of [³H]dopamine from rat brain in
vitro: effects of opioids on release induced by
potassium, nicotine and L-glutamic acid. Can. J.
Physiol. Pharmacol., 61, 43-60.

McBean, G.J. and Roberts, P.J. (1981). Glutamate
preferring receptors regulate release of D-aspartate
from rat hippocampal slices. Nature, 291, 593-594.

Recasens, M., Saadoun, F., Maitre, M., Baudry, M.,
Lynch, G., Mandel, P. and Vincendon, G. (1983).
Cysteine sulphinate receptors in rat CNS. In CNS
Receptors - From Molecular Pharmacology to Behaviour
(eds. P. Mandel and F.V. DeFeudis), Raven Press, New
York.

Roberts, P.J. and Anderson, S.D. (1979). Stimulatory
effect of L-glutamate and related amino acids on [³H]
dopamine release from rat striatum: an in vitro model
for glutamate actions. J. Neurochem., 32, 1539-1545.

Roberts, P.J., McBean, G.J., Sharif, N.A. and Thomas,
E.M. (1982). Striatal glutamatergic function:
modifications following specific lesions. Brain Res.,
235, 83-91.

Schwartzkroin, P.A. (1975). Characteristics of CA1
neurones recorded intracellularly in the hippocampal in
vitro slice preparation. Brain Res., 85, 423-426.

Segal, M. (1981). The actions of glutamic acid
on neurones in the rat hippocampal slice. In Glutamate
as a Neurotransmitter. (eds. G. Di Chiara and G.L.
Gessa). Raven Press, New York.

Skerritt, J.H., Johnston, G.A.R. and Willow, M. (1981).
Amino acid release from rat brain slices: evidence for
autoreceptors for GABA but not for D-aspartate. Proc.
Aus. Physiol. Pharmacol. Soc., 12, 131P.

Starke, K. (1979). Presynaptic regulation of release in
the central nervous system. In The Release of catecholamines from Adrenergic Neurones (ed. D.M. Paton)

Pergamon Press, Oxford.   pp 143-183.

Storm-Mathisen, J. and Wold, J.E. (1981).   In vivo
uptake and axonal transport of D-aspartate in excitatory
neurones.   Brain Res., 230, 427-433.

14

Effects of Kainic Acid on Different Populations of Cerebellar Interneurones and Astrocytes Cultured *in vitro*

V. Gallo and G. Levi

The potent toxic action of kainic acid has been initially attributed to its excitatory effects on neuronal cells, but numerous following studies showed that the mechanism of kainate toxicity is more complex (Coyle, 1983). Several hypotheses have been formulated in order to explain the mechanism of action of kainate, since the discovery of specific receptors for this excitotoxin in the central nervous system (London and Coyle, 1979; Monaghan and Cotman, 1982). Some authors have evidenced that the presence of the glutamatergic afferent fibers is necessary for kainate to be effective in some brain areas (Biziere and Coyle, 1978, 1979; Panula, 1980),suggesting a cooperative interaction between glutamate and injected kainate to cause neuronal degeneration. On the other hand, other reports suggest that kainic acid neurotoxic effects can be manifested also in the absence of the glutamatergic input (Mc Lennan, 1980). More recently, kainate has been shown to affect neurotransmitter release and uptake in different tissue preparations of several brain areas. For example, the excitotoxin caused glutamate and aspartate release from brain synaptosomes (Pastuszko et al., 1984), cerebellar, hippocampal and striatal slices (Ferkany et al.,1982) and also stimulated GABA release from isolated, perfused chick retina (Tapia and Arias, 1982). Kainic acid inhibited high affinity uptake of glutamate in brain and striatal slices (Potashner and Gerard, 1983) and in brain synaptosomal preparations (Pastuszko et al., 1984). These studies would be suggestive of a presynaptic action of kainic acid and some authors have recently postulated the presence of kainate presynaptic receptors localized on nerve cell terminals releasing glutamate and aspartate (Ferkany et al., 1982; Ferkany and Coyle, 1983).

In studies <u>in vivo</u> or using heterogeneous preparations such as brain slices, it is difficult to discriminate between direct and indirect effects of kainate, to study the primary site of its action and to understand why certain classes of neurones appear to be more sensitive than others to kainic acid. Furthermore, the possible involvement of non neuronal cells in the anatomical and biochemical alterations caused by kainate can not be easily assessed using such preparations. This type of problems has been approached by the use of primary cultures of neural cells which can be maintained in vitro for several days. In these preparations neuronal and non neuronal cells can be distinguished morphologically and with immunological techniques (Giacobini et al., 1980; Raff et al., 1979); furthermore different neuronal subtypes can be identified either morphologically or on the basis of the presence of neurotransmitter-synthesizing enzymes. An obvious advantage of the use of cultured neural cells is that molecules possessing a known physiological effect can be tested in vitro in reproducible conditions and at controlled concentrations.

The cytotoxicity of kainic acid has been tested in several systems of cultured dissociated cells (Messer and Maskin, 1980) or "microexplants" (Seil and Woodward, 1980; Whetsell et al., 1979; Gibbs et al., 1982) of nervous tissue. These studies confirmed that various neuronal subtypes are differently vulnerable to kainic acid (Seil and Woodward, 1980; Gibbs et al., 1982). Postnatal rat cerebellar cultures offer several advantages when used as a model in vitro for studying the mechanism of action of kainate. In fact they comprise: i) GABAergic inhibitory interneurones, which are vulnerable to kainic acid (Messer and Maskin, 1980); ii) excitatory interneurones, granule cells, scarcely sensitive to kainate and releasing their neurotransmitter glutamate (Gallo et al., 1982) and iii) astroglial cells, which can reveal whether the toxic effect of kainate is specifically directed towards neurones.

CHARACTERISTICS OF THE CELLS PRESENT IN CEREBELLAR PRIMARY CULTURES

Granule cells

Cerebellar cultures obtained from 8-day-old rats consist essentially of granule cells, when cytosine arabinofuranoside, a mitotic inhibitor, is added to the culture medium (Thangnipon et al., 1983; Kingsbury et al., 1985). The small excitatory interneurones represent more than 95% of the total cells present in the cultures (Kingsbury et al., 1985; Levi et al.,1984) and differentiate morphologically and biochemically in vitro during the 2 week cultivation time (Kingsbury et al., 1985; Gallo et al., 1982; Levi et al.,

1984; Webb et al., 1985). Granule cells start to emit fibers, positive for tetanus toxin, a few hours after plating on poly-L-lysine
(Thangnipon et al., 1983). During maturation of the cells in culture
fibers tend to fasciculate into thick bundles and cells migrate to
form aggregates (Thangnipon et al., 1983; Kingsbury et al., 1985).
Electron microscopic observations of the fibers have shown that several presynaptic-like structures are distributed along them. These
varicosities are filled with synaptic vesicles, but only very rarely they form a "classic" synaptic structure with synaptic cleft and
synaptic thickening (Kingsbury et al., 1985).

Cultured granule cells express their neurotransmitter-associa-
ted properties and progressively acquire the capacity of releasing
glutamate (endogenous and neosynthesized from the precursor glutamine) in response to a depolarizing stimulus. Endogenous and newly
synthesized glutamate release evoked by high K^+ was small and not affected by extracellular Ca^{2+} at 2 days in vitro (DIV), but it steadly
increased after this age and became clearly Ca^{2+}-dependent (up to
85%) at later stages (4, 8 and 12 DIV) (Gallo et al., 1982). During
the same culture period veratridine became able to evoke glutamate
release in a tetrodotoxin-sensitive manner (Gallo et al., 1982).
These observations not only indicate that cerebellar granule cells
differentiate in culture, but also strongly support the concept that
glutamate is the neurotransmitter of the small granular neurones of
the cerebellar cortex. The glutamate analog ^{3}H-D-aspartate is accumulated by cultured differentiated granule cells and their processes as
shown by light-microscopy autoradiography (Levi et al., 1984; see also Fig.5). In the same cultures Ca^{2+}-dependent, high K^+-induced release and tetrodotoxin-sensitive, veratridine-induced release was
detectable (Levi et al., 1984). Depolarizing stimuli did not evoke
glutamate release from astrocyte-enriched cultures, comprising 80-
90% of GFAP (glial fibrillary acidic protein)- positive cells, although glutamate was present and synthesized from the precursor glutamine in these cells (Gallo et al., 1982).

Together with the maturation of the neurotransmitter release machinery in the cultured granule cells we also studied immunocytochemically the expression of a neurone-specific phosphorylated protein,
Synapsin I, which has been shown to be associated with the cytoplasmic surface of synaptic vesicles (De Camilli et al., 1983a; De Camilli et al., 1983b) and which might be related to the neurotransmitter release process. Interestingly, Synapsin I was poorly expressed
in granule cells at stages in culture at which glutamate release was
still limited, but was present in large amounts at later stages

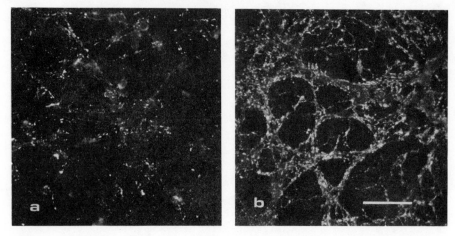

Fig. 1. Cerebellar neurone-enriched cultures at 3(a) and 9(b) DIV
stained with anti-Synapsin I antibody. Cultures were fixed with 2%
paraformaldehyde, permeabilized with acetone and then stained with
anti-Synapsyn I antibodies and rhodamine-conjugated goat anti-rabbit
IgG. Scale bar = 75 μm.

of maturation in vitro (Fig. 1).

Inhibitory interneurones

Another population of interneurones present in the cerebellar
cultures has been identified and well characterized (Currie and
Dutton, 1980; Aloisi et al., 1985). These cells, presumably stellate and basket cells, account for about 2% of the total cells
growing in the cultures and can be identified by indirect immunofluorescence using antibodies raised against the GABA-synthesizing
enzyme glutamate decarboxylase, often used as a marker for GABAergic neurones (Aloisi et al., 1985). Cells cultured for 5 days or
longer appeared morphologically differentiated with highly branched
processes, characterized by the presence of numerous varicosities
along them (Aloisi et al., 1985). At 5 DIV or later GABAergic interneurones and their long processes could be visualized also by
^{3}H-GABA autoradiography (see Aloisi et al., 1985 also for other
references). A large Ca^{2+}-dependent release of ^{3}H-GABA could be
evoked by high K^+ and a tetrodotoxin-sensitive, veratridine stimulated release was also demonstrated (Aloisi et al., 1985).

Astroglial cells

In cerebellar interneurone-enriched cultures astroglial cell proliferation was prevented by the addition of cytosine arabinofuranoside to the culture medium. In these conditions astroglial cells, evidenced by indirect immunofluorescence with anti-GFAP antibodies, accounted for about 5% of the total cells at 2 DIV, but their number decreased to 0.5% at 12 DIV (Levi et al., 1983). Two different types of GFAP-positive cells could be distinguished in these cultures: one with a stellate morphology, the other polygonal or elongated in its shape (Wilkin et al., 1983). In 12 DIV neurone-enriched cultures stellate astrocytes accounted for 70-80% of the astroglial cell population present (Levi et al., 1983). Interestingly, while the L-glutamate analog ^{3}H-D-aspartate was avidly taken up by both types of astrocytes, as evidenced by autoradiography, ^{3}H-GABA was substantially accumulated only by the stellate GFAP-positive cells (Wilkin et al., 1983; Johnstone et al., submitted). ^{3}H-GABA transport into these cells was inhibited by GABA analogs which have been reported to interfere selectively with the neuronal uptake of the amino acid, such as ACHC and DABA, but was scarcely influenced by β-alanine, a molecule considered to be specific for the glial cell transport system (Levi et al., 1983; Johnstone et al., submitted). In astrocyte-enriched cultures, in which no antimitotic was present, the stellate cell population capable of ^{3}H-GABA uptake progressively disappeared, being overwhelmed by the more rapidly proliferating polygonal astrocytes (Wilkin et al., 1983).

KAINIC ACID EFFECT ON ^{3}H-D-ASPARTATE AND ^{3}H-GABA UPTAKE AND RELEASE IN NEURONE-ENRICHED CULTURES

We have previously shown (Levi et al., 1984) that cultured cerebellar granule cells acquire the property of accumulating and releasing ^{3}H-D-aspartate once they have reached an advanced degree of morphological and biochemical differentiation, after 5 DIV. Therefore, we tested the effect of kainic acid on ^{3}H-D-asparate release and uptake in 10-11 DIV cultures. When kainate was added at concentrations above 10-15 μM to the incubation medium, it caused a significant increase on ^{3}H-D-aspartate release (Fig. 2). The kainate effect was concentration-dependent (in the range of 10-100 μM) and it was only partially antagonized (20-30%) by the kainate receptor antagonist 2,3-cis-piperidine dicarboxylic acid (PDA) (Fig. 2). The ef-

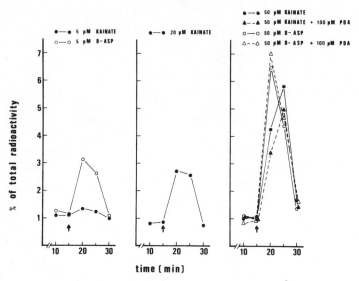

Fig. 2. Effect of kainic acid and D-aspartate on ^{3}H-D-aspartate release from cerebellar neurone-enriched cultures. Cells cultured for
10 DIV were incubated in a Krebs-Ringer medium with 1 μCi/ml of [3]H-
D-aspartate (0.1 μM) and then subjected to 5 min washes. Kainic acid,
unlabelled D-aspartate and PDA were applied starting from the time
indicated by an arrow, for 10 min. The radioactivity recovered in
each wash was expressed as a percent of that present in the cells
before the washes. Averages of 3 duplicate experiments are shown.

fect of the glutamate analog was unaltered when Ca^{2+} ions were omit-
ted from the incubation medium (data not shown). N-methyl-DL-aspar-
tic acid (100 μM) a glutamate agonist for a receptor different from
kainate receptor, did not produce any significant effect on the release of labelled ^{3}H-D-aspartate. Since kainic acid has been reported to interfere with the acidic amino acid transport system (see
introduction and also following paragraph), the releasing effect
observed, and in particular the PDA-insensitive component, could be
due to a heteroexchange process (Levi and Raiteri, 1976). Exchange
processes should be abolished in the absence of extracellular Na^+
ions, and indeed this was the case for the ^{3}H-D-aspartate release
induced by unlabelled D-asp (cfr. Fig. 3 with Fig. 2). However, kai-

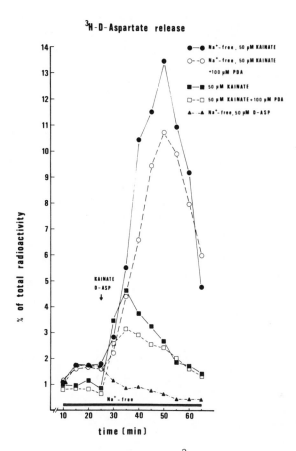

Fig. 3. Effect of kainic acid and PDA on ^{3}H-D-aspartate release from cerebellar neurone-enriched cultures in the absence of Na[+] ions. Cells cultured for 10 DIV were incubated in a Krebs-Ringer, Na[+]-containing medium with 1 μCi/ml of ^{3}H-D-aspartate (1 μM) and then subjected to 5 min washes. After the second wash cells were incubated in a medium in which sucrose replaced NaCl. Kainic acid, D-aspartate and PDA were applied from the time indicated by the arrow to the end of the experiment. One representative experiment is shown. Results expressed as in Fig. 1.

nic acid appeared to be much more effective in releasing ^{3}H-D-asp in the absence of Na[+], although the PDA-sensitive component remained

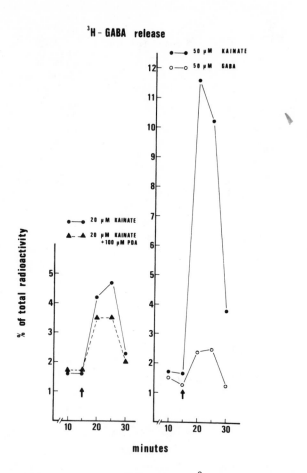

Fig. 4. Effect of kainic acid and GABA on ^{3}H-GABA release from cerebellar neurone-enriched cultures. Cells cultured for 10 DIV were incubated in a Krebs-Ringer containing 10 µM aminooxyacetic acid with 1 µCi/ml of ^{3}H-GABA (0.1 µM) and then subjected to 5 min washes. Kainic acid, GABA and PDA were applied starting from the time indicated by an arrow, for 10 min. Averages of 3 experiments run in duplicate are shown. Results expressed as in Fig. 1.

unchanged (Fig. 3).

Exposure of the cells to kainate produced also a significant effect on preaccumulated ^{3}H-GABA release (Fig. 4): the glutamate

Table 1.   Kainate inhibition of ^{3}H-D-aspartate and ^{3}H-GABA
uptake in neurone-enriched cerebellar cultures.

	IC$_{50}$ (molar)	
Inhibitor	^{3}H-D-aspartate	^{3}H-GABA
Kainate	8×10^{-5}	3×10^{-5}
D-aspartate	2×10^{-5}	–
GABA	–	4×10^{-6}

Neurone-enriched cultures (8 DIV) were incubated in a Krebs
Ringer medium at 37°C for 5 min and then in the same medium with 1 μCi/ml of ^{3}H-D-aspartate or ^{3}H-GABA (1 μM) for
10 min in the presence or in the absence of inhibitors.
Uptake in these conditions was linear up to 20 min. Radioactivity and protein values were determined in NaOH (0.1 N)
extracts of the cells. Averages of 3 experiments are Shown.

analog appeared to be more effective on ^{3}H-GABA than on ^{3}H-D-asparta-
te release (cfr. Fig. 4 with Fig. 2). Kainic acid stimulated ^{3}H-GABA
release in a partially PDA-sensitive (Fig. 4), Na$^+$-independent way
(data not shown).

In order to examine more directly whether kainic acid would interfere with ^{3}H-D-aspartate and ^{3}H-GABA transport, we tested kainate
as an inhibitor of ^{3}H-D-aspartate and ^{3}H-GABA uptake into neurone-
enriched cultures. Table I summarizes such experiments. Kainic acid
showed an IC$_{50}$ value for ^{3}H-D-aspartate uptake which was almost 3
times higher than that for ^{3}H-GABA uptake. Furthermore, D-aspartate
and GABA were more potent than kainate as inhibitors of ^{3}H-D-asparta-
te and ^{3}H-GABA uptake respectively.

The biochemical results described above were confirmed by light
microscopy autoradiography (Fig. 5): kainate completely inhibited
^{3}H-GABA uptake into neurones (Fig. 5a and b), whereas it was less
effective on ^{3}H-D-aspartate uptake (Fig. 5c and d). The inhibitory
effect of kainic acid on ^{3}H-GABA and ^{3}H-D-aspartate uptake was not
due to an acute irreversible toxic action of the drug. In fact, when
cells were preexposed to kainate for 10 min and then washed with a

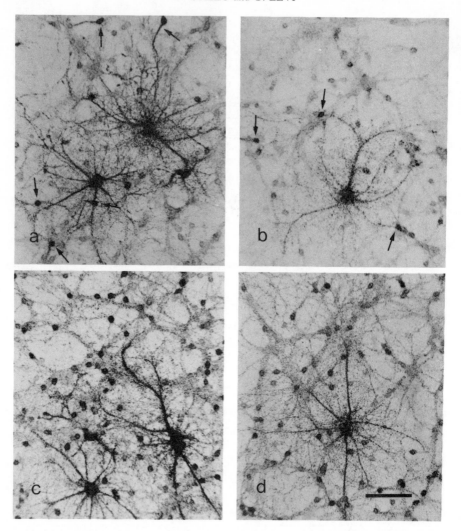

Fig. 5. Effect of kainic acid on the accumulation of ^{3}H-GABA and ^{3}H-
-D-aspartate by cerebellar neurone-enriched cultures at 10 DIV.
Autoradiograms after exposure of the cells for 10 min to 4 μCi/ml of
^{3}H-GABA (1 μM) (a,b) or ^{3}H-D-aspartate (1 μM) (c,d), with (b,d) or with
out (a,c) kainic acid. Kainate concentration was 20 μM (b) or 100 μM
(d). ^{3}H-D-Aspartate labels granule cells, ^{3}H-GABA labels GABAergic interneurones (arrows) and both amino acids label stellate astrocytes.
The radioactivity accumulated by neurones and astrocytes is decreased
when kainic acid was present in the labelling period, the effect being more pronounced with ^{3}H-GABA. Scale bar = 75 μm.

kainate-free medium the subsequent accumulation of radioactivity, detected by autoradiography, was comparable to that seen in control cultures.

DIFFERENTIAL EFFECTS OF KAINATE ON ^{3}H-D-ASPARTATE AND ^{3}H-GABA UPTAKE AND RELEASE IN ASTROCYTE-ENRICHED CULTURES

The autoradiographic studies shown in Fig. 5 evidenced that the effect of kainic acid was not limited to neurones: the autoradiographic accumulation of radioactivity was decreased also in stellate astrocytes, present in low number in neurone-enriched cerebellar cultures and, again, the effect was much more pronounced in the case of ^{3}H-GABA. Therefore, we decided to investigate whether the excitotoxin would affect ^{3}H-GABA and ^{3}H-D-aspartate transport into glial cells, using cerebellar astrocyte-enriched cultures, completely devoid of neurones. Table II summarizes the results of ^{3}H-D-aspartate and ^{3}H-GABA uptake experiments performed in such cultures at different ages in vitro, in order to compare the effect of kainate in systems differently enriched in the two subpopulations of cerebellar astrocytes described in the introduction. Kainic acid was much less effective in inhibiting ^{3}H-D-aspartate than ^{3}H-GABA uptake at 5 and 8 DIV. Furthermore, kainate IC_{50} values for ^{3}H-GABA uptake drastically increased with time in vitro, along with the disappearance of stellate astrocytes from the cultures (Wilkin et al., 1983). Interestingly, also the GABA IC_{50} values for ^{3}H-GABA uptake diminished with the age of the cultures. Moreover, GABA and to a much greater extent D-aspartate appeared to be more effective than kainate at any time in vitro in inhibiting ^{3}H-GABA and ^{3}H-D-aspartate uptake respectively. Autoradiographic studies performed on 5 DIV astrocyte-enriched cultures, showed that whereas GABA uptake into stellate astrocytes was markedly affected by kainate, kainic acid did not influence ^{3}H-D-aspartate accumulation in an appreciable way (data not shown). The potency of kainate in inhibiting ^{3}H-GABA uptake into stellate astrocytes was comparable to that of ACHC (data not shown).

In view of the unexpected effect of kainic acid on stellate astrocytes, we tested the ability of kainate to release preaccumulated labelled amino acids from astrocyte-enriched cultures. Figure 6 summarizes such experiments, performed at two different ages of the cultures. It is evident that the excitotoxin had a stimulatory effect only on ^{3}H-GABA release at 5 DIV, an age at which a maximal number of stellate astrocytes was present in the cultures, and that

Table 2. Kainate inhibition of ^{3}H-D-aspartate and ^{3}H-GABA uptake in astrocyte-enriched cerebellar cultures.

| | IC$_{50}$ (molar) | | | | |
| | ^{3}H-D-aspartate | | ^{3}H-GABA | | |
Inhibitor	5 DIV	8 DIV	5 DIV	8 DIV	14 DIV
Kainate	$> 10^{-4}$	$> 10^{-3}$	4×10^{-5}	4×10^{-4}	$> 10^{-3}$
D-aspartate	3×10^{-5}	6×10^{-5}	-	-	-
GABA	-	-	5×10^{-6}	9×10^{-6}	$> 10^{-4}$

Astrocyte-enriched cultures were incubated in a Krebs-Ringer medium at 37°C for 5 min and then in the same medium with 1 μCi/ml of ^{3}H-D-aspartate or ^{3}H-GABA (1 μM) for 10 min in the presence or in the absence of inhibitors. Uptake in these conditions was linear up to 20 min. Radioactivity and protein values were determined in NaOH (0.1 N) extracts of the cells. Averages of 3 experiments are shown.

the effect was concentration-dependent. At 14 DIV the effect of kainate on ^{3}H-GABA release completely disappeared.

CONCLUSIONS

It has been shown that chronic treatment of cerebellar cultured neurones with kainic acid produces a degeneration of inhibitory interneurones (Messer and Maskin, 1980) or Purkinje cells (Gibbs et al., 1982) in dissociated cell cultures or in "microexplant"cultures, respectively. In contrast, granule cells appeared to be unaffected after such treatment. Several studies have proposed that at least part of kainate toxic effects in vivo could be mediated by an enhancement of excitatory neurotransmitter amino acids release and/or an inhibition of their reuptake into nerve cells (Pastuszko et al., 1984; Ferkany et al., 1982). Besides the effects on excitatory amino acids, other experiments have evidenced a stimulatory action of kainate on GABA release from superfused chick retina (Tapia and Arias,

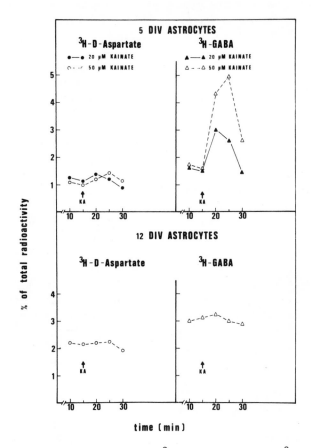

Fig. 6. Effect of kainic acid on ^{3}H-D-aspartate and ^{3}H-GABA release from cerebellar astrocyte-enriched cultures. Astrocytes cultured for 5 DIV (top panels) or for 12 DIV (bottom panels) were incubated in a Krebs-Ringer medium with 1 μCi/ml of ^{3}H-D-aspartate or ^{3}H-GABA, and then subjected to 5 min washes. Kainic acid was applied starting from the time indicated by the arrow, for 10 min. The radioactivity recovered in each wash was expressed as a percent of that present in the cells before the washes. Averages of 3 eperiments are shown.

1982), suggesting an interference of the glutamate analog with the GABA transport system. The rationale of our studies was that kainic

acid might exert a multiple action on various cell types and on different neurotransmitter systems and that all these effects may be
involved in its neurotoxic action.

A combination of biochemical and autoradiographic experiments
evidenced that kainate influences amino acid neurotransmitter release and uptake processes in both glutamatergic and GABAergic cerebellar interneurones. The partial antagonism by PDA and the Na^+ independence of kainate releasing effects may be indicative of a receptor-mediated mechanism. The existence of PDA-sensitive kainate
receptors coupled to a guanylate cyclase system has been recently
demonstrated in cerebellar neurone-enriched cultures (Novelli and
Guidotti, personal communication). On the other hand, the large PDA-
insensitive component in the release evoked by kainate, the inhibitory effect of kainic acid on ^{3}H-D-aspartate and ^{3}H-GABA uptake and
the ability of unlabelled D-aspartate and GABA to release the pre-
accumulated labelled amino acids in cultured cerebellar neurones by
homoexchange would suggest that kainate can utilize the transport
systems of GABA and glutamate and can release these amino acids by
heteroexchange. It has to be noted that the stimulatory effect of
unlabelled GABA on ^{3}H-GABA release was very modest compared to that
of kainate and to that of D-aspartate on ^{3}H-D-aspartate release.
Since kainate did not appear to be more potent than GABA in inhibiting ^{3}H-GABA uptake, its releasing activity should be largely independent on an interference with the GABA transport system. It is
difficult at present to assess to what extent the interaction of kainic acid with a specific receptor is responsible for the effects described on ^{3}H-GABA and ^{3}H-D-aspartate uptake and release. Although
several observations suggest that kainate may cause ^{3}H-D-aspartate
and ^{3}H-GABA release through a heteroexchange process, preliminary
results indicate that ^{3}H-kainic acid is very poorly transported into
cultured neurones and astrocytes. An intriguing aspect concerning the
stimulation of ^{3}H-D-aspartate release by kainate is that the effect
of the excitotoxin was amplified in the absence of extracellular Na^+
ions. It can be suggested that kainate binding to its receptors and,
therefore, its biological effects are enhanced in a Na^+-free medium.

The experimental approach utilized in the present study allowed
us to reveal a so far unexplored site of kainate action. In fact,
the excitotoxin strongly inhibited ^{3}H-GABA uptake into and stimulated its release from a subpopulation of cerebellar astrocytes exhibiting characteristic morphological, antigenic and functional features (Wilkin et al., 1983; Johnstone et al., submitted). In the
same cells, the effect of kainic acid on ^{3}H-D-aspartate uptake and

release was minimal. We are currently investigating the possibility that stellate astrocytes possess specific receptors for kainic acid.

Acknowledgements. The skillful and enthusiastic collaboration of Mrs. M.T. Ciotti is gratefully acknowledged. We thank Dr. P. De Camilli for the generous gift of anti-Synapsin I antibodies and Dr. A. Guidotti for the generous gift of PDA. This work was partially supported by a grant of the Italian National Research Council (Progetto Finalizzato MPR-SP3) and was performed at the Institute of Cell Biology, C.N.R., Rome, Italy.

REFERENCES

Aloisi, F., Ciotti, M.T. and Levi, G. (1985). Characterization of GABAergic neurons in cerebellar primary cultures and selective neurotoxic effect of a serum fraction. J. Neurosci., in press.

Biziere, K. and Coyle, J.T. (1978). Influence of cortico-striatal afferents on striatal kainic acid neurotoxicity. Neurosci. Lett., 8, 303-310.

Biziere, F. and Coyle, J.T. (1979). Effects of cortical ablation on the neurotoxicity and receptor binding of kainic acid in striatum. J. Neurosci. Res., 4, 383-398.

Coyle, J. T. (1983). Neurotoxic action of kainic acid. J. Neurochem., 41, 1-11.

Currie, D.N. and Dutton, G.R. (1980). $[^3H]$-GABA uptake as a marker for cell type in primary cultures of cerebellum and olfactory bulb. Brain Res., 199, 473-481.

De Camilli, P., Cameron, R. and Greengard, P. (1983a). Synapsin I (Protein I), a nerve terminal-specific phosphoprotein. I. Its general distribution in synapses of the central and peripheral nervous system demonstrated by immunofluorescence in frozen and plastic sections. J. Cell Biol., 96, 1337-1354.

De Camilli, P., Harris, S.M., Jr., Huttner, W.B. and Greengard, P. (1983b). Synapsin I (Protein I), a nerve terminal-specific phosphoprotein. II. Its specific association with synaptic vescicles demonstrated by immunocytochemistry in agarose-embedded synaptosomes. J. Cell Biol., 96, 1355-1373.

Ferkany, J.W. and Coyle, J.T. (1983). Kainic acid selectively stimulates the release of endogenous excitatory acidic amino acids.

J. Pharmacol. Exp. Ther., 225, 399-406.

Ferkany, J.W., Zaczek, R. and Coyle, J.T. (1982). Kainic acid stimulates excitatory amino acid neurotransmitter release at presynaptic receptors. Nature, 298, 757-759.

Gallo, V., Ciotti, M.T., Coletti, A., Aloisi, F. and Levi, G. (1982).
Selective release of glutamate from cerebellar granule cells
differentiating in culture. Proc. Natl. Acad. Sci. (USA), 79,
7919-7923.

Giacobini, E., Vernadakis, A. and Shahar, A. (1980) Tissue Culture
in Neurobiology. Raven Press, New York.

Gibbs, W., Neale, E.A. and Moonen, G. (1982). Kainic acid sensitivity of mammalian Purkinje cells in monolayer cultures. Dev.
Brain Res., 4, 103-108.

Johnstone, S.R., Levi, G., Wilkin, G.P., Schneider, A. and Ciotti,
M.T. Subpopulations of rat cerebellar astrocytes in primary
culture: morphology, cell surface antigens and $[^3H]$-GABA
transport. Dev. Brain Res., submitted.

Kingsbury, A.E., Gallo, V., Woodhams, P.L. and Balazs, R. (1985)
Survival, morphology and adhesion properties of cerebellar
interneurones cultured in chemically defined and serum-
supplemented medium. Dev. Brain Res., 17, 17-25.

Levi, G., Aloisi, F., Ciotti, M.T. and Gallo, V. (1984). Autoradiographic localization and depolarization-induced release of
acidic amino acids in differentiating cerebellar granule cell
cultures. Brain Res., 290, 77-86.

Levi, G. and Raiteri, M. (1976). Synaptosomal transport processes.
Int. Rev. Neurobiol., 19, 51-74.

Levi, G., Wilkin, G.P., Ciotti, M.T. and Johnstone, S. (1983) Enrichment of differentiated, stellate astrocytes in cerebellar
interneurone cultures as studied by GFAP immunofluorescence
and autoradiographic uptake patterns with $[^3H]$-D-aspartate and
$[^3H]$-GABA. Dev. Brain Res., 10, 227-241.

London, E.D. and Coyle, J.T. (1979). Specific binding of $[^3H]$-kainic
acid to receptor sites in rat brain. Mol. Pharmacol., 15, 492-
505.

Mc Lennan, H. (1980). The effect of decortication on the excitatory
amino acid sensitivity of striatal neurones. Neurosci. Lett.,
18, 313-316.

Messer, A. and Maskin, P. (1980). Short-term effects of kainic acid on rat cerebellar cells in monolayer cultures. Neurosci. Lett., 19, 173-177.

Monaghan, D.T. and Cotman, C.W. (1982). The distribution of $[^3H]$-kainic acid binding sites in rat CNS as determined by auto-radiography. Brain Res., 252, 91-100.

Panula, P.A.J. (1980). A fine structure and histochemical study of the effect of kainic acid on cultured neostriatal cells. Brain Res., 181, 185-190.

Pastuszko, A., Wilson, D.F. and Erecinska, M. (1984). Effects of kainic acid in rat brain synaptosomes: the involvement of Ca^{2+}. J. Neurochem., 43, 747-754.

Potashner, S.J. and Gerard, D. (1983). Kainate-enhanced release of D-$[^3H]$Aspartate from cerebral cortex and striatum: reversal by baclofen and pentobarbital. J. Neurochem., 40, 1548-1557.

Raff, M.C., Fields, K.L., Hakomori, S., Mirsky,R., Pruss, R.M. and Winter, J. (1979). Cell type specific markers for distinguishing and studying neurones and the major classes of glial cells in culture. Brain Res., 174, 283-308.

Seil, F.J. and Woodward, W.R. (1980). Kainic acid neurotoxicity in granuloprival cerebellar cultures. Brain Res., 197, 285-289.

Thangnipon, W., Kingsbury, A.E., Webb, M. and Balazs, R. (1983). Observation on rat cerebellar cells in vitro: influence of substratum, potassium concentration and relationship between neurones and astrocytes. Dev. Brain Res., 11, 177-189.

Tapia, R. and Arias, C. (1982). Selective stimulation of neuro-transmitter release from chick retina by kainic and glutamic acids. J. Neurochem., 39, 1169-1178.

Webb, M., Gallo, V., Schneider, A. and Balazs, R. (1985). The expression of concavalin A-binding glycoproteins during the development of cerebellar granule neurones in vitro. Int. J. Dev. Neurosci., 3, 199-208.

Whetsell, W.O., Jr., Ecob-Johnston, M.S. and Nicklas, W.J. (1979). Studies of kainate-induced caudate lesions in organotypic tissue culture. In Huntington's Disease, Advances in Neurology, vol. 23. (eds. T.N. Chase, N.S. Wexler and A. Barbeau). Raven Press, New York, pp. 645-654.

Wilkin, G.P., Levi, G., Johnstone, S.R. and Riddle, P.N. (1983).

Cerebellar astroglial cells in primary culture: expression of different morphological appearances and different ability to take up $[^3H]$ D-aspartate and $[^3H]$-GABA.   Dev.   Brain Res., 10, 265-277.

15

Some Comparative Aspects of the Pharmacology of Excitatory Amino Acids in Spinal Cord and Hippocampus

H. McLennan, K. Curry, M.J. Peet and D.S. Magnuson

EXTRACELLULAR STUDIES

The original work demonstrating the depolarizing and excitatory actions of the naturally occurring acidic amino acids was conducted on spinal cord neurones (Curtis et al., 1960), and subsequent studies which first showed the more powerful and often more prolonged actions of many other natural and non-natural compounds possessing the same general pattern of functional groups were also conducted in this tissue (Curtis and Watkins, 1960, 1963; Biscoe et al., 1976; McLennan and Wheal, 1978; Krogsgaard-Larsen et al., 1980). Many of the early descriptions of the actions of antagonists of the excitatory amino acids also derived from observations made in spinal cord (Haldeman and McLennan, 1972; Biscoe et al., 1978; McLennan and Hall, 1978), and it has come to be recognized that several subsets of receptors mediate the responses to the various compounds. Thus it was observed that the first amino acid antagonist described, L-glutamate diethylester (GDEE), could be used to rank the excitants in order of their susceptibility to blockade, that D-α -aminoadipate (DAA) blocked actions which were resistant to GDEE, and that the effects produced by a few compounds were relatively unaffected by either antagonist (Hicks et al., 1978; McLennan and Lodge, 1979). On the basis of these data and the effects of the more potent and specific DAA-like substances D-2-amino-5-phosphonovalerate (DAPV) and D-2-amino-7-phosphonoheptanoate (DAPH) (Evans et al., 1982; McLennan and Liu, 1982), it is now commonly accepted that three receptor types exist, characteristically activated by N-methyl-D-aspartate (NMDA), quisqualate and kainate (for refs. see McLennan, 1983).

Of these the NMDA receptors are specifically antagonized by DAPV, the quisqualate variety by GDEE, while kainate receptors are are only affected by substances which also have appreciable actions against one or other or both of the NMDA and quisqualate types (Davies and Watkins, 1981; Davies et al., 1982; Jones et al., 1984).

There is also a consensus that the great majority of central
neurones possess all three types of receptor, although there were
early indications that quantitatively there may be differences
between various groups of cells (McLennan et al., 1968).
Nevertheless the pharmacological pattern of responsiveness for the
most part has been found consistent in thalamic (Hicks et al.,
1978), cortical (Stone et al., 1981), hippocampal (Collingridge et
al., 1983a) and other neurones with what had been developed from
spinal cord studies. However it is now becoming apparent that some
differences do exist.

A problem vexing to an understanding of the physiological
significance of NMDA receptors has been the identification of a
possible endogenous reactant. The majority of powerful agonists
and antagonists of these receptors are compounds with a
D-configuration, yet such do not occur in the CNS. Nevertheless
such substances as DAPV do affect synaptic processes in the cord
(Evans et al., 1982) and elsewhere (e.g. Crunelli et al., 1983),
and the apparently anomalous chirality of the receptor was
difficult of explanation.

A possible solution was presented in 1981 when Stone and
Perkins described the excitatory action of quinolinic acid
(2,3-pyridine dicarboxylic acid) on cortical neurones, and showed
that its effects were blocked by DAPV. Quinolinate, a catabolite
of tryptophan, is planar and achiral, and thus the problem of the
conformational preference of the receptor would be obviated. A new
complication was soon recognized however, namely that there
appeared to be considerable regional variations in the sensitivity
of different groups of neurones to quinolinate (Perkins and Stone,
1983a,b); and specifically that spinal neurones were very much less
responsive than were those of the cortex. This result was confirmed
in a later study (McLennan, 1984), which also adduced evidence that
the same spinal cells which were less affected by quinolinate were
as responsive as those in the cortex to NMDA. Nevertheless it
appeared that both quinolinate and NMDA were reacting with
qualitatively the same population of receptors. One is left with
the result that in the cortex those receptors are somehow
relatively more accessible to quinolinate while access by NMDA is
the same in both regions.

Other differences appear when the actions of several molecules
related to quinolinate are examined. Perkins and Stone (1982)
reported that, in cortex, kynurenic acid reduced the excitatory
effects of quinolinate, NMDA and, surprisingly, quisqualate. We
have recently confirmed this observation, but have now also shown
that kynurenate is active in the spinal cord equally against
quisqualate, NMDA and kainate, and that in the cord it is
considerably more potent as an antagonist than is the case at
cortical cells. A related compound, acridinic acid, shows a
similar differential effect between cord and cortex (Table I).

However when these compounds were tested upon hippocampal pyramidal cells of the CA1 region another difference was revealed. Quinolinate is a powerful excitant of these neurones (see below), and its effects are prevented by kynurenate and DAPV; but in this tissue acridinate is completely without effect.

Finally, it is interesting to note that hydrogenation of the aromatic ring of quinolinate yields 2,3-piperidine dicarboxylate, the cis isomer of which (PDA) was noted earlier as a relatively unspecific antagonist of amino acid-induced excitations in the cord (Davies et al., 1982; McLennan and Liu, 1982). Yet this substance retains some of the excitatory properties of its "parent", for in the hippocampus it activates cells sufficiently to render its use there as a pharmacological antagonist very difficult (Collingridge et al., 1983a; Crunelli et al., 1983).

One is thus left with a collection of facts not all of which are easily reconcilable. (1) The division of excitatory amino acid receptors into three subgroups seems to be generally applicable throughout the various regions of the CNS. (2) Both quisqualate and NMDA types appear to be involved in synaptic processes (for refs. see McLennan, 1983; Watkins, 1984). (3) The endogenous activator of quisqualate receptors is probably L-glutamate. (4) Quinolinate is the only naturally occurring compound recognized as activating NMDA receptors; however there are marked regional variations in the sensitivity of individual neurones to this substance. (5) Compounds related to quinolinate (kynurenate, acridinate, PDA) are predominately blockers both of NMDA and of other receptor types; however they too show regional variation in their effectiveness and PDA can exhibit excitatory actions in some areas. It is thus clear that for NMDA receptors certainly but possibly for the others as well, there are subtle and at present unexplained differences in reactivity to selected agonists and antagonists.

INTRACELLULAR STUDIES

The pioneering work on the actions of glutamic and aspartic acids on the transmembrane potentials was also conducted by Curtis et al. (1960), who showed that these compounds applied extracellularly depolarized motoneurones. Since at that time it was believed that all active amino acids reacted with the same three-point attachment site (Curtis and Watkins, 1960), it was assumed that a common ionic mechanism must underlie this effect.

Table I. Percent reduction of amino acid-induced
excitations by kynurenic and acridinic acids.

| | SPINAL CORD | | |
	Quisqualate	NMDA	Kainate
Kynurenate Mean ± SD (n)	64±28 (21)	82±24 (24)	73±26 (13)
Acridinate	55±20 (29)	69±19 (26)	74±20 (17)

| | CORTEX | |
	Quisqualate	NMDA/quinolinate
Kynurenate Mean ± SD (n)	25±11 (7)	43±20 (10)
Acridinate	26±24 (5)	18±28 (9)

    Since then the picture has become more complex and not
completely consistent. Thus it has been reported that glutamate
can activate two distinct membrane processes, one increasing and
the other reducing membrane conductance (MacDonald and Wojtowicz,
1980, 1982), presumably due to changes respectively in Na and K
ionic permeabilities: at low doses the net effect is an increase
in conductance (Bernardi et al., 1972; Lambert et al., 1981).
The latter authors have shown that quisqualate gives a similar
response. Other excitants however have very different effects.
Compounds reacting with NMDA receptors have been reported to have
as their principal action a marked decrease in conductance on spinal
(Lambert et al., 1981; MacDonald and Wojtowicz, 1982) and cortical
(Flatman et al., 1983) neurones, while kainate evokes only a large
increase on spinal cells. These various data therefore support
the separate existence of the three classes of receptor.

    Whether all of these seeming variations in qualitative effects
are indeed real may however be questioned. Thus in the hippocampus
Peet et al. (1985) have found that under conditions where
voltage-related changes in the membrane are precluded, both
quisqualate and NMDA increase the membrane conductance; however
beyond that similarity the actions of the two are quite different.
Quisqualate induces a relatively rapid depolarization which reaches
a plateau upon which are superimposed simple action potentials
which appear when the membrane potential has been reduced by about
15mV (Fig. 1A). By contrast the depolarization induced by NMDA is
slower but continues to rise throughout the period of
administration, and at 5-6mV of depolarization sudden rhythmically

occurring depolarizing shifts of the membrane potential are
observed: upon these are superimposed bursts of spikes (Fig. 1B).
As membrane depolarization continues there is often a sudden "step"
which initiates the generation of action potentials and the
rhythmical shifts of membrane potential disappear. These latter
changes occur at about 15mV of depolarization (Fig. 1C). In the

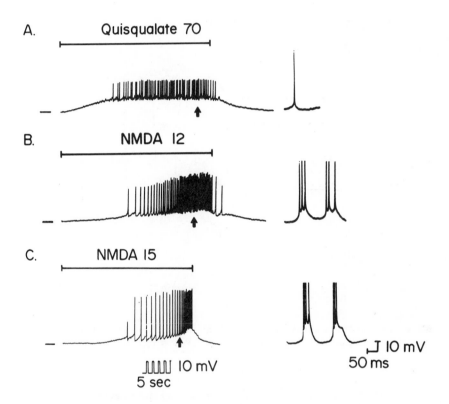

Fig. 1. Intracellularly recorded responses of CA1 hippocampal
neurones to the iontophoretic administration of amino acids into
the adjacent stratum pyramidale. This and subsequent figures show
chart records with sample oscilloscope traces taken at the arrows.
A,B: from the same cell compare the effects of quisqualate (70nA
ejecting current from a solution of 5mM in 150mM NaCl) and NMDA
(12nA; 50mM in 100mM NaCl). C: 15nA NMDA on a different cell
illustrates the "step" depolarization following which simple action
potentials only were observed.

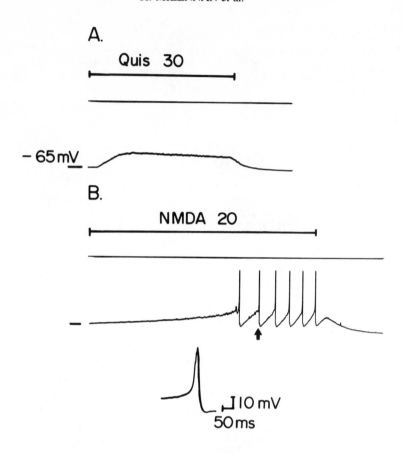

Fig. 2.   Intracellular responses of an hippocampal neurone to A:
quisqualate (30nA) and B: NMDA (20nA) in a slice superfused with
1μM TTX.

presence of TTX the simple spikes induced by quisqualate are
blocked but the underlying depolarization is unaffected; and the
NMDA-evoked actions are changed by the elimination of the bursts
superimposed on the depolarizing shifts.   Instead longer duration
spikes are revealed (Fig. 2), which disappear as the membrane
continues to depolarize.   DAPV prevents all of the actions of NMDA
but does not alter the quisqualate responses.

There is now convincing evidence that the spike potentials
induced by NMDA in TTX-poisoned hippocampal cells depend upon the
activation of a voltage-dependent increase in calcium conductance
(Dingledine, 1983; Peet et al., 1985), and it is probable that the
depolarizing shifts giving rise to action potentials in untreated
tissue have a similar origin since they too are eliminated in the

presence of cobalt.

All of the effects of NMDA are mimicked by its specific
agonists including quinolinate. Many of the other amino acids
tested, among them kainate, L-glutamate and L-aspartate, resemble
quisqualate in that only TTX-sensitive action potentials are
generated, and at no level or rate of change of depolarization are
calcium-dependent potentials evoked. A few compounds, most notably
D-glutamate which from other studies (Hicks et al., 1978) is
capable of reaction with both NMDA and quisqualate receptors, on
approximately equal numbers of cells showed either the quisqualate
or the NMDA type of response. Finally it has been found that all of
the actions of NMDA and quinolinate in addition to being
antagonized by DAPV are powerfully blocked by kynurenate (see also
Ganong et al., 1983), which had little effect on
quisqualate-evoked responses.

Thus in the hippocampus at least, the following conclusions
can be reached. (1) NMDA and quisqualate both depolarize neurones
by mechanisms which result in a net increase in membrane
conductance. (2) NMDA-induced depolarizations of about 5mV give
rise to rhythmic, calcium-dependent shifts of the membrane potential
upon which bursts of action potentials are superimposed. In the
presence of TTX these potential shifts are seen to be caused by
large slow spikes which are followed by pronounced
hyperpolarizations. (3) The actions of NMDA are mimicked by
quinolinate, and are blocked by DAPV and kynurenate.
(4) Quisqualate-induced depolarizations only trigger the
generation of TTX-sensitive action potentials. (5) Under no
circumstances can quisqualate or kainate activate the calcium
conductance and its sequellae. (6) The co-existence of different
types of amino acid receptors upon hippocampal cells which here
can be shown to elicit qualitatively distinct changes in membrane
conductance is unequivocal.

PHYSIOLOGICAL IMPLICATIONS

In the spinal cord there is now rather good evidence that
polysynaptic reflex activity induced by stimulation of dorsal root
afferent fibres is selectively prevented by NMDA receptor
antagonists (Davies and Watkins, 1983); while not quite such
compelling data imply that the glutamate/quisqualate receptor may
be implicated in monosynaptic pathways activated by low threshold
afferents (Haldeman and McLennan, 1972; Davies and Watkins, 1985).
The likely involvement of one or other of these two, and possibly
of kainate receptors as well, at other synaptic sites in the CNS
has been inferred (McLennan, 1983; Watkins, 1984).

However what is not specifically known is whether NMDA elicits
changes in the calcium conductance of spinal neurones and thus acts
through a mechanism analogous to that occurring in the hippocampus.

The failure of cadmium ions to modify the NMDA responses of spinal
cells (Mayer et al., 1984) suggests that it may not, and that again
a qualitative difference between spinal and supraspinal neurones
is revealed.

In the hippocampus it is now accepted also that monosynaptic
excitatory responses are not dependent upon NMDA receptors
(Collingridge et al., 1983b; Harris et al., 1984), yet the fact that
less specific amino acid antagonists such as γ-D-glutamylglycine
(Collingridge et al., 1983b; Crunelli et al., 1983) or kynurenate
(Ganong et al., 1983) are effective suggests that one of the other
receptor types may be involved.

What is however true is that the NMDA antagonists are able to
prevent the prolonged increase in synaptic response which follows
a brief high frequency tetanus of the afferent input ("long-term
potentiation") (Collingridge et al., 1983b; Harris et al., 1984).
By way of explanation it has been proposed that the high frequency
stimulation results in sufficient transmitter release to activate
the NMDA receptors and their attendant calcium influx, whereas a
single stimulus is unable to do so. The alternative explanations
therefore appear to be these. (1) Glutamate is the single
endogenous substance whose release in small amounts activates the
quisqualate (or less probably the kainate) receptor, but which in
larger concentration, i.e. following tetanic stimulation,
additionally reacts with NMDA receptors to induce calcium responses.
Although glutamate is an effective competitor at presumed NMDA
binding sites (Monaghan et al., 1984), the fact that under no
circumstances have we been able to induce calcium spikes in
hippocampal cells with quisqualate (see above), glutamate or
kainate does not support this hypothesis. (2) Two transmitters,
possibly glutamate and quinolinate, are co-released from the
afferent nerve terminals to react with two separate populations of
receptors, again with the tetanic stimulation providing a
sufficient local concentration of agonist that the consequences of
NMDA receptor activation become manifest. Although no direct
evidence in favour of this possibility presently exists, the
pharmacological data would seem to favour it.

REFERENCES

Bernardi, G., Zieglgänsberger, W., Herz, A. and Puil, E.A. (1972).
Intracellular studies on the action of L-glutamic acid on spinal
neurones of the cat. Brain Res. 39, 523-525.

Biscoe, T.J., Davies, J., Dray, A., Evans, R.H., Martin, M.R. and
Watkins, J.C. (1978). D-α-aminoadipate, α,ε-diaminopimelic acid and
HA-966 as antagonists of amino acid-induced and synaptic excitation
of mammalian spinal neurones in vivo. Brain Res. 148, 543-548.

Briscoe, T.J., Evans, R.H., Headley, P.M., Martin, M.R. and Watkins, J.C. (1976). Structure-activity relations of excitatory amino acids on frog and rat spinal neurones. Br. J. Pharmac. 58, 373-382.

Collingridge, G.L. Kehl, S.J. and McLennan, H. (1983a). The antagonism of amino acid-induced excitations of rat hippocampal CA1 neurones in vitro. J. Physiol. 334, 19-31.

Collingridge, G.L., Kehl, S.J. and McLennan, H. (1983b). Excitatory amino acids in synaptic transmission in the Schaffer collateral-commissural pathway of the rat hippocampus. J. Physiol. 334, 33-46.

Crunelli, V., Forda, S. and Kelly, J.S. (1983). Blockade of amino acid-induced depolarizations and inhibition of excitatory post-synaptic potentials in rat dentate gyrus. J. Physiol. 341, 627-640.

Curtis, D.R., Phillis, J.W. and Watkins, J.C. (1960). The chemical excitation of spinal neurones by certain acidic amino acids. J. Physiol. 150, 656-682.

Curtis, D.R. and Watkins, J.C. (1960). The excitation and depression of spinal neurones by structurally related amino acids. J. Neurochem. 6, 117-141.

Curtis, D.R. and Watkins, J.C. (1963). Acidic amino acids with strong excitatory actions on mammalian neurones. J. Physiol. 166, 1-14.

Davies, J., Evans, R.H., Francis, A.A., Jones, A.W., Smith, D.A.S. and Watkins, J.C. (1982). Conformational aspects of the actions of some piperidine dicarboxylic acids at excitatory amino acid receptors in the mammalian and amphibian spinal cord. Neurochem. Res. 7, 1119-1133.

Davies, J. and Watkins, J.C. (1981). Differentiation of kainate and quisqualate receptors in the cat spinal cord by selective antagonism with γ-D(and L)-glutamylglycine. Brain Res. 206, 172-177.

Davies, J. and Watkins, J.C. (1983). Role of excitatory amino acid receptors in mono- and polysynaptic excitation in the cat spinal cord. Exp. Brain Res. 49, 280-290.

Davies, J. and Watkins, J.C. (1985). Depressant actions of γ-D-glutamylaminomethyl sulphonate (GAMS) on amino acid-induced and synaptic excitation in the cat spinal cord. Brain Res. 327, 113-120.

Dingledine, R. (1983). N̲-methyl aspartate activates voltage-dependent calcium conductance in rat hippocampal pyramidal cells. J. Physiol. 343, 385-405.

Evans, R.H., Francis, A.A., Jones, A.W., Smith, D.A.S. and Watkins, J.C. (1982). The effects of a series of ω-phosphonic α-carboxylic amino acids on electrically evoked and excitant amino acid-induced responses in isolated spinal cord preparations. Br. J. Pharmac. 75, 65-75.

Flatman, J.A., Schwindt, P.C., Crill, W.E. and Stafstrom, C.E. (1983). Multiple actions of N-methyl-D-aspartate on cat neocortical neurons in vitro. Brain Res. 266, 169-173.

Ganong, A.H., Lanthorn, T.H. and Cotman, C.W. (1983). Kynurenic acid inhibits synaptic and acidic amino acid-induced responses in the rat hippocampus and spinal cord. Brain Res. 273, 170-174.

Haldeman, S. and McLennan, H. (1972). The antagonistic action of glutamic acid diethylester towards amino acid-induced and synaptic excitations of central neurones. Brain Res. 45, 393-400.

Harris, E.W., Ganong, A.H. and Cotman, C.W. (1984). Long-term potentiation in the hippocampus involves activation of N-methyl-D-aspartate receptors. Brain Res. 323, 132-137.

Hicks, T.P., Hall, J.G. and McLennan, H. (1978). Ranking of the excitatory amino acids by the antagonists glutamic acid diethylester and D-α-aminoadipic acid. Can. J. Physiol. Pharmacol. 56, 901-907.

Jones, A.W., Smith, D.A.S. and Watkins, J.C. (1984). Structure-activity relations of dipeptide antagonists of excitatory amino acids. Neuroscience 13, 575-581.

Krogsgaard-Larsen, P., Honoré, T., Hansen, J.J., Curtis, D.R. and Lodge, D. (1980). New class of glutamate agonist related to ibotenic acid. Nature 284, 64-66.

Lambert, J.D.C., Flatman, J.C. and Engberg, I. (1981). Actions of excitatory amino acids on membrane conductance and potential in motoneurones. In Glutamate as a Neurotransmitter. (eds.,G. DiChiara and G.L. Gessa). Raven Press, New York. pp. 205-216.

MacDonald, J.F. and Wojtowicz, J.M. (1980). Two conductance mechanisms activated by applications of L-glutamic, L-aspartic, DL-homocysteic, N̲-methyl-D-aspartic, and DL-kainic acids to cultured mammalian central neurones. Can. J. Physiol. Pharmacol. 58, 1393-1397.

MacDonald, J.F. and Wojtowicz, J.M. (1982). The effect of
L-glutamate and its analogues upon the membrane conductance of
central murine neurones in culture. Can. J. Physiol. Pharmacol.
60, 282-296.

Mayer, M.L., Westbrook, G.L. and Guthrie, P.B. (1984).
Voltage-dependent block by Mg^{2+} of NMDA responses in spinal cord
neurones. Nature 309, 261-263.

McLennan, H. (1983). Receptors for the excitatory amino acids in
the mammalian central nervous system. Progr. Neurobiol. 20,
251-271.

McLennan, H. (1984). A comparison of the effects of
N-methyl-D-aspartate and quinolinate on central neurones of the rat.
Neurosci. Lett. 46, 157-160.

McLennan, H. and Hall, J.G. (1978). The action of
D-α-aminoadipate on excitatory amino acid receptors of rat thalamic
neurones. Brain Res. 149, 541-545.

McLennan, H., Huffman, R.D. and Marshall, K.C. (1968). Patterns of
excitation of thalamic neurones by amino-acids and by
acetylcholine. Nature 219, 387-388.

McLennan, H. and Liu, J.R. (1982). The action of six antagonists
of the excitatory amino acids on neurones of the rat spinal cord.
Exp. Brain Res. 45, 151-156.

McLennan, H. and Lodge, D. (1979). The antagonism of amino
acid-induced excitation of spinal neurones in the cat. Brain Res.
169, 83-90.

McLennan, H. and Wheal, H.V. (1978). A synthetic, conformationally
restricted analogue of L-glutamic acid which acts as a powerful
neuronal excitant. Neurosci. Lett. 8, 51-54.

Monaghan, D.T., Yao, D., Olverman, H.J., Watkins, J.C. and Cotman,
C.W. (1984). Autoradiography of D-2-{^{3}H}
amino-5-phosphonopentanoate binding sites in rat brain. Neurosci.
Lett. 52, 253-258.

Peet, M.J., Gregersen, H. and McLennan, H. (1985).
2-Amino-5-phosphonovaleric acid selectively blocks the
Ca^{2+}-dependent excitation of rat CA1 pyramidal neurones by
N-methyl-D-aspartic acid. Submitted for publication.

Perkins, M.N. and Stone, T.W. (1982). An iontophoretic
investigation of the actions of convulsant kynurenines and their
interaction with the endogenous excitant quinolinic acid. Brain
Res. 247, 184-187.

Perkins, M.N. and Stone, T.W. (1983a). Quinolinic acid: regional variations in neuronal sensitivity. Brain Res. 259, 172–176.

Perkins, M.N. and Stone, T.W. (1983b). Pharmacology and regional variations of quinolinic acid-evoked excitations in the rat central nervous system. J. Pharm. exp. Ther. 226, 551–557.

Stone, T.W. and Perkins, M.N. (1981). Quinolinic acid: a potent endogenous excitant at amino acid receptors in CNS. Eur. J. Pharmacol. 72, 411–412.

Stone, T.W., Perkins, M.N., Collins, J.F. and Curry, K. (1981). Activity of the enantiomers of 2-amino-5-phosphono-valeric acid as stereospecific antagonists of excitatory amino acids. Neuroscience 6, 2249–2252.

Watkins, J.C. (1984). Excitatory amino acids and central synaptic transmission. Trends in Pharmacol. Sci. 5, 373–376.

ACKNOWLEDGEMENT

The authors' work was supported by the Medical Research Council of Canada.

16

Pharmacology of Excitatory Amino Acids on Spinal Cord Neurones*

A. Nistri and A. King

INTRODUCTION

There is little doubt that there are distinct receptor classes for excitatory amino acids on vertebrate spinal neurones (Watkins and Evans 1981; McLennan, 1983). Much of this classification has been derived from experiments on isolated CNS preparations, in particular the amphibian spinal cord in vitro. This preparation was first described nearly a quarter of a century ago as a very useful model system to screen the basic pharmacological properties of neuroactive amino acids (Curtis et al, 1961). Subsequent work on this tissue has mainly employed extracellular recording of motoneuronal activity elicited by various amino acid agonists and antagonists (Watkins and Evans, 1981). We wished to explore in more detail the motoneuronal membrane responses to excitatory amino acids using the direct and inevitably more complex technique of intracellular recording with glass micropipettes during continuous superfusion of test substances. In this report we describe how the effects of excitatory amino acids on frog spinal neurones were characterized in terms of neuronal specificity, ionic dependence and receptor classification.

METHODS

Experiments were carried out on parasagittal spinal cord slices of the frog (R. temporaria). The preparation was mounted in a small three-chambered bath in which pairs of ventral and dorsal lumbar roots were connected to miniature suction electrodes for stimulation. Intracellular recordings, usually carried out with 3M KCl filled microelectrodes, were amplified through a high impedance electrometer with facilities for bridge balancing, capacity neutralization and current injection. Data were displayed on a storage oscilloscope, stored on magnetic tape and recorded on a pen recorder. The preparation was superfused with Mg^{2+}-free Ringer solution (maintained at 7°C) to which drugs were added. Further details of the

* The financial support of the Joint Research Board of St. Bartholomew's Hospital and of the MRC is gratefully acknowledged.

experimental procedure and solutions have been published (Nistri
and Arenson, 1983a; Nistri et al, 1985).

RESULTS

Effects of Glutamate on Motoneurones and Interneurones

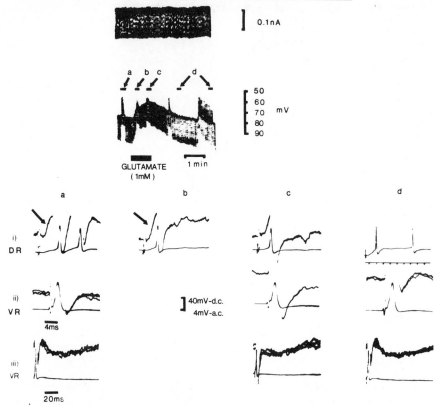

Fig. 1. Effects of glutamate on spinal motoneurone. Top: chart
records of injected current (above) and membrane potential (below)
with hyperpolarizing electrotonic potentials (downward deflections)
and EPSPs with spikes (upward deflections). Bottom: oscilloscope
records (i, ii and iii) taken during periods (a, b, c and d) shown
by lines above the chart record of membrane potential. DR and VR
denote dorsal root (orthodromic) and ventral root (antidromic)
evoked responses respectively. Time marks for i are in i,d (10 ms).
Time calibrations for ii and iii are in iia and iiia respectively.
In each insert top trace is high gain a.c. and bottom trace is low
gain d.c. Note EPSPs marked by arrows. Row iii (slow sweep) shows
antidromic spike afterhyperpolarizations.

Fig. 1 shows a typical response of a lumbar motoneurone to 1 mM
glutamate. The top two tracings are chart records of intracellularly

injected current and membrane potential. Application of glutamate
(black horizontal bar) produced a depolarization with little variation in motoneuronal conductance (note only slight change in amplitude of electrotonic potentials - downward deflections - even at the
peak of the depolarization). After wash the membrane potential
returned to control level but there was a sustained conductance
decrease (larger electrotonic potentials). Changes in the spike
potentials before, during and after glutamate application are shown
in the bottom half of Fig. 1 (i, ii and iii, where DR and VR denote
dorsal or ventral root evoked spikes respectively). During the
glutamate-induced depolarization the EPSP was larger and faster
(Fig. 1, ib) and both the DR (ib and ic) and VR (iic) spikes had
reduced latency. The spike afterhyperpolarization was also enhanced
(iiic). All these data were thus consistent with a strong excitatory action of glutamate, although two features of this response
should be pointed out: the lack of both a significant conductance
increase and of repetitive action potential firing during the
depolarization. In some experiments in which the glutamate depolarization was accompanied by a conductance increase this phenomenon
was thought to be an indirect action as it was coupled to a notice-
able voltage noise presumably caused by activation of afferent
interneurones. This was confirmed by exposing the preparation to
tetrodotoxin (6 μM) which blocked interneuronal activity, voltage
noise and associated conductance increase (Nistri et al, 1985). In
approximately 30% of motoneurones the depolarization induced by
glutamate was preceded initially by a transient (5-15s) enhancement
of the antidromic action potential afterdepolarization and subsequently by a small hyperpolarization (associated with a large conductance rise; Nistri and Arenson, 1983b; see also Fig. 2). Since
these early components of the glutamate response persisted even
after block of chemical synaptic transmission by Mn^{2+} (1.5-2 mM),
they were unlikely caused by release of unidentified endogenous
transmitters and probably represented direct actions of glutamate
on the motoneuronal membrane.

On the relatively small population of interneurones we successfully impaled in the ventral horn of the spinal slice, the depolarizing activity of glutamate was closely comparable to that found on
motoneurones. Nevertheless, interneurones always displayed a large
conductance increase (not accounted for by membrane rectification)
and sustained firing (Nistri and Arenson, 1983a).

Ionic Dependence of Glutamate-Induced Effects on Motoneurones

The precise mechanism responsible for glutamate hyperpolarizations is still unclear. In fact these responses persisted after
treatment with tetrodotoxin, Mn^{2+} or intracellularly injected Cs^+
(a K^+ channel blocker) and had the same polarity of Cl^- dependent
GABA-induced effects. These observations suggest but do not prove
that glutamate might be activating a Cl^- permeability increase perhaps analogous to that found on invertebrate muscle (Cull-Candy and
Usherwood, 1973). The infrequent occurrence of glutamate hyperpolarizations has so far prevented more systematic studies.

In the case of glutamate-induced dpolarizations studies with ion-sensitive microelectrodes have suggested that an increased permeability to Na^+ and secondarily to K^+ was responsible for such an effect (Bührle and Sonnhof, 1983). It was however felt necessary to investigate the role of Na^+ with a different approach, namely to substitute it with impermeable cations, such as glucosamine or choline, and to test the responsiveness of motoneurones to glutamate and other excitatory amino acids such as N-methyl-D-aspartate (NMDA) and quisqualate. Zero Na^+ media were often poorly tolerated by motoneurones so that most experiments were conducted in 14% Na^+ Ringer. When glucosamine was the substitute ion, the depolarizations evoked by glutamate or quisqualate were blocked with recovery on return to control Ringer (Fig. 2). Similar data were obtained for NMDA responses. When Na^+ was replaced by choline (plus 1 μM atropine to abolish the cholinomimetic activity of this substance), glutamate and NMDA responses were blocked whereas the quisqualate depolarizations were not attenuated. However, addition of Mn^{2+} (2 mM) depressed the quisqualate response in choline Ringer (Nistri et al, 1985). These results thus demonstrated a difference in the motoneuronal sensitivity to quisqualate dependent on the impermeant cation used. It was also observed that, unlike glucosamine, choline strongly blocked the antidromic spike afterhyperpolarization (Fig.3), which is a phenomenon caused by K^+ permeability rise. We therefore suggested that choline was probably a blocker of K^+ efflux from frog motoneurones. The following hypothesis was then formulated to account for these data (Nistri et al, 1985): Na^+ was likely to be the main ion normally responsible for the depolarizing activity of glutamate, NMDA and quisqualate. In special circumstances, for instance when K^+ conductances were blocked by choline, an additional (probably Ca^{2+} dependent) mechanism involved in the depolarization evoked by quisqualate was unmasked. We found however no evidence that this Ca^{2+} component is activated by glutamate or NMDA or thatit is prominent in generating quisqualate responses in a normal medium (Mn^{2+} did not block quisqualate in control Ringer).

Separate experiments were carried out to test whether the depolarizing activity of excitatory amino acids might be caused by a block of K^+ conductances (cf. Puil, 1981). Bath applications of K^+ channel blockers such as tetraethylammonium (TEA; 3-5 mM), and 4-aminopyridine (4-AP; 0.5 - 1 mM) or intracellular injections of Cs^+ produced distinctive changes in motoneuronal action potentials and excitability consistent with a strong block of K^+ conductances (Arenson and Nistri, 1985). Neither TEA nor 4-AP decreased the effects of glutamate, NMDA or quisqualate, while a degree of blockade was seen after Cs^+ injections but considered to be not specifically related to K^+ channel block (Arenson and Nistri, 1985).

Pharmacology of Excitatory Amino Acid Antagonists

Early studies (reviewed by Watkins and Evans, 1981; McLennan 1983) indicated that some synthetic substances, D-aminoadipic acid (DAA) and glutamic acid diethylester (GDEE), were apparently preferential blockers of NMDA and quisqualate responses respectively. These data thus provided the first indication of distinct receptor

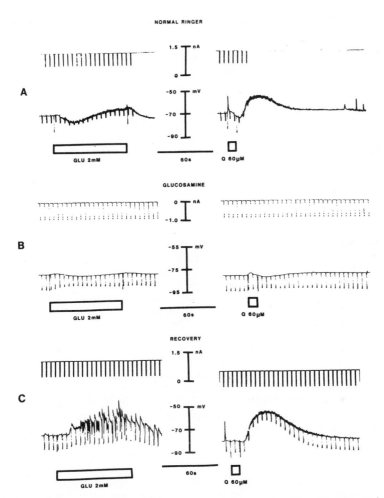

Fig. 2. Chart records (current-top, membrane potential-bottom) of
the effects of glutamate (GLU) or quisqualate (Q) in control Ringer
(A), in 14% Na+ (replaced by glucosamine; B) and after recovery in
control Ringer (C).

classes for excitatory amino acids. We have confirmed that, with
intracellular recording from motoneurones, DAA was a relatively
potent antagonist of NMDA (and glutamate) without a concomitant
depression of quisqualate (Arenson et al, 1984).   More recent studies
by Evans et al, (1982) showed that D-aminophosphonovaleric acid
(D-APV) was a rather more potent and selective antagonist than DAA
for NMDA receptors. We therefore decided to examine directly the
action of D-APV and compare it with that of GDEE on motoneurones.

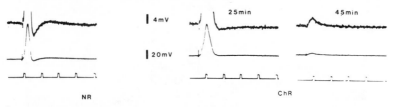

Fig. 3. Oscilloscope traces of antidromic spikes in control Ringer
(NR), and during superfusion with 14% Na/86% choline Ringer (ChR).
Note that after 25 min in ChR the spike displays little afterhyper-
polarization. At 45 min loss of spike. Time marks: 10 ms.

    D-APV (0.5 - 10 µM) often induced a shallow slow hyperpolari-
zation of the motoneuronal membrane without significant conductance
changes (see Fig. 4 top). The origin of this small hyperpolarization (found also in tetrodotoxin solutions) is not yet clear.
After 15-30 min of 10 µM D-APV superfusion and stabilization of the
cell membrane potential, glutamate responses were attenuated while
NMDA responses were abolished (Fig. 4 bottom). Even much smaller
concentrations (0.5 µM) of D-APV strongly reduced the NMDA response,
depressed the action of glutamate and left unchanged that of quisqualate (Fig. 5). Table 1 summarizes the selectivity of D-APV
antagonism towards the depolarizations evoked by different amino
acids. Distinct receptor sites for NMDA and quisqualate were
clearly demonstrated with the use of D-APV. Glutamate and aspartate appeared to act on both receptor types as their effects were
reduced but not blocked even by higher D-APV concentrations. More
recently, single electrode voltage clamp studies on motoneurones
have revealed an inward somatic current elicited by glutamate or
quisqualate; again the quisqualate current was unaffected by D-APV
(10 µM) while that produced by glutamate was reduced by one third
(King and Nistri, 1985). There is therefore strong evidence for a
"mixed agonist" action of glutamate binding to quisqualate and NMDA
receptors.
    It has recently been proposed that on mouse cultured neurones
D-APV, while blocking the NMDA-like activity of glutamate, would
reveal a considerable conductance increased evoked by glutamate via
activation of quisqualate receptors (Mayer and Westbrook, 1984). In
our studies with combined administrations of tetrodotoxin and D-APV
we were unable to detect a significant conductance increase during
glutamate applications. Furthermore, the average conductance
increase elicited by quisqualate was only about 30% in tetrodotoxin
solution and not significantly affected by D-APV (A. Nistri and
A. King, unpublished). The discrepancy between our results and
those of Mayer and Westbrook might be ascribed to a variety of
factors such as animal species, tissue cultures vs. slices, and
different concentrations of divalent cations in the bathing solutions (known to affect amino acid responses; cf. Watkins and Evans,
1981).

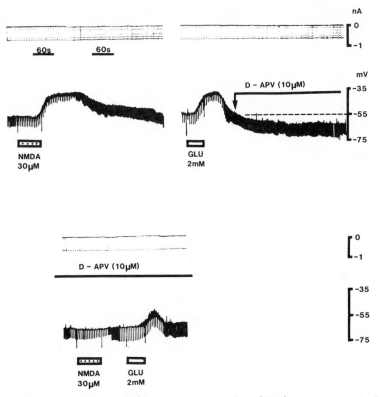

Fig. 4. Chart records of NMDA and glutamate (GLU) responses before (top) and after (bottom) application of D-APV. Note hyperpolarization (with respect to control membrane potential indicated by dashed line) induced by D-APV.

Table 1 Effect of D-APV (1 μM) on excitatory amino acid responses

Agonist	No of Cells	$\Delta V \pm$ s.e. (mV) in Control Ringer	$\Delta V \pm$ s.e. (mV) in D-APV	% of Control
glutamate	7	10.1 ± 2.3	6.8 ± 1.5	67
quisqualate	6	8.8 ± 1.4	9.2 ± 1.3	104
NMDA	7	11.6 ± 2.9	1.5 ± 0.5	13
aspartate	6	13.5 ± 2.2	8.6 ± 1.9	63

Attempts to demonstrate a selective antagonism by GDEE of quisqualate responses were unsuccessful. In a dose range of 0.1 – 1 mM GDEE did not block motoneuronal responses to quisqualate, glutamate, NMDA or DL-α-amino-3-hydroxy-5-methyl-4-isoxazole-propionic acid (AMPA) (Fig. 6). The latter substance is a synthetic glutamate

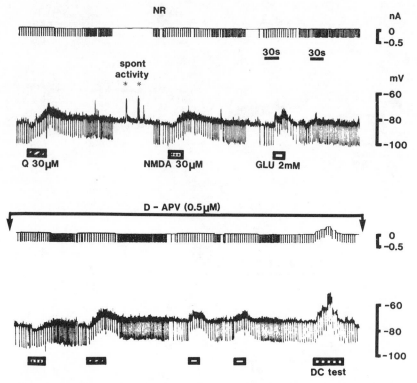

Fig. 5. Chart records of quisqualate (Q), NMDA and glutamate (GLU) responses in control Ringer (NR; top) and in D-APV solution (bottom). DC test indicates injection of steady d.c.-current to check for cell membrane rectification.

analogue presumably acting via quisqualate receptors (Krogsgaard-Larsen and Honoré, 1983) and unaffected by DAA or D-APV (King et al, 1985). High concentrations of GDEE actually produced some local anaesthetic effects by depressing the antidromic action potential of motoneurones. Hence we suggest that at least in the case of spinal motoneurones in vitro GDEE was ineffective as an amino acid antagonist. These findings reinforce the need for potent and selective antagonists for quisqualate receptors in order to probe the distribution and function of these sites in the central nervous system.

Pharmacological Antagonism of Excitatory Postsynaptic Potentials - EPSPs

Mono and polysynaptic EPSPs can be recorded from motoneurones following dorsal root electrical stimulation (cf. Fig. 1, ia). These EPSPs can be differentiated on the basis of stimulus strength, latency, amplitude and timecourse. Since D-APV was a potent blocker of NMDA receptors, and the identity of the EPSP transmitter is still

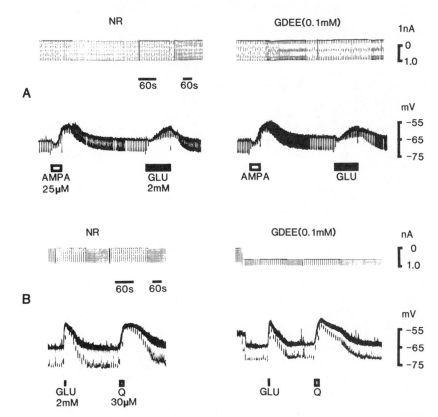

Fig. 6. Chart records of the effects of AMPA, glutamate (GLU) and quisqualate (Q) in control Ringer (NR) and in GDEE solutions. A and B are from different cells.

unknown, we tested whether D-APV sensitive sites might be involved in the generation of mono or polysynaptic EPSPs. This work was also encouraged by our early observation that DAA reduced monosynaptic EPSPs (Arenson et al, 1984). D-APV (0.5 - 10 μM) significantly and reversibly decreased the amplitude of monosynaptic EPSPs by approximately 60% while it did not significantly change the amplitude of polysynaptic EPSPs. The time course of the latter was however reduced by 38%, a finding which suggested that D-APV-sensitive sites might be involved in the origin of later components of the polysynaptic EPSPs.

The sensitivity of monosynaptic EPSPs to D-APV closely parallels that of glutamate and aspartate responses, implying that one or both of these amino acids may be the transmitter of fast excitatory signals to motoneurones. Since dorsal root electrical stimulation preferentially elicits release of glutamate rather than aspartate from the frog spinal cord in vitro (Takeuchi et al, 1983), it seems likely that glutamate subserves the transmitter function for the monosynaptic EPSP of motoneurones.

CONCLUSIONS
    Our intracellular studies on frog spinal neurones in vitro
confirm the usefulness of these cells for a characterization of
amino acid receptor mechanisms. We have been able to identify
distinct membrane responses of motoneurones and interneurones to
glutamate, to show the predominant involvement of Na^+ in excitatory
amino acid-induced depolarizations and the existence of separate
receptor sites for NMDA and quisqualate on motoneurones. This work
has also provided experimental support for the role of glutamate as
the transmitter of the monosynaptic EPSP on motoneurones. We
suggest that the amphibian spinal cord in vitro, even twenty-four
years after its introduction as an important model system for CNS
pharmacology, continues to fulfill the expectations raised in those
pioneering studies (Curtis et al, 1961).

Arenson, M.S. and Nistri, A. (1985). The effects of potassium
channel blocking agents on the responses of in vitro frog motoneurones to glutamate and other excitatory amino acids: an intracellular study. Neurocience, 14, 317-325.

Arenson, M.S., Berti, C., King, A. and Nistri, A. (1984). The
effect of D-α-aminoadipate on excitatory amino acid responses
recorded intracellularly from motoneurones of the frog spinal cord.
Neurosci. Lett., 49, 99-104.

Bührle, C.P. and Sonnhof, U. (1983). The ionic mechanisms of
excitatory action of glutamate upon the membranes of motoneurones
of the frog. Pflügers Arch., 396, 154-162.

Cull-Candy, S.G. and Usherwood, P.N.R. (1973). Two populations of
L-glutamate receptors on locust muscle fibres. Nature New Biol.,
246, 62-64.

Curtis, D.R., Phillis, J.W. and Watkins, J.C. (1961). Actions of
amino-acids on the isolated hemisected spinal cord of the toad.
Br. J. Pharmac., 16, 262-283.

Evans, R.H., Francis, A.A., Jones, A.W., Smith, D.A.S. and Watkins,
J.C. (1982). The effects of a series of ω-phosphonic α-carboxylic
amino acids on electrically evoked and excitant amino acid-induced
responses in isolated spinal cord preparations. Br. J. Pharmac.,
75, 65-75.

King, A. and Nistri, A. (1985). Current and voltage clamp studies
on the mixed agonist action of L-glutamate and its antagonism by
D-aminophosphonovalerate on spinal motoneurones in vitro. Soc.
Neurosci. Abstr., 11, in press.

King, A.E., Nistri, A. and Rovira, C. (1984). The excitation of frog motoneurones in vitro by the glutamate analogue, DL-α-amino-3-hydroxy-5-methyl-4-isoxazole-propionic acid (AMPA) and the effect of amino acid antagonists. Neurosci. Lett., 55, 77-82.

Krogsgaard-Larsen, P. and Honoré, T. (1983). Glutamate receptors and new glutamate agonists. Trends Pharmac. Sci., 4, 31-33.

Mayer, M. and Westbrook, G.L. (1984). Mixed-agonist action of excitatory amino acids on mouse spinal cord neurones under voltage clamp. J. Physiol., 354, 29-53.

McLennan, H. (1983). Receptors for the excitatory amino acids in the mammalian central nervous system. Prog. Neurobiol., 20, 251-271.

Nistri, A. and Arenson, M.S. (1983a). Differential sensitivity of spinal neurones to amino acids: an intracellular study on the frog spinal cord. Neuroscience, 8, 115-122.

Nistri, A. and Arenson, M.S. (1983b). Multiple postsynaptic responses evoked by glutamate on in vitro spinal motoneurones. Adv. Biochem. Psychopharmac., 37, 229-236.

Nistri, A., Arenson, M.S. and King, A. (1985). Excitatory amino acid-induced responses of frog motoneurones bathed in low Na^+ media: an intracellular study. Neuroscience, 14, 921-927.

Puil, E. (1981). S-Glutamate: its interaction with spinal neurons. Brain Res. Rev., 3, 229-322.

Takeuchi, A., Onodera, K. and Kawagoe, r. (1983). The effects of dorsal root stimulation on the release of endogenous glutamate from the frog spinal cord. Proc. Jap. Acad. B., 59, 88-92.

Watkins, J.C. and Evans, R.H. (1981). Excitatory amino acid transmitters. A. Rev. Pharmac. Toxicol., 21, 165-204.

17

L–Glutamic Acid Receptors on Locust Muscle: Pharmacological Properties and Single Receptor Channel Studies

M.S.P. Sansom and P.N.R. Usherwood

INTRODUCTION

Although evidence that L-glutamic acid is a neurotransmitter in the vertebrate central nervous system (CNS) is incomplete it is generally accepted that this amino acid is the mediator of transmission at arthropod nerve-muscle junctions (Usherwood, 1978, 1980a,b,1981; Nistri & Constanti, 1979). Given the complexities of vertebrate central nervous organization compared with the arthropod nerve-muscle system it is perhaps unwise to compare their pharmacological properties without major qualifications. Recent electrophysiological evidence suggests the presence of three types of amino acid receptor for glutamate in the CNS; a quisqualate-responsive site: a N-methyl-D-aspartate-responsive site and a kainate-responsive site (Watkins & Evans, 1981). This conclusion is generally supported by the results of ligand binding studies (e.g. Foster & Roberts, 1978). The quisqualate-responsive receptor is considered by some to be the main neurotransmitter receptor for L-glutamic acid in the CNS. The pharmacology of glutamatergic systems in arthropods is also complicated and there are differences between species (Usherwood, 1978; Nistri & Constanti, 1979; Piek, 1985). Amongst these arthropod systems, the glutamate receptors of locust leg muscle which form the subject of this review have been the most thoroughly investigated both electrophysiologically and biochemically. At excitatory junctions on locust muscle where L-glutamate is the chemical mediator and where the postjunctional receptor gates a cationic (Na, K, Ca) channel (Anwyl & Usherwood, 1974), the receptor sites are mainly quisqualate-responsive. However, at a minority of junctions ibotenate- and aspartate-responsive sites co-exist with quisqualate-responsive receptors (Usherwood, 1978; Gration et al., 1979a; Clark et al., 1979a). In addition autoreceptors for L-glutamic acid with, as yet, undefined pharmacological properties, are present on the nerve terminals of the excitatory motoneurones where they contribute to a negative feedback system for regulating transmitter release (Usherwood & Machili, 1968; Dowson & Usherwood, 1972).

The occurrence on innervated locust leg muscle of extra-junctional receptors for L-glutamic acid delineates this insect glutamatergic system from its arthropod counterparts. These extrajunctional receptors co-exist as two pharmacologically dis-tinct populations, the H-receptors and D-receptors which gate anionic (Cl) and cationic (Na, K, Ca) channels respectively (Usherwood & Cull-Candy, 1973; Cull-Candy & Usherwood, 1973; Lea & Usherwood, 1973; Kits et al., 1985). The D-receptors are quisqualate-sensitive whereas the H-receptors are responsive to ibotenic acid. Following denervation a marked increase in population density of D-receptors, but not H-receptors occurs (Usherwood, 1969; Clark et al., 1979c; Gration et al., 1979b). On the basis of extensive pharmacological studies of locust muscle it has been concluded that the quisqualate-responsive sites probably bind L-glutamic acid in partially folded conforma-tion, whereas the ibotenate-responsive sites probably bind L-glutamic acid in its fully extended conformation (Lea & Usherwood, 1973; Usherwood & Cull-Candy, 1975). The D-receptors and junctional D-receptors are only weakly responsive to kainic acid (Daoud & Usherwood, 1978).

All of the glutamate receptor populations on locust muscle are subject to desensitization (Usherwood & Machili, 1968; Clark et al., 1979b; Gration et al., 1980; Anis et al., 1981). Desensitization of the extrajunctional D-receptors, but not the H-receptors, can be inhibited by prior treatment of the muscle with concanavalin A (> 10^{-6}M) (Mathers & Usherwood, 1976, 1978). However, application of this lectin to a desensitized muscle does not reverse desensitization (Anis et al., 1979; Evans & Usherwood, 1985). After recovery following prolonged desensiti-zation of junctional and extrajunctional D-receptors with a high concentration of agonist (e.g. L-glutamic acid or L-quisqualic acid) the sensitivity of the muscle to later applications of agonist is reduced, suggesting the possibility of receptor down-regulation.

The crystal lattice of the nitrogen atom of quisqualic acid, which joins the heterocyclic system to the amino acid residue in this molecule has a pyramidal configuration comparable with the tetrahedral carbon in the corresponding position in L-glutamic acid. Whether this configuration is retained in solution or is rapidly interconverting remains to be established, but it may have an important bearing upon the pharmacological properties of this amino acid. Electrophysiological studies in our laboratory by Boden et al. (1985) suggest that interconversion is likely since D-quisqualic acid is active (only 10x less than L-quisqualic acid) on locust muscle, unlike D-glutamic acid (Usherwood & Machili, 1968; Clements & May, 1974). The inacti-vity of D-glutamic acid may stem from its inability to undergo suitable conformational adjustments when locking at the receptor; D-quisqualic acid, with its ability to change readily from a trigonal to a pyramidal arrangement may have much greater mobi-lity for docking.

SINGLE CHANNEL STUDIES OF EXTRAJUNCTIONAL D-RECEPTORS

The absence of extensive connective tissue on locust leg muscle makes possible the study of channel gating by single extrajunctional D-receptors using the patch clamp technique (Patlak et al., 1979).

CHANNEL GATING

Inhibition of receptor-channel desensitization by concanavalin A (Mathers & Usherwood, 1976,1978) greatly facilitates investigation of channel gating kinetics by allowing recordings to be made from single membrane patches for extended periods of time. Such recordings have been analysed in an attempt to develop models of the kinetics of receptor-channel gating by following the stochastic theory of single channel behaviour as discussed by Colquhoun & Hawkes (1983) and, more recently, by Fredkin et al. (1985).

Initial investigations centred about single channel data obtained at a single glutamate concentration (10^{-4}M) (Ashford et al., 1984; Kerry et al., 1985a). Data obtained from different preparations showed good site-to-site agreement. The single channel recordings were reduced to lists of channel opening and closing times using the threshold crossing detection method of Gration et al. (1981). Such event lists were the basis of further analysis. The first part of the analysis consisted of the construction of open and closed time histograms and the determination of the number of exponential terms required to fit the observed open and closed time distributions. As discussed by, for example, Colquhoun & Hawkes (1983), the numbers of such components give minimum estimates of the number of kinetically distinct states of the receptor channel. So, for single channel data obtained in the presence of 10^{-4}M glutamate, 3 exponential terms were required to fit the open time distribution and 4 exponential terms to fit the closed time distribution. Thus one may conclude that, at 10^{-4}M glutamate, there are at least 3 distinct open states of the receptor-channel and at least 4 distinct closed states. The next stage of the analysis involved investigation of the dependence of successive channel dwell (open or closed) times. This was achieved by estimation of channel dwell time autocorrelation functions, a means of analysis recently applied by Labarca et al. (1985) to the single channel data obtained from the nicotinic acetylcholine receptor Torpedo electroplax. A significant correlation between successive channel dwell times was demonstrated (Kerry et al., 1985c), indicative of more than one pathway via which the open states of the receptor-channel may isomerize to the closed states. Further analysis revealed at least 3 such pathways.

How do these results relate to our ideas concerning channel gating by the glutamate receptor? The demonstration of more than a single pathway linking the open and closed states of the channel excludes from further consideration linear gating schemes such as the following:-

$$2A + C \rightleftharpoons A + AC \rightleftharpoons A_2O$$

(where A represents the agonist; C the closed channel and O the open channel). Instead, we can express the results in terms of a minimal model for gating of the glutamate receptor-channel:-

$$
\begin{array}{ccccccc}
C_1 & \rightleftharpoons & C_2 & \rightleftharpoons & C_3 & \rightleftharpoons & C_4 \\
\updownarrow & & \updownarrow & & \updownarrow & & \\
O_1 & \rightleftharpoons & O_2 & \rightleftharpoons & O_3 & &
\end{array}
$$

Such a scheme does not include details of the agonist binding steps. Clearly to obtain information on such steps channel kinetics must be studied as a function of agonist concentration. Single channel recordings have been obtained over a range (10^{-6}M to 10^{-4}M) of glutamate concentrations and such analysis is underway. It is hoped that the end result of this will be an objective assessment of alternative channel gating schemes.

AGONIST POTENCY

Much of the early pharmacology of locust excitatory nerve-muscle junctions was determined by comparing the effects of different putative agonists and antagonists on whole nerve-muscle preparations using muscle twitch tension as an index of receptor activation/inactivation (Usherwood & Machili, 1968; Clements & May, 1974). However, twitch tension in this context belongs to the category of 'high-order effects' which do not necessarily bear a simple relationship to drug-receptor activation and cannot be used, as such, as a quantitative measure of agonist potency (Werman, 1969). With locust nerve-muscle preparations and with vertebrate CNS preparations the interpretation of data obtained following application of compounds to whole tissues, even when single cell recordings are made from such tissues, are complicated by the fact that amino acids may interact with spatially (and possibly pharmacologically) undefined receptor populations whose activities are modified by receptor desensitization, and the quantitative distribution both in time and space of these chemicals may be influenced by the presence of amino acid sequestration sites in these tissues (Usherwood & Machili, 1968; Lea & Usherwood, 1973; Cull-Candy & Usherwood, 1973; Clark et al., 1979b,1980; Gration et al., 1980; Anis et al., 1981). One advantage of the patch clamp technique is that it eliminates most, if not all, of these complications by limiting drug-receptor interactions to a single receptor channel protein complex. In other words it provides a way of directly studying molecular pharmacology and channel biophysics, since contact between the contents of the patch pipette and the receptor protein(s) isolated in the membrane patch under its tip result ideally in interactions of single molecules of ligand with a

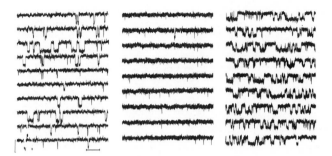

Fig. 1 Sample recordings of single channels obtained from adult locust muscle using mega-ohm seals, with 10^{-4}M L-glutamate (A), 10^{-3}M L-cysteine sulphinate (B) and 10^{-4}M L-quisqualate (C) in the patch pipette. All recordings were made with the muscle fibres voltage clamped at -100mV and at a bandwidth of 0-2.5KHz. Note obvious differences in the life-times or open times of the channels gated by these three amino acids. Calibration: 10pA; 25ms. (From Gration, Lambert, Ramsey & Usherwood, unpublished data).

single receptor molecule, each of which may lead to the gating of a channel.

Gration and Usherwood (1980) and Gration et al. (1981) demonstrated the usefulness of single channel recordings in obtaining information on agonist potency. By studying the three agonists L-glutamic acid, L-quisqualic acid and L-2-amino-3-sulphinopropionic acid (L-cysteine sulphinic acid) they showed that the conductance of the channel gated by the D-receptor of locust muscle is not influenced by agonist identity but that the mean open time of the channel and percentage of time that it spends in the open state is agonist dependent. Examples of channels gated by these amino acids are presented in Fig. 1. According to these criteria L-quisqualate is the most potent agonist of the three that were studied and the weakest is L-cysteine sulphinate, conclusions which accord with those arrived at previously by a variety of authors using whole muscle preparations (see reviews by Usherwood & Cull-Candy, 1975; Usherwood 1978). Similar conclusions about agonist potency derived from single channel studies of extrajunctional D-receptors were subsequently published by Cull-Candy et al. (1981).

The results of a more extensive study of agonist potency at the single channel level are presented in Table 1 which contains concentrations required to keep the channel open for 50% of the recording time.

Table 1. Relative potencies and apparent dissociation constants of L-glutamate and six analogues determined from analysis of single channel percentage open times.

	Potency relative to L-glutamate	Apparent K_D
L-Glutamate	1	4×10^{-4}M
L-Quisqualate	2	2×10^{-4}M
L-4-Methylene glutamate*	0.5	8×10^{-4}M
DL-4-Fluoroglutamate*	0.07	6×10^{-3}M
L-Cysteate	0.04	1×10^{-2}M
L-allo-4-Hydroxyglutamate	0.02	2×10^{-2}M
L-Cysteine sulphinate	0.013	3×10^{-2}M

* Mixture of isomers

CHANNEL BLOCKERS

Several positively charged drugs have been shown to block the locust glutamate receptor-channel, including curare (Cull-Candy & Miledi, 1983; Yamamoto & Washio, 1979). Single channel recordings from intact locust muscle has been used further to investigate the channel blocking activities of a variety of these agents, e.g. chlorisondamine (Ashford et al., 1984). Of these compounds, curare has been the most intensively investigated (Kerry et al., 1985b,c). Single channel kinetics have been investigated as a function of the relative concentrations of blocker (curare) and agonist (glutamate) in an attempt to probe the details of the channel blocking mechanism.

In the first series of experiments, the curare concentration in the patch pipette was held at 5×10^{-4}M, whilst the glutamate concentration was varied between 10^{-5}M and 10^{-3}M. The mean channel open time (m_o) was found to be constant, at about 0.25ms. In the absence of curare, m_o increased from 0.74ms at 10^{-5}M glutamate to 10.7 ms at 10^{-3}M glutamate.

In the second series of experiments, the glutamate concentration was fixed at 10^{-4}M and the curare concentration increased from 5×10^{-6}M to 5×10^{-4}M. The m_o decreased, in a dose dependent manner, from 0.73 ms (5×10^{-6}M curare) to 0.26 ms (5×10^{-4}M curare). Furthermore, increasing concentrations of curare also decreased the correlation between successive channel open times and between successive channel closed times, such correlations being a characteristic feature of the kinetics of the glutamate receptor-channel (Ball et al., 1985).

Taken together, these results are consistent with a scheme in which the channel blocker, curare, interacts with the open channel and dissociates relatively slowly once bound. There remains the possibility of some additional interaction between the closed channel and the blocker, a point which is currently under investigation.

CULTURED LOCUST MUSCLE

Current single channel analysis of the glutamate receptor employs mega-ohm seals with intact locust muscle. To extend kinetic studies to a finer time resolution and to probe the ion selectivity of the glutamate receptor channel in more depth requires application of giga-ohm seal techniques to locust muscle channels. So far, only limited progress has been made in applying these methods to the intact adult muscle. However, development of methods for growth of embryonic locust muscle in tissue culture has resulted in considerable progress.

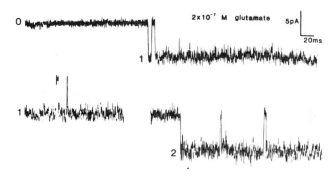

Fig. 2 A-C Single channel recording from an inside-out patch of membrane excised from a locust embryonic myofibre. Patch pipette held at +90mV. Note the 2 conductance states, which may indicate the presence of more than one channel in the membrane patch and the increased noise level when the channel is open (1; 2) compared with that seen when it is closed (o). The patch pipettee contained 2×10^{-7} M L-glutamate and 10^{-7} M Ca^{2+} (buffered with EGTA). The bulk solution was standard locust saline (Clark et al., 1980). (From Cook & Usherwood, unpublished data).

Embryonic locust muscle is grown in vitro as described by Cook et al. (1985), and after 1 month giga-ohm seal recordings may be obtained. The cell attached patch, inside-out patch and outside-out patch configuration may all be achieved.

With > 5 x 10^{-9}M glutamate in the patch pipette, channels of 90–125pS conductance may be obtained from most recording sites. Glutamate receptors appear to be uniformly distributed over the surface of the myofibre. However, even with the lowest concentration of glutamate, a minimum of 2 equal conductance states was usually seen (Fig. 2). The channel currents reversed at a membrane potential of 0 ± 10mV (S.D.) (n = 50). The noise level appeared to be higher when the channel was open. In the absence of concanavalin A the channels appeared in clusters separated by long closed times; application of concanavalin A resulted in continuous channel activity. This suggests that, as in the adult system, concanavalin A inhibits receptor-channel desensitization (Mathers & Usherwood, 1976). Preliminary kinetic analysis of single channel data from concanavalin A treated cells suggests that the embryonic channel may exist in multiple open and closed states, as does the adult channel.

These preliminary data suggest that embryonic locust muscle in vitro may be a useful system for further single channel analysis of the locust muscle glutamate receptor.

ACKNOWLEDGEMENTS: These studies were supported by grants from the SERC.

REFERENCES

Anis, N.A., Clark, R.B., Gration, K.A.F. and Usherwood, P.N.R. (1981). Influence of agonist on desensitization of glutamate receptors on locust muscle. J. Physiol. 312, 345–364.

Anis, N.A., Evans, M. and Usherwood, P.N.R. (1979). Plant lectins and desensitization of locust glutamate receptors. J. Physiol. 291, 47P.

Anwyl, R. and Usherwood, P.N.R. (1974). Voltage clamp studies of glutamate synapse. Nature, 252, 591–593.

Ashford, L.J., Boden, P., Ramsey, R.L., Shinozaki, H. and Usherwood, P.N.R. (1984). Voltage-dependent block by chlorisond-amine of glutamate channels in Schistocerca gregaria muscle fibres. J. Physiol. 357, 65P.

Ball, F., Sansom, M.S.P. and Usherwood, P.N.R. (1985). Clustering of single glutamate receptor-channel openings recorded from locust (Schistocerca gregaria) muscle. J. Physiol. 360, 66P.

Boden, P., Bycroft, B.W., Chhabra, S.R., Chiplin, J., Crawley,

P.J., Grant, R.J., King, T.J., McDonald, E., Raferty, P. and Usherwood, P.N.R. (1985). Structure-activity studies of quisqualic acid and related analogues at a defined glutamatergic synapse. (In preparation).

Clark, R.B., Gration, K.A.F. and Usherwood, P.N.R. (1979a). Responses to DL-ibotenic acid at locust glutamatergic neuromuscular junctions. Br. J. Pharmac. 66, 267-273.

Clark, R.B., Gration, K.A.F. and Usherwood, P.N.R. (1979b). Desensitization of glutamate receptors on innervated and denervated locust muscle fibres. J. Physiol. 290, 551-568.

Clark, R.B., Gration, K.A.F. and Usherwood, P.N.R. (1979c). Relative trophic influences of excitatory and inhibitory innervation of locust skeletal muscle fibres. Nature, 280, 679-682.

Clark, R.B., Gration, K.A.F. and Usherwood, P.N.R. (1980). Influence of glutamate and aspartate on the time course of decay of excitatory synaptic currents at locust neuromuscular junctions. Brain Res. 192, 205-216.

Clements, A.N. and May, T. (1974). Pharmacological studies on locust neuromuscular preparation. J. exp. Biol. 61, 421-442.

Colquhoun, D. and Hawkes, A.G. (1983). On the stochastic properties of single ion channels. Proc. Roy. Soc. Lond. B. 199, 205-235.

Cook, J., Ramsey, R.L. and Usherwood, P.N.R. (1985). Giga-ohm recordings of glutamate receptor-gated channels from locust muscle in vitro. J. Physiol. 361, 16P.

Cull-Candy, S.G., Miledi, R. and Parker, I. (1981). Single glutamate-activated channels recorded from locust muscle fibres with perfused patch electrodes. J. Physiol. 321, 195-210.

Cull-Candy, S.G. and Miledi, R. (1983). Block of glutamate-activated synaptic channels by curare and galamine. Proc. R. Soc. B. 218, 111-118.

Cull-Candy, S.G. and Usherwood, P.N.R. (1973). Two populations of L-glutamate receptors on locust muscle fibres. Nature 246, 62-64.

Daoud, M.A.R. and Usherwood, P.N.R. (1978). Desensitization and potentiation during glutamate application to locust skeletal muscle. Comp. Biochem. Physiol. 59C, 105-110.

Dowson, R. and Usherwood, P.N.R. (1972). The effect of low concentrations of L-glutamate and L-aspartate on transmitter release at the locust excitatory nerve-muscle synapses. J. Physiol. 229, 13-14P.

Evans, M.L. and Usherwood, P.N.R. (1985). The effect of lectins
on desensitization of locust muscle glutamate receptors. Brain
Res. (In press).

Foster, A.C. and Roberts, P.J. (1978) Affinity L-[^{3}H] glutamate
binding to postsynaptic receptor sites on rat cerebellar membranes. J. Neurochem. 31, 1467-1477.

Fredkin, D.R., Montal, M. and Rice, J.A. (1985). Identification
of aggregated Markovian models: Application to the nicotinic
acetylcholine receptor. Proc. of the Berkeley Conference in
Honour of Jerzy Neyman and Jack Kiefer Wadoworth Publ. Co. (In
press).

Gration, K.A.F., Clark, R.B. and Usherwood, P.N.R. (1979a).
Three types of L-glutamate receptor on junctional membrane of
locust muscle. Brain Res. 171, 360-364.

Gration, K.A.F., Clark, R.B. and Usherwood, P.N.R. (1979b).
Denervation of insect muscle: A comparative study of the changes
in L-glutamate sensitivity on locust retractor unguis and ex-
tensor tibiae muscle. Neuropharmac. 18, 201-208.

Gration, K.A.F., Lambert, J.J. and Usherwood, P.N.R. (1980). A
comparison of glutamate single-channel activity at desensitizing
and non-desensitizing sites. J. Physiol. 310, 49P.

Gration, K.A.F. and Usherwood, P.N.R. (1980). Interactions of
glutamate with amino acid receptors on locust muscle. Verh.
Dtsch. Zool. Ges. 1980, 122-132.

Gration, K.A.F., Lambert, J.J., Ramsey, R.L., Rand, R.P. and
Usherwood, P.N.R. (1981). Agonist potency determination by patch
clamp analysis of single glutamate receptors. Brain Res. 230,
400-405.

Kerry, C.J., Kits, K.S., Ramsey, R.L., Sansom, M.S.P. and
Usherwood, P.N.R. (1985a). Single channel kinetics of a glutamate receptor. (Submitted to Biophys. J.).

Kerry, C.J., Ramsey, R.L., Sansom, M.S.P., Usherwood, P.N.R. and
Washio, H. (1985b). The effect of tubocurarine on single glutamate-activated channels in locust muscle. J. Physiol. 365, 82P.

Kerry, C.J., Ramsey, R.L., Sansom, M.S.P., Usherwood, P.N.R. and
Washio, H. (1985c). Single channel studies of the action of
δ-tubocurarine on locust muscle glutamate receptors. (Submitted
to Br. J. Pharmac.)

Kits, K., Ramsey, R.L., Sansom, M.S.P. and Usherwood, P.N.R.
(1985). Ionic properties of channels gated by glutamate receptors

of locust muscle studied at the single channel level. (In preparation).

Labarca, P., Rice, J.A., Fredkin, D.R. and Montal, M. (1985). Kinetic analysis of channel gating: application to the cholinergic receptor channel and the chloride channel from Torpedo californica. Biophys. J. A7 469-478.

Lea, T. and Usherwood, P.N.R. (1973). The site of action of ibotenic acid and the identification of two populations of glutamate receptors on insect muscle fibres. Comp. gen. Pharmac. 4, 333-350.

Mathers, D. and Usherwood, P.N.R. (1976). Concanavalin A blocks desensitization of glutamate receptors of insect muscle fibres. Nature, 259, 409-411.

Mathers, D. and Usherwood, P.N.R. (1978). Effects of concanavalin A on junctional and extrajunctional L-glutamate receptors on locust skeletal muscle fibres. Comp. Biochem. Physiol. 59C, 151-155.

Nistri, A. and Constanti, A. (1979) Pharmacological characterization of different types of GABA and glutamate receptors in vertebrates and invertebrates. Progr. Neurobiol. 13, 117-235.

Patlak, J.B., Gration, K.A.F. & Usherwood, P.N.R. (1979). Single glutamate-activated channels in locust muscle. Nature, 278, 643-645.

Piek, T. (1985). Neurotransmission and neuromodulation of skeletal muscles In Comprehensive insect physiology, biochemistry and pharmacology. Eds. Kerkut, G.A. and Gilbert, L.T., Pergamon Press, Oxford. pp. 55-118.

Usherwood, P.N.R. (1969). Glutamate sensitivity of denervated insect muscle fibres. Nature, 203, 411-413.

Usherwood, P.N.R. (1978). Amino acids as neurotransmitters. Adv. Comp. Physiol. Biochem. 7, 227-309.

Usherwood, P.N.R. (1978). Glutamate receptors in eucaryotes. Adv. Pharmac. Ther. 1, 107-116.

Usherwood, P.N.R. (1980a). Neuromuscular transmitter receptors of insect muscle. In Receptors for Neurotransmitters, Hormones and Pheromones in Insects. Eds. D.B. Sattelle, L.M. Hall and J.G. Hildebrand. Elsevier/North Holland, Amsterdam and New York. pp. 141-152.

Usherwood, P.N.R. (1980b). Peripheral glutamatergic synapses in insects. In <u>Ontogenesis and Functional Mechanisms of Peripheral Synapses</u>. Ed. J. Taxi, Elsevier/North Holland, Amsterdam and New York. pp. 367–383.

Usherwood, P.N.R. (1981). Glutamate synapses and receptors on insect muscle. In <u>Glutamate as a Neurotransmitter</u>. Ed. Di Chiara, G. and Gessa, G.L., Raven Press, New York. pp. 183–193.

Usherwood, P.N.R. and Cull-Candy, S.G. (1973). Distribution of glutamate sensitivity on insect muscle fibres. Neuropharmac. <u>13</u>, 455–461.

Usherwood, P.N.R. and Cull-Candy, S.G. (1975). Pharmacology of somatic muscle. In <u>Insect Muscle</u>. Ed. P.N.R. Usherwood, Academic Press, New York, London. pp. 207–280.

Usherwood, P.N.R. and Machili, P. (1968). Pharmacological properties of excitatory neuromuscular organisms in the locust. J. Exp. Biol. <u>49</u>, 341–361.

Watkins, J.C. and Evans, R.H. (1981). Evaluating amino acid transmitters. Ann. Rev. Pharmac. <u>21</u>, 165–204.

Werman, R. (1969). An electrophysiological approach to drug-receptor mechanisms. Comp. Biochem. Physiol. <u>30</u>, 997–1017.

Yamamoto, D. and Washio, H. (1979). Curare has a voltage-dependent blocking action at the glutamate synapse. Nature, <u>281</u>, 372–373.

18

Discrimination of Excitatory Amino Acid Receptor Sub-types Using Radioligand Binding Techniques

G.E. Fagg, T.H. Lanthorn[1], A.C. Foster[2], L. Maier[3], J. Dingwall[3], J.D. Lane[4] and A. Matus

INTRODUCTION

Radioligand binding techniques have formed the basis of many investigations of the nature and properties of neurotransmitter receptors. The first attempts to exploit these techniques to label receptors for the excitatory amino acids were reported in 1974 by Roberts and by Michaelis and his coworkers. Using synaptic membrane fractions isolated from the rat brain, these authors demonstrated the presence of a population(s) of L-[^{14}C]glutamate binding sites which exhibited moderately high affinity for the radioligand and which could be blocked by other neuroexcitatory amino acids. Since that time, many papers have described the binding of L-[^{3}H]glutamate and related radioligands to brain membranes (reviewed by Foster & Fagg, 1984). Until recently, however, a consistent problem had been that, based on their ligand selectivity, none of the binding sites examined could be equated conclusively with the physiologically-defined receptor classes.

The past three years have witnessed major progress in the techniques of excitatory amino acid receptor binding and receptor autoradiography, such that it is now possible to label and study each of the three principal receptor types characterized in functional analyses (see next section). With respect to membrane binding, this has been achieved largely by (1) using subcellular fractions highly enriched in postjunctional membranes, (2) elucidating the effects of detergent and freeze-thaw pre-treatments and the effects of ions on the properties of L-[^{3}H]glutamate binding and (3) the introduction of more selective radioligands. This article will overview some of the key observations which have

_____
Present addresses: [1] NINCDS, National Institutes of Health, Bethesda, MD 20205, USA; [2] Neuroscience Research Centre, Merck, Sharp & Dohme Ltd., Harlow CM20 2QR, U.K.; [3] Ciba-Geigy AG, CH-4002 Basel, Switzerland; [4] Department of Pharmacology, Texas College of Osteopathic Medicine, Fort Worth, TX 76107, USA.

led to the identification of excitatory amino acid receptor subtypes in isolated brain membranes, indicate the radioligands
available to label individual sites, and present some preliminary
biochemical data on the nature of the postjunctional receptors. In
addition, a nomenclature is suggested to clarify reference to these
sites in future studies. The reader is referred to the chapters by
Cotman and Young and their collaborators (this volume) for summaries
of the literature on receptor autoradiography.

ACIDIC AMINO ACID RECEPTOR SUB-TYPES: FUNCTIONALLY-DEFINED SITES

Receptor studies using radioligand binding techniques are of
value only if it has been demonstrated that the ligand labels a
receptor of physiological or pharmacological significance. It is
therefore pertinent to briefly consider the receptor categories
which have been defined on the basis of functional studies.

The concept of multiple receptors for the excitatory amino acids
gained acceptance following the extensive pharmacological analyses
conducted in the late 1970s and early 1980s by McLennan and Watkins
and their collaborators (see chapters by these authors in this
volume; also Watkins & Evans, 1981; McLennan, 1983). Based on
studies of a large number of acidic amino acid analogues, these
investigators proposed the existence of three distinct excitatory
receptors, for which N-methyl-D-aspartate (NMDA), quisqualate and
kainate were potent and selective agonists. Rank ordering of the
potencies of a series of excitants indicated an approximately
inverse selectivity between the NMDA- and quisqualate-preferring
sites (McLennan & Lodge, 1979), whereas kainate-preferring receptors
displayed a distinct and more restricted specificity (Davies et al.,
1982).

The most well characterized excitatory amino acid receptor is
the NMDA-preferring sub-type. A number of antagonists for this site
have been described, of which the most potent and selective are the
ω-phosphonic acid derivatives, D-2-amino-5-phosphonopentanoate (D-
AP5) and D-2-amino-7-phosphonoheptanoate (D-AP7) (see Evans et al.,
1982). The relative potencies of this family of ω-phosphonic acid
homologues (3-8 carbon atoms) provide a valuable 'marker' for the
NMDA-preferring receptor category, and these compounds have been
utilized both to evaluate the role of this site in defined
physiological responses (e.g., Evans et al., 1982; Harris et al.,
1984) and to determine its presence in radioligand binding studies
(see later). Specific antagonists for the quisqualate- and kainate-
preferring receptors are currently unavailable, although two
peptides (γ-D-glutamylaminomethylsulphonate and γ-D-glutamyl-
taurine) recently have been reported to exhibit selectivity for
these sites (as opposed to the NMDA-preferring category) (Jones et
al., 1984), and L-glutamic acid diethylester has been used
successfully by some investigators to antagonize responses evoked by
quisqualate (e.g., McLennan & Lodge, 1979; Krogsgaard-Larsen et al.,

1980; but see Watkins & Evans, 1981).

The existence of additional excitatory amino acid receptor sub-types remains open to question. Koerner & Cotman (1981) have postulated a fourth category on the basis of the potent antagonist action of L-2-amino-4-phosphonobutanoate (L-AP4) at a sub-population of excitatory amino acid synapses (also see Slaughter & Miller, 1981; Evans et al., 1982; Lanthorn et al., 1984), and other authors have proposed additional sites based on the resistance of glutamate/aspartate-evoked responses to concentrations of antagonists which abolish NMDA, quisqualate and kainate-induced responses (Teichberg et al., 1984; Surtees & Collins, 1985; see chapters by Teichberg and Collins, this volume). More extensive analyses are required to determine the inter-relationship of these sites, as well as their precise locations and roles in the neuronal membrane.

ACIDIC AMINO ACID BINDING SITES: RELATIONSHIP TO FUNCTIONAL EXCITATORY RECEPTORS

The first evidence that Na^+-independent L-[^{3}H]glutamate binding sites in synaptic membranes are heterogeneous came with the discovery that these sites may be separated into two pharmacologically-distinct populations based on their Cl^--dependence (Fagg et al., 1982). Subsequent studies showed that these two sites may be distinguished further on the basis of their differential modulation by mono- and divalent cations (Mena et al., 1982; 1984; Fagg et al., 1984) and their lability to freeze-thaw (Fagg et al., 1983b) and detergent treatments (Fagg & Matus, 1984). Computer-assisted analyses of inhibitor dose-response curves also provided indirect support for the existence of more than one type of L-[^{3}H]glutamate binding site in synaptic membranes (Werling & Nadler, 1982; Slevin et al., 1982; Werling et al., 1983).

Recently, this heterogeneity has been resolved by using a subcellular fraction enriched in postsynaptic densities and essentially devoid of pre- and extrasynaptic membranes (Fagg & Matus, 1984). The postsynaptic density is a proteinaceous component of the postjunctional membrane, and is thought to provide a structural framework for transmitter receptors and other molecules involved in generating the postsynaptic neuronal response (see Matus, 1981). It is an especially prominent feature of excitatory (type I) synapses (Gray, 1969). Studies employing postsynaptic density preparations may therefore be expected to yield valuable information on the nature and properties of postjunctional excitatory amino acid receptor types.

The following sections summarize data on each of the populations of L-[^{3}H]glutamate binding sites which have been characterized; studies using more specific radioligands are discussed as appropriate, and the relationship (or not) of each site to the three principal receptor categories defined on the basis of functional experiments is clearly indicated (for additional details, see Foster & Fagg, 1984).

Cl⁻-Dependent Binding Sites

The Cl⁻-dependent population of L-[³H]glutamate binding sites was originally characterized by virtue of its high sensitivity to inhibition by the L-isomer of AP4 (Fagg et al., 1982; see Table 1). More recently, DL-[³H]AP4 has been employed as a selective radioligand for this site (Monaghan et al., 1983a; Butcher et al., 1983, 1984). The pharmacological profile, modulation by Ca^{2+} and Na^+, and lability to freeze-thawing (Fagg et al., 1983a,b; 1984; Mena et al., 1982, 1984; Monaghan et al., 1983a; Butcher et al., 1983, 1984) all indicate that this site is the principal binding site examined in many investigations of L-[³H]glutamate binding

Table 1. Inhibition of Cl⁻-dependent L-[³H]glutamate binding by ω-phosphono and ω-phosphino derivatives: lack of correspondence with L-AP4-sensitive synaptic receptors.

$HOOC(NH_2)CH(CH_2)_nPO_2H.R$

Compound	Structure	L-Glu Binding K_i, μM	Synaptic response IC_{50}, μM
A) Varying chain length, R = OH			
DL-AP3	n = 1	>1000[a]	5000[b]
L-AP4	= 2	5[a]	2.5[b]
D-AP4	= 2	75[a]	100[b]
DL-AP5	= 3	39[a]	250[b]
B) Varying ω-terminal, n = 2[c]			
	R = H	3.1	>100
DL-AP4	= OH	7.3	2.7
	= CH_3	3.4	>100
	= CH_2COOH	7.5	19
	= $CH_2CH_2CH(NH_2)COOH$	38	NT
	= $CH_2.C_6H_4Br$	2.4	>100
	= $CH_2.C_6H_3Cl_2$	3.0	NT

K_i values were derived by Scatchard analyses of inhibition data and IC_{50} values from log concentration-inhibition curves. The synaptic response examined was lateral perforant path-evoked field potentials recorded extracellularly in the outer molecular layer of the rat dentate gyrus *in vitro*. Data are from [a]Fagg et al. 1982; [b]Koerner & Cotman, 1981; [c]Fagg & Lanthorn, submitted. NT, not tested.

(e.g., Foster & Roberts, 1978; Baudry & Lynch, 1979, 1980, 1981; Lynch et al., 1982; Mewett et al., 1983; Mamounas et al., 1984; Siman et al., 1985). This is because, when freshly-prepared synaptic membranes are incubated in the presence of Cl^- (e.g., Tris-HCl buffer) $\pm$ Ca^{2+}, the AP4-sensitive population represents 70-80% of L-[^{3}H]glutamate specifically bound (Fagg et al., 1982).

What is the physiological role of Cl^--dependent L-glutamate (AP4) binding sites? Consideration of their overall ligand specificity indicates that these sites cannot be equated readily with either the NMDA-, quisqualate- or kainate-preferring receptors identified in physiological experiments (see Foster & Fagg, 1984). Initial studies, however, suggested that this population might correspond to the L-AP4-sensitive receptors described by Koerner & Cotman (1981). AP4 and a number of related analogues displayed essentially the same potencies and stereoselectivity (1) as inhibitors of Cl^--dependent L-glutamate binding and (2) as antagonists of lateral perforant path-evoked synaptic field potentials in the rat dentate gyrus in vitro (Fagg et al., 1982; see Table 1A). However, more extensive analyses have shown that this correspondence was probably fortuitous. In our recent investigations, a series of α- and ω-phosphono and phosphino analogues of glutamate have been found to exhibit disparate activities between Cl^--dependent binding sites and lateral perforant path-granule cell synaptic responses (Fagg & Lanthorn, manuscript submitted). Six of these analogues are shown in Table 1B. Whereas all ω-phosphino analogues are good to moderately potent inhibitors of Cl^--dependent L-glutamate binding, only one (the ω-carboxymethylphosphino derivative) is an effective antagonist in electrophysiological experiments. Moreover, compounds which are active in the binding assay but inactive as antagonists do not reverse the synaptic depressant effect of AP4, demonstrating that these compounds are neither agonists nor antagonists at AP4-preferring receptors (data not shown). Similar observations, but using a different series of glutamate analogues, recently have been made by Robinson et al. (1985). Hence, it appears that Cl^--dependent L-glutamate (AP4) binding sites in synaptic membranes are not the receptors through which AP4 mediates its neuropharmacological actions.

If Cl^--dependent L-glutamate binding sites are not AP4-preferring receptors, what is their function? A recent report suggests that the 'binding' observed in synaptic membrane fractions may in fact be a Cl^--driven uptake of L-glutamate into resealed plasma membrane vesicles (Pin et al., 1984), and several lines of evidence support this hypothesis. Firstly, a number of membrane transport systems, including those for glutamate and aspartate in neural tissue (Kuhar & Zarbin, 1978; Marvizon et al., 1981; Waniewski & Martin, 1984) exhibit a Cl^--requirement. Furthermore, the anion specificity of these systems is similar to that for Cl^--dependent L-glutamate binding (Mena et al., 1982, 1984; Butcher et al., 1983, 1984), and corresponds to those anions known to permeate the plasma membrane (and hence able to establish a transmembrane electrochemical gradient). Secondly, the high density of Cl^--dependent L-glutamate (AP4) binding sites, and their temperature- and freeze-thaw-sensitivity (see Foster & Fagg, 1984; also Monaghan et al., 1983a; Butcher et al., 1983; Robinson et al., 1985), are characteristics generally associated with membrane transport processes. The number of binding sites, in particular, is greatly in excess of those determined for other neurotransmitter receptors in synaptic membranes, including receptors expected to be present in abundance, such as those for GABA (see Olsen, 1982). Thirdly, Cl^--dependent L-glutamate binding sites are absent from isolated postsynaptic densities (Fagg & Matus, 1984; see next section). Hence, the weight of evidence currently available favours the idea that the 'binding' of L-[^{3}H]glutamate and DL-[^{3}H]AP4 measured in the presence of Cl^- largely does not represent a receptor interaction, but rather corresponds to the sequestration of radioligand in resealed synaptic plasma membrane vesicles.

Cl^--Independent Binding Sites

NMDA-preferring receptors. The first data leading to the positive identification of NMDA-preferring receptors using radioligand binding techniques was gained from studies of the Cl^--independent population of L-[^{3}H]glutamate binding sites in synaptic membranes (Fagg et al., 1983a). The absolute and relative inhibitory potencies of the ω-phosphonic acid homologues, and the high potencies of L-aspartate and NMDA (relative to Cl^--dependent binding sites), suggested that a sub-component of these sites corresponded to the NMDA-preferring receptor defined by electrophysiological analyses. This was further supported by experiments involving frozen and thawed synaptic membranes. In these preparations, the Cl^--dependent population of binding sites is absent (see previous section), and the inhibitory activity of NMDA is even more apparent (Fagg et al., 1983b).

Recently, two groups have succeeded in labeling NMDA-preferring receptors in rat brain membrane fractions, either by employing preparations enriched in postsynaptic densities (Fagg & Matus, 1984; see above) and/or by using specific radioligands (Olverman et al., 1984; Fagg & Foster, 1985). In isolated postsynaptic densities, the Cl^--dependent population of L-[^{3}H]glutamate binding sites is not detectable, and the predominant site appears to correspond to the NMDA-preferring receptor (Fagg & Matus, 1984). Thus, in contrast to observations in synaptic membranes, Cl^- has little effect on the binding of L-[^{3}H]glutamate to postsynaptic densities, and AP4 is a weak inhibitor both in the presence and absence of this anion. Examination of the complete series of ω-phosphonic acid homologues reveals the characteristic potency profile expected for the NMDA-preferring receptor, with peaks of activity at the 5-carbon (AP5, K_i 1.0 µM) and 7-carbon (AP7, K_i 4.6 µM) members of the series. In addition, N-methyl-DL-aspartate (NMA) itself inhibits L-[^{3}H]gluta-

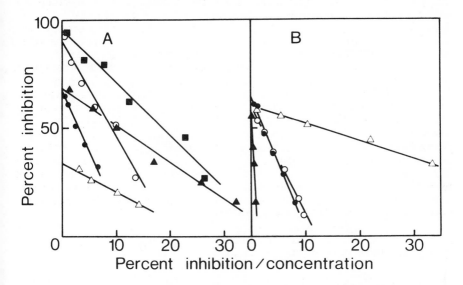

Fig. 1. Scatchard plots of the inhibition of L-[³H]glutamate binding to postsynaptic densities by (A) ●NMA, ○ibotenate, ▲L-aspartate, △quisqualate, ■L-glutamate (abscissa scale is x 0.1) and (B) ●NMA, ○DL-AP7, ▲quinolinate and△DL-AP5. NMA, AP5, AP7 and quinolinate act at one subpopulation of L-glutamate binding sites (see B) and quisqualate at another (see A). Some compounds bind to both populations. Data from Fagg & Matus (1984).

mate binding with a K_i in the low micromolar range. Scatchard analyses demonstrate that NMA, AP5 and AP7 all interact with the same sub-population of L-glutamate binding sites in postsynaptic densities (Fig. 1), and the overall ligand selectivity of this site confirms that this is the NMDA-preferring receptor identified in physiological studies (for details, see Fagg & Matus, 1984).

Investigations using other radioligands have yielded similar results. Thus, D-[³H]AP5 (Olverman et al., 1984), as well as DL-[³H]AP5, D-[³H]aspartate and [³H]NMDA (Fagg & Foster, 1985; see Table 2) all label a site with pharmacological properties essentially the same as those of NMDA-sensitive L-[³H]glutamate binding sites, and corresponding to those expected of an NMDA-preferring receptor. However, the affinity of these ligands for the receptor is somewhat lower than that of L-[³H]glutamate, with the consequence that the ratio of specific to non-specific binding is less favourable for accurate analyses of binding and inhibition kinetics. In crude postsynaptic densities, specific L-[³H]glutamate binding represents 87% of the total binding, whereas the respective

Table 2. Comparative properties of the binding of L-[^{3}H]glutamate and [^{3}H]NMDA to crude postsynaptic densities[a].

Compound	Concentration μM	Percent inhibition of binding L-[^{3}H]glutamate	[^{3}H]NMDA
DL-AP4	50	22	22
DL-AP5	50	72 (K_i 1.0 μM)	97 (IC_{50} 5.6 μM)
DL-AP6	50	30	17
DL-AP7	50	65	71
L-glutamate	10	96 (K_i 0.2 μM)	99 (IC_{50} 0.12 μM)
L-aspartate	10	85	87
NMA	10	53 (K_i 7.2 μM)	55 (IC_{50} 9.5 μM)
Quisqualate	10	25	33
Kainate	50	16	18

[a]Data are from Fagg & Foster (1985) and unpublished observations.

values for D-[^{3}H]aspartate, DL-[^{3}H]AP5 and [^{3}H]NMDA are 61%, 37% and 29% (Fagg & Foster, 1985). It seems that these four substances together bracket the borderline of affinity for useful radioligands at NMDA-preferring receptors, with L-[^{3}H]glutamate (K_d 0.1-0.3 μM) at the useable end of the spectrum and [^{3}H]NMDA (K_d 2-5 μM, as estimated from inhibition studies) at the other extreme.

Quisqualate-preferring receptors. The most significant advance towards the development of a receptor binding assay for quisqualate-preferring receptors was the synthesis of the heterocyclic compound, α-amino-3-hydroxy-5-methyl-4-isoxazolepropionic acid (AMPA) (Krogsgaard-Larsen et al., 1980) and its subsequent radiolabeling (Honore et al., 1982). AMPA is a weak inhibitor of Cl$^-$-dependent L-[^{3}H]glutamate (AP4) binding (Werling et al., 1983; Butcher et al., 1983) and this, coupled with the low density of [^{3}H]AMPA binding sites in synaptic membranes and their resistance to freeze-thawing (Honore et al., 1982), suggests that the sites labeled by this radioligand form part of the Cl$^-$-independent population of L-[^{3}H]glutamte binding sites. Studies of the binding of L-[^{3}H]glutamte to postsynaptic densities support this proposal (Fagg & Matus, 1984; also see Fig. 1).

Electrophysiological experiments indicate that AMPA acts primarily at the quisqualate-preferring type of excitatory amino acid receptor (Krogsgaard-Larsen et al., 1980), and the ligand selectivity of [^{3}H]AMPA binding sites is in accord with this idea (Honore et al., 1982). Thus, quisqualate, AMPA and L-glutamate are powerful inhibitors of binding, and a series of ibotenate analogues (which act at quisqualate-preferring receptors) show parallel potencies in the binding assay and as neuronal excitants.

Interestingly, kainate also is an inhibitor of [³H]AMPA binding (Honore et al., 1982; see Fagg & Matus, 1984), although its potency at this site is 2-3 orders of magnitude lower than at kainate-preferring receptors (Simon et al., 1976; London & Coyle, 1979; Foster et al., 1981).

Few reports of [³H]AMPA binding have been published to date, and several groups have experienced difficulty with this assay. One possibility is that this is due to the rather low density of these receptors in synaptic membranes (see Honore et al., 1982). Recently, however, Honore et al. (1985) have reported that chaotropic ions enhance the affinity of [³H]AMPA for its recognition site, and this should facilitate investigations of the quisqualate-preferring receptor type in future studies.

Kainate-preferring receptors. Saturable binding of [³H]kainate to rat brain membrane preparations was first reported by Simon et al. (1976), and has since been described by a number of authors (e.g., London & Coyle, 1979; Foster et al., 1981; Slevin et al., 1983). Like [³H]AMPA recognition sites (see previous section), [³H]kainate binding sites are present at low density and are unaffected by Cl⁻ or by freeze-thawing. In addition, they are enriched in synaptic junction fractions (Foster et al., 1981). For these reasons, kainate sites do not appear to be part of the Cl⁻-dependent population of L-[³H]glutamate binding sites.

The selectivity of the [³H]kainate recognition site is distinct from and more restricted than that of other acidic amino acid binding sites. With the exception of kainate itself and the structurally-analogous domoate, only quisqualate, α-ketokainate and L-glutamate exhibit high potency (Simon et al., 1976; London & Coyle, 1979; Slevin et al., 1983). Furthermore, structure-activity analyses using a series of kainate analogues have demonstrated the importance of the isopropenyl side-chain, a substituent which has no homology in the glutamate molecule (Slevin et al., 1983). A similar pharmacological profile is seen both in neuroexcitatory and neurotoxicity tests (Coyle et al., 1981; Davies et al., 1982) and this, together with the similar regional distributions of kainate binding sites and neurons susceptible to kainate-induced toxicity (see Foster & Fagg, 1984), constitutes firm evidence that [³H]kainate labels kainate-preferring receptors as defined by functional criteria.

Unlike the NMDA- and quisqualate-preferring receptor types described in the previous sections, membrane binding sites with high affinity (K_i 5-50 nM) for kainic acid have not currently been labeled using L-[³H]glutamte. This may be due to the very low density of these sites in whole brain membranes (such that they form an indetectable proportion of L-[³H]glutamate binding sites) since, using autoradiographic methods, a discrete population of kainate-sensitive L-glutamate binding sites has been identified (Monaghan et al., 1983b) which corresponds to those sites labeled by [³H]kainate

(Monaghan & Cotman, 1982).

RECEPTOR CHARACTERIZATION AND CLASSIFICATION: A SUMMARY AND SOME PROPOSALS

Recent years have seen a convergence of data from membrane binding, autoradiographic and functional analyses of excitatory amino acid receptor activity, with the result that the three principal receptor categories characterized physiologically have been identified using radioligand binding techniques (Table 3). An

Table 3. Acidic amino acid receptor sub-types.

Nomenclature[a]	A1	A2	A3
Trivial name	NMDA	Quisqualate	Kainate
A) Functional analyses			
Most potent and selective agonists	NMDA Ibotenate	Quisqualate AMPA	Kainate Domoate
Most potent and selective antagonists	D-AP5 D-AP7	(GAMS) (Glu-tau)	
B) Direct receptor assays			
Most potent ligands	L-glutamate D-AP5 D-aspartate	Quisqualate AMPA L-glutamate	Kainate Quisqualate L-glutamate
Most selective ligands	D-AP5	AMPA	Kainate
Most suitable radioligands[b]	L-glutamate D-AP5	AMPA L-glutamate	Kainate

[a]Nomenclature according to Foster & Fagg (1984). [b]Other radioligands (e.g., L-aspartate, L-cysteine sulphinate, L-cysteate, DL-AP7, N-acetyl-aspartyl-glutamate; see Foster & Fagg, 1984; Koller & Coyle, 1984) have not been demonstrated to label sites with characteristics expected of currently defined receptor types. Abbreviations are as in the text except for GAMS: γ-D-glutamylamino methylsulphonate and Glu-tau: γ-D-glutamyltaurine.

important finding from the membrane binding studies is that both the NMDA- and quisqualate-preferring receptor types are associated with purified postsynaptic densities (Fagg & Matus, 1984) and hence, in agreement with electrophysiological investigations in a number of pathways (see refs. in Watkins & Evans, 1981; Fagg & Matus, 1984), are probably directly involved in the generation of postsynaptic excitatory responses. The kainate-preferring sub-type is enriched in synaptic junction fractions (which comprise pre- and postsynaptic membrane structures, see Foster et al., 1981) but its presence specifically in the postsynaptic membrane has not been examined.

Qualitatively and, to some extent, quantitatively, each of the identified binding sites shows a ligand selectivity similar to that of its corresponding physiological receptor; indeed, this has formed a primary criterion for receptor identification _in vitro_. However, receptor studies in intact tissue inevitably are complicated by other cellular processes (of which ligand uptake is certainly an important factor, see Garthwaite, 1985; Foster & Fagg, 1984), and a number of differences between receptor binding and functional analyses are now emerging. For example, electrophysiological investigations suggest that NMDA is the most powerful agonist at NMDA-preferring receptors, and that L-aspartate is more potent than L-glutamate (Watkins & Evans, 1981; McLennan, 1983); in contrast, direct binding assays demonstrate that L-glutamate is the most potent ligand currently known at this receptor (Olverman et al., 1984; Fagg & Matus, 1984; Fagg & Foster, 1985). Similarly, quisqualate and kainate are each thought to be highly selective receptor agonists. Radioligand binding assays, however, have shown that kainate interacts moderately well with 'quisqualate receptors' (Honore et al., 1982, 1985; Fagg & Matus, 1984) and that quisqualate is as potent at 'kainate receptors' as it is at its own recognition site (compare Simon et al., 1976; London & Coyle, 1979; Slevin et al., 1983 with Honore et al., 1982, 1985; Fagg & Matus, 1984). Quisqualate, in fact, appears to be a relatively non-specific ligand, binding with high affinity not only to quisqualate- and kainate-preferring receptors, but also to Cl^--dependent L-glutamate 'binding' sites (see Foster & Fagg, 1984) and (less potently) to NMDA-sensitive recognition sites (Greenamyre et al., 1985; Fagg & Foster, unpublished observations).

These observations raise a number of points. Most obviously, it is clear that precise measures of the potency and selectivity of ligand-receptor interactions can only be gained by eliminating other cellular influences; this situation is found under the conditions employed for radioligand binding assays, and partially when assaying receptor function using dissociated neurons (Garthwaite, 1985). Secondly, the widely-used terms 'NMDA receptors', 'quisqualate receptors' and 'kainate receptors' are misnomers. Excitatory amino acid receptors can be neither defined nor identified on the basis of their 'sensitivity' to NMDA, quisqualate or kainate (in fact, this is likely to yield misleading results), but rather on the basis of a broader profile of ligand selectivity. In view of the correspondence now achieved between radioligand binding and physiological techniques regarding receptor sub-types, this seems to be an appropriate time to propose the wider use of a nomenclature which avoids connotations of single ligand selectivity. Such a nomenclature was previously introduced (Foster & Fagg, 1984), and is presented again in Table 3. Thus, in accordance with schemes utilized for other neurotransmitter receptors, the three principal

Acidic amino acid receptors are termed the **A1** (NMDA-preferring), **A2** (quisqualate-preferring) and **A3** (kainate-preferring) sub-classes. The A4 site (AP4-preferring, Foster & Fagg, 1984) is eliminated from this scheme since agreement between binding and functional analyses is lacking (moreover, in view of the proposed transport function of Cl^--dependent binding sites, investigations of this site may require reinterpretation; see above).

FUTURE DIRECTIONS

The identification of excitatory amino acid receptor sub-types using membrane binding and autoradiographic techniques opens up a number of questions for future research. For example, what changes occur in receptor type, number and distribution during development, ageing and disease, and how are these regulated? What is the molecular structure and relationship of receptor sub-types? Our preliminary biochemical investigations indicate that the postjunctional receptors are destroyed by high temperature and by trypsin and chymotrypsin, but are little affected by calpains I and II, concanavalin A or by the thiol reagent p-hydroxymercuribenzoate (Table 4). Hence, these sites appear to be proteinaceous in nature, but are distinct from the concanavalin A-sensitive 'glutamate binding protein' isolated by Michaelis and his coworkers (Michaelis et al., 1974, 1983; this volume) and, in contrast to the Cl^--dependent L-glutamate 'binding' (transport?) sites (Baudry & Lynch, 1980; Baudry et al., 1981; Siman et al., 1982, 1985), are not affected by Ca^{2+}-activated proteases. Honore and his colleagues (this volume), using the irradiation inactivation method, have provided preliminary data on the molecular size of individual excitatory amino acid receptor sub-types and evidence that each site

Table 4. Biochemical properties of L-[^{3}H]glutamate binding sites in crude postsynaptic densities.

Treatment	Percent binding remaining
60°C	1
Trypsin, 1%	9
Chymotrypsin, 1%	24
Calpain I, 1%	89
Calpain II, 1%	82
Concanavalin A, 0.5 mg/ml	76
p-Hydroxymercuribenzoate, 0.1 mM	95

Data are from Fagg, Baud & Matus, unpublished. Calpain I and II were kindly provided by Dr. T. Murachi.

is a distinct molecular entity. Further studies in these directions promise to yield exciting new data on the molecular and regulatory properties of acidic amino acid receptors and their involvement in neuronal function.

REFERENCES

Baudry, M. & Lynch, G. (1979). Nature (Lond.) 282, 749-750.
Baudry, M. & Lynch, G. (1980). Proc. Natl. Acad. Sci. (USA), 77, 2298-2302.
Baudry, M. & Lynch, G. (1981). J. Neurochem., 36, 811-820.
Baudry, M., Bundman, M.C., Smith, E.K. & Lynch, G. (1981). Science, 212, 937-938.
Butcher, S.P., Collins, J.F. & Roberts, P.J. (1983). Br. J. Pharmacol., 80, 355-364.
Butcher, S.P., Roberts, P.J. & Collins, J.F. (1984). J. Neurochem., 43, 1039-1045.
Coyle, J.T. (1981). Neurosci. Res. Prog. Bull., 19, 354-360.
Davies, J., Evans, R.H., Jones, A.W., Smith, D.A. & Watkins, J.C. (1982). Comp. Biochem. Physiol., 72, 211-224.
Evans, R.H., Francis, A.A., Jones, A.W., Smith, D.A. & Watkins, J.C. (1982). Br. J. Pharmacol., 75, 65-75.
Fagg, G.E., Foster, A.C., Mena, E.E. & Cotman, C.W. (1982). J. Neurosci., 2, 958-965.
Fagg, G.E., Foster, A.C., Mena, E.E. & Cotman, C.W. (1983a). Eur. J. Pharmacol., 88, 105-110.
Fagg, G.E., Mena, E.E., Monaghan, D.T. & Cotman, C.W. (1983b). Neurosci. Lett., 38, 157-162.
Fagg, G.E. & Matus, A. (1984). Proc. Natl. Acad. Sci. (USA), 81, 6876-6880.
Fagg, G.E., Riederer, B. & Matus, A. (1984). Life Sci., 34, 1739-1745.
Fagg, G.E. & Foster, A.C. (1985). J. Physiol. (Lond.) (Abstract), in press.
Foster, A.C. & Roberts, P.J. (1978). J. Neurochem., 31, 1467-1477.
Foster, A.C., Mena, E.E., Monaghan, D.T. & Cotman, C.W. (1981). Nature (Lond.), 289, 73-75.
Foster, A.C. & Fagg, G.E. (1984). Brain Res. Rev., 7, 103-164.
Garthwaite, J. (1985). Br. J. Pharmacol., 85, 297-307.
Gray, E.G. (1969). Prog. Brain Res., 31, 141-155.
Greenamyre, J.T., Olson, J.M.M., Penney, J.B. & Young, A.B. (1985). J. Pharmac. exp. Ther., 233, 254-263.
Harris, E.W., Ganong, A.H. & Cotman, C.W. (1984). Brain Res., 323, 132-137.
Honore, T., Lauridsen, J. & Krogsgaard-Larsen, P. (1982). J. Neurochem., 38, 173-178.
Honore, T. & Nielsen, M. (1985). Neurosci. Lett., in press.
Jones, A.W., Smith, D.A. & Watkins, J.C. (1984). Neuroscience, 13, 573-581.
Koerner, J.F. & Cotman, C.W. (1981). Brain Res., 216, 192-198.

Koller, K.J. & Coyle, J.T. (1984). Eur. J. Pharmacol., **104**, 193–194.

Krogsgaard-Larsen, P., Honore, T., Hansen, J.J., Curtis, D.R. & Lodge, D. (1980). Nature (Lond.), **284**, 64–66.

Kuhar, M.J. & Zarbin, M.A. (1978). J. Neurochem., **31**, 251–256.

Lanthorn, T.H., Ganong, A.H. & Cotman, C.W. (1984). Brain Res., **290**, 174–178.

London, E.D. & Coyle, J.T. (1979). Molec. Pharmacol., **15**, 492–505.

Lynch, G., Halpain, S. & Baudry, M. (1982). Brain Res., **244**, 101–111.

Mamounas, L.A., Thompson, R.F., Lynch, G. & Baudry, M. (1984). Proc. Natl. Acad. Sci.. (USA), **81**, 2548–2552.

Marvizon, J.G., Mayor, F., Aragon, M.C., Gimenez, C. & Valdivieso, F. (1981). J. Neurochem., **37**, 1401–1406.

Matus, A. (1981). Trends in Neurosci., **4**, 51–53.

McLennan, H. & Lodge, D. (1979). Brain Res., **169**, 83–90.

McLennan, H. (1983). Prog. Neurobiol., **20**, 251–271.

Mena, E.E., Fagg, G.E. & Cotman, C.W. (1982). Brain Res., **243**, 378–381.

Mena, E.E., Whittemore, S.R., Monaghan, D.T. & Cotman, C.W. (1984). Life Sci., **35**, 2427–2433.

Mewett, K.N., Oakes, D.J., Olverman, H.J., Smith, D.A.S. & Watkins, J.C. (1983). Adv. Biochem. Psychopharmacol., **37**, 163–174.

Michaelis, E.K., Michaelis, M.L. & Boyarski, L.L. (1974). Biochim. Biophys. Acta, **367**, 338–348.

Michaelis, E.K., Michaelis, M.L., Stormann, T.M., Chittenden, W.L. & Grubbs, R.D. (1983). J. Neurochem., **40**, 1742–1753.

Monaghan, D.T. & Cotman, C.W. (1982). Brain Res., **252**, 91–100.

Monaghan, D.T., McMills, M.C., Chamberlin, A.R. & Cotman, C.W. (1983a). Brain Res., **278**, 137–144.

Monaghan, D.T., Holets, V.R., Toy, D.W. & Cotman, C.W. (1983b). Nature (Lond.), **306**, 176–179.

Olsen, R.W. (1982). Ann. Rev. Pharmacol. Toxicol., **22**, 245–277.

Olverman, H., Jones, A.W. & Watkins, J.C. (1984). Nature (Lond.), **307**, 460–462.

Pin, J.-P., Bockaert, J. & Recasens, M. (1984). FEBS Lett., **175**, 31–36.

Roberts, P.J. (1974). Nature (Lond.), **252**, 399–401.

Robinson, M.B., Crooks, S.L., Johnson, R.L. & Koerner, J.F. (1985). Biochemistry, **24**, 2401–2405.

Siman, R., Baudry, M. & Lynch, G. (1982). Soc. Neurosci. Abstr., **8**, 251.24.

Siman, R., Baudry, M. & Lynch, G. (1985). Nature (Lond.), **313**, 225–228.

Simon, J.R., Contrera, J.F. & Kuhar, M.J. (1976). J. Neurochem., **26**, 141–147.

Slaughter, M.M. & Miller, R.F. (1981). Science, **211**, 182–185.

Slevin, J., Collins, J.F., Lindsley, K. & Coyle, J.T. (1982). Brain Res., **249**, 353–360.

Slevin, J.T., Collins, J.F. & Coyle, J.T. (1983). Brain Res., **265**, 169–172.

Surtees, L. & Collins, G.G.S. (1985). Brain Res., **334**, 287–295.

Teichberg, V.I., Tal, N., Goldberg, O. & Luini, A. (1984). Brain Res., 291, 285-292.
Waniewski, R.A. & Martin, D.L. (1984). J. Neurosci., 4, 2237-2246.
Watkins, J.C. & Evans, R.H. (1981). Ann. Rev. Pharmacol. Toxicol., 21, 165-204.
Werling, L.L. & Nadler, J.V. (1982). J. Neurochem., 38, 1050-1062.
Werling, L.L., Doman, A. & Nadler, J.V. (1983). J. Neurochem., 41, 586-593.

19

Anatomical Organization of NMDA, Kainate and Quisqualate Receptors

D.T. Monaghan and C.W. Cotman

The anatomical localization of excitatory amino acid receptors is necessary for evaluating their functional role in specific pathways and systems in the CNS. Autoradiography allows a quantitative description of anatomical and pharmacological properties of binding sites in specific CNS locations which can then be correlated to function. Receptors in a single dendritic field can be both biochemically characterized and rigorously studied by electrophysiological techniques. It is also possible, once the sites are defined, to study their development, the effects of lesions, and the consequences of treatments upon receptors in discrete pathways.

Autoradiographic analysis using radiolabelled excitatory amino acids reveals three classes of binding sites whose pharmacological profiles correspond to the three physiologically-identified receptor classes. Under the appropriate conditions, binding sites can be detected which appear to correspond to the N-methyl-D-aspartate (NMDA), kainate (KA), and quisqualate (QA) receptor subpopulations. These sites can be reliably assessed by either the selective radioligands, D-$[^3H]$-2-amino-5-phosphonopentanoate (D-$[^3H]$-AP5), $[^3H]$-KA and $[^3H]$-amino-3-hydroxy-5-methyl-4-isoxazolepropionate ($[^3H]$-AMPA) respectively, or by the relatively nonselective ligand, L-$[^3H]$-glutamate. These sites each show a marked anatomical specificity, and an overall distribution and pharmacological profile appropriate for excitatory amino acid-mediated transmission.

In the presence of Ca^{++} and Cl^- ions, L-$[^3H]$-glutamate binds at two additional sites, neither corresponding to the 2-amino-4-phosphonobutyrate-sensitive Ca^{++}/Cl^--dependent site observed in membrane preparations. In general, there is good agreement between autoradiographic studies after differing conditions are taken into account. Thus, it becomes apparent that autoradiography can now serve as a powerful experimental method for studying excitatory amino acid receptors in discrete areas of the CNS. The goal of this review is to illustrate that autoradiography is an appropriate method for characterizing excitatory amino acid binding sites which reflect the physiologically appropriate receptors.

AUTORADIOGRAPHY OF NMDA RECEPTORS

<u>NMDA-Sensitive L-[^{3}H]-Glutamate Binding Sites</u>

Autoradiographic visualization of the NMDA receptor was first described using L-[^{3}H]-glutamate as the radioligand (Monaghan et al., 1983a). In whole brain, NMDA-sensitive L-[^{3}H]-glutamate binding displays rapid and reversible binding. Determination of the dissociation constant at 0°C by analysis of the rate constants indicates a Kd of 1.0 μM (Monaghan and Cotman, 1985). Binding is not facilitated by increases in temperature, rather binding levels are actually higher when measured at 0°C than when measured at 30°C. This difference may be due to a lower affinity at 30°C. Using equilibrium or saturation analysis (Scatchard plot) we obtained a Kd of 0.6 μM at 30°C, and a Kd of 0.3 μM at 0°C.

Although L-[^{3}H]-glutamate labels several binding sites, we have found it possible to study the anatomical distribution of NMDA site binding selectively by using NMDA as a displacer. To study the pharmacological characteristics of the NMDA binding site, we have assessed the properties of L-[^{3}H]-glutamate binding sites in regions enriched in NMDA sites. NMDA agonists and antagonists interact with this binding site with the appropriate potency, while acidic amino acids which are not potent at this receptor are relatively ineffective (Monaghan et al., 1983a; 1985a; Fig. 1a). Thus, the NMDA antagonist D-α-aminoadipate and the NMDA agonists L-homocysteate, L-glutamate, L-aspartate, D-aspartate, D-glutamate, ibotenate, and NMDA are potent displacers of the NMDA-sensitive L-[^{3}H]-glutamate binding. Compounds which do not act at the NMDA receptor, most notably KA, QA, AMPA, and AP4, are poor displacers of NMDA-sensitive L-[^{3}H]-glutamate binding.

Of particular value in the identification of NMDA receptors is the phosphonic acid series with increasing carbon chain length from 2-amino-4-phosphonobutyrate (AP4) to 2-amino-8-phosphonoctanoate (AP8). The 5 and 7 carbon analogues (AP5 and 2-amino-7-phosphonoheptanoate, AP7) exhibit potent and selective NMDA receptor antagonism while the 4, 6, and 8

Figure 1. Pharmacological Specificity of Three Distinct L-[^{3}H]-Glutamate Binding Sites. L-[^{3}H]-Glutamate binding in stratum radiatum of hippocampus (A), stratum lucidum (B), and stratum radiatum in the presence of 100 μM NMDA (C) exhibit acidic amino acid specificities corresponding to NMDA, KA, and AMPA receptors respectively. Inhibition by compounds tested at multiple concentrations were expressed as % inhibition of maximal inhibition (approximately 80%, 80%, and 50% of total specific binding in the three cases respectively). Compounds tested at 100 μM were expressed as % inhibition of total specific binding. Displacement curves are also shown for cold ligand displacement of D-[^{3}H]-AP5 and [^{3}H]-AMPA binding to stratum radiatum in (A) and (C). [^{3}H]-KA binding (100 - % of Bmax) to whole tissue sections is shown in (B). Abbreviations: D-GLU, D-glutamate; D-AA, D-α-aminoadipate; IBO, ibotenate; L-ASP, L-aspartate; D-ASP, D-aspartate; L-SOS, L-serine-O-sulfate; L-HC, L-homocysteate. Data from Monaghan and Cotman, 1982; Monaghan et al., 1983a; 1984a; 1984b; 1985a).

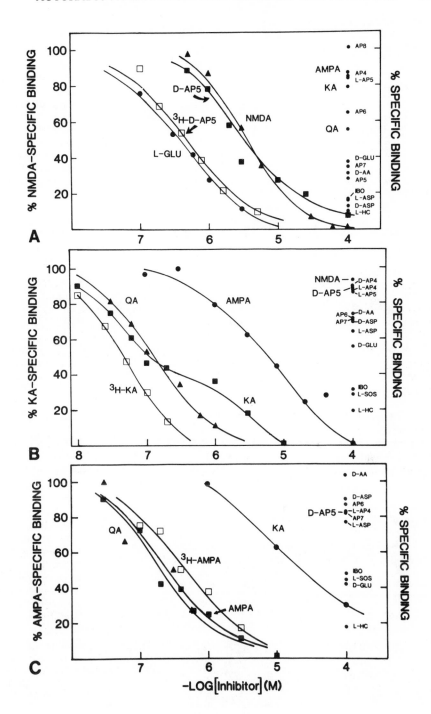

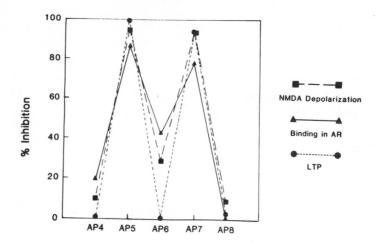

Figure 2. NMDA Excitation and Binding Exhibit Similar Pharmacological Profiles. Inhibition by a series of ω-phosphono acidic amino acid analogs of NMDA-induced focal depolarizations, LTP, and NMDA-sensitive L-[³H]-glutamate binding shows a similar potency in each system. AP5, AP6, AP7, and AP8 are the longer chain derivatives (pentanoate, hexanoate, heptanoate, and octanoate) of 2-amino-4-phosphonobutyrate (AP4) (data from Harris et al., 1984; and Monaghan et al., 1985a). All measurements are from the stratum radiatum of hippocampus and tested with 100 μM inhibitor.

carbon analogues are poor NMDA antagonists (Davies and Watkins, 1983; Watkins, 1984). Within a single, well-defined dendritic field (hippocampal CA1 stratum radiatum), we have characterized the potency of these compounds as 1) antagonists of NMDA-induced depolarizations, 2) antagonists of the formation of long term potentiation (LTP), and 3) inhibitors of NMDA-sensitive L-[³H]-glutamate binding (Harris et al., 1984; Monaghan et al., 1985a). As shown in Figure 2, each of these measures displays the characteristic profile of NMDA receptors. Furthermore, in each of these systems it is the D rather than the L isomer of AP5 which has NMDA site activity. Together, the pharmacological data strongly support the NMDA receptor identification of this binding site. In apparent confirmation of this identification is the demonstration that L-glutamate has a similar high affinity for NMDA receptors labelled by D-[³H]-AP5 (Olverman et al., 1984). Furthermore, Greenamyre et al., (1985) have recently confirmed the presence of this NMDA-sensitive L-[³H]-glutamate binding site, and they find that the NMDA antagonist β-D-aspartyl-aminomethylphosphonate is also a potent displacer of this binding site.

The NMDA-sensitive L-[³H]-glutamate binding site has not been extensively studied in membrane preparations. Initially, we reported that a major proportion of L-[³H]-glutamate binding sites observed in the absence of Cl- ions has properties similar to that of the NMDA receptor (Fagg et al., 1983). This site has been described in postsynaptic density preparations

(Fagg and Matus, 1984); and under these conditions, this site exhibits properties similar to the site observed in autoradiography (see also the chapter by Fagg in this volume).   More recently, we have studied the pharmacological profile of these binding sites in membrane preparations using conditions optimal for this class of binding site.   With these conditions, the resulting pharmacological profile is indistinguishable from that obtained in autoradiographic preparations and is appropriate for that of the NMDA receptor (Monaghan et al., 1985b).

NMDA-sensitive L-[^{3}H]-glutamate binding sites are found predominately within the telencephalon (Monaghan and Cotman, 1985; Monaghan et al., 1985a). Highest levels of binding are found in the stratum radiatum and stratum oriens of the hippocampal CA1 field. The

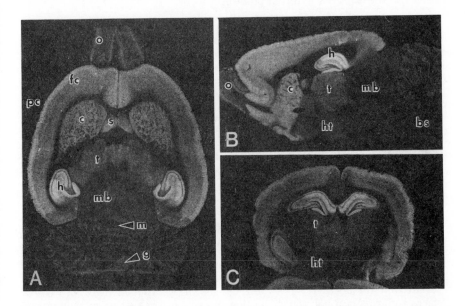

Figure 3.   Distribution of NMDA-Sensitive L-[^{3}H]-Glutamate Binding Sites. Tritium sensitive film autoradiograms of L-[^{3}H]-glutamate binding in rat brain using conditions which result in predominately NMDA-sensitive site labelling (cf Fig. 6a, 6b). Data from Monaghan and Cotman, (1985). Planes of section were A. horizontal, B. parasaggital, C. coronal. Abbreviations for Figures 3, 5, 6, 7, and 8: o, olfactory bulb; fc, frontal cortex; pc, parietal cortex; tc, temporal cortex; ec, entorhinal cortex; cc, cingulate cortex; c, caudate/putamen; s, septum; h, hippocampus; t, th, thalamus; mb, midbrain; m, cerebellar molecular layer; g, cerebellar granule cell layer; ht, hypothalamus; bs, brain stem; 1-3, layers I, II, and III of cerebral cortex; r, reticular nucleus of thalamus; 5, layer V of cerebral cortex; me, median eminence; pt, pituitary; ls, lateral septum; p, pyramidal cell layer of hippocampus; Cb, cerebellum; Cx, cerebral cortex; 5a, layer Va of cerebral cortex; L, stratum lucidum of hippocampus; ml, molecular layer of the dentate gyrus.

corresponding layers in CA3 have moderately high levels of binding as does the inner layer of the dentate gyrus molecular layer. Within the hippocampus, relatively low levels of binding are found in the hilus and in the stratum lucidum. Cerebral cortex displays moderate to high levels of binding with higher densities found in layers I, II, III, and Va. Frontal, pyriform, anterior cingulate, and perirhinal regions contain higher levels of binding than do the parietal, posterior cingulate, and entorhinal corticies. Within the basal ganglia, the nucleus accumbens has the highest levels of NMDA sites, closely followed by the caudate nucleus. These regions have considerably higher levels of binding than does the globus pallidus. The anterior olfactory nuclei, olfactory accessory bulb, olfactory tubericles, and nucleus of the lateral olfactory tract all have relatively high levels of binding sites. Within the olfactory bulb, the highest concentration of these sites is found within the external plexiform layer.

In general, thalamic regions have moderate levels of binding sites whereas the hypothalamus has low levels. Within the thalamus the anterior dorsal, and certain midline nuclei (e.g. rhomboid, and reuniens) have higher binding levels than do the reticular nucleus and the zona incerta. Midbrain and brain stem have an overall low density of NMDA-sensitive sites with certain regions having higher densities. These include the superficial gray and intermediate gray layers of the superior colliculus, dorsal medial inferior colliculus, dorsal raphe nucleus, central gray, cuniform nucleus, granule cell layer of the cochlear nucleus, medial vestibular nucleus, parabrachial nucleus, nucleus of the solitary tract, and the substantia gelatinosa of the spinal cord. The only ventral, brain stem structure preferentially labelled was the inferior olive. Low levels are also found within the cerebellum, with the granule cell layer exhibiting more binding than does the molecular layer (for quantitative values see Monaghan and Cotman, 1985).

In the areas studied by Greenamyre et al. (1985) (hippocampus, cerebral cortex and cerebellar cortex) a similar distribution of NMDA-sensitive sites was reported. However, they describe the layers of higher binding within neocortex as layers I, II, and IV, whereas our studies indicate that layers I, II, III, and Va exhibit the higher levels of binding. It is perhaps relevant that the cortical autoradiogram shown by Greenamyre et al., (1985) appears to be from a horizontal section of area 2, a region of relatively thin layers I, II, and III (Krieg, 1946).

D-[^{3}H]-AP5 Binding Sites

NMDA receptors may also be labelled by D-[^{3}H]-AP5 (Olverman et al., 1984). Although D-[^{3}H]-AP5 rapidly dissociates from its binding site (dissociation half-life of 5 seconds), autoradiograms prepared with this ligand display the appropriate pharmacological profile and a distinct anatomical specificity (Monaghan et al., 1984b). Consistent with the properties of the NMDA receptor, D-[^{3}H]-AP5 binding sites exhibit higher affinity for L-glutamate, L-homocysteate, L-aspartate, D-AP5, and NMDA than for QA, L-AP4, KA, and AMPA. Since L-glutamate appears to have the same high affinity for the D-[^{3}H]-AP5 binding site as it does for the NMDA-sensitive L-[^{3}H]-glutamate binding site, and since the binding site density is also

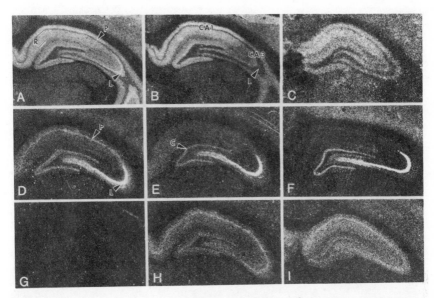

Figure 4. D-AP5-, KA-, and AMPA-sensitive L-[³H]-Glutamate Binding Sites and the D-[³H]-AP5, [³H]-KA, and [³H]-AMPA Binding Sites Have Similar Distributions Within the Hippocampus. Radioligands: L-[³H]-glutamate, A, B, D, E, G, and H; D-[³H]-AP5, C; [³H]-KA, F; [³H]-AMPA, G. (A) represents total L-[³H]-glutamate binding (in the absence of Ca^{++}/Cl^- ions); other L-[³H]-glutamate incubations contained the following displacers: B, 10 µM KA; D, 100µM NMDA; E, 100 µM NMDA and 10 µM L-serine-O-sulfate (to remove the AMPA-sensitive sites); G, 500 µM L-glutamate; and H, 100 µM NMDA and 1µM KA. Note the similar distributions obtained with different radioligands in B and C; E and F; and H and I. Data from Monaghan et al., 1983a; 1984a; 1984b.

comparable, it seems likely that these two ligands are binding to the same site. This conclusion is reinforced by the observation that both ligands have their highest binding site density within the hippocampus and that within this structure they have identical distributions (compare Figs. 4B and 4C). In both autoradiographic and membrane fraction ligand binding experiments, D-AP5 more potently displaces D-[³H]-AP5 than NMDA-sensitive L-[³H]-glutamate binding (Fagg et al., 1983; Monaghan et al., 1983a; 1984b; 1985a; 1985b; Olverman 1984).

AUTORADIOGRAPHY OF KA RECEPTORS

[³H] Kainate Binding Sites

The [³H]-KA binding site was the first excitatory amino acid binding site to be characterized by autoradiographic techniques. Our initial observation of a strikingly high density of [³H]-KA binding sites within the

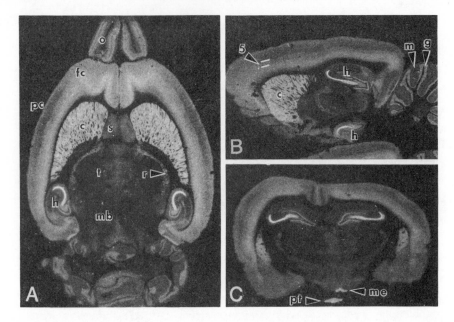

Figure 5. Distribution of [³H]-KA Binding Sites in Rat Brain. A. horizontal plane, B. parasaggital plane, and C. coronal plane. Abbreviations listed in legend to Fig. 3.. Autoradiographic procedure modified from Monaghan and Cotman, 1982.

stratum lucidum of the hippocampus (Foster et al., 1981) has now been confirmed by various studies (Monaghan and Cotman, 1982; Unnerstall and Wamsley, 1983; Berger and Ben-Ari, 1983; Berger et al., 1984). [³H]-KA binding sites exhibit high affinity, reversible, specific binding. Pharmacologically, these binding sites are similar to those described in membrane preparations (Simon et al., 1976; London and Coyle, 1979). The [³H]-KA sites observed in autoradiography are displaced by low concentrations of KA, QA, and L-glutamate, whereas dihydrokainate, D-glutamate, and L-aspartate were less effective (Monaghan and Cotman, 1982; Unnerstall and Wamsley, 1983). Together with the similar anatomical distributions obtained for [³H]-KA binding sites in the two preparations, it is likely that the same binding site is being labelled.

[³H]-KA binding sites are found predominately within the telencephalon (Monaghan and Cotman, 1982; Unnerstall and Wamsley, 1983; see Figure 5). The caudate/putamen has the highest binding levels of major structures (in agreement with dissection/membrane-fraction binding experiments), and the hippocampal stratum lucidum has the highest binding site density altogether.   Binding within the cerebral cortex exhibits variations among both layers and regions. As with other excitatory amino acid binding sites, higher levels are found in the frontal cortex, anterior olfactory nuclei, and perirhinal cortex, and lower levels are found in the parietal, temporal, and entorhinal corticies. Particular to [³H]-KA sites is the higher density of binding in the insular cortex than in the pyriform cortex (Fig. 5c) This latter observation is interesting in view of the greater sensitivity of the pyriform cortex to KA excitotoxic damage (Schwob et al., 1980). Also in contrast to the other excitatory amino acid binding sites is the relatively high density of [^{3}H]-KA sites in layer V and yet higher binding site density in layer VI of neocortex (Fig. 5b). In the thalamus and hypothalamus the distribution is again complimentary to that found for the NMDA sites. That is, the hypothalamus, the thalamic reticular nucleus, and the zona incerta have higher levels than the remaining portions of the thalamus. Both receptors exhibit moderate densities in the midline thalamic nuclei. Within the rat cerebellum, both NMDA and KA sites are found in higher concentrations in the granule cell layer than in the molecular layer. A particularly interesting finding is the strikingly high concentration of [^{3}H]-KA binding sites in the posterior pituitary and median eminence reported by Unnerstall and Wamsley (1983), see Fig. 5c.

Lower brain regions contain low levels of [^{3}H]-KA binding. Using the recently available high specific activity [^{3}H]-KA, we have found that KA sites display significant anatomical specificity within the lower brain regions (Monaghan and Cotman, manuscript in preparation). Similar to NMDA-sensitive L-[^{3}H]-glutamate binding sites, [^{3}H]-KA binding sites are found in the nucleus of the solitary tract, the granule cell layer of the dorsal cochlear nucleus, and the medial vestibular nucleus. In contrast to NMDA sites, the pontine nucleus contains relatively high levels of [^{3}H]-KA binding sites.

Few studies have been perfomed in species other than rat. We find that [^{3}H]-KA binding sites in human brain have a distribution similar to that found for rat (Geddes et al., 1985). In pigeon brain, Henke et al., (1981) have reported a high density of [^{3}H]-KA binding sites within the molecular layer of the cerebellum.

KA-Sensitive L-[^{3}H]-Glutamate Binding Sites

L-[^{3}H]-Glutamate may also be used to label KA binding sites. L-Glutamate is known to have a high affinity for [^{3}H]-KA binding sites (Simon et al., 1976; London and Coyle, 1979). However, given the relatively low maximal binding site density displayed by KA sites compared to the sum of the other glutamate binding sites, KA would not be expected to displace a significant amount of L-[^{3}H]-glutamate binding in membrane preparations. Using quantitative autoradiography we have demonstrated that L-[^{3}H]-glutamate binding sites in the stratum lucidum of the rat hippocampus are predominately KA binding sites (Monaghan et al., 1983a; Monaghan et al., 1985a). As with [^{3}H]-KA binding sites in membrane preparations, KA-sensitive L-[^{3}H]-glutamate binding sites are more potently displaced by KA, QA, and L-glutamate than by D-glutamate and AMPA (Fig. 1b). Furthermore, these sites exhibit little displacement by 100 μM concentrations of NMDA, D-AP5, L-APB, L-aspartate, D-aspartate, and D-α-aminoadipate. These sites exhibit the same distribution as do [^{3}H]-KA binding sites. Highest concentrations are found in the stratum lucidum of the hippocampus, and relatively high levels are also found in the

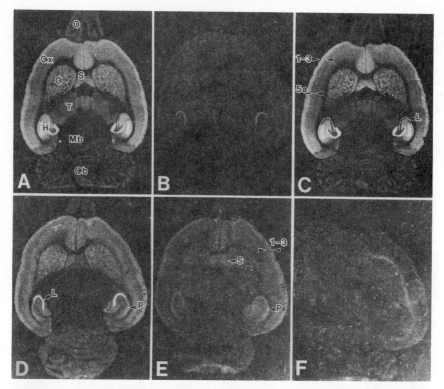

Figure 6.   Distribution of NMDA-, KA-, and AMPA-Sensitive L-[³H]-Glutamate Binding Sites. L-[³H]-glutamate was incubated in the presence of the following compounds: A, no displacers; B and D, 100 μM NMDA; C, 1 μM KA; E, 100 μM NMDA and 1 μM KA; F, 500 μM L-glutamate. For clarity, Figures D, E, and F are printed with more contrast and brighter than A, B, and C. Figures A and B show NMDA displacement; Figures A and C, and D and E show KA displacement; and the binding in Figure E is AMPA sensitive as shown in Figures 8c and 8d. Abbreviations listed in legend to Figure 3.

caudate/putamen, dentate gyrus inner molecular layer, and deep cerebral cortical layers (compare Figure 5a with Figures 6d and 6e). Relatively low levels are found in the septum, outer cortex, and lower brain regions.

Multiple KA binding sites

There are two related issues to consider when comparing the [³H]-KA and KA-sensitive L-[³H]-glutamate binding sites observed in autoradiographic preparations to those described using membrane preparations: The effect of ions upon KA binding sites, and the presence of high and low affinity KA binding sites. Beaumont et al, (1979), have reported that Ca++ and other ions will readily prevent [³H]-KA binding in membrane preparations. Others, however, have not found inhibition by Ca++ (discussed in Foster and Fagg, 1984). Using L-[^{3}H]-glutamate as a ligand in autoradiographic preparations, we find L-[^{3}H]-glutamate binding to KA sites to be potently inhibited by Ca++ ions (Monaghan et al., 1983a); however, when [^{3}H]-KA is the ligand, Ca++ is relatively ineffective (Monaghan et al., 1985b). These discrepencies appear to be accounted for by the observation (Monaghan et al., 1985b; Dr. Honore and colleagues, personal communication) that Ca++ selectively inhibits binding at the high affinity [^{3}H]-KA sites. Since the KA-sensitive L-[^{3}H]-glutamate binding sites are quite Ca++-sensitive, whereas the [^{3}H]-KA (at 100 nM) are mostly Ca++-insensitive, these results suggest that L-[^{3}H]-glutamate predominately labels high affinity sites, and that [^{3}H]-KA (100 nM) binding in Tris-citrate buffer represents predominately low affinity sites. This conclusion is consistent with the known affinities of KA and L-glutamate for the high and low affinity [^{3}H]-KA binding sites (London and Coyle, 1979). Since L-glutamate has lower affinities for the KA binding sites, 100 nM L-[^{3}H]-glutamate results in mostly high affinity binding. If K+ acts like Ca++ to inhibit the high affinity sites, this might explain the result in our earlier report (Monaghan and Cotman, 1982) that the [^{3}H]-KA binding in the presence of a K+-containing buffer was predominately of low affinity character. Furthermore, Unnerstall and Wamsley (1983) used a Tris-citrate buffer (which does not affect high affinity binding sites) and reported the presence of a high affinity component.

Another relevant factor for the interpretation of the [^{3}H]-KA binding is the relative dissociation rates of the [^{3}H]-KA binding sites. It has been reported that high affinity [^{3}H]-KA binding sites exhibit much slower dissociation than the low affinity sites (London and Coyle, 1979). Thus, longer rinsing of the tissue sections in autoradiography should result in a selective labelling of the high affinity site. However, in our experiments using EGTA-washed membranes (Monaghan et al., 1985b) we find that following partial dissociation (10 minutes after adding excess unlabelled KA) there is only a modest increase in Ca++-sensitivity. This result is consistent with the Scatchard analysis of Unnerstall and Wamsley, (1983) where after 3 minutes of dissociation the proportion of high and low affinity sites is not greatly different than when measured under equilibrium conditions. Likewise, Berger et al., (1984) have also observed a sizable low affinity component following 5 minutes of dissociation and have proposed the presence of a slow-dissociating low affinity [^{3}H]-KA binding site. Thus, our results are in agreement with Berger et al., (1984), and suggest that there are at least 3 kinetically-distinct KA sites (or states). Consequently, experiments assessing the regional localization of the differing KA sites (e.g. Berger and Ben-Ari, 1983) will be important for determining the significance of the multiple sites (or states).

AUTORADIOGRAPHY OF QUISQUALATE (AMPA) RECEPTORS

[^{3}H]-AMPA Binding Sites

The third major class of excitatory amino acid receptor is preferentially activated by QA and AMPA. The binding site corresponding to this receptor appears to be selectively labelled in membrane (Honore et al, 1982) and autoradiographic preparations by $[^3H]$-AMPA (Monaghan et al., 1984a; Rainbow et al., 1984). $[^3H]$-AMPA exhibits saturable, reversible, high affinity binding to sites located throughout the brain. In either membrane fraction or autoradiographic methods, $[^3H]$-AMPA binding exhibits the same characteristic pharmacology. Binding is most potently displaced by AMPA, QA, and L-glutamate (inhibition constants between 0.1 and 1 μM). KA has a moderate affinity (7-10 μM) for this site; this affinity is 100 fold lower than displayed by $[^3H]$-KA for KA binding sites (Simon et al., 1976; London and Coyle, 1979). D-Glutamate has a still lower affinity, while L- and D- aspartate, NMDA, D-AP5, L-AP4 cause little displacement at a 100 μM concentration. The identification of this binding site as that of the QA receptor is based upon the similar order of potency exhibited by various compounds between QA receptor activation and $[^3H]$-AMPA binding inhibition (Honore et al., 1982). A more definitive identification of this binding site could be made if potent and selective QA receptor antagonists were available.

It is important to note that the property of having a high affinity for QA is shared by other excitatory amino acid binding or uptake sites (the Cl—dependent L-AP4-sensitive L-$[^3H]$-glutamate or $[^3H]$-AP4 apparent binding site, Fagg et al., 1983; Butcher et al., 1983; Monaghan et al., 1983b; the Cl—dependent L-$[^3H]$-glutamate apparent binding site of the cerebellar molecular layer observed in autoradiography, Greenamyre et al., 1984; 1985; see Fig. 8; the Cl—independent, AP4-sensitive binding site recently described by Nadler et al., 1985; the KA binding site, Simon et al., 1976; London and Coyle, 1979; and the AMPA binding site, Honore et al., 1982). Thus, QA-sensitivity, in itself, does not prove that an apparent binding site is the QA receptor (or a subtype of QA receptor). Conversely, there are at least two sites observed in autoradiography which are relatively insensitive to QA (the NMDA site and the hippocampal Cl-/Ca++ dependent site observed in autoradiography, Fagg and Matus, 1984; Monaghan et al., 1983a). Thus, QA-insensitivity is also not selective. Because of the lack of specificity displayed by QA we have focused upon the properties of the distinct class of binding sites preferentially displaced by AMPA.

Highest densities of $[^3H]$-AMPA binding sites are found over the hippocampal pyramidal cells and over the cell bodies of the indusium grisium, suggesting a neurochemical similarity between these developmentally related cells. High levels are found in the CA1 stratum radiatum and stratum oriens of the hippocampus (Fig. 7). Outer cortical layers (I, II, and III) have higher levels of binding than layers V and VI while layer IV has lower levels. In posterior sections, there is a dense band of binding found in layer V, probably corresponding to layer Vb. The caudate/putamen and the nucleus accumbens have more dense binding than does the globus pallidus, substantia innominata, and ventral pallidum. Basolateral, lateral, and posterior amygdaloid nuclei have higher levels of binding than the adjacent central, medial, and anterior cortical nuclei. Mid brain and brain stem have considerably lower levels of binding.

This distribution of $[^3H]$-AMPA binding sites is largely similar to that of the NMDA sites with the exception of relatively higher concentrations of

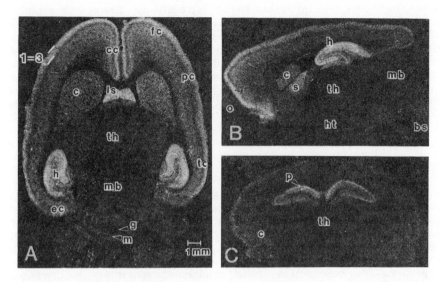

Figure 7.   Distribution of [^{3}H]-AMPA Binding Sites. Rat brain tissue sections were incubated with [^{3}H]-AMPA and processed for autoradiography.   Shorter autoradiographic film exposure time (C) emphasizes higher binding levels in the hippocampal pyramidal cell layer. A. horizontal plane, B. parasaggital plane, and C. coronal plane. Abbreviations listed in legend to Fig. 3. Data from Monaghan et al., 1984b.

[^{3}H]-AMPA binding in the hippocampal pyramidal cell layer, the induseum griseum, the lateral septum, the dentate gyrus hilus, and the layer V binding observed in posterior cortex.   Also distinct from the NMDA-sensitive L-[^{3}H]-glutamate binding is the relatively lower levels of [^{3}H]-AMPA binding found in the thalamus, the Va layer of anterior cortex, the external plexiform layer of the olfactory bulb, and the granule cell layer of the cerebellum.   Although there is a striking qualitative and quantitative agreement between the study of Monaghan et al., (1984a) and Rainbow et al., (1984), the latter study found relatively lower levels of binding over the hippocampal pyramidal cell layer.   This could possibly be the result of the latter group's use of thicker tissue sections (32 versus 6 µm) or the use of a longer rinse time (10 versus 0.5 minutes).

AMPA-Sensitive L-[^{3}H]-Glutamate Binding Sites

As with NMDA sites, the first report of AMPA site autoradiography was with the use of L-[^{3}H]-glutamate (Monaghan et al., 1983a).   After displacing the NMDA-sensitive L-[^{3}H]-glutamate binding, one can observe the AMPA-sensitive L-[^{3}H]-glutamate binding sites over the pyramidal cell layer of the hippocampus (Figs. 4 and 6).   The pharmacological profile of these sites is similar to that found for the [^{3}H]-AMPA binding sites (Monaghan et al., 1985a; see figure 1c).   AMPA, QA, and L-glutamate each have a high affinity; kainate and D-glutamate have an intermediate affinity; and D- and L-aspartate, NMDA, D-AP5, and L-AP4 have relatively low affinities for this site (little displacement at 100 μM). After displacement of the NMDA- and KA- sensitive L-[^{3}H]-glutamate binding sites, the remaining binding is AMPA-displacable and is found in the outer cortical layers, the anterior cingulate cortex, the lateral septum, and the hippocampus. Since AMPA-sensitive L-[^{3}H]-glutamate binding sites and the [^{3}H]-AMPA binding sites exhibit a similar pharmacological profile, distribution, and binding site density, it is most likely that the two ligands are labelling the same binding site population.

EVIDENCE EXCLUDING LABELLING DUE TO UPTAKE OR METABOLISM

In studying the binding of acidic amino acids to whole brain tissue (as in autoradiographic preparations) it is conceivable that apparent binding could be some form of sequestration or uptake, binding to enzymes, or metabolism. Four lines of evidence indicate that these problems do not account for the binding at the three pharmacologically-identified binding sites described above. 1) The NMDA, KA, and AMPA sites correspond very well to the sites described in purified membrane preparations. Furthermore, the NMDA and AMPA sites have been described in purified postsynaptic densities (Fagg and Matus, 1984), while the NMDA and KA sites have also been described in purified synaptic junctions (Foster et al., 1981; Yao et al., 1984). 2) Binding at these sites occurs in the absence of Na$^+$ and Cl$^-$ ions which are thought to be required for acidic amino acid transport (Wheeler et al., 1978; Waniewski and Martin, 1983; 1984). 3) Binding to these sites is not significantly increased by an increase in temperature. 4) L-[^{3}H]-glutamate, [^{3}H]-KA, D-[^{3}H]-AP5, and [^{3}H]-AMPA bound to tissue sections, then extracted, comigrated with authentic ligand upon thin layer chromatography plates. In some studies of L-[^{3}H]-glutamate binding at 30°C, some of the radioactive L-glutamate appeared to be metabolized; up to 20% of the radioactivity no longer comigrated with L-glutamate carrier. No metabolism was detected at 0°C.

In addition to the pharmacological evidence which indicates that these binding sites correspond to the excitatory amino acid receptors, their distribution is also consistent with physiological evidence. NMDA has greater activity in the outer cortical layers (Pumain et al., 1984); and has little activity in the cerebellum (Crepel, et al., 1982). Likewise, KA sites have their most potent activity in the region of their greatest density (Robertson and Deadwyler, 1981).

AUTORADIOGRAPHY OF Cl-/Ca++-DEPENDENT BINDING (?) SITES

Many of the inconsistancies between L-[^{3}H]-glutamate binding studies in both isolated membrane fractions and autoradiographic preparations are accounted for by the presence of Cl$^-$/Ca^{++}-dependent binding or uptake of L-[^{3}H]-glutamate. In autoradiographic preparations, incubation of L-[^{3}H]-glutamate with rat brain tissue sections in the presence of Cl$^-$ ions (especially with the addition of Ca^{++} ions) results in

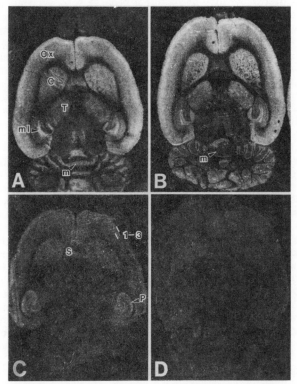

Figure 8.   Two Populations of Ca⁺⁺/Cl⁻-dependent L-[³H]-Glutamate Sites in Autoradiography.   L-[³H]-glutamate was incubated with horizontal sections of rat brain in the presence of 2.5 mM Ca⁺⁺, 20 mM Cl⁻, and 100 μM NMDA in Tris-acetate buffer as described in Monaghan et al. (1983a). A. no other displacers present; B. 100 μM QA ; C. 200 μM SITS; D. 200 μM SITS and 1 μM AMPA. Abbreviations listed in legend to Fig. 3. Data from Monaghan et al., (1985b) and Monaghan and Cotman unpublished observations.

apparent binding at two distinct sites, neither of which corresponds to the major Cl⁻/Ca⁺⁺-dependent binding site observed in membrane preparations. In the hippocampus, the Cl⁻/Ca⁺⁺-dependent binding sites observed by autoradiography are relatively insensitive to nearly all of the excitatory amino acid analogs (e.g. QA, AMPA, L-AP4, NMDA, and KA). In contrast, the Cl⁻/Ca⁺⁺-dependent glutamate binding site of the cerebellar molecular layer is sensitive to QA (Greenamyre et al., 1984; 1985; see figure 8b). These sites are also insensitive to L-AP4 and NMDA, and weakly sensitive to AMPA and KA (Monaghan and Cotman, unpublished observations). It has been suggested (Halpain et al., 1984, Greenamyre et al., 1984) that in autoradiographic experiments L-[³H]-glutamate binds to the same sites previously described in membrane preparations (Foster and Roberts, 1978; Baudry and Lynch, 1981; Fagg et al., 1982; and Werling et al., 1983). However, the site described in membrane preparations differs from those described in autoradiography by having a greater sensitivity to L-AP4, D-α-aminoadipate, and AP6 (Monaghan et al., 1983a, Monaghan et al., 1985b). Furthermore, the site described in membrane preparations is quite sensitive to Na^+ ions (Baudry and Lynch, 1981; Fagg et al., 1984; Mena et al., 1985), whereas the sites assessed in autoradioghraphy are not Na^+-sensitive (at 5 mM Na^+, Monaghan et al., 1985b).

It has also been suggested that the Cl^-/Ca^{++}-dependent L-[^{3}H]-glutamate binding site of the cerebellar molecular layer corresponds to QA receptors (Greenamyre et al., 1985). While this is possible, these sites have not yet been shown to exhibit a pharmacological profile appropriate for QA receptors other than QA-sensitivity itself. Therefore, given the lack of specificity displayed by QA for excitatory amino acid binding sites, one cannot yet conclude that this site represents the QA receptor.

Some of the properties of the Cl^-/Ca^{++}-dependent binding site in membrane preparations are suggestive of an uptake system (Pin et al., 1984; Mena et al., 1984; Fagg 1985). Given the similarity in ion dependency observed for the sites in autoradiography, it is possible that the Cl^-/Ca^{++}-dependent binding represents either uptake or binding to an uptake carrier site. This interpretation is consistent with the observation that the Cl^- channel inhibitor 4-acetamido-4'-isothiocyano-2,2'-disulfonic acid (SITS) is a potent inhibitor of 1) Cl^--dependent L-[^{3}H]-glutamate transport into astrocytes (Waniewski and Martin; 1983), 2) Cl^--dependent L-[^{3}H]-glutamate binding to purified astrocyte membranes (Bridges et al., 1985), and 3) Cl^-/Ca^{++}-dependent L-[^{3}H]-glutamate binding in autoradiography (Monaghan et al., 1985b). Consequently, with the evidence available, one cannot yet conclude that the Cl^-/Ca^{++}-dependent binding observed in autoradiography corresponds to a postsynaptic neurotransmitter receptor for L-glutamate.

It is clear, however, that the Cl^-/Ca^{++}-dependent binding in the cerebellum is distinct from the [^{3}H]-AMPA/AMPA-sensitive L-[^{3}H]-glutamate binding site. In addition to the differing pharmacological profiles and anatomical distributions (Monaghan et al. 1983a; 1984a; 1985a; Halpain et al., 1984; Rainbow et al., 1984), the Cl^-/Ca^{++}-dependent sites (both QA sensitive and insensitive) are potently inhibited by SITS, while the AMPA-sensitive sites are not (see Figure 8a, 8c). Indeed, in the presence of Cl^- and Ca^{++} ions, it is necessary to include SITS to remove the Cl^-/Ca^{++}-sites (and NMDA to remove the NMDA sites) in order to visualize the AMPA-sensitive L-[^{3}H]-glutamate binding site. That these two QA-sensitive sites are different accounts for the apparent discrepency of weak displacement by L-serine-O-sulfate against the cerebellar Cl^-/Ca^{++}-dependent L-[^{3}H]-glutamate site (Greenamyre et al., 1985) and the more potent displacement by L-SOS against the AMPA site (Monaghan et al., 1983a; 1984a; 1985a).

Some initial reports of L-[^{3}H]-glutamate binding indicated that the compounds NMDA and KA are not potent inhibitors of L-[^{3}H]-glutamate binding in autoradiographic preparations (Halpain et al., 1984; Greenamyre et al., 1983). A major reason for this apparent discrepency is that Cl^- ions were included in the incubation buffers of these studies, thus resulting binding stimulated by Cl^-. Another reason is that these studies were

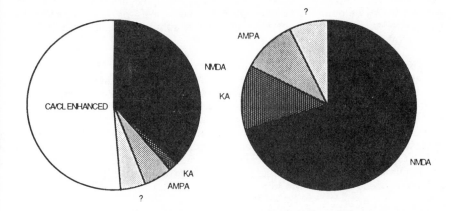

Figure 9.   Relative Proportions of the L-[³H]-Glutamate Binding Sites. L-[³H]-glutamate binding to whole rat brain tissue sections which are selectively displaced by NMDA, KA, and AMPA under differing incubation conditions.   Left: 30 minutes, 30°C, 2.5 mM Ca⁺⁺, 20 mM Cl⁻, 50 mM Tris-acetate, pH 7.2.   Right: Optimal conditions for L-[³H]-glutamate binding to tissue sections, 10 minutes, 0°C, 50 mM Tris-acetate, pH 7.2..

performed at 23-37°C.   The Cl^-/Ca^{++}-dependent L-[³H]-glutamate sites display a partial temperature dependency (Monaghan, Yao, and Cotman, unpublished observations), with higher levels of binding observed at 30°C than at 0°C.   In contrast, the levels of NMDA-sensitive binding are lower at 30°C than at 0°C (Monaghan and Cotman, 1985).   Thus, total L-[³H]-glutamate binding levels are not greatly affected by changes in temperature, but the proportions of the differing subtypes are changed. KA and AMPA populations represent a relatively small percentage of the L-[³H]-glutamate binding, especially KA sites when Ca^{++} is present. Thus, in whole tissue sections, these compounds do not displace the majority of the binding.   (See Figure 9 for a quantitative representation.)   After consideration of the effects of temperature and Cl^- and Ca^{++} ions upon L-[³H]-glutamate binding, the results obtained in the various autoradiographic studies are consistent.

CONCLUSIONS

The three major classes of excitatory amino acid receptors may be visualized by autoradiographic techniques.   The corresponding binding sites are differentially distributed throughout the central nervous system of the rat and are localized in discrete dendritic fields.   NMDA sites have been studied with D-[³H]-AP5 or L-[³H]-glutamate and appropriate specific analogues.   NMDA   binding   sites   are   found   primarily   within   the telencephalon, particularly the hippocampus.   Cerebral cortex is heavily labelled with outer layers of frontal, pyriform, anterior cingulate and perirhinal regions showing highest binding.   The pharmacological profile of these sites in the stratum radiatum of the hippocampus parallels that derived from neurophysiological studies. Overall, the results on NMDA receptors show excellent agreement between autoradiographic, membrane binding and electrophysiological data.

Kainic acid binding sites can be resolved using either [^{3}H]-kainate or [^{3}]-glutamate with selective displacers. The highest binding levels are in the caudate/putaman and the terminal field of the mossy fibers in the hippocampus. In the cerebral cortex the deep layers contain the highest density of binding sites. The overall distribution corresponds reasonably well with areas where kainate has its most potent excitotoxic action. There are at least two kinetically distinct kainate binding sites, distinguishable on the basis of affinity constants and calcium-sensitivity. [^{3}H]-Glutamate (100 nM) labels predominantly the high affinity sites whereas [^{3}H]-kainate (100 nM) labels predominantly the low affinity sites.

The third class of excitatory amino acid receptor is preferentially activated by QA and AMPA. It is now clear, however, that in binding studies, QA is less selective than AMPA. In general, there are parallels between the anatomical organization of AMPA and NMDA sites. The highest density AMPA sites are in the hippocampus and outer cortical layers. Midbrain and brain stem have considerably lower levels. The pharmacological properties of AMPA sites and their distribution is consistent with the electrophysiological evidence. Thus, the three major classes of excitatory amino acid receptors can be studied by way of quantitative autoradiography. A striking feature of the binding is that the pharmacological properties correspond to the receptor classes, and thus confirms this receptor classification scheme previously defined by electrophysiological methods.

In the absence of calcium and chloride ions, and when incubated at low temperatures, these three sites account for over 90 percent of the labelling by L-[^{3}H]-glutamate. The majority of the labelling occurs to the NMDA site. Many of the anomolies in the autoradiography literature and the discrepancies with respect to membrane binding appear to be due to the presence or absence of calcium and chloride ions in the incubation medium. The presence of these ions gives rise to an increase in [^{3}H]-glutamate binding throughout the brain. It seems that the Cl^-/Ca^{++}-enhanced sites may correspond to uptake or binding to a carrier site. After consideration of the effects of ions, as well as temperature, the results obtained in autoradiography appear quite consistent and in accord with the hypothesis that the method is accurate for studying NMDA, KA, and AMPA receptor populations within the CNS.

Acknowledgements: This work was supported by grants DAMD 17-83-C-3189 and DAAG 29-82-K-0194.

REFERENCES

Baudry, M. and Lynch, G. (1981). Characterization of two [^{3}H] glutamate binding sites in rat hippocampal membranes. J. Neurochem. 36: 811-

820.

Beaumont, K., Y. Maurin, T.D. Reisine, J.Z. Fields, E. Spokes, E.D. Bird, and H.I. Yamamura (1979). Huntington's disease and its animal model: alterations in kainic acid binding. Life Sci. 24 809–816.

Berger, M.L. and Ben-Ari, Y. (1983). Autoradiographic visualization of [^{3}H] kainic acid receptor subtypes in the rat hippocampus. Neurosci. Lett. 39: 237–242.

Berger, M.L., Tremblay, M., Nitecka, L., and Ben-Ari, Y. (1984). Maturation of kainic acid seizure-brain damage syndrome in the rat. III. Postnatal development of kainic acid binding sites in the limbic system. Neurosci. 13: 1095–1104.

Bridges, R.J., Nieto-Sampedro, M., and Cotman, C.W. (1985). Stereospecific binding of L-glutamate to astrocyte membranes. Soc. Neurosci. Abstr. (in press).

Butcher, S.P., J.F. Collins, and P.J. Roberts (1983). Characterization of the binding of DL-[^{3}H] -2-amino-4-phosphonobutyrate to L-glutamate-sensitive sites on rat brain synaptic membranes. Brit. J. Pharmacol. 80: 355–364.

Crepel, F., S.S. Ahanjal, and T.A. Sears (1982). Effect of glutmate, aspartate, and related derivatives on cerebellar Purkinje cell dendrites in the rat: an in vitro study. J. Physiol. (Lond.) 329: 297–317.

Davies, J. and Watkins, J.C. (1983). Role of excitatory amino acid receptors in mono- and polysynaptic excitation in the cat spinal cord. Exp. Brain Res. 49 280–290.

Fagg, G.E. (1985). L-glutamate, excitatory amino acid receptors and brain function. Trends Neurosci. 8: 207–210.

Fagg, G.E., Foster, A.C., Mena, E.E., and Cotman, C.W. (1982). Chloride and calcium ions reveal a pharmacologically distinct population of L-glutamate binding sites in synaptic membranes: correspondence between biochemical and electrophysiological data. J. Neurosci. 2: 958–965.

Fagg, G.E., Foster, A.C., Mena, E.E., and Cotman, C.W. (1983). Chloride and calcium ions separate L-glutamate receptors in synaptic membranes. Eur. J. Pharmacol. 88: 105–110.

Fagg, G.E. and Matus, A. (1984). Selective association of N-methyl-aspartate and quisqualate types of L-glutamate receptor with brain postsynaptic densities. Proc. Natl. Acad. Sci. U.S.A. 81: 6876–6880.

Fagg, G.E., Riederer, B., and Matus, A., (1984). Sodium ions regulate a specific population of acidic amino acid receptors in synaptic membranes. Life Sciences. 34: 1739–1745.

Foster, A.C. and Fagg, G.E. (1984). Acidic amino acid binding sites in mammalian neuronal membranes: Their characteristics and relationship to synaptic receptors. Brain Res. Rev. 7 103–164.

Foster, A.C. and P.J. Roberts (1978) High-affinity L-^{3}H-glutamate binding to postsynaptic receptor sites on rat cerebellar synaptic membranes. J. Neurochem. 31: 1467–1477.

Foster, A.C., Mena, E.E., Monaghan, D.T., and Cotman, C.W. (1981). Synaptic localization of kainic acid binding sites. Nature 281: 73–75.

Geddes, J.W., Monaghan, D.T., Lott, I.T., Chui, H., Kim, R., and Cotman, C.W. (1985). Soc. Neurosci. Abstr. (in press).

Greenamyre, J.T., Olson, J.M., Penny, J.B., and Young, A.B. (1985). Autoradiographic characterization of N-methyl-D-aspartate-, quisqualate-, and kainate–sensitive glutamate binding sites. J. Pharm. Exp. Therap. 233: 254–263.

Greenamyre, J.T., A.B. Young, and J.B. Penny (1983).   Quantitative autoradiography of L-[^{3}H]-glutamate binding to rat brain. Neurosci. Lett. 37: 155–160.

Greenamyre, J.T., A.B. Young, J.B. Penny (1984).   Quantitative autoradiographic distribution of L-[^{3}H]-glutamate binding sites in rat central nervous system. J. Neurosci. 4: 2133–2144.

Halpain, S., C.M. Wieczorek, and T.C. Rainbow (1984). Localization of L-glutamate receptors in rat brain by quantitative autoradiography. J.Neurosci. 4: 2247–2258.

Harris, E.W., Ganong, A.H., and Cotman, C.W. (1984).   Long-term potentiation in the hippocampus involves activation of N-methyl-D-aspartate receptors. Brain Res. 323: 132–137.

Henke, H., Beaudet, A., and Cuenod, M. (1981).   Autoradiographic localization of specific kainic acid binding sites in pigeon and rat cerebellum. Brain Res. 219: 95–105.

Honore, T., Lauridsen, J. and Krogsgaard-Larsen, P. (1982). The binding of [3H] AMPA, a structural analogue of glutamic acid, to rat brain membranes. J. Neurochem. 38: 173–178.

Krieg, W.J.S. (1946). Connections of the cerebral cortex, J. Comp. Neurol., 84: 221–275.

London, E.D., and Coyle, J.T. (1979). Specific binding of [^{3}H]-kainic acid to receptor sites in rat brain. Molec. Pharmacol. 15 492–505.

Mena, E.E., D.T. Monaghan, S.R. Whittemore and C.W. Cotman (1985). Cations differentiate subtypes of L-glutamate binding sites in rat forebrain. Brain Res., 329: 319–322.

Mena, E.E., Whittemore, S.R., Monaghan, D.T., and Cotman, C.W. (1984). Ionic regulation of glutamate binding sites. Life Sciences 35: 2427–2433.

Monaghan, D.T. and Cotman, C.W. (1982). Distribution of [^{3}H]-kainic acid binding sites in rat CNS as determined by autoradiography. Brain Res. 252: 91–100.

Monaghan, D.T. and Cotman, C.W. (1985). Distribution of NMDA–sensitive L-[^{3}H]-glutamate binding sites in rat brain as determined by quantitative autoradiography. J. Neuroscience (in press).

Monaghan, D.T., Holets, V.L., Toy, D.W., and Cotman, C.W. (1983a). Anatomical distributions of four pharmacologically distinct [^{3}H]-L-glutamate binding sites. Nature 306: 176–179.

Monaghan, D.T., McMills, M.C., Chamberlin, A.R., and Cotman, C.W. (1983b).   Synthesis of ^{3}H 2-amino-4-phosphonobutyric acid and characterization of its binding to rat brain membranes: a selective ligand for the chloride/calcium–dependent class of L-glutamate binding sites. Brain Res. 278: 137–144.

Monaghan, D.T., Yao, D., and Cotman, C.W. (1984a). Distribution of ^{3}H-AMPA binding sites in rat brain as determined by quantitative autoradiography. Brain Res. 324: 160–164.

Monaghan, D.T., Yao, D., and Cotman, C.W. (1985a).   L-[^{3}H]-glutamate binds to kainate-, NMDA-, and AMPA–sensitive binding sites: an autoradiographic analysis. Brain Res. 340: 378–383.

Monaghan, D.T., Yao, D., Nguyen, L., and Cotman, C.W. (1985b). Excitatory amino acid binding sites: Correspondence between autoradiographic and membrane fraction preparations. Soc. Neurosci. (in press).

Monaghan, D.T., Yao, D., Olverman, H.J., Watkins, J.C., and Cotman, C.W. (1984b). Autoradiography of D-[^{3}H]-2-amino-5-phosphonopentanoate binding sites in rat brain. Neurosci. Lett. 52: 253-258.

Nadler, J.V., Wang, A., and Werling, L.L. (1985). Binding sites for L-[^{3}H]-glutamate on hippocampal synaptic membranes: Three populations differentially affected by chloride and calcium ions. J. Neurochem. 44: 1791-1798.

Olverman, H.J., Jones, A.W., and Watkins, J.C. (1984). L-glutamate has higher affinity than other amino acids for ^{3}H-D-AP5 binding sites in rat brain membranes. Nature 307: 460-462.

Pin, J-P., Bockaert, J., and Recasens, M. (1984). The Ca^{++}/Cl$^-$ dependent L-[^{3}H] glutamate binding: a new receptor or a particular transport process? FEBS Letters 175: 31-36.

Pumain, R., Kurcewicz, I., Louvel, J., and Heinemann, U. (1984). Electrophysiological evidence for a differential localization of excitatory amino acid receptors in the rat neocortex. Neurosci. Lett. (Supplement) 18: S433.

Rainbow, T.C., Wieczorek, C.M., and Halpain, S. (1984). Quantitative autoradiograpy of binding sites for [^{3}H] AMPA, a structural analogue of glutamic acid. Brain Res. 309: 173-177.

Robertson, J.H. and Deadwyler, S.A. (1981). Kainic acid produces depolarization of CA3 pyramidal cells in the in vitro hippocampal slice. Brain Res. 221: 117-127.

Schwob, J.E., Fuller, T., Price, J.L. and Olney, J.W. (1980). Widespread patterns of neuronal damage following systemic or intracerebral injections of kainic acid: a histological study. Neurosci. 5: 991-1014.

Simon, J.R., Contrera, J.F., and Kuhar, M.J. (1976). Binding of [^{3}H]-kainic acid, an analogue of L-glutamate to brain membranes. J. Neurochem. 26: 141-147.

Unnerstall, J.R. and Wamsley, J.K. (1983) Autoradiographic localization of high-affinity [^{3}H]-kainic acid binding sites in the rat forebrain. Eur. J. Pharmacol. 86: 361-371.

Waniewski, R.A., and Martin, D.L. (1984). Selective inhibition of glial versus neuronal uptake of L-glutamate by SITS. Brain Res. 268: 390-394.

Waniewski, R.A., and Martin, D.L. (1984). Characterization of L-glutamic acid transport by glioma cells in culture: evidence for sodium-independent, chloride dependent high affinity influx. J. Neurosci. 4: 2237-2246.

Watkins, J.C., (1984). Excitatory amino acids and central synaptic transmission. Trends in Pharmacol. Sci. 84 373-376.

Werling, L.L., A. Doman, and J.V. Nadler (1983). L-[^{3}H]-glutamate binding to hippocampal synaptic membranes: two binding sites discriminated by their differing affinities for quisqualate. J. Neurochem. 41: 586-593.

Wheeler, D.D. (1979). A model of high affinity glutamic acid transport by rat cortical synaptosomes- a refinement of the originally proposed model. J. Neurochem. 33: 883-894.

Yao, D., Monaghan, D.T., Ganong, A.H., Harris, E.W., and Cotman, C.W. (1984). NMDA receptors in the rat brain. I. Subcellular and anatomical distribution. Soc. Neurosci. 10: 419.

20

Functions and Regulations of Glutamate Receptors

M. Baudry

The existence of multiple classes of receptors is now well-documented for a large number of neurotransmitters or neuromodulators, and is becoming a generalized property of chemical messengers (Snyder, 1984). Why this rule has developed during the evolution of biological systems is not yet understood, although it clearly results in an increase in the number of ways cells interact with each other, especially if different receptors perform different functions or are localized on different cellular elements. This could allow the same neurotransmitter to elicit different cellular responses or to interact with specific cell types or cell domains depending on the conditions under which it is released. The cellular responses can be brief and rapid when the receptor is associated (or contains in itself) an ionic channel or long-lasting and slowly developing when the interaction neurotransmitter-receptor results in the intracellular generation of a second messenger (calcium, cyclic nucleotides, or phosphatidylinositol derivatives). The multiplicity of receptor types also provides for the possibility of independent and multiple regulations of the different types of receptors, thus further amplifying the flexibility of chemical transmission. Here also the regulation can be short-lasting and reversible as in the case of receptor desensitization or long-lasting as in situations resulting in changes in the rates of synthesis or degradation of receptor molecules. Although most of these latter regulatory mechanisms are expressed under pathological or pharmacological conditions, it remains that they reflect the existence of fundamental properties of receptors which could be used to adjust chemical transmission under various physiological or pharmacological situations.

Another emerging general property of neurotransmitter receptors is for the same receptor molecule to exhibit different states characterized by different affinities for the neurotransmitter, the transition between the states being regulated by a variety of factors, such as the concentration of the agonist, the membrane potential, various ions and possibly various types of interactions

301

with cytoskeletal elements. In the case of the nicotinic cholinergic receptor, it has been shown that the reversible transitions between different states were in fact likely to be responsible for various desensitization mechanisms (Feltz and Trautmann, 1982), and it has even been proposed that these short-lasting transitions could eventually become long-lasting via some types of covalent modifications of the receptors (Heidmann and Changeux, 1982). On the basis of theoretical considerations at least, this was shown to provide a simple and plausible mean to modify the efficacy of synaptic transmission which could potentially account for a wide range of adaptive phenomena such as homosynaptic and heterosynaptic facilitation as well as classical conditioning. The recently elucidated structure of the nicotinic cholinergic receptor suggests indeed that this receptor is an allosteric protein, providing strong support for the above described mechanism (Changeux et al., 1984).

Glutamate is now considered to be a major excitatory neurotransmitter in vertebrate central nervous system and several recent reviews have summarized the evidence in favor of such a role for this acidic amino acid (Fagg and Foster, 1983; Fonnum 1984). Electrophysiological as well as biochemical studies have also revealed the existence of multiple classes of receptors for glutamate, indicating that it shares the general property of neurotransmitters to possess multiple receptor types (Foster and Fagg, 1984). The present review intends to show that, in addition, glutamate receptors exhibit several of the general properties outlined above, namely that different classes of receptors participate in various functions and exhibit different forms of regulation. Moreover, I will also discuss the possibility that some forms of glutamate receptor regulations might be responsible for some fundamental properties of brain and in particular the ability to store large amounts of information.

CLASSIFICATION OF GLUTAMATE RECEPTORS

Several experimental approaches have been used to characterize glutamate receptors. Electrophysiological techniques have shown, using both extra- and intracellular recordings, that excitatory amino acids depolarize neuronal and, more recently, glial cells (Watkins et al., 1981; Bowman and Kimelberg, 1984). In general this depolarization results from the opening of sodium channels although it has been suggested that some agonists may also be associated with calcium channels (Dingledine, 1983a). On the basis of the results obtained with a large number of agonists and antagonists, three classes of glutamate receptors have been tentatively identified, which are generally defined as:
- an NMA receptor type preferentially stimulated by N-methyl-D-aspartate and antagonized by -phosphono-derivatives.
- a quisqualate receptor type stimulated by quisqualate.
- a kainate receptor type, stimulated by kainate. For the latter

two types of receptors no specific antagonist has been so far proposed. However, it must be stressed that this classification has been mainly determined from studies using the spinal cord and that there does not exist a general agreement between the classification obtained in this system as compared to that derived from studies using the hippocampus or the cerebellum for instance (Crepel et al., 1982; Fagni et al., 1983). In particular, Fagni et al. (1983) proposed the existence of 4 types of glutamate receptors on the basis of the differential apparent desensitization which results from the successive applications of certain agonists and the selective antagonism of responses elicited by various agonists. Two of these 4 receptor types, namely the NMA and the kainate types, satisfy the general properties reported in other systems, while the other two could correspond to 2 subclasses of the quisqualate receptors. One of these latter two types appears to represent the synaptic receptor (at least in the Schaffer-commissural system) and the other likely represents an extra-synaptic receptor preferentially stimulated by L-glutamate.

Biochemical techniques have also been employed to provide a description of the types of receptors involved in the effects of excitatory amino acids. Binding techniques have been extensively used in the last 5 years, involving either labeling of synaptic membranes with various radiolabeled ligands or autoradiographic studies and localizations in tissue sections. Although agreement between various laboratories is still not achieved, a somewhat coherent picture is starting to emerge from these numerous studies, which is better described with the existence of 4 different types of receptors. Using ^{3}H-L-glutamate as a ligand, several groups have shown that three types of receptors can be evidenced in membrane preparations (Baudry and Lynch, 1981a; Mena et al., 1982; Werling and Nadler, 1983; Koller and Coyle, 1984). Two of these sites, corresponding to the NMA receptor and a subtype of a quisqualate receptor, are labeled when the binding is performed in the absence of chloride ions (Fagg and Matus, 1984). In the presence of chloride ions, a third type of receptor is revealed which might correspond to the synaptic receptor at least in hippocampus (Baudry and Lynch, 1981b). ^{3}H-Kainic acid labels a small population of sites which exhibit the characteristics of the kainate receptor of the electrophysiologists (Slevin et al., 1983). This subdivision fits relatively well with results obtained using a variety of different radiolabeled ligands. Thus, ^{3}H-D-amino-5-phosphono-valerate (^{3}H-D-AP5) labels a site with a pharmacological specificity identical to that of the NMA receptor (Olverman et al., 1984), while ^{3}H-α-amino-3-hydroxy-5-methyl-4-isoxazol-propionic acid (^{3}H-AMPA) labels a subtype of the quisqualate receptor (Honore et al., 1982). ^{3}H-amino-phosphonobutyrate (^{3}H-APB) labels a site with properties similar to the Cl-dependent ^{3}H-glutamate binding sites and could correspond to the synaptic receptor at some synapses in the hippocampus (Monaghan et al., 1983a). Autoradiographic studies have provided results in relatively good agreement with the membrane binding studies and a better description of the regional localization of the different types of receptors. Thus ^{3}H-L-glutamate has been shown to label 4 types of receptors exhibiting different pharmacological profiles and localizations. As was found in membrane preparations, the NMA receptor and a subtype of quisqualate receptor are labeled when the binding is performed in the absence of chloride ions. This latter site also seems to be labeled by ^{3}H-AMPA while ^{3}H-D-amino-5-phosphonovalerate (^{3}H-D-AP5) binding is associated with the NMA receptor (Monaghan et al., 1983b; Halpain et al., 1984, Monaghan et al., 1984a, b; Greenamyre et al., 1985a). In addition a small population of kainate receptors is also labeled by ^{3}H-L-glutamate. The presence of chloride ions reveals a fourth type of site which unfortunately does not seem to correspond to the chloride-dependent site observed in membrane fractions since binding to this site is not affected by APB and since no other ligand than L-glutamate has been found to compete with (Monaghan et al, 1983b). The nature of this latter site is therefore quite puzzling at present and these results indicate that some caution must be taken in interpreting data obtained with this technique. They also suggest that the pattern of labeling might be critically dependent on different parameters used in the preparation and handling of the samples. As far as the regional localization of these sites is concerned there is a general tendency for their preferential association with telencephalic structures, although striking differences are also observed within these structures for some of these sites (see Cotman and Monaghan, this volume).

Another biochemical approach to determine the classification of excitatory amino acid receptors has consisted in evaluating the ability of various agonists to stimulate the rate of sodium efflux in brain slices preloaded with ^{22}Na (Luini et al., 1981; Baudry et al., 1983a). Using slices prepared from both the striatum and the hippocampus, this technique has provided data which are also best interpreted with the existence of 4 different types of receptors. Luini et al. (1981) reached this conclusion on the basis of the different pharmacological profiles of various agonists, while we reached a similar conclusion on the basis of the differential degree of desensitization following successive applications of agonists as well as the relative inhibition by some antagonists (Baudry et al., 1983a). Thus, we identified a G1 receptor (tentatively associated with a synaptic glutamate receptor), a G2 receptor (an extrasynaptic glutamate receptor), an NMA receptor, and a kainate receptor. The only difference between this classification and the more generally accepted subdivision of glutamate receptors into 3 classes resides in the introduction of a partition of the quisqualate class into 2 subtypes (G1 and G2 in our nomenclature). Several reports have also indicated that, in certain brain regions or cells in culture, excitatory amino acids are able to stimulate the generation of cyclic nucleotides, but this technique has not been widely used to characterize the different types of receptors (Garthwaite and Balazs, 1980).

This brief summary of the different approaches used so far to characterize excitatory amino acid receptors indicates that, while important progress has been achieved in the last 5 years, we are far from having a definitive and clear-cut classification of the different types of receptors. Several factors contribute to the confusion and discrepancies noted between different approaches and laboratories. First is the lack of more specific agonists and antagonists which could discriminate better different types of receptors. Hopefully, research directed at the synthesis of such compounds will remedy this situation. Second is the fact that the conditions employed in the different approaches are radically different. Thus, binding studies are generally performed at equilibrium and under ionic conditions which have no similarity with those required to observe physiological responses. Since various ions have marked effects on the properties of the binding of various ligands, it must be the case that the properties of the receptors are critically dependent upon the ionic environment. It remains that with the current knowledge of the properties of the different glutamate receptors, it becomes possible to ask questions concerning the possible functions and regulations of these different classes.

FUNCTIONS OF EXCITATORY AMINO ACID RECEPTORS

As previously mentioned, excitatory amino acid receptors are associated with cationic channels and while patch-clamp studies of the effects of excitatory amino acids on central mammalian neurons are still in their infancy, it is already apparent that different types of receptors are associated with different ionic channels (Nowak and Ascher, 1984). Thus the quisqualate receptor might be associated with different channels (or alternatively quisqualate receptors represent multiple classes of receptors) with small unitary conductance. The kainate receptor is also associated with a cationic channel with small unitary conductance and short open time. NMA receptor-associated channels have a larger unitary conductance as well as a longer open time. In addition, the latter channels exhibit a voltage-dependent blockade by magnesium ions which is released when the neurons are depolarized (Nowak et al., 1984; Mayer et al., 1984) (see below). This effect is probably responsible for the apparent increase in membrane resistance elicited by NMA agonists noted by several authors (Hablitz, 1982; Dingledine, 1983b; Flatman et al., 1983). Although these studies suggested that the channels associated with these different classes of receptors were non-selective cationic channels, it has been proposed that the NMA receptor-associated channel was relatively selective for calcium ions since the effect of NMA could be blocked by cadmium and various calcium blockers (Dingledine, 1983a). However, this result has been recently challenged and it seems more likely that NMA agonists activate various voltage-dependent ionic channels directly or indirectly as a result of neuronal depolarization (Hablitz, 1985).

Whatever may be the precise mechanisms underlying the ionic responses, it remains that excitatory amino acids elicit rapid depolarization in the majority of neurons in the mammalian CNS. However, the duration of the depolarization is limited by a number of factors including receptor desensitization, depolarization-block and efficient uptake mechanisms. Although desensitization has been commonly observed in invertebrate preparations (Takeuchi and Takeuchi, 1964) and in patch-clamp studies (Nowak and Ascher, 1984), it has only rarely been reported in studies using iontophoretic application of agonists. In our own studies, we consistently observed a prominent apparent desensitization to the depolarizing effects of various agonists following successive bath-applications using the in vitro hippocampal slice preparation (Fagni et al., 1982, 1983) (see below).

A complicating factor in analyzing the effects of excitatory amino acids is due to the fact that the large depolarization they elicit results in the inactivation of voltage-dependent sodium channels, thus preventing neuronal firing (depolarization-block); in addition, by decreasing the electrical driving force, excitatory amino acids elicit apparent decreases in synaptic potentials, thus appearing to act as synaptic antagonists (Nistri and Constantini, 1979). Potent uptake mechanisms normally eliminate L-glutamate from extracellular spaces (Fonnum, 1984). These uptake systems are generally sodium-dependent, exhibit high-affinity for L-glutamate and have been shown to be localized in glial as well as neuronal elements (Hosli and Hosli, 1978). So far, only a few compounds have been described which might selectively inhibit glutamate uptake without by themselves depolarizing neurons; in these few cases, the depolarizing effect of glutamate was prolonged as predicted if uptake represents an efficient mechanism to lower glutamate concentration in the extracellular space (Johnston et al., 1980). An interesting aspect of the uptake process is the coupling of the glutamate carrier with both sodium and potassium. Thus while glutamate is transported across cell membranes after binding to a sodium-carrier complex, the carrier requires intracellular potassium to relocate to the extracellular side of the membrane (Kanner and Bendahan, 1982). This implies that the ratio of sodium to potassium concentrations regulates the rate of glutamate transport, suggesting that under certain circumstances resulting in imbalance between these two cations, significant reductions of glutamate transport can be observed (Kramer and Baudry, 1984).

In addition to rapid ionic responses, excitatory amino acid receptors appear to be associated with more slowly occurring cellular responses. As previously mentioned, several excitatory amino acids have been shown to stimulate the synthesis of cyclic GMP (Garthwaite and Balazs, 1980). Moreover, as a result of depolarization, excitatory amino acids might, directly or indirectly (through voltage-dependent calcium channels), activate calcium-dependent kinases and thereby modulate the state of phosphorylation of various proteins. Mechanisms of this type would allow a wide range of interaction between neurotransmitters acting by altering the levels of calcium and/or cyclic nucleotides (Nestler et al., 1984). More recently, a large number of reports have indicated that various neurotransmitters interact with their target cells by increasing the turnover of membrane phospholipids and in particular of inositol phospholipids (Berridge and Irvine, 1984; Nishizuka, 1984). This increase in turnover has been shown to generate two breakdown products, diacylglycerol which activates a calcium-dependent phospholipid-dependent protein kinase, kinase C, and inositol-triphosphate which releases calcium from intracellular stores. In a recent publication, Costa's group reported that ibotenate was able to stimulate phosphatidylinositol turnover in rat hippocampal slices (Nicoletti et al., 1985). This effect was potently antagonized by aminophosphonobutyric acid but not by other antagonists. In addition, kainate, glutamate and NMA were inactive.

In our laboratory, we confirmed that these latter agonists did not stimulate phosphatidylinositol turnover; if anything they tended to inhibit the turnover. Moreover, all these agonists markedly inhibited the stimulation of phosphatidylinositol turnover elicited by cholinergic or catecholaminergic agonists (Baudry et al., 1985). This effect did not simply result from the depolarization that they elicit since potassium-induced depolarization by itself stimulated phosphatidylinositol turnover and did not modify the stimulatory effect of carbachol. The response to glutamate receptor agonists was mediated by specific glutamate receptors since, for instance, the NMA response was blocked by amino-phosphonovalerate or amino-adipate while the kainate response was not affected by these compounds. This effect also did not appear to be dependent on the presence of extracellular calcium since it was still present in slices incubated in the absence of calcium. Rather it seemed to be due to the large influx of sodium elicited by excitatory amino acids. The opposite effect of ibotenate reported by Costa's group is rather puzzling but might be related to the observation that physiological experiments have often revealed some inhibitory responses specifically elicited by this agonist in various preparations (Nistri and Constantini, 1979).

Although the mechanism underlying the effect of excitatory amino acids on phosphatidylinositol turnover remains obscure for the moment, it appears clearly that this cellular response offers a new and potentially important way by which neurotransmitters can interact at the level of target cells. In preliminary experiments, we observed that the effects of excitatory amino acids on carbachol-induced stimulation of phosphatidylinositol turnover are very long-lasting even after a very brief application. Moreover, it has been shown that, following high-frequency stimulation of the perforant path, phosphatidylinositol turnover was first reduced and after several minutes increased for long periods of time (Bar et al., 1984). This effect was not observed following low frequency stimulation suggesting that certain patterns of electrical activity in glutamatergic neurons are required in order for the endogenously released neurotransmitter to trigger this cellular response.

The unresolved question underlying most of these studies concerns the nature of the synaptic receptors. Is it the case that, depending on the type of synapses, each of the 4 classes of excitatory amino acid receptors represents a synaptic receptor and that each brain region exhibit differential densities of 4 types of excitatory synapses, or rather, that there are certain classes of receptors which play only a modulatory role in synaptic transmission, but do not participate directly in synaptic transmissionπ Several arguments suggest that the latter is more likely, although it remains possible that the roles of the different receptors might be different in different brain regions. So far, there seems to be little doubt that the NMA receptor is probably not the synaptic receptor activated by low-frequency electrical stimulation. Although some effects of NMA receptor antagonists have been described at various central synapses, which could be attributed to blockade of synaptic receptors (Thompson et al., 1985), the evidence that the affected potentials were monosynaptic were never very convinc ing, considerably weakening the argument. The kainate receptors could be synaptic receptors at some synapses where they seem to be highly concentrated, such as the mossy fiber terminals (Monaghan and Cotman, 1982); in most other regions, their density is relatively low and there are a number of studies showing that they might be localized presynaptically where they could have some role in the regulation of release of various neurotransmitters (Ferkany et al., 1982). This would leave quisqualate, or a subclass of quisqualate receptors, as the predominant class of synaptic receptors. Again the lack of specific antagonists of this class of receptors makes this conclusion speculative at the moment.

REGULATION OF GLUTAMATE RECEPTORS

Supersensitivity during the postnatal period and following denervation

Using the excitatory amino acid-induced stimulation of ^{22}Na efflux in hippocampal slices allowed us to study the changes in sensitivity of glutamate receptors under various experimental conditions (Baudry et al., 1983b). For all the agonists tested, the stimulation of ^{22}Na efflux was much larger in immature than in adult rats. Since this effect was due to an increase in their maximal responsiveness without changes in their apparent affinities, it seems more likely that this supersensitive response to agonists was due to a greater density of the receptors rather than changes in uptake, or distribution of the agonists. This supersensitivity was specially marked with L-glutamate which caused a 6-fold greater stimulation of ^{22}Na efflux in neonatal than in adult rats. This agrees well with the previously reported supersensitivity to the stimulation by L-glutamate of cyclic GMP accumulation in neonatal cerebellar slices (Garthwaite and Balazs, 1978). This suggests that the regulation of neurotransmitter receptors in neurons could be similar to what is observed in muscles, namely that developing dendritic trees would exhibit large amounts of receptors and that the formation of synapses would result in the elimination of a large number of receptors (Pumplin and Fambrough, 1982). Indeed the time-course for the postnatal changes in sensitivity of the L-glutamate-induced stimulation of ^{22}Na efflux rate (between postnatal days 11 and 30) coincides with that observed for the formation of synapses in the hippocampal formation.

One week following unilateral hippocampal aspiration which removes the hippocampal commissural projections, the maximal responsiveness of L-glutamate on the ^{22}Na efflux rate was also increased by 50%, without significant changes in the half-maximal concentration. This effect appeared to be rather specific for the receptor stimulated by L-glutamate (G2 in our nomenclature) since the response to agonists of other classes of receptors were in fact decreased following hippocampal aspiration. This latter result implies that some of these receptors would be localized presynaptically (Baudry et al., 1983b).

Thus, under these two conditions, following lesions of afferent pathways and during the developmental period, one class of glutamate receptor (G2) appears to obey the same general rules demonstrated for cholinergic receptors in skeletal muscles, suggesting that the synthesis of these receptors is regulated by their state of activity. However, different classes of receptors and in particular the synaptic receptor (G1) do not seem to exhibit the phenomenon of denervation supersensitivity indicating that this receptor might be subjected to different mechanisms of regulation.

Desensitization Following Successive Applications of Agonists

Although it was commonly thought that desensitization to glutamate does not occur in mammalian brain, we observed that successive and short bath-application of various agonists results in marked decreases in their depolarizing effects (Fagni et al., 1982, 1983). This effect was not due to a modification in the uptake process since the same desensitization was observed for amino acids which are not substrates of the high-affinity uptake system, such as D-glutamate or N-methylaspartate. Only 2 of the 4 classes of receptors exhibited this apparent desensitization (this differential desensitization to the effects of various agonists was, as mentioned above, used to classify the different classes of receptors), namely the NMA receptors and the L-glutamate G2 receptors. The characteristics of this apparent desensitization phenomenon clearly differ from those observed with other neurotransmitters (i.e. rapid onset and recovery) (Feltz and Trautmann, 1982) since it took several applications of the agonists for several minutes and its reversibility required hours. This pecu-

liar form of regulation suggests however that these receptors exist
under different configurations with different affinities for their
agonists, the transition between states being regulated by slowly-
occurring events (phosphorylation or other covalent modification
for instance); alternatively the activation of these receptors
might result in long-lasting modifications of the ionic channels
with which they are associated. Interestingly the synaptic receptors stimulated by the transmitter released under low-frequency
electrical stimulation did not seem to exhibit desensitization even
following repeated episodes of potassium-induced deplorization. A
paradoxical result provided by these experiments consists in the
fact that L-glutamate, even at high concentrations, does not seem
to stimulate the synaptic receptors, since it is possible to
totally suppress the depolarizing effects of L-glutamate without
affecting synaptic transmission. This would imply that either the
postsynaptic receptor has a very low affinity for L-glutamate, or
bath-applied L-glutamate does not reach the synaptic receptors.
Possibly a combination of efficient uptake mechanisms and low
affinity of the post-synaptic receptors for glutamate could account
for this paradox. This in turn raises the question of the nature
of the high-affinity binding sites for ^{3}H-L-glutamate and of their
relationships with postsynaptic receptors. As mentioned previously,
binding experiments are performed under conditions far remote from
the ionic conditions prevailing in situ and we know that a variety
of ions have profound effects on ^{3}H-glutamate binding; in particular low concentrations of Na+ ions totally inhibit the Cl-dependent
^{3}H-glutamate binding (Baudry and Lynch, 1979). Possibly then, some
classes of high-affinity binding sites for ^{3}H-glutamate represent
either receptors in a desensitized state or the absence of critical
ions in the binding assay alters the receptor configuration and
confers them a high-affinity for ^{3}H-glutamate (see below).

Magnesium regulation of the NMA receptor

As mentioned above, the NMA receptor exhibits a very peculiar
regulation, which consists in a voltage-dependent blockade of its
associated ionic channel by magnesium ions. At normal resting
membrane potential the channel is blocked by magnesium ions and the
responses elicited by NMA agonists are very small. With depolarization the magnesium blockade would be released and the response to
agonists would increase. Varying magnesium concentrations in
slices or in cells in culture indicated that this effect occurs at
relatively low concentrations of magnesium (Dingledine, 1983a).
Removal of magnesium from the incubation medium revealed more
conventional voltage-current relationships of the channel which is
a non-selective cationic channel (Nowak et al., 1984; Mayer et al.,
1984). Although the mechanism underlying this effect is not yet
known it provides a powerful way of regulating the depolarizing
effect of glutamate or of the transmitter which normally stimulates
this receptor. Several studies have shown that specific NMA receptor antagonists have little effect on synaptic potentials elicited

by electrical stimulation performed at low frequency (Collingridge et al., 1983; Harris et al,, 1984). However, these compounds totally prevented the induction of long-term potentiation, a long lasting increase in synaptic efficacy which follows high frequency stimulation in several telencephalic pathways, and which has been extensively studied since its discovery by Bliss and Lomo (1973) as it represents a potential mechanism for the storage of information in mammalian CNS (Swanson et al., 1982, Eccles, 1983). The general interpretation of these results is as follows (Fig. 1):

- At low frequency stimulation the postsynaptic target cells never get depolarized enough to release the channels associated with the NMA receptor from the magnesium blockade, which accounts for the lack of effect of NMA antagonists on synaptic transmission.

- At high frequency stimulation, more synaptic receptors are activated, the postsynaptic membrane gets depolarized enough to release the channels from the magnesium blockade, allowing further depolarization. This would result, directly or indirectly through the opening of voltage-dependent calcium channels, in an increase in intracellular calcium, which has been shown to be necessary to induce long-term potentiation (Lynch et al., 1983). By blocking the NMA receptors, NMA receptor antagonists would prevent this effect and thus block the induction of long-term potentiation. In addition to preventing the induction of long-term potentiation, NMA receptor antagonists have been shown to exhibit potent anticonvulsant and anti-epileptic properties (Croucher et al., 1982; Jones et al., 1984) as well as to prevent neuronal degeneration induced by hypoxia (Simon et al., 1984). A similar explanation is likely to account for all the properties of these antagonists, namely that NMA receptors are activated under unusual conditions requiring massive release of endogenous transmitter.

Calcium regulation of glutamate receptors

As discussed earlier, ^{3}H-L-glutamate has been widely used as a ligand to study various types of glutamate receptors and several subclasses were defined partly on the basis of their ionic requirements. Our laboratory studied extensively the properties of the Cl-dependent ^{3}H-glutamate binding in synaptic membranes from rat hippocampus, a structure where several intrinsic and extrinsic pathways use glutamate as a neurotransmitter (Baudry and Lynch, 1984). Whereas monovalent cations inhibited ^{3}H-glutamate binding, several divalent cations and in particular calcium, induced a marked stimulation of the binding, by increasing the maximal number of sites, without changing the apparent affinity for glutamate, nor the pharmacological profile of the binding (Baudry and Lynch, 1979). The concentration of calcium eliciting half-maximal stimulation was found to be about 20-30 µM and the maximal effect

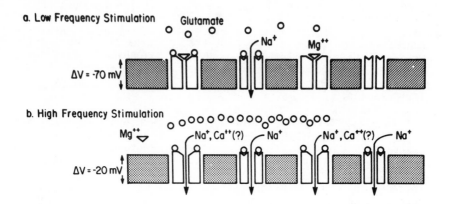

Figure 1: Schematic representation of the mechanism of activation of NMA receptors. At low frequency stimulation, the ionic channel associated with NMA receptors is blocked by magnesium. Following high frequency stimulation and/or large depolarization, the magnesium blockade of the channel is released and the channel allows large influx of cations (see text).

obtained at 100-250 µM calcium. In addition, the calcium-induced increase in glutamate binding was irreversible, and by quenching the effect of calcium with the calcium chelator EGTA, we determined that the increase in binding was half maximal at about 1 minute of exposure of membranes with calcium, and maximal at about 5 minutes (Baudry et al., 1983c). This suggested that calcium activated some enzymatic process which results in a rapid unmasking of additional ^{3}H-glutamate binding sites which could not normally be detected by the binding assay (because they were shielded from ^{3}H-glutamate or they existed in a state of low-affinity for ^{3}H-glutamate not measurable with the binding technique). A variety of inhibitors of calcium-dependent proteases were found to totally prevent the calcium-induced increase in binding without affecting basal ^{3}H-glutamate binding (Baudry and Lynch, 1980; Baudry et al., 1981). Moreover, the potency of these compounds to inhibit the calcium-induced increase in binding paralleled relatively closely their potency to inhibit calcium-dependent protease activity (Baudry and Kramer, submitted). In addition, exogenous proteases such as chymotrypsin also increased the number of ^{3}H-glutamate binding sites (Baudry and Lynch, 1980). This leaves little doubt that calcium increases the number of ^{3}H-glutamate binding sites by activating a membrane-associated calcium-dependent protease. Such an enzyme was indeed identified in synaptic membrane preparations (Siman et al., 1983), and shown to belong to the general class of calcium-dependent proteases found in a variety of tissues and given the generic name of calpain by Murachi et al.(1981). The substrates of this category of enzymes consist of most of the proteins forming the cell cytoskeleton, such as the microtubule-associated proteins, the neurofilament proteins, tubulin, actin and a high molecular weight protein called fodrin by and Weber, 1978; Pant and Gainer, 1980; Malik et al., 1981; Zimmerman and Schlaepfer, 1982) (see Goodman and Zagar, 1985, for a review) In synaptic membranes it appears that fodrin is the main substrate for calpain (Baudry et al., 1981; Siman et al., 1984) and that calpain-mediated degradation of fodrin is responsible for the calcium-induced increase in ^{3}H-glutamate binding (Siman et al., 1985). All these results were integrated into a model of the organization of glutamate receptors in synaptic membranes and their regulation via the interaction with the underlying cell cytoskeleton (Baudry and Lynch, 1984). We proposed that this category of glutamate receptors exist in synaptic membranes under two configurations with different affinities for ^{3}H-glutamate, and that the transition between these two states is regulated by constraints imposed on their spontaneously-occuring movements by their interaction with the cytoskeleton mediated through the fodrin network. Recent data indicate that the same mechanism could be implicated in the regulation of another class of ^{3}H-glutamate binding sites, namely a subclass of the Cl-independent binding sites. In the absence of chloride ions, calcium also increases the number of ^{3}H-glutamate binding sites The subpopulation of these sites which is sensitive to NMA and NMA receptor agonists is not affected while the number of quisqualate-displacable binding sites is increased two- to threefold. This effect is also blocked by inhibitors of calpain and is irreversible (Baudry, in preparation). Thus, we have to assume that either several types of glutamate receptors are regulated by a similar mechanism, or that the distinction of 2 classes of receptors according to their chloride sensitivity requires some revision. It remains indeed possible that, as sodium ions can change the pharmacological characteristics of various neurotransmitter receptors (Pert and Snyder, 1974), chloride ions can similarly modify the pharmacological properties of a subclass of glutamate receptors. Purification of the receptors or obtention of antibodies might be required to provide definitive answers to these questions. It remains that the number and properties of certain types of glutamate receptors can be irreversibly altered by a mechanism triggered by relatively modest concentrations of calcium and involving the transient breakdown of the cytoskeletal network, in a way not unlike what has been suggested for the cholinergic nicotinic receptors (Heidmann and Changeux, 1982). The functional implications of the existence of such a regulatory mechanism in a large number of excitatory synapses will be discussed later.

Regulation of Glutamate Binding by Glutamate and Tyrosylglutamate

Another regulatory mechanism has also recently been discovered; the number of the Cl-dependent binding sites is increased several-fold following exposure of membranes or hippocampal slices to high concentrations of glutamate, glutamate analogs or the dipeptide tyrosyl-glutamate (Ferkany et al., 1984; Kessler et al., 1985). This mechanism is independent of calcium and in fact, once increased by glutamate analogs, the binding can be further increased by calcium. The fact that the concentrations of glutamate or of glutamate analogs required to increase the number of binding sites are a thousand fold higher than those required to inhibit the binding suggests the possibility that the same glutamate receptor exists under two interconvertible states with widely different affinities for glutamate and glutamate analogs. Prolonged occupation of the low affinity sites by a variety of ligands (agonists as well as compounds classically considered as antagonists) would result in a transition to a high affinity state for these ligands. The peculiar aspect of this transition is that the reverse transition to the low affinity state would be very slow (several hours) except in the presence of sodium ions which can rapidly trigger it. An alternative interpretation is that the binding site would be part of a multicomponent complex involving several sites with different affinities for glutamate or glutamate analogs. Evidence for the existence of such a multicomponent complex has in fact been provided on the basis of experiments using radiation inactivation of the binding site for ^{3}H-glutamate (Bardsley and Roberts, 1985) or ^{3}H-AMPA (Honore and Nielsen, 1985). Again the functional significance of such a regulatory mechanism is not clear at the moment due to the lack of correspondency between the conditions used in the binding assays and those used to determine functional responses to glutamate agonists. It is interesting to note that, as is the case for calcium regulation of ^{3}H-L-glutamate binding, this same regulatory mechanism is also found for the quisqualate-sensitive component of the Cl-independent binding.

CONCLUSIONS

Although we are far from having a clear understanding of the properties of glutamate receptors, it remains that they satisfy the general rules outlined in the introduction, i.e. that there exist different classes of receptors, which perform different functions and exhibit different mechanisms of regulation. With their association with cationic channels, these receptors mediate fast responses which are likely to represent excitatory potentials in a large number of synapses, considering the wide range of neuronal pathways which have been proposed to use glutamate or aspartate as neurotransmitters. In addition, these receptors can also mediate

changes in cyclic nucleotides thereby participating in the regulation of the state of phosphorylation of various proteins, which
would represent slower-occurring and longer-lasting modifications
of the properties of cells expressing these receptors. Finally,
these receptors are also implicated in the regulation of phosphatidylinositol turnover, an important step in cellular function
since the breakdown products of phosphatidylinositol, diacylglycerol and inositol-triphosphate, have been shown to elicit a
wide range of intracellular effects from calcium mobilization to
activation of protein kinase C. In addition, since several chemical
signals interact at the level of phosphatidylinositol turnover, the
inhibitory effect of excitatory amino acids on this response considerably expands the complex interactions between different neurotransmitters which can occur in a given target cell.

Various regulatory mechanisms of glutamate receptors have been
discussed and although the functional significance of several of
these mechanisms remains quite obscure, some of them have already
been shown to be of particular importance in a variety of brain
functions. First the voltage-dependent magnesium blockade of NMA
receptors makes this receptor an ideal amplifying mechanism of
depolarisation. It is therefore not surprising that antagonists of
this class of receptors have been found to be potent anticonvulsant
and antiepileptic as well as effective agents to prevent cell
damage induced by anoxia, conditions under which large amounts of
glutamate are likely to be released. Moreover, these antagonists
were also found to prevent the induction of long-term potentiation
in vitro as well as in vivo and in recent experiments to produce
selective impairment of certain forms of learning (Morris et al.,
1985). Thus AP-5 causes a dramatic impairment of place navigation
without affecting visual discrimination learning. This result not
only reinforces the idea that long-term potentiation is a mechanism
which is used at a variety of excitatory synapses to store specific
information, but also provides further support for the existence of
different forms of learning and memory underlied by different
cellular mechanisms. In this regard it will be of particular
interest to use this type of compounds to delineate what forms of
learning depend on the activation of glutamate receptors.

Second, we developed elsewhere the idea that the calcium
regulation of glutamate receptor distribution provided an ideal
mechanism to store information in the central nervous system by
modifying the efficacy of excitatory synapses as a result of either
high-frequency stimulation or simultaneous activation of various
converging inputs on the same target cell (Lynch and Baudry, 1984).
In our hypothesis, the large influx of calcium resulting from these
stimulation conditions activates a calcium-dependent protease, inducing the breakdown of cytoskeletal proteins and an increase in
the number of glutamate receptors. This hypothesis is supported by
experimental data showing that an inhibitor of calcium-dependent
protease, the peptide leupeptin, induces selective impairment of
learning in an 8-arm radial maze as well as in a smell discrimina-

tion task without affecting the learning of active or passive shock
avoidance tasks (Staubli et al, 1984, 1985). It is also supported
by the fact that the number of glutamate receptor binding sites are
increased in the rabbit hippocampus following classical conditioning (Mamounas et al., 1985). Finally, it has recently been shown
that in Alzheimer's disease the number of glutamate receptors is
dramatically decreased in cortical structures, suggesting that
changes in the regulation of glutamate receptors might also be
involved in pathological conditions (Greenamyre et al., 1985).

Acknowledgements. This work was supported by grant BNS 81-12156
from the National Science Foundation. The author wishes to acknowledge Markus Kessler for numerous and fruitful discussions and
Jackie Porter for preparing the manuscript.

REFERENCES
Bar, P.R., Wiegant, F., Lopes da Silva, F.H. and Gispen, W.H.
(1984). Tetanic Stimulation Affects the Metabolism of Phosphino-
sitides in Hippocampal Slices. Brain Res., 321, 381-385.
Bardsley, M.E. and Roberts, P.J. (1985). Molecular Size of the
High-affinity Glutamate Binding Site on Synaptic Membranes from Rat
Brain. Biochem. Biophys. Res. Comm., 126, 227-232.
Baudry, M. and Lynch, G. (1979). Regulation of Glutamate Receptors
by Cations. Nature, 282, 748-750.
Baudry, M. and Lynch, G. (1980). Regulation of Hippocampal Gluta-
mate Receptors. Evidence for the Involvement of a Calcium-
activated Protease. Proc. Nat. Acad. Sci. (U.S.A.), 77, 2298-2302.
Baudry, M. and Lynch, G. (1981a). Characterization of Two ^{3}H-
Glutamate Binding Sites in Rat Hippocampal Membranes. J. Neurochem.,
36, 811-820.
Baudry, M. and Lynch, G. (1981b). Hippocampal Glutamate Receptors.
Mol. Cell. Biochem., 38, 5-18.
Baudry, M., Bundman, M., Smith, E. and Lynch, G. (1981) Microlar
Levels of Calcium Stimkulate Proteolytic Activity and Glutamate
Receptor Binding in Rat Brain Synaptic Membranes. Science, 212,
937-938.
Baudry, M., Kramer, K., Fagni, L., Recasens, M. and Lynch, G.
(1983a). Classification and Properties of Excitatory Amino Acid
Receptors in Hippocampus. II. Biochemical Studies Using the
Sodium Efflux Assay. Mol. Pharmacol., 24, 222-228.
Baudry, M., Kramer, K. and Lynch, G. (1983b). Classification and
Properties of Acidic Amino Acid Receptors in Hippocampus. III.
Supersensitivity during the Post-natal Period and Following Dener-
vation. Mol. Pharmacol., 24, 229-234.
Baudry, M., Kramer, K., and Lynch, G. (1983c). Irreversibility
and Time-course of the Calcium-stimulation of ^{3}H-Glutamate Binding
to Rat Hippocampal Membranes. Brain Res., 270, 142-145.
Baudry, M., Evans, J. and Lynch, G. (1985). Excitatory Amino Acids
Inhibit Stimulation of Phosphatidylinositol Turnover by Cholinergic
Agonists in Rat Hippocampal Slices. Submitted.

Berridge,.M. J. and Irvine, R. F. (1984). Inositol Triphosphate, A Novel Second Messenger in Cellular Signal Transduction. Nature, 312, 315-321.

Bliss, T.V.P. and Lomo, T. (1973) Long-lasting Potentiation of Synaptic Transmission in the Dentate Area of the Anesthetized Rabbit Following Stimulation of the Perforant Path. J. Physiol. (London), 232, 331-356.

Bowman, C.L. and Kimelberg, H.K. (1984). Excitatory Amino Acids Directly Depolarize Rat Brain Astrocytes in Primary Culture. Nature, 311, 656-659.

Changeux, J.P. Devillers-Thiery, A. and Chemouilli, P. (1984). Acetylcholine Receptor: An Allosteric Protein. Science, 225, 1335-1345.

Collingridge, G.L., Kehl, S.J. and McLennan, H. (1983). The Antagonism of Amino-acid-induced Excitation of Rat Hippocampal CA1 Neurones In Vitro. J. Physiol., 334, 19-31.

Crepel, F., Dhanjal, S.S. and Sears, T.A. (1982). Effect of Glutamate, Aspartate and Related Derivatives on Cerebellar Purkinje Cell Dendrites in the Rat: An In Vitro Study. J. Physiol., 329, 297-327.

Croucher, M.J., Collins, J.F. and Meldrum, B.S. (1982). Anticonvulsant Action of Excitatory Amino Acid Antagonists. Science, 216, 899-901.

Dingledine, R. (1983)a. N-Methylaspartate Activates Voltage-dependent Calcium Conductance in Rat Hippocampal Pyramidal Cells. J. Physiol., 343, 385-405.

Dingledine, R. (1983)b. Excitatory Amino Acids: Modes of Action on Hippcampal Pyramidal Cells. Fed. Proceed., 42, 2881-2885.

Eccles, J. C. (1983). Calcium in Long-term Potentiation as a Model for Memory. Neurosci., 10, 1071-1081.

Fagg, G.E., Foster, A.C., Mena, E.E. and Cotman, C.W. (1982). Chloride and Calcium Ions Reveal A Pharmacologically Distinct Population of L-glutamate Binding Sites in Synaptic Membranes: Correspondence between Biochemical and Electrophysiological Data. J. Neurosci., 2, 958-965.

Fagg, G.E. and Foster, A.C. (1983). Amino Acid Neurotransmitters and Their Pathways in the Mammalian Central Nervous System. Neurosci., 9, 701-719.

Fagg, G.E. and Matus, A. (1984). Selective Association of N-Methylaspartate and Quisqualate Types of L-glutamate Receptor with Brain Postsynaptic Densities. Proc. Nat. Acad. Sci. (U.S.A.), 81, 6876-6880.

Fagni, L., Baudry, M. and Lynch, G. (1982). Desensitization to Glutamate Does Not Affect Synaptic Transmission in Rat Hippocampal Slices. Brain Res., 1983, 261, 167-171.

Fagni, L., Baudry, M. and Lynch, G. (1983). Classification and Properties of Excitatory Amino Acid Receptors in Hippocampus. I. Electrophysiological Studies of an Apparent Desensitization and Interactions with Drugs Which Block Transmission. J. Neurosci., 3, 1538-1546.

Feltz, A. and Trautmann, A. (1982). Desensitization at the frog neuromuscular junction: a biphasic process. J. Physiol., 322, 257-272.

Ferkany, J.W., Zaczek, R. and Coyle, J.T. (1982). Kainic Acid Stimulates Excitatory Amino Acid Neurotransmitter Release at Presynaptic Receptors. Nature, 298, 757-759.

Ferkany, J., Zaczek, R., Tharkl, A. and Coyle, J.J. (1984). Glutamate-containing Dipeptides Enhance Specific Binding at Glutamate Receptors and Inhibit Specific Binding at Kainate Receptors in Rat Brain. Neurosci. Lett., 44, 281-286.

Flatman, J.A., Schwindt, P.C., Crill, W.E. and Stafstrom, C.E. (1983). Multiple Actions of N-Methyl-D-Aspartate on Cat Neocortical Neurons In Vitro. Brain Res., 266, 169-173.

Fonnum, F. (1984). Glutamate: A Neurotransmitter in Mammalian Brain. J. Neurochem., 42, 1-11.

Foster, A.C. and Fagg, G.E. (1984). Acid amino acid binding sites in mammalian neuronal membdranes; their characteristics and relationships to synaptic receptors. Brain Res. Rev., 7, 103-164.

Garthwaite, J. and Balazs, R. (1978). Supersensitivity tio the Cyclic GMP Response to Glutamate During Cerebellar Maturation. Nature, 275, 328-330.

Garthwaite, J. and Balazs, R. (1980). Excitatory Amino Acid-induced Changes in Cyclic GMP Levels in Slices and Cell Suspensions from the Cerebellum. In Glutamate as a Transmitter. (eds. G. D. Chiara and G.L. Gena). Raven Press, New York.

Goodman, S.R. and Zabon, I.S. (1985). Brain spectrin: a review. Brain Res. Bull., 13, 813-832.

Greenamyre, J.T., Olson, J.J.J., Penney, J.P. and Young, A.B. (1985a). Autoradiographic characterization of N-Methyl-D-aspartate-, quisqualate- and kainate-sensitive glutamate binding sites. J. Pharm. Exp. Therap., 233, 254-263.

Greenamyre, J.T., Penney, J.B., Young, A.B., D'Amato, C., Hicks, S.P. and Shoulson, I. (1985b). Alterations in L-glutamate Binding in Alzheimer's and Huntington's Diseases. Science, 227, 1496-1499.

Hablitz, J.J. (1982). Conductance Changes Induced by DL-Homocysteric Acid and N-Methy-DL-Aspartic Acid in Hippocampal Neurons. Brain Res., 247, 149-153.

Hablitz, J.J. (1985). Action of Excitatory Amino Acids and Their Antagonists in Hippocampal Neurons. J. Neurophys. (in press)

Halpain, S., Wieczorek, C.M. and Rainbow, T. C. (1984). Localization of L-glutamate receptors in rat brain by quantitative autoradiography. J. Neurosci., 4, 2247-2258.

Harris, E.W., Ganong, A.H. and Cotman, C.W. (1984). Long-term potentiation in the hippocampus involves activation of N-methyl-D-aspartate receptors. Brain Res., 323, 132-137.

Heidmann, T. and Changeux, J.P. (1982). Un Modele Moleculaire de Regulation d'efficacite au Niveau Postsynaptique d'une Synapse Chimique. C.R. Acad. Sci. Paris, 295, 665-670.

Honore, T., Lauridsen, J. and Krogsgaard-Larsen, P. (1982). J. Neurochem., 38, 173-178.

Honore, T. and Nielsen, M. (1985). Complex Structure of Quisqualate-sensitive Glutamate Receptors in Rat Cortex. Neurosci. Letters, 54, 27-32.

Hosli, E. and Hosli, L. (1978). Autoradiographic Localization of the Uptake of ^{3}H-GABA and ^{3}H-L-glutamate Acid in Neurons and Glial Cells of Cultured Dorsal Root Ganglia. Neurosci. Lett., 1, 173-176.
Johnston, G.A.R., Lodge, D., Bornstein, J.C. and Curtis, D. R. (1980). Potentiation of L-glutamate and L-aspartate Excitation of Cat Spinal Neurones by the Stereoisomers of Threo-3hydroxyaspartate. J. Neurochem, 34, 241-243.
Kanner, B.I. and Bendahan, A. (1982). Binding Order of Substrates to the Sodium and Potassium Ion Coupled L-glutamic and Transporter from Rat Brain.
Kessler, M., Baudry, M., Cummins, J.T. and Lynch, G. (1985). Induction of Glutamate Binding Sites in Hippocampal Membranes by Transient Exposure to High Concentrations of Glutamate or Glutamate Analogs. J. Neurosci. (in press)
Koller, K.J. and Coyle, J.T. (1984). Characterization of the Interactions of N-Acetylaspartyl Glutamate with ^{3}H-L-Glutamate Receptors. Europ. J. Pharmac., 98, 193-199.
Kramer, K. and Baudry, M. (1984). Low Concentrations of Potassium Inhibit the Na-dependent ^{3}H-glutamate Binding to Rat Hippocampal Membranes. Europ. J. Pharmac., 102, 155-158.
Luini, A., Goldberg, D., and Teichberg, V. (1981). Distinct Pharmacological Properties of Excitatory Amino Acid Receptors in the Rat Stratus: Study by the Na+ Efflux Assay. Proc. Nat. Acad. Sci. (U.S.A.), 78, 3250-3254.
Lynch, G., Larson, J., Kelso, S., Barrionuevo, G. and Schottler, F. (1983). Intracellular Injections of EGTA Block the Induction of Hippocampal Long-term Potentiation. Nature,, 305, 719-721.
Lynch, G. and Baudry, M. (1984). The Biochemistry of Memory: A New and Specific Hypothesis. Science, 224, 1057-1063.
Malik, M.N., Meyers, L.A., Iqbal, K., Sheikh, A.M., Scotto, L. and Wisniewski, H.M. (1981). Calcium Activated Proteolysis of Fibrous Proteins in Central Nervous System. Life Sci., 29, 795-802.
Mamounas, L., Thompson, R.F., Lynch, G. and Baudry, M. (1984). Classical Conditioning of the Rabbit Eyelid Response Increases Glutamate Receptor Binding in Hippocampal Synaptic Membranes. Proc. Nat. Acad. Sci. (U.S.A.), 81, 2478-2482.
Mayer, M.L., Westbrook, G.L. and Guthrie, P.B. (1984). Voltage-dependent block by Mg^{++} of NMDA responses in spinal and neurons. Nature, 309, 261-263.
Mena, E.E., Fagg, G.E. and Cotman, C.W. (1982). Chloride Ions Enhance L-glutamate Binding to Rat Brain Synaptic Membranes. Brain Res., 243, 378-381.
Monaghan, D.T. and Cotman, C.W. (1982). The distribution of ^{3}H-Kainic and Binding Sites in Rat CNS as Determined by Autoradiography. Brain Res., 252, 91-100.
Monaghan, D., Holets, V.R., Toy, D.W. and Cotman, C.N. (1983). Anatomical distributions of four pharmacologically distinct ^{3}H-L-glutamate binding sites. Nature, 306, 176-178.
Monaghan, D.T., Yao D., and Cotman, C.W. (1984a). Distribution of ^{3}H-AMPA in Rat Brain as Determined by Quantitative Autoradiography. Brain Res., 324, 160-164.
Monaghan, D.T., Yao, D., Olverman, H.J., Watkins, J.C. and Cotman,

C.W. (1984b). Autoradiography of D-2-³H-amino-5-phosphonopentanoate Binding Sites in Rat Brain. Neurosci. Letters, 52, 253-258.

Morris, R.G.M., Anderson, E., Lynch, G. and Baudry, M. (1985). Selective Impairment of Learning and Blockade of Long-term Potentiation by an N-methyl-D-aspartate Receptor Antagonist, APV. Submitted.

Nestler, E.J., Walaas, S.I. and Greengard, P. (1984). Neuronal Phosphoproteins: Physiological and Clinical Implications. Science, 225, 1357-1364.

Nicoletti, F., Meek, J.L., Chuang, D.M., Iadarola, M. Roth, B.L. and Costa, E. (1985). Ibotenic Acid Stimulates Inositol Phospholipid Turnover in Rat Hippocampal Slices: An Effect Mediated by APB-sensitive Receptors. Abstract FASEB Meeting, 1985, 493.

Nishizuka, Y. (1984). Turnover of Inositol Phospholipids and Signal Transduction. Science, 225, 1364-1370.

Nistri, A. and Constantini, A.V. (1979). Pharmacological Characterization of Different Types of GABA and Glutamate Receptors in Vertebrates and Invertebrates. Prog. Neurobiol., 13, 117-235.

Nowak, L.M. and Ascher, P. (1984). N-Methyl-D-aspartic, Kainic and Quisqualic Acids Evoked Currents in Mammalian Central Neurons. Abstract, Society for Neuroscience, 10, 23.

Nowak, L., Bregestovski, P., Ascher, P., Herbet, A., and Prochiantz, A. (1984). Magnesium Gates Glutamate-activated Channels in Mouse Central Neurons. Nature, 307, 462-465.

Olverman, H.J., Jones, A.W. and Watkins, J.C. (1984). L-Glutamate Has Higher Affinity Than Other Amino Acids for ³H-D-AP5 Binding Sites in Rat Brain Membranes. Nature, 307, 460-462.

Pant, H.C. and Gainer, H. (1980). Properties of a Calcium-Activated Protease in Squid Axoplasm Which Selectively Degrades Neurofilament Proteins. J. Neurobiol., 11, 1-12.

Pert, C. and Snyder, S.H. (1974). Opiate Receptor Binding of Agonists and Antagonists Affected Differentially by Sodium. Mol. Pharmacol., 10, 866-879.

Pumplin, D.W. and Fambrough, D.M. (1982). Turnover of Acetylcholine Receptors in Skeletal Muscle. Ann. Rev. Physiol., 44, 319-335.

Sandoval, I.V. and Weber, K. (1978). Calcium-induced Inactivation of Microtubule Formation in Brain Extracts. Eur. J. Biochem., 92, 463-470.

Siman, R., Baudry, M. and Lynch, G. (1983). Purification from Synaptosomal Plasma Membranes of Calpain I, A Thiol-protease Activated by Micromolar Calcium Concentrations. J. Neurochem., 41, 950-956.

Siman, R., Baudry, M. and Lynch, G. (1984). Brain Fodrin: Substrate for the Endogenous Calcium-activated Protease Calpain I., Proc. Nat. Acad. Sci (U.S.A.), 81, 3276-32680.

Siman, R., Baudry, M. and Lynch, G. (1985). Regulation of glutamate receptor binding by cytoskeletal protein fodrin. Nature, 1985, 313, 225-227.

Simon, R.P., Swan, J.H., Griffiths, T. and Meldrum, B.S. (1984). Blockade of N-Methyl-D-aspartate receptors may protect against ischemic damage in the brain. Science, 226, 850-852.

Slevin, J.T., Collins, J.F. and Coyle, J. T. (1984). Analogue interactions with the brain receptor labeled by ^{3}H-kainic acid. Brain Res., 265, 169-172.

Snyder, S.H. (1984). Drug and Neurotransmitter Receptors in the Brain. Science, 224, 22-31.

Staubli, U., Baudry, M. and Lynch, G. (1984). Leupeptin, A Thiol -proteinase Inhibitor, Causes A Selective Impairment of Spatial Maze Performance in Rats. Behav. and Neural Biol. 40, 58-69.

Staubli, U., Baudry, M. and Lynch, G. (1985). Olfactory Discrimination Learning is Blocked by Leupeptin, A Thiol-Proteinase Inhibitor. Brain Res. (in press).

Swanson, L.W., Teyler, T. J. and Thompson, R.F. (1982). Neurosci. Res. Prog. Bull., 20, 5.

Takeuchi, A. and Takeuchi, N. (1964). The Effect on Crayfish Muscle of Iontophoretically Applied Glutamate. J. Physiol, 170, 296.

Thomson, A., West, D.C. and Lodge, D. (1985). An N-Methylaspartate Receptor-mediated Synapse in Rat Cerebral Cortex: A Site of Action of Betamine. Nature, 313, 479-481.

Watkins, J.C., Davies, J., Evans, R.H., Francis A.A. and Jones, A. (1981). Pharmacology of Receptors for Excitatory Amino Acids. In Glutamate As a Neurotransmitter. (eds G. Di Chiara and G. L. Gena). Raven Press, New York.

Werling, L. and Nadler, V.J. (1983). Multiple Binding Sites for L-^{3}H-glutamate in Hippocampal Synaptic Membranes: Effects of Calcium and Chloride Ions. Soc. Neuro. Abst., 9, 1186.

Zimmerman, V.J.P. and Schlaepfer, W.W. (1982). Characterizations of a Brain Calcium-activated Protease that Degrades Neurofilament Proteins. Biochemistry, 21, 3977-3982.

21

Functional Reconstitution of Binding-ion Channel Activity and Immunochemical Characterization of Brain Synaptic Membrane Glutamate Binding Protein

E.K. Michaelis, T.M. Stormann, S. Roy and H.H. Chang

INTRODUCTION

The molecular characterization of plasma membrane receptors for neurotransmitters should include the development of approaches for the definition of the ligand recognition sites in the receptor complex, the characterization of the ion channel or macromolecular array involved in the signaling of transmitter-receptor interaction, and the isolation and biochemical characterization of the molecular entities that comprise the receptor complex. The investigations that have been conducted in our laboratory have involved all of these approaches in an effort to delineate the molecular and supramolecular characteristics of the receptor sites in neuronal membranes with which the excitatory neurotransmitter L-glutamic acid interacts to produce its effects on central nervous system neurons (e.g., Michaelis et al., 1974; Michaelis, 1975; Chang and Michaelis, 1980; 1981; 1982; Michaelis et al., 1981; Michaelis et al., 1983; Michaelis et al., 1984a).

Our initial efforts involved attempts to identify the ligand recognition sites for L-glutamic acid in brain neuronal preparations. Since L-glutamic acid causes excitation following application only in the extracellular environment of neurons (Coombs, et al., 1955), it was assumed that the ligand recognition sites for these receptors would be localized on plasma membranes of neurons, most probably membranes obtained from the synaptic region. Indeed such localization of L-glutamate binding sites in isolated synaptic plasma membranes was observed (Michaelis et al., 1974; Foster and Roberts, 1978; Foster et al., 1981), and these membranes were used as the starting material for the purification and characterization of the glutamate binding sites (Michaelis, 1975; Michaelis et al., 1981; Michaelis et al., 1983; Michaelis, et al., 1984a).

A small molecular weight (M_r = 14300) glutamate binding
protein was purified from brain synaptic membranes and was
biochemically characterized. This glutamate binding protein,
hereafter referred to as **GBP**, is an intrinsic synaptic plasma—
membrane acidic glycoprotein (Michaelis, 1975; Michaelis et al.,
1980; Michaelis et al., 1983; Michaelis et al., 1984a). The
binding affinity and selectivity of its binding site for L-glutamic
acid and glutamate analogs are very similar to the binding affinity
and selectivity of sodium-independent, calcium or chloride-
independent binding sites in synaptic membranes (Michaelis et al.,
1981; Michaelis et al., 1983; Michaelis et al., 1984a). Further
similarities between this isolated protein and the synaptic plasma
membrane glutamate-binding sites referred to above include the
nearly identical sensitivity of the protein and the membrane
binding sites to the metal-chelating agents NaN_3, KCN, and
o-phenanthroline, and to the plant lectin concanavalin A
(Michaelis, 1975; Michaelis, 1979; Michaelis et al., 1982). Not
only were the similarities between the isolated protein and the
glutamate binding sites on synaptic membranes strong but, in
addition, the protein that was isolated from either the rat or
bovine brain neuronal membranes exhibited no glutamate-metabolizing
enzymatic activity, indicating that it did not represent any of the
known glutamate-metabolizing enzymes (Michaelis, 1975; Michaelis
et al., 1984a). In order to assess the role of these synaptic
membrane glutamate binding sites, and especially of the **GBP**, in the
physiology of central nervous system synapses, we chose to develop
the methodology for simultaneously monitoring the activity of
glutamate-activated ion channels, presumably receptor-related ion
channels, and of the glutamate transport carriers in these
membranes. The activity of glutamate transport carriers in
synaptic plasma membranes was selected since it represented the
other major class of membrane-associated, glutamate-interacting
macromolecules in the synaptic region of neurons. It became clear
that both the glutamate binding sites that were being detected by
our assays and the **GBP** were not components of the glutamate
transport system in synaptic membranes. This conclusion was based
on the observations that (1) the glutamate binding activity of
these entities was being measured in the absence of an existing Na^+
gradient across the synaptic plasma membrane, as a gradient is
necessary for normal activity of the carriers, and (2) there were
marked differences between the sensitivity of the carriers and the
binding sites to various chemical manipulations. A listing of the
differential characteristics between the glutamate binding sites of
synaptic membranes and the synaptic membrane glutamate transport
carriers is presented in Table 1.

The demonstration of differences between synaptic membrane
glutamate binding sites or the **GBP** and the glutamate transport
carriers with regard to their sensitivity to inhibitors is not
sufficient evidence that the binding protein or the synaptic
membrane binding sites are components of the physiologic glutamate

TABLE 1   Differential Characteristics of Glutamate Binding and
Glutamate Transport Sites in Synaptic Membranes

Agent	L-Glutamate Binding	L-Glutamate Transport	References
Concanavalin A	Inhibited	No Effect[a]	(Michaelis et al., 1974) (Michaelis et al., 1975)
GDEE	Inhibited	No Effect[a]	(Michaelis et al., 1981)
Metal Ligands (KCN, NaN₃, etc.)	Inhibited	No Effect[a]	(Michaelis et al., 1982)
Amphipathic Drugs	Inhibited	No Effect[a]	(Michaelis et al., 1984b)

[a] The L-glutamate transport activity was not sensitive to a 10-fold or even
greater concentration of these agents.

receptors. Partial confirmation of such a relationship can be
obtained by demonstrating that the effects of various glutamate
receptor agonists and antagonists on the binding to synaptic
membranes or the purified protein are analogous to their observed
electrophysiological effects on the glutamate receptors. However,
the pharmacology of glutamate–activated receptor sites of intact
neurons is still deficient in terms of agents which are highly
selective in their actions on specific subpopulations of excitatory
amino acid receptors (Watkins and Evans, 1981; Foster and Fagg,
1984; Monaghan et al., 1983). Furthermore, as McLennan (1983) has
already pointed out, the correlation between the effects of these
agents on ligand binding sites and their actions on physiologic
receptors is still somewhat indefinite.

Additional criteria needed to be developed for evaluating the
relationship of the isolated **GBP** to the macromolecular complex of
the physiologic glutamate receptors. One would expect that if
the **GBP** is involved in the function of some physiologic glutamate
receptors, then one should be able to demonstrate that this protein
is necessary for obtaining ligand–induced activation of the
excitatory amino acid receptor–ion channel complexes. Furthermore,
the distribution of the protein at the cellular and subcellular
levels should correspond to the location of putative glutamate-
sensitive synaptic sites. Our attempts to examine the first level
of activity of the **GBP**, i.e., its involvement in ion channel
activation, led us to the development of methods for the study of
chemically–sensitive ion channels in isolated–resealed synaptic
membranes and in reconstituted proteoliposomes. In order to probe

the distribution of this protein in neuronal preparations and in
intact brain tissue we proceeded with the development of specific
antibodies against the **GBP**, that we could use in both immunochemical and immunocytochemical studies.

FUNCTIONAL PROPERTIES OF THE **GBP**

Both rat brain synaptosomal preparations as well as isolated
synaptic plasma membranes that are resealed in the presence of a
selected ionic environment, exhibit both specific L-glutamate
binding sites and glutamate-sensitive sodium ion (Na^+) flux
activity (Chang and Michaelis, 1980, 1981). Micromolar and
submicromolar concentrations of L-glutamate, L-aspartate, cysteine
sulfinic acid, and kainic acid brought about an increase above the
baseline $^{22}Na^+$ flux in these preparations. The L-glutamate-
stimulated Na^+ influx into synaptosomes and synaptic plasma
membrane vesicles was not related either to an inactivation of the
plasma membrane ($Na^+ + Ka^+$)-ATPase or to an increase in the ion
permeability through voltage-dependent Na^+ channels (Chang and
Michaelis, 1980). Furthermore, this glutamate-stimulated ion flux
process exhibited the following properties: (1) it was
bidirectional, i.e., dependent on the direction of the Na^+
concentration gradient; (2) it could be activated by L-glutamate in
the trans location across the membrane with respect to the ion
gradient; (3) the Na^+ was moving into an osmotically sensitive
space; and (4) the activity was not eliminated by lowering the
incubation temperatures to 4°C (Chang and Michaelis, 1981). These
characteristics of the glutamate-sensitive ion flux activity in
synaptic membranes are indicative of the activation of an ion
channel rather than of a transport carrier.

More recently we have shown that the ion flux processes in
synaptic membranes, which are enhanced by glutamate and other
excitatory amino acids, are electrogenic and cause the appearance
of a transient intravesicular positive potential difference which
averages between 5 and 13 mV (Chang and Michaelis, 1982; Chang
et al., 1984). The electrogenic properties of the activation of
Na^+ flux by the excitatory amino acids were detected by measuring
the distribution of the lipophilic ion $[^{35}S]$ SCN^-, a probe of
transmembrane potential. All excitatory amino acids and amino acid
analogs tested have been found to increase the flux of SCN^- across
the synaptic membrane in a concentration-dependent manner, whereas
the inhibitory amino acid γ-amino butyric acid caused a small
decrease in SCN^- flux. Half-maximal activation by L-glutamic acid
of either $^{22}Na^+$ or SCN^- flux across synaptic membranes was observed
at concentrations of this amino acid that ranged between 0.05 and
0.3 μM (Chang and Michaelis, 1981, 1982). Finally, this
glutamate-stimulated ion flux was inhibited by the glutamate
receptor antagonist L-glutamate diethyl ester, a glutamate analog
which has no effect on the glutamate transport carriers. These

observations indicate that synaptic plasma membranes isolated from neurons of mammalian brain tissue contain both the recognition and ion channel components of the excitatory amino acid receptors.

Solubilization of synaptic membranes with either 2% (w/v) n-octylglucoside or 2% Triton X-100 in the presence of asolectin led to the extraction and reconstitution into liposomes of a glutamate-sensitive Na^+ flux response (Stormann et al., 1984). The L-glutamate stimulation of Na^+ flux detected following solubilization and reconstitution of the extracted synaptic membrane proteins was consistently observed in almost all preparations, but represented a small net Na^+ uptake above that observed under conditions of passive Na^+ flux (Table 2). Nevertheless, this glutamate-sensitive Na^+ flux process was dependent on the concentration of L-glutamate introduced into the assays and was blocked by L-glutamate diethyl ester, indicating it had characteristics very similar to those of the glutamate-initiated ion flux detected in intact synaptic plasma membrane preparations.

Solubilized synaptic membrane proteins were processed through the steps previously used in the purification of the **GBP,** including affinity batch chromatography through glass fiber with co-reticulated L-glutamate and through concanavalin A sepharose (Michaelis, 1975, Michaelis et al., 1983). With each successive enrichment of the preparations in the **GBP,** there was a progressive enhancement in the maximal stimulating action of L-glutamate on Na^+ flux in the reconstituted proteoliposomes (Table 2).

The purification scheme that was used to obtain the various fractions subsequently reconstituted into liposomes was nearly identical to the scheme employed previously for the purification of **GBP** from Triton X-100-solubilized synaptic membrane proteins (Michaelis 1975; Michaelis et al., 1983). The affinity of the L-glutamate binding sites in the reconstituted proteoliposomes obtained following protein purification through either the first step, with the glutamate-loaded glass fiber, or the second step with concanavalin A sepharose, were nearly identical. L-glutamate binding to these proteoliposomes exhibited characteristics of ligand binding to two sets of sites, as has also been observed with the purified **GBP** under conditions of high protein concentration in the assays (Michaelis et al., 1983) and with synaptic membranes (Michaelis et al., 1981; Werling and Nadler, 1982). The estimated dissociation constants (K_D) for ligand binding to these sites were $K_{D1} = 0.05-0.07$ µM and $K_{D2} = 1.0-1.2$ µM.

The comparative purity of the isolated and liposome-reconstituted fractions is currently being determined by means of electrophoretic and immunoblot transfer methods. Our preliminary studies indicate that there is an enrichment of the reconstituted proteoliposomes from both steps of purification with

Table 2   Stimulation of ^{22}Na$^+$ Flux in Solubilized-Reconstituted
Protein Liposomes During the Purification of the **GBP**

Fraction[a]	Baseline Flux[b]	Glutamate-Stimulated[b]	Δ Flux
Synaptic Membrane (Total Solubilized)	67.53±7.4 (n=5)	72.83±2.3 (n=5)	5.3
Glass-Fiber Non-Bound	62.73±11.1 (n=5)	73.86±8.6 (n=5)	11.1
Glass-Fiber Bound	63.27±12.2 (n=5)	91.74±6.0 (n=5)	28.5
Con-A Non-Bound	55.74±9.1 (n=6)	67.00±6.4 (n=6)	11.3
Con-A Bound	53.16±2.9 (n=4)	93.68±4.9 (n=4)	40.5

[a] Synaptic membranes were solubilized in the presence of 2% (v/v)
Triton X-100 and 15 mg/ml asolectin. Reconstitution of the proteins
following centrifugation was accomplished by incubating the solubilized extract with Bio-Beads SM2 to remove the detergent.

[b] The ^{22}Na flux activities shown represent the mean (±S.E.M.) from
duplicate or triplicate determinations from the number of purification
and reconstitution experiments shown in parentheses. Each value
represents ^{22}Na uptake at 120 sec of incubation at 24°C expressed
in nmol/mg protein.

immunologically detectable **GBP**. The protein fraction obtained
following incubation with and selective elution from the glass
fiber with co-reticulated glutamate has given the most consistent
ion flux responses to L-glutamate, while the fraction obtained
following concanavalin A sepharose chromatography has given very
strong responses in some preparations and less so in others. These
differences may be due either to the loss of some associated
proteins that co-purify with the **GBP** through the glass fiber step
or to a change in the state of the protein following the last
purification step of concanavalin A sepharose chromatography.

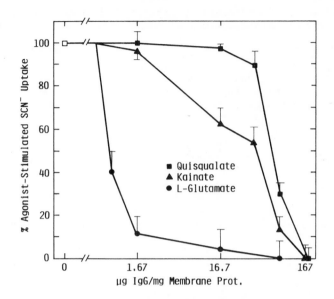

FIG. 1.   Inhibition by increasing concentrations of anti-GBP IgG of the excitatory amino acid stimulation of SCN⁻ influx into synaptic membrane vesicles.   The synaptic plasma membranes were preincubated for 1 h at 0°C with the amounts of IgG indicated.   Complete kinetic determinations (5-90 s) of SCN⁻ influx into the synaptic membranes were obtained under the following conditions: in membranes preincubated in buffer (controls) and assayed in the absence and presence of 5 µM concentration of each amino acid or amino acid analog, and in membranes preincubated with each of the amounts of IgG shown and assayed in the absence or presence of each amino acid or analog.   The values shown represent the percentage of net control SCN⁻ uptake measured in the presence of each amino acid analog at 45 s of incubation under each condition.   Each point represents the mean (±SEM) of six to nine determinations from three membrane preparations (from Roy et al., 1985).

IMMUNOLOGICAL CHARACTERISTICS OF THE **GBP** AND IMMUNOHISTOCHEMICAL DETECTION OF ITS DISTRIBUTION IN BRAIN TISSUE

The second approach we have followed in exploring the activity of **GBP** in mammalian brain tissue has been to develop antibodies (Abs) to the protein and to use these Abs in probing the localization and function of the **GBP**.   The initial set of Abs developed in rabbits were against the **GBP** purified from bovine brain cerebral cortex and cerebellum (Roy and Michaelis, 1984). These Abs exhibited a very high degree of immune reactivity and specificity for the **GBP** from bovine as well as from rat brain.   A glutamate-aspartate binding protein which forms part of the dicarboxylic amino acid transport in bacteria (Schellenberg and Furlong, 1977), and the enzymes glutamate dehydrogenase, glutamine synthetase, and γ-glutamyl transpeptidase from mammalian sources had little or no cross-reactivity with the anti- GBP Abs (Roy and Michaelis, 1984). Only the enzyme glutamic acid decarboxylase from either bacteria (Roy and Michaelis, 1984) or bovine brain cortex (Roy, Galton and Michaelis, submitted for publication) exhibited a moderate to low reactivity with the anti- GBP Abs. Pre-immune sera obtained from the rabbits prior to immunization with the GBP did not react with the GBP. The specificity and strength of immuno-reactivity of the Abs against the various proteins were determined by the highly sensitive enzyme-linked immunosorbent assay (ELISA) (Roy and Michaelis, 1984).

Three physiological processes in which the GBP could be involved were examined for their sensitivity to the anti- GBP Abs: the excitatory amino acid stimulation of Na^+ flux in synaptic membranes, L-glutamic acid transport across these membranes, and depolarization-induced L-glutamate release form synaptic membranes. Only the amino acid-induced changes in ion flux were affected by the anti- GBP Abs (Roy et al., 1985).

The stimulation of SCN^- influx into synaptic membranes by the excitatory amino acids and amino acid analogs L-glutamic acid, kainic acid, quisqualic acid, and N-methyl-D-aspartic acid (NMDA) was used as a measure of receptor-activated ion flux and depolarization. The L-glutamate activation of ion flux was at least 40 times more sensitive to the effects of the anti- GBP Abs than was the stimulation of ion flux by kainate and 60 times more sensitive than that induced by quisqualate (Fig. 1). The stimulation of SCN^- influx by NMDA was unaffected by the introduction of anti- GBP Abs at concentrations as high as 83 μg IgG/mg protein (Roy et al., 1985). The specificity of the inhibition of L-glutamate-stimulated SCN^- influx into synaptic membranes was demonstrated by the fact that pre-immune sera from the same animals had no effect on the activation of ion flux by L-glutamate. Finally, the actions of the Abs were highly selective towards one type of function in synaptic membranes, the activation of ion flux by the excitatory amino acids; they had no effect on either the Na^+-gradient dependent transport of L-glutamic acid or the depolarization-induced release of L-glutamic acid from preloaded synaptic plasma membranes (Roy et al., 1985).

The Abs used in these studies are polyclonal Abs developed against the purified, solubilized GBP and are therefore likely to interact with several immunogenic determinants, some of which may be buried within the synaptic plasma membrane matrix when the protein is associated with the membranes. The glutamate binding site of the protein does not appear to be one of the primary immunogenic determinants since pre-exposure of synaptic membranes to these Abs produces very low levels of inhibition of glutamate

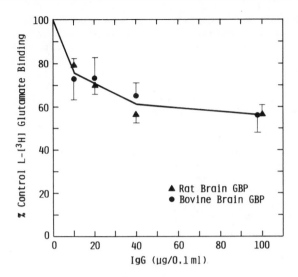

FIG. 2. Inhibition by L-[^{3}H]glutamate binding to rat and bovine brain GBP by anti-GBP antibodies (IgG fraction). All binding assays were performed at 24°C in the presence of 120 nM L-[^{3}H]glutamate in 0.1 ml of 10 mM potassium phosphate pH 7.4 according to the procedures described in Michaelis et al. (1983). The GBP samples were treated with varying amounts of IgG for 30 min at 24°C prior to the initiation of the binding assay. Samples that functioned as controls were exposed to the same conditions, but received either equivalent amounts of bovine serum albumin or phosphate-buffered saline. Each data point represents the mean (±SEM) of 8-12 determinations (from Roy et al., 1985).

binding to the synaptic membranes (usually on the order of 10-15%). In addition, large amounts of the Abs are required to bring about moderate inhibition of L-glutamate binding to the active site of the soluble, purified protein (Fig. 2). The weak inhibitory effect of the Abs on the ligand binding activity of the purified **GBP** in solution is probably the result of interactions of the Abs with domains of the protein that are not normally exposed when this protein is embedded in the synaptic membranes. Nevertheless, although the anti- **GBP** Abs weakly inhibit glutamate binding to its recognition sites in synaptic membranes, they are strong inhibitors of the changes in macromolecular conformations that lead to activation of an ion channel by glutamate.

In initial studies of the distribution of **GBP** immunoreactivity in brain tissue samples using the ELISA procedure, we have found that the hippocampus has the highest immunoreactivity per mg of tissue protein of any brain region that we have examined, followed by that found in the cerebral cortex, cerebellar cortex, and

caudate-putamen (Roy and Michaelis, unpublished observations).
Furthermore, there is progressive enrichment in immunochemically
reactive proteins as one proceeds from whole brain homogenate to
the purified synaptic plasma membranes. These membranes have the
highest immunoreactivity of any brain subcellular fraction that we
have examined. We have recently initiated a more detailed
examination of the immunocytochemical localization of the protein
at the light and electron microscopic level. These studies are not
yet completed, but the results obtained at the light microscopic
level by means of the unlabeled peroxidase-anti-peroxidase method
are indicative of a localized distribution of immunoreactive
products in discrete, punctate or "knob-like" structures localized
primarily on brain neuronal dendrites. The hippocampus and
temporal cerebral cortex appear to be areas of considerable
enrichment in terms of the appearance of these small, immunoreactive loci on neuronal processes.

CONCLUSIONS

    The evidence that we have gathered with regard to the possible
function of the glutamate binding protein from brain synaptic
membranes is suggestive of a role for this protein in the activity
of some excitatory amino acid-sensitive receptors in brain
synapses. Since we have purified this protein from synaptic plasma
membranes and have demonstrated by means of immunochemical
techniques that the synaptic membranes obtained from brain neurons
are most highly enriched with GBP immunoreactivity, it seems
reasonable to conclude that on the basis of its localization
the GBP may play a role in synaptic physiology. Its activity is
probably not related either to enzymatic functions involved in
glutamate metabolism, or to neuronal uptake and release processes
for L-glutamic acid and related amino acids.

    The fact that the reconstitution of solubilized, GBP- enriched
protein fractions led to the preservation of both glutamate
recognition function and glutamate-activated ion channel responses
suggests that this protein may indeed be related to the function of
some synaptic plasma membrane glutamate receptor complexes. The
additional observation that antibodies raised against
the GBP blocked the glutamate stimulation of ion fluxes across
synaptic membranes can be considered as further evidence that this
protein is a component of the supramolecular complex of a receptor
activated by L-glutamate. Our observations also indicate that
the GBP may be a component of or may be closely interacting with
the receptor complexes that are activated by kainic acid and
quisqualic acid. This idea is being advanced on the basis of the
observed inhibition of the kainate and quisqualate-activated ion
fluxes in synaptic plasma membranes by the anti- GBP Abs. It is
also supported by our observations in previous studies that the ion
fluxes in synaptic membranes that were activated by glutamate and

kainate were not additive as one might have expected of agonists stimulating distinct receptor complexes (Chang et al., 1984). The definition of the exact relationship of the GBP to the kainate and quisqualate receptor complexes in neuronal membranes will have to await the further characterization of the macromolecular species that comprise those entities.

Acknowledgments

This work was supported by grants from NIAAA (AA 04732), from the AHA, Kansas Affiliate (KS-83-73 and KS-84-G21), and from U.S. ARO (DAAG29-83-K0065). We acknowledge the support provided by the Center for Biomedical Research - The University of Kansas. We thank Ms. Kristin Adams for typing this manuscript.

REFERENCES

Coombs, J.S., Eccles, J.C., and Fatt, P. (1955). The specific ionic conductances and ionic movements across motoneuronal membrane that produce the inhibitory postsynaptic potential. J. Physiol. (Lond.), 130, 326-373.

Chang, H.H. and Michaelis, E.K. (1980). Effects of L-glutamic acid on synaptosomal and synaptic membrane Na^+ fluxes and (Na^+-K^+)-ATPase. J. Biol. Chem., 255, 2411-2417.

Chang, H.H. and Michaelis, E.K. (1981). L-Glutamate stimulation of Na^+ efflux from brain synaptic membrane vesicles. J. Biol. Chem., 256, 10084-10087.

Chang, H.H. and Michaelis, E.K. (1982). L-Glutamate effects on electrical potentials of synaptic plasma membrane vesicles. Biochim. Biophys. Acta, 688, 285-294.

Chang, H.H., Michaelis, E.K. and Roy, S. (1984). Functional characteristics of L-glutamate, N-methyl-D-aspartate and kainate receptors in isolated brain synaptic membranes. Neurochem. Res., 9, 901-913.

Foster, A.C., Mena, E.E., Fagg, G.E., and Cotman, C.W. (1981). Glutamate and aspartate binding sites are enriched in synaptic junctions isolated from rat brain. J. Neurosci., 1, 620-625.

Foster, A.C. and Fagg, G.E. (1984). Acidic amino acid binding sites in mammalian neuronal membranes: Their characteristics and relationship to synaptic receptors. Brain Res. Rev., 7, 103-164.

Foster, A.C. and Roberts, P.J. (1978). High affinity L-[^{3}H]glutamate binding to postsynaptic receptor sites on rat cerebellar membranes. J. Neurochem., 31, 1467-1477.

Grubbs, R.D. and Michaelis, E.K. (1980). Characterization of the glutamate receptor-like protein: Reconstitution into liposomes. Neurosci. Abstr., 6, 254.

McLennan, H. (1983). Receptors for the excitatory amino acids in the mammalian central nervous system. Prog. Neurobiol., 20, 251-271.

Michaelis, E.K. (1975). Partial purification and characterization of a glutamate-binding glycoprotein from rat brain. Biochem. Biophys. Res. Commun., 65, 1004-1012.

Michaelis, E.K. (1979). The glutamate receptor-like protein of brain synaptic membranes is a metalloprotein. Biochem. Biophys. Res. Commun., 87,106-113.

Michaelis, E.K., Michaelis, M.L., and Boyarsky, L.L. (1974). High affinity glutamic acid binding to brain synaptic membranes. Biochim. Biophys. Acta, 367, 338-348.

Michaelis, E.K., Michaelis, M.L., and Grubbs, R.D. (1980). Distinguishing characteristics between glutamate and kainic acid binding sites in brain synaptic membranes. FEBS Lett., 118, 55-57.

Michaelis, E.K., Michaelis, M.L., Chang, H.H., Grubbs, R.D., and Kuonen, D.R. (1981). Molecular characteristics of glutamate receptors in the mammalian brain. Mol. Cell. Biochem., 38, 163-179.

Michaelis, E.K., Belieu, R.M., Grubbs, R.D., Michaelis, M.L., and Chang, H.H. (1982). Differential effects of metal ligands on synaptic membrane glutamate binding and uptake systems. Neurochem. Res., 7, 417-430.

Michaelis, E.K., Michaelis, M.L., Stormann, T.M., Chittenden, W.L., and Grubbs, R.D. (1983). Purification and molecular characterization of the brain synaptic membrane glutamate binding protein. J. Neurochem., 40, 1742-1753.

Michaelis, E.K. Chittenden, W.L., Johnson, B.E., Galton, N., and Decedue, C. (1984a). Purification, biochemical characterization, binding activity, and selectivity of the glutamate binding protein purified from bovine brain. J. Neurochem., 42, 397-406.

Michaelis, E.K., Magruder, C.D., Lampe, R.A., Galton, N., Chang, H.H., and Michaelis, M.L. (1984b). Effects of amphipathic drugs on L-[^{3}H]glutamate binding to synaptic membranes and the purified binding protein. Neurochem. Res., 9, 29-44.

Monaghan, D.T., Hotels, V.R., Toy, D.W., and Cotman, C.W. (1983). Anatomical distributions of four pharmacologically-distinct L-[^{3}H]glutamate binding sites. Nature (Lond.), 306, 176-179.

Roy, S. and Michaelis, E.K. (1984). Antibodies against the bovine brain glutamate binding protein. J. Neurochem., 42, 838-841.

Roy, S., Galton, N., and Michaelis, E.K. (1985). Effects of anti-glutamate-binding protein antibodies on synaptic membrane ion flux, glutamate transport and release, and L-glutamate binding activities. J. Neurochem., in press.

Schellenberg, G.D. and Furlong, C.E. (1977). Resolution of the multiplicity of the glutamate and aspartate transport systems of Escherichia coli. J. Biol. Chem., 252, 9055-9064.

Stormann, T.M., Chang, H.H., Johe, K., and Michaelis, E.K. (1984). Functional reconstitution of the synaptic membrane glutamate-binding protein: Glutamate receptor-like activity in liposomes. Neurosci. Abstr., 10, 958.

Watkins, J.C. and Evans, R.H. (1981). Excitatory amino acid transmitters. Annu. Rev. Pharmacol. Toxicol., 21, 165-204.

Werling, L.L. and Nadler, J.V. (1982). Complex binding of L-[^{3}H]glutamate to hippocampal synaptic membranes in the absence of sodium. J. Neurochem., 38, 1050-1062.

22

Analysis of Excitatory Amino Acid Receptor Function with $^{22}Na^+$ Fluxes in Brain Slices

V.I. Teichberg and O. Goldberg

INTRODUCTION

A large body of evidence supports today the views that glutamate (Glu) mediates the majority of the neuroexcitatory transactions in the CNS and is a possible contributing factor to several neuropathological states and neurological disorders (DiChiara and Gessa, 1981; Meldrum, 1985). The elucidation of the various neuroexcitatory and neurotoxic properties of glutamate and the development of ways to prevent or modulate its deleterious actions have therefore become increasingly important research issues. The study of the glutamate receptors has received particular emphasis since the latter clearly play a pivotal role in mediating the various effects of glutamate. The interaction of glutamate with its neuronal receptors has been studied by both electrophysiological and biochemical methods. The first make use of intracellular or extracellular recording devices to measure either the changes in membrane potential or spike frequency produced by topically applied glutamate or excitatory amino acids analogs (Watkins and Evans, 1981). The biochemical methods include measurements of: radioactive ligand (glutamate or analogs) binding to brain membranes (Fagg and Foster, 1984), excitatory amino acids induced elevation of cyclic nucleotide in cerebellar slices (Foster and Roberts, 1980), excitatory amino acid induced release of acetylcholine in striatal slices (Scatton and Lehman, 1982), the stimulatory effect of glutamate on radioactive Ca^{++} uptake by cortical slices (Ichida et al., 1982) and the excitatory amino acid induced increases in $^{22}Na^+$ efflux from $^{22}Na^+$-preloaded brain slices (Luini et al., 1981; Teichberg et al., 1981). The purpose of this chapter is to review the contributions of the $^{22}Na^+$ efflux receptor assay to the elucidation of the pharmacological properties and synaptic function of the excitatory amino acid receptors, to the screening and detection of new antagonists of excitation and to the development of potential anticonvulsant drugs.

2. $^{22}Na^+$ EFFLUX RECEPTOR ASSAY: PRINCIPLE, ADVANTAGES AND DISADVANTAGES.

The $^{22}Na^+$ efflux receptor assay is based on the following premises: 1. Glutamate and its analogs depolarize CNS neurons by increasing their membrane permeability to Na^+ ions (Puil, 1981);

2. Excitatory amino acids increase the movements of Na^+ ions in brain slices (Bradford and McIlwain, 1966; Harvey and McIlwain, 1968);

3. The glutamate induced depolarization of neurons is prevented in a Na^+ free medium (Hosli et al., 1976).

Since during the excitatory amino acid stimulated increase in membrane permeability to Na^+ ions, an increased traffic of Na^+ ions is expected to take place in both directions across the membrane, increases in both Na^+ influx and efflux can be used as indices of the interactions of excitatory amino acids with membrane receptors. With brain slices, however, the measurements of $^{22}Na^+$ influxes present some technical problems. In particular, the amplitude of the signal (amount of tracer ions taken up by brain tissue upon exposure to an effector) depends on the amount of tissue used, and since the background radioactivity is high, the signal/noise ratio is often very small. Receptor assays based on $^{22}Na^+$ influx measurements cannot, thus, be carried out in a routine way if brain slices are used. In contrast, measurements of $^{22}Na^+$ efflux are not affected by these problems.

The protocol of the $^{22}Na^+$ efflux receptor is described in figure 1. Brain slices are preloaded with the radioactive tracer ion: $^{22}Na^+$. The radioactivity is then washed out by repetitive transfers of the slices into wash-out vessels containing a non-radioactive physiological solution.

The direct relationship between the effector (glutamatergic agonist) induced-increase in specific $^{22}Na^+$ efflux rate ($\triangle$) and the interaction of the effector with receptors controlling the membrane permeability has been established without ambiguity (Teichberg et al., 1980; Luini et al., 1981; Teichberg et al., 1981). This phenomenon is not affected by the presence of 10^{-7} M tetrodotoxin, a specific inhibitor of the action potential Na^+ channel, by the use of physiological solutions containing low Ca^{++} (0.02 mM) and high Mg^{++} (10 mM) concentrations - to block polysynaptic activation, or by decreasing the temperature from $37^\circ C$ to $15^\circ C$

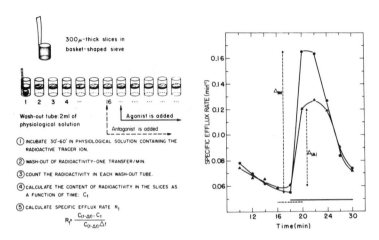

Figure 1:   Scheme of the tracer ion efflux assay.
C_t is the tracer ion content of the slices of time t, and
$C_{(t-\Delta t)}$ is the content at $(t-\Delta t)$. The graph on the right
shows the evolution of the specific efflux rate R_t as the
function of time. The full and the broken horizontal bars indicate the time of exposure to the agonist and to the antagonist respectively. The stimulatory effect of an effector on
the specific efflux rate is expressed in terms of Δ, a parameter equal to the difference between the average values of R_t
measured during the two minutes before and the two minutes after the exposure to the effector. This procedure allows a
standardization of the results as they become independent of
the number of slices used or the actual amount of radioactivity released (According to Teichberg et al., 1981).

which markedly inhibits the activity of Na$^+$/K$^+$ activated
ATPases. Moreover, the effector-induced increase in specific
^{22}Na$^+$ efflux rate is inhibited by several compounds known to
block selectively amino acid excitation both in vivo and in
vitro (Luini et al., 1981; Teichberg et al., 1981).
     The ^{22}Na$^+$ efflux method displays several attractive
properties: It is reliable since it duplicates results
obtained by electrophysiological methods. It is reproducible
since the signal measured does not vary from experiment to
experiment by more than 20%. It is quantitative since dose-
response curves can be established. It is fast since
experiments under 18 different conditions in triplicate can be
carried out routinely in one day by a single experimenter. It
is versatile since manipulations of the ionic composition of

the medium can be made. It is applicable in any laboratory with simple precautions regarding the use of radioactivity.

The limitations of the $^{22}Na^+$ efflux assay are the following: The effects of excitatory amino acids or of their antagonists are not being studied on a real time scale. Diffusion barriers, uptake mechanisms or degrading enzymes may theoretically affect the concentration of every ligand (agonists and antagonists) and cause it to differ at the receptor level from its value in the bath. Thus, a displacement towards higher concentrations may affect the dose-response curve of some ligands. The depletion of radioactive ions from ligand-sensitive cell pools is a factor that prohibits the use of very high agonist concentrations.

3. IDENTIFICATION OF AGONISTS AND ANTAGONISTS WITH THE $^{22}Na^+$ EFFLUX ASSAY.

Using the $^{22}Na^+$ efflux assay, we have carried out a very large screening study to detect compounds capable of either increasing the $^{22}Na^+$ efflux rate (agonists) or blocking the excitatory amino acid induced increase in $^{22}Na^+$ efflux rate (antagonists). As a result, three categories of compounds were identified: agonists, antagonists and compounds devoid of either activity. Tables 1-3 list the tested compounds according to these three categories. The agonists are ranked in table 1 according to their relative potency taking L-glutamate as a standard. Table 2 lists the effectors that were found to antagonize the increase in $^{22}Na^+$ efflux produced in preloaded striatum slices by at least one of the following 5 agonists: L-Glu, L-aspartate (L-Asp), N-methyl-D-aspartate (NMDA), quisqualate (Quis) and kainate (KA). Some of these antagonists such as γ-D-Glu-Gly are competitive antagonists (Tal et al., 1982) whereas others such as the barbiturates, alcohols and local anaesthetics are noncompetitive antagonists (Teichberg et al., 1984).

Table 3 lists 77 compounds that were tested and found not to possess the ability to increase $^{22}Na^+$ efflux from $^{22}Na^+$-preloaded striatum slices. The twenty-three compounds, labeled with an asterisk, which were further tested as potential antagonists of NMDA, Quis, Glu and KA turned out to be devoid of any activity.

4. DETECTION OF VENOMS AND TOXINS WITH EXCITATORY AMINO ACID ANTAGONIST ACTIVITY.

Table 3 includes also a list of 40 venoms that were tested for their potential ability to block the increase in $^{22}Na^+$ efflux produced in $^{22}Na^+$-preloaded striatum slices by NMDA, Quis, Glu or KA. In these experiments, the $^{22}Na^+$-preloaded slices were exposed to the venom for 20 min before the addition of the agonist. The venom phospholipase activities were inhibited by the use of a physiological medium

Table 1:   Relative efficacies of effector agonists inducing an increase

in ^{22}Na$^+$ efflux from ^{22}Na$^+$-preloaded rat striatum slices.

N-methyl-D-aspartate	50.0	cephalosporin C	0.5
DL-homocysteate	12.0	L-aminoadipate	0.4
kainate (KA)	8.0	cis-2,3-piperidinedicarboxylate	0.3
quisqualate (Quis)	6.4	L-cysteine	0.2
kainate methylketone	5.2	2-aminophosphonobutyrate	0.2
ibotenate	5.0	hydroxykainate	0.2
allokainate methyl ketone	4.6	L-proline	0.1
carboxykainate	2.5	L-glutamate-γ-hydroxamate	0.1
carboxyallokainate	2.3	L-serine	0.1
L-cysteine sulfinate	2.1	phenylthiohydroxykainate	0.1
quinolinate	2.1	dihydroxykainate	0.1
D-glutamate	1.2	2-aminophosphonopropionate	0.1
L-cysteate	1.2	α-L-Glu-L-Ala	0.1
L-glutamate (Glu)	1.0	N-(DL)γ-Glu-anilide	0.1
D-aspartate	0.8	γ-L-Glu-L-Glu	0.1
L-aspartate (Asp)	0.6	methyltetrahydrofolate	0.05
dihydrokainate	0.6		

The efficacy was established by calculating for each effector the inverse ratio between the bath concentration of the effector eliciting a response $\triangle$ and the concentration of L-glutamate eliciting the same response (example: 600μM Glu produced an increase in ^{22}Na$^+$ efflux ($\triangle$) identical to that produced by 50 μM DL-homocysteate). All compounds were tested in triplicates at, at least, two concentrations.

Table 2:   List of effectors antagonizing the excitatory amino acid

induced increase in ^{22}Na$^+$-preloaded rat striatum slices.

DL-aminoadipate	kainate iodolactone
DL-aminosuberate (DLAS)	kainate phenylthiolactone
2-amino-5-phosphonovalerate (APV)	γ-kainyl-Gly
amobarbital	γ-kainyl-Ala
butanol	γ-kainyl-GABA
butobarbital	γ-kainyl-Glu
chloral hydrate	γ-kainyl-Tyr
chloroform	γ-kainyl-kainate
diethyl glutamate	γ-kainyl-amide
ethanol	kynurenate
γ-D-Glu βAla	methanol
γ-L-Glu-Gly	γ-methylkainate
γ-D-Glu-Gly (γDGG)	pentanol
γ-D-Glu-L-Phe	pentobarbital
γ-D-Glu-D-Phe	phenobarbital
γ-D-Glu-L-Pro	procaine
3-hydroxy-2-quinoxalinecarboxylate (HQC)	propanol
hexobarbital	secobarbital
kainate hydroxylactone	thiopentone
kainate lactone	

The data concerning the effectiveness of these antagonists have been presented in the following articles: Luini et al., 1981; Teichberg et al., 1981; Tal et al., 1982; Teichberg et al., 1984; Erez et al., 1985; Goldberg and Teichberg, 1985a, b.

Table 3: List of compounds devoid of the ability to increase the $^{22}Na^+$

efflux from $^{22}Na^+$-preloaded brain slices

N-acetyl-L-aspartate; N-acetyl-DL-aspartate; N-acetylkainate; N-acetylaspartylglutamate*; adenosine; γ-aminobutyric acid; 7-aminocephalosporanic acid; 6-aminopenicillamic acid; 2-amino ethanol*; aspartylglutamate; β-L-aspartylglycine; apomorphine; ATP; betaine; carbamyl choline; carnosine*; chelidamic acid; chlorpheniramine maleate; chlorpromazine; cyclic AMP; L-cystine*; diazepam*; L-dihydroxyphenylalanine; diphenylhydantoin*; dipropylacetamide*; dopamine; dithiothreitol; ergothioneine; ethosuccimide; folic acid; L-glutamate γ-hydrazide; α-L-glutamyl-L-Ala; γ-L-glutamyl-L-His; γ-L-glutamyl-L-Tyr; glutamine; glutathione; glutarate; glycine; hydrocortisone*; 1-hydroxy-3-aminopyrrolidone*; 6-hydroxynicotinic acid; hydroxyproline*; 8-hydroxyquinoline-5-sulphonic acid*; hypoxanthine; IMP; L-kynurenine; L-lysine; mephenesin*; L-methionine; DL-α-methylglutamate; metrazole; morphine*; naloxone; norepinephrine; orotic acid*; penicillin G; penicillin V; pentylglycineamide*; picrotoxin; L-pipecolic acid; phosphocreatine; O-phosphoserine; L-phenylalanine; L-proline-proline; 2,3-pyrazinedicarboxylic acid*; 2,4-pyridinedicarboxylic acid*; 2,5-pyridinedicarboxylic acid*; 2,6-pyridinedicarboxylic acid*; serotonine; substance P; L-taurine*; tetrodotoxine*; theophylline; L-threonine; L-tyrosine; uracil*; uric acid*; xanthurenic acid*; venoms: androctonus mauritanicus mauritanicus*; apis mellifera*; atractaspis engaddensis*; agkistrodon bilineatus*; agkistrodon contortrix*; agkistrodon halys*; agkistrodon piscivorus*; bitis arietans*; bitis gabonica*; bitis nasicornus*; bothrops atrox*; bothrops cotiara*; bothops jararaca*; bothrops jararacussa*; bothrops numnifer*; bufo arenarum*; bufo marinus*; buthus judaicus*; causus rhombeatus*; crotalus adamanteus*; crotalus atrox*; crotalus basiliscus*; crotalus viridis helleri*; crotalus ruber*; crotalus durrisus*; crotalus horridus*; crotalus horridus horridus*; crotalus molossus molossus*; dendroaspis augusticus*; dendroaspis viridis*; echis colorata*; leptochelis*; maurus palmatus*; ophiophagus hannah*; pardachirus marmoratus*; sepedon hemachatus*; trimeresurus popearum*; vipera amodytes*; vipera palestinae*: walternissia aegiptia*.

All the above compounds were tested in triplicates at, at least two concentrations, the highest being 1mM for the fully soluble substances. The poorly soluble substances were tested at the concentration corresponding to their highest solubility. Compounds with asterisks (*) were tested also in the presence of each of the following agonists (Glu, Asp, Quis, KA, NMDA) and found not to affect their responses.
The venoms were tested at the concentration of 25 $\mu g/ml$ using the procedure described in figure 1. Brain slices were transferred in a series of venom containing washout tubes in order to be exposed to the venom for 20 min before the addition of one of the following agonists: NMDA, Glu, Quis and KA. The venom and the slices were incubated in a physiological solution containing 2 μM Ca^{++}, 1.3 mM Mg^{++} and 10^{-4} M EDTA to inhibit as much as possible the venom phospholipase activities.

containing 2 μM Ca^{++}, 1.3 mM Mg^{++} and 0.1 mM EDTA (Shipolini et al., 1971). These experimental conditions did not affect the response produced by the excitatory amino acids. With the exceptions of the venoms from various naja species, from bungarus multicinctus and from the scorpion leiurus quinquestriatus which were found to inhibit by 40-50% the response to the four agonists tested, none of the venoms listed in table 3 produced any effect.

4a. β—Bungarotoxin blockade of the response to selected excitatory amino acids.

Chromatographic separation and analysis of the various proteins composing the bungarus multicinctus venom revealed that the excitatory amino acid blocking activity of the venom was entirely due to β-bungarotoxin (βT). Studying the effects of pure βT, we found that its excitatory amino acid blocking activity could be markedly increased by using a physiological solution containing 2 mM Ca^{++}. Under these conditions, βT was found to block the responses to excitatory amino acids in the following order: Quis>KA>NMDA>>Glu. At 100nM βT, all the excitatory amino acid responses were totally blocked. Since βT has been established to possess an intrinsic Ca^{++}-dependent phospholipase A2 activity (Strong et al., 1976), we investigated whether the excitatory amino acid antagonist activity of βT could be blocked by substituting in the incubation medium Sr^{++} for Ca^{++}. This substitution has been reported to reduce the phospholipase A2 activity of βT by 99% (Abe et al., 1977).

Table 4 shows that the replacement of Ca^{++} by Sr^{++} in the incubation medium did not affect the response to NMDA while it reduced the blocking action of βT, but not completely (65%). Moreover, it was found that the incubation of the slices with pure phospholipase A2 (from N. Mozambica) affected only slightly the response to NMDA. These results indicate that the NMDA blocking action of βT is not entirely due to its phospholipase A2 activity. The nature of the phospholipase-independent antagonist action of βT on excitatory amino acid responses has not yet been clarified.

Table 4: Inhibition of the NMDA response by β-bungarotoxin and phospholipase A2.

	% Response to NMDA	
	2mM Ca^{++}	2mM Sr^{++}
Control	100$\pm$ 9	96$\pm$12
βT (1μg/ml)	3$\pm$ 2*	65$\pm$10*
Phospholipase A2 (1μg/ml)	82$\pm$ 5*	N.T.

* $p < 0.01$, N.T. not tested.

4b. Modulation by phospholipases of the response to selected excitatory amino acids.

In view of the rather selective effect produced in the presence of 2 mM Ca^{++} by βT on the response to excitatory amino acids, i.e. blocking Quis>KA>NMDA>Glu, we investigated whether other phospholipases would produce similar results. We therefore incubated brain slices with one of the following enzymes used at 1 μg/ml: The phospholipase A2 from either N. Mozambica or Maurus palmatus, phospholipase D (from cabbage) and phospholipase C (from Bacillus welchii). After a 20 min incubation at $37^{\circ}C$ to facilitate the phospholipase activity, the slices were washed, loaded with $^{22}Na^{+}$ and then processed as described in Fig. 1. In spite of the experimental conditions favouring the phospholipase activity, the responses to the various excitatory amino acids were found not to be markedly inhibited (no more than 43% inhibition) but were nevertheless differentially affected: whereas the phospholipases A2 blocked the response to NMDA>Quis>KA>Glu, the phospholipases C and D affected the KA response but not that of Quis, Glu or NMDA. The quantitative data are included in figure 2.

4c. Antagonism of excitatory amino acids by scorpion toxins.

The chromatographic separation and analysis of the proteins composing the venom of the scorpion leiurus quinquestriatus revealed that the excitatory amino acid blocking activity of this venom was due to the so-called neurotoxin V (Miranda et al., 1970), a well-known ligand of the voltage-dependent Na^{+} channel (Catterall, 1977). When brain slices were incubated for 20 min with 25 μg/ml neurotoxin V, washed extensively, loaded with $^{22}Na^{+}$ and further processed as shown in figure 1, the increase in $^{22}Na^{+}$ efflux stimulated by excitatory amino acids was totally abolished. Similar resuts were obtained using the pure neurotoxin II isolated from the venom of the scorpion Androctonus australis which has an amino acid sequence very similar to that of neurotoxin V (Kopeyan et al., 1978). A fifty percent inhibition of the response to either Quis, Glu, NMDA or KA was obtained with 0.1 μg/ml neurotoxin II. The inhibitory effects of neurotoxin II were abolished if brain slices were incubated concomitantly with 10^{-7} M tetrodotoxin and duplicated upon incubation with 10^{-4} M veratridine.

We have not been able to explain the mechanisms responsible for the above effects. Using the same paradigm as outlined above, we found that preincubation of the brain slices in depolarizing conditions (with either 30 μM NMDA or 50 mM K^{+}) did not ultimately affect the extent of $^{22}Na^{+}$ uptake into the slices during $^{22}Na^{+}$-loading or of the excitatory amino acid-stimulated increase in $^{22}Na^{+}$ efflux. This rules out the possibility that a neuronal depolarization is at the origin of the effects of the scorpion toxins. The fact that their action is blocked by tetrodotoxin strongly suggest, however, that the scorpion toxins do not directly interact with the excitatory amino acid receptors.

5. PHARMACOLOGICAL PROPERTIES OF EXCITATORY AMINO ACID RECEPTORS.

Using the ^{22}Na$^+$ efflux assay, we have carried out a systemic study of the effects produced by the antagonists listed in table 2 on the increase in ^{22}Na$^+$ efflux stimulated in ^{22}Na$^+$-preloaded striatum slices by excitatory amino acids. The results of these experiments led to a classification of the agonists and of the antagonists into groups displaying specific pharmacological properties. Four classes of agonists were detected, the most representative compound of each class being respectively NMDA, Kainate, Quisqualate and L-Glu (Teichberg et al., 1981; Luini et al., 1981). Thus, DL homocysteate, quinolinate, L-glutamate, L-aminoadipate, and L-serine belong to the class of NMDA-like agonists whereas L-cysteine sulfinate and D-aspartate to that of Glu-like agonists (Luini et al., 1984a). Ibotenate and the various kainate derivatives listed in table 1 behave as KA-like agonists whereas quisqualate is so far the only representative of its group.

Figure 2 shows the profiles of inhibition produced by 31 antagonists on the responses to NMDA, KA, Quis and Glu. These profiles differ from each other in many respects. Antagonists 1-5 (alcohols) and 12-16 (barbiturates) differentiate very well between NMDA-like and Quis-KA. The antagonists 20-21-22 (γ-dipeptides derived from D-Glutamate) are specific NMDA antagonists whereas the lactones derived from KA are both NMDA and KA antagonists. Antagonists 28-31 are rather specific NMDA antagonists with the exception of γ-D-Glu-Gly which also blocks KA to some extent.

The differences observed in the inhibition profiles have been taken as evidence of the existence of four distinct classes of excitatory amino acid receptors, namely, the NMDA-receptor, KA-receptor, Quis-receptor and Glu-receptor (Luini et al., 1981; Teichberg et al., 1981). Similar studies using electrophysiological techniques have suggested, however, the presence of only three classes of receptors, namely, the NMDA-receptor, KA-receptor and Quis-receptor and found little evidence for the existence of a separate Glu-receptor (Watkins and Evans, 1981). Having been able to demonstrate that by combining two antagonists, secobarbital and 2-amino-5-phosphonovalerate, it is possible to block completely the responses to NMDA, Quis and KA while affecting

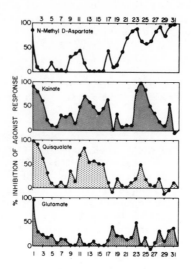

Figure 2: Antagonism of the increase in specific $^{22}Na^+$ efflux rate produced by NMDA, KA, Quis and Glu on $^{22}Na^+$-preloaded striatum slices The excitants were used at concentrations eliciting sub-maximal responses of identical amplitudes (30 μM NMDA, 100 μM Quis, 100 μM KA and 0.5 mM Glu). The following antagonists were used: 1. Pentanol; 2. Butanol; 3. Propanol; 4. Ethanol; 5. Methanol; 6. Barbital; 7. Phospholipase D; 8. Phospholipase C; 9. Phospholipase A2 (N. Mozambica); 10. Phospholipase A2 (Maurus Palmatus); 11.β bungarotoxin; 12. Secobarbital; 13. Pentobarbital; 14. Amobarbital; 15. Butobarbital; 16. Phenobarbital; 17. Kynurenate; 18. Diethyl-Glu; 19. γ-D-Glu-β L-Ala; 20. γ-D-Glu-L-Phe; 21. γ-D-Glu-D-Phe; 22. γ-D-Glu-L-Pro; 23. 3-hydroxy-2-quinoxalinecarboxylic acid; 24. Kainate phenylthiolactone; 25. Kainate iodolactone; 26. Kainate hydroxylactone; 27. Kainate lactone; 28. DL-aminodipate; 29. DL-aminosuberate; 30. γ-D-Glu-Gly; 31. 2-amino-5-phosphonovalerate. The data are collected from the following papers: Teichberg et al., 1981; Luini et al., 1981; Goldberg et al., 1981; Tal et al., 1982; Teichberg et al., 1984; Erez et al., 1985.

only slightly the response to Glu (Teichberg et al., 1984), we believe that the classification into four distinct receptors is compelling.

The Glu-receptor has properties which clearly distinguish it from the other receptors. It is not very much affected by the presence of aliphatic alcohols, barbiturates or phospholipases which are known to perturb the membrane lipid bilayer. This observation tends to suggest that the Glu receptor is imbedded in a very different lipidic environment than the other receptors. The Glu-receptor is also refractory to the inhibitory action of most antagonists.

The NMDA-receptor, in contrast, is sensitive to a large number of antagonists, including 2-amino-5-phosphonovalerate, DL-aminosuberate, various γ-D-glutamyl dipeptides and several γ-dipeptides of kainic acid (Goldberg and Teichberg, 1985b) (table 5). The NMDA-receptor is a protein distinct from the other excitatory amino acid receptors since it is not present in the cerebellum, a region which contains the Glu-, KA- and Quis-receptors (Luini et al., 1983). Although the NMDA-receptor is not sensitive to the presence of aliphatic alcohols or barbiturates, it is affected by phospholipases and β-bungarotoxin. This property further differentiates it from the other receptors.

The KA-receptor and the Quis-receptor are both sensitive to the presence of aliphatic alcohols, barbiturates, phospholipases and β-bungarotoxin but the Quis-receptor markedly differs from the KA-receptor in its sensitivity to antagonists derived from chemical modifications of glutamate or kainic acid.

Table 5: Response to agonists in the presence of γ-kainyl derivatives.

		Antagonist Activity		(% Response to Agonist)	
		NMDA 30 μM	KA 0.1 mM	Glu 0.5 mM	Quis 0.1 mM
R=NHCH$_2$CO$_2$H,	γKA-Gly	46±18***	91±9	96±23	81±7**
R=NHCH$_2$CH$_2$CO$_2$H,	γKA-βAla	76±28	92±22	86±16	81±26
R=NHCHCO$_2$H (L), CH$_2$-⟨O⟩-OH	γKA-Tyr	48±25*	87±14	72±16*	70±20
R=NHCHCO$_2$H (L), CH$_2$CH$_2$CO$_2$H	γKA-Glu	16±9***	84±5	90±31	69±11**
R= (ring), CH$_2$CO$_2$H,	γKA-KA	4±13***	86±11	80±30	71±17

*:p<0.02; **:p<0.01; ***:p<0.001. All antagonists were tested at 1 mM.

6. ROLE OF EXCITATORY AMINO ACID RECEPTORS IN SYNAPTIC NEUROTRANSMISSION.

Using $^{22}Na^+$-preloaded brain slices, it has been possible to study the role of excitatory amino acid receptors in synaptic neurotransmission. The principle of the method is illustrated in figure 3. $^{22}Na^+$-preloaded brain slices are exposed to a physiological solution in which part of the Na^+ ions has been substituted with K^+ ions. The resulting K^+ depolarization produces the release of neurotransmitters from nerve terminals (right panel). Among the released neurotransmitters, are the excitatory neurotransmitters which act on postsynaptic receptors controlling the membrane permeability to Na^+ ions, and produce therefore an increase in $^{22}Na^+$ efflux. The latter effect can be blocked by 99% if one incubates the slices in low Ca^{++} concentrations (middle panel).

Using $^{22}Na^+$-preloaded slices from various brain regions, we have compared the effects produced by excitatory amino acid antagonists on the K^+-evoked Ca^{++} dependent $^{22}Na^+$ efflux and on the agonist-dependent $^{22}Na^+$ efflux (Luini et al., 1984b). Figure 4 illustrates the results. In the left part of the figure, NMDA, Glu and KA are each tested on striatal slices in the absence (hatched column) and presence of specific antagonists (open columns) and the amplitude of the responses is measured. In the middle and right part of the figure, the responses to the excitatory neurotransmitters endogenously released by K^+ ions from various brain areas are measured as well as the extent of their inhibition by the same antagonists used above. On the basis of the similarity of the profiles of inhibition shown in the upper panel of figure 4, one can conclude that the endogenous excitatory neurotransmitters released upon depolarization of nerve terminals in the cortex and in the striatum are acting mainly on subsynaptic NMDA-receptors. Using the same argumentation, one may propose that in the hippocampus and cerebellum, the excitatory neurotransmitters would act mainly on subsynaptic Glu-receptors and in the substantia nigra and hypothalamus, on KA-receptors. The data do not rule out the participation of other receptors in the excitatory transactions that take place in the above brain areas but the assaying method does not allow their detection.

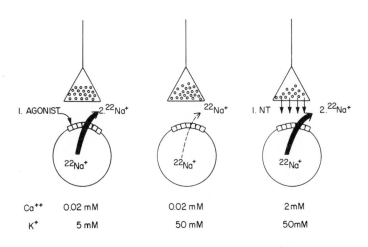

Figure 3: Agonist-dependent and depolarization-dependent $^{22}Na^+$ efflux from $^{22}Na^+$-preloaded brain slices. The agonist-dependent $^{22}Na^+$ efflux (left panel) is observed when slices are bathed in a physiological solution containing inter alia 0.02mM Ca^{++}, 10mM Mg^{++} and 5mM K^+. When the slices are depolarized by increasing K^+ concentration, a very small $^{22}Na^+$ efflux is observed at low Ca^{++} concentration (middle panel). With 2mM Ca^{++}, 50mM K^+ depolarize nerve terminals and trigger the release of excitatory neurotransmitters (NT) which control the neuronal membrane permeability to Na^+ ions (right panel).

7. USE OF THE $^{22}Na^+$ EFFLUX RECEPTOR ASSAY IN THE DETECTION OF ANTICONVULSANT ANTAGONISTS OF EXCITATION.

Since the excitatory amino acid receptors, as studied by the $^{22}Na^+$ efflux assay, are involved, as shown above, in synaptic excitatory neurotransmission, one expects that the antagonists of excitatory amino acids detected with this assay will also display anticonvulsive properties. This expectation is based on the studies of Meldrum and his colleagues (Croucher et al., 1982; Meldrum et al., 1983a,b) who demonstrated that the excessive brain excitation which produces epileptic seizures is mediated by excitatory amino acid receptors and can be blocked by NMDA antagonists.

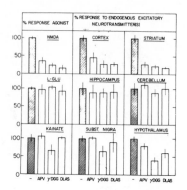

<u>Figure 4</u>: Antagonism of the increase in specific $^{22}Na^+$ efflux
rate produced by excitatory amino acids on striatal slices
(left   panel)   and   by   the   endogenous   excitatory
neurotransmitters released   upon   exposure   of   slices   of
different brain regions to a physiological solution containing
40mM K$^+$ (center and right panels; from Luini <u>et</u> <u>al</u>., 1984b).

We have therefore tested the anticonvulsant effects of
compounds   that   were   identified   as   excitatory   amino   acid
antagonists using solely the $^{22}Na^+$ efflux receptor assay.
Figure   5A   shows   that   the   intracerebroventricular   (icv)
administration of increasing doses of γ-kainyl-glutamate
(Goldberg and Teichberg, 1985a), a newly synthesized NMDA
antagonist (see table 5) increasingly protects mice from
picrotoxin-induced convulsions.   Figure 5B shows the dose-
dependent   delay   produced   by   icv   injection   of
3-hydroxy-2-quinoxalinecarboxylic   acid   in   the   time   of
appearance of picrotoxin-induced convulsions in mice (Erez <u>et</u>
<u>al</u>., 1985).   In both cases, it has been possible to correlate
the anticonvulsive potency of these compounds with their
ability to antagonize the effects of NMDA on $^{22}Na^+$-preloaded
brain slices (Erez <u>et</u> <u>al</u>., 1985; Goldberg and Teichberg,
1985a).

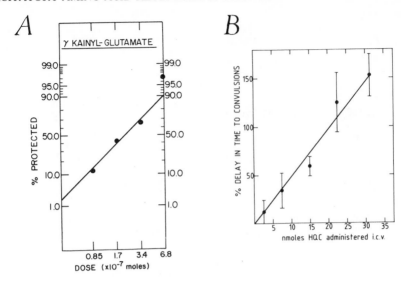

Figure 5 A: Dose-dependent protection from picrotoxin-induced convulsions in CD1 mice by intracerebroventricular injections of γ-kainyl-glutamate. Five microliters of the various doses indicated were injected. Each point corresponds to 10 mice. The results are calculated by the method of Litchfield and Wilcoxon (1949). 5B: Dose-dependent delay produced by intracerebroventricular injection of HQC in the time to picrotoxin induced convulsions in CD1 mice. The vertical bar indicates standard error (5B from Erez et al., 1985).

SUMMARY.

The ^{22}Na$^+$ efflux receptor assay is a biochemical method enabling the study of excitatory amino acid receptors in their functional state. Its relevance and usefulness having been proven for the screening, identification and evaluation of novel excitatory amino acid agonists and antagonists, this assay can now assist in the development of new anticonvulsant agents and possibly of drugs protecting from anoxic/ischaemic brain damage (Meldrum, 1985).

ACKNOWLEDGEMENTS

This work is supported by Grants from the Israel Commission for Basic Research, the Wills Foundation and the Esther A. and Joseph Klingenstein Fund. V.I.T. holds the Louis and Florence Katz-Cohen professorial chair in neuropharmacology.

REFERENCES

Abe, T., Alema, S. and Miledi, R. (1977) Isolation and characterization of presynaptically acting neurotoxins from the venom of Bungarus snakes. Europ. J. Biochem. 80, 1-12.

Bradford, H.F. and McIlwain, H. (1966) Ionic basis for the depolarization of cerebral tissues by excitatory acidic amino acids. J. Neurochem. 13, 1163-1177.

Catterall, W.A. (1976) Purification of a toxic protein from scorpion venom which activates the action potential Na^+ ionophore. J. Biol. Chem. 251, 5528-5536.

Croucher, M.J., Collins, J.F. and Meldrum, B.S. (1982) Anticonvulsant action of excitatory amino acid antagonists. Science 216, 899.

Di Chiara, G. and Gessa, G.L., eds. (1981) "Glutamate as a Neurotransmitter" Raven Press, New York.

Erez, U., Frenk, H., Goldberg, O., Cohen, A. and Teichberg, V.I. (1985) Anticonvulsant properties of 3-hydroxy-2-quinoxalinecarboxylic acid, a newly found antagonist of excitatory amino acids. Europ. J. Pharmacol. 110, 31-39.

Foster, A.C. and Fagg, G.E. (1984) Acidic amino acid binding sties in mammalian neuronal membranes: their characteristics and relationship to synaptic receptors. Brain Res. Rev. 7, 103-164.

Foster, G.A. and Roberts, P.J. (1980) Pharmacologoy of excitatory amino acid receptors mediating the stimulation of rat cerebellar cyclic GMP levels. Life Sci. 27, 215-221.

Goldberg, O., Luini, A. and Teichberg, V.I. (1981) Lactones derived from kainic acid: novel selective antagonists of amino acid induced Na^+ fluxes in rat striatum slices. Neurosci. Lett. 23, 187-191.

Goldberg, O. and Teichberg, V.I. (1985a) A dipeptide derived from kainic acid and L-glutamic acid: a selective antagonist of amino acid-induced neuroexcitation with anticonvulsant properties. J. Medic. Chem (in press).

Goldberg, O. and Teichberg, V.I. (1985b) Peptides derived from kainic acid as antagonists of N-methyl-D-aspartate-induced neuroexcitation. (submitted).

Harvey, J.A. and McIlwain, H. (1968) Excitatory acidic amino acids and the cation content and sodium ion flux of isolated tissues from the brain. Biochem. J. 108, 269-274.

Hosli, L., Andres, P.F. and Hosli, E. (1976) Ionic mechanisms associated with the depolarization by glutamate and aspartate on human and rat spinal neurons in tissue culture. Pflugers Arch. 363, 43-48.

Ichida, S., Tokunaga, H., Moriyama, M., Oda, Y., Tanaka, S. and Kita, T. (1982) Effects of neurotransmitter candidates on ^{45}Ca uptake by cortical slices of rat brain: stimulatory effect of L-glutamic acid. Brain Res. 248, 305-311.

Kopeyan, C., Martinez, G. and Rochat, H. (1978) Amino acid sequence of neurotoxin V from the scorpion. FEBS Lett. 89, 54-58.

Litchfield, J.T. and Wilcoxon, F. (1949) A simplified method of evaluating dose-effects experiments. J. Pharmacol. Ther. 96, 99-113.

Luini, A., Goldberg, O. and Teichberg, V.I. (1981) Distinct pharmacological properties of excitatory amino acid receptors in the rat striatum: study by Na$^+$ efflux assay. Proc. Natl. Acad. Sci. USA 78, 3250-3254.

Luini, A., Goldberg, O. and Teichberg, V.I. (1983) Differential sensitivity of selected brain areas to excitatory amino acids. Neurosci. Lett. 41, 307-312.

Luini, A., Tal, N., Goldberg, O. and Teichberg, V.I. (1984a) An evaluation of selected brain constituents as putative excitatory neurotransmitters. Brain Res. 324, 271-277.

Luini, A., Goldberg, O. and Teichberg, V.I. (1984b) Differential sensitivity of selected brain areas to excitatory neurotransmitters released by K$^+$ depolarization. Neurosci. Lett. 49, 325-330.

Meldrum, B.S., Croucher, M.J., Czuczwar, S.J., Collins, J.F., Curry, K., Joseph, M. and Stone, T.W. (1983a) A comparison of the anticonvulsant potency of (+) 2-amino-5-phosphonopentanoic acid and (+) 2-amino-7 phosphonohepthanoic acid. Neurosci. 9, 925-930.

Meldrum, B.S., Croucher, M.J., Badman, G. and Collins, J.F. (1983b) Antiepileptic action of excitatory amino acid antagonists in the photosensitive baboon, Papio papio. Neurosci. Lett. 39, 101-105.

Meldrum, B.S. (1985) Excitatory amino acids and anoxic/ischaemic brain damage. Trends Neurosci. 8, 47-48.

Miranda, F., Kopeyan, C., Rochat, M., Rochat, C. and Lissitzky, S. (1970) Purification of animal neurotoxins. Europ. J. Biochem. 16, 514-523.

Puil, E. (1981) S. Glutamate: its interactions with spinal neurons. Brain Res. Rev. 3, 229-433.

Scatton, B. and Lehmann, J. (1982) N-methyl-D-aspartate-type receptors mediate striatal ^{3}H acetylcholine release evoked by excitatory amino acids. Nature, Lond. 297, 422-424.

Shipolini, R.A., Callewaert, G.L., Cottrel, R.C., Doonan, S., Vernon, C.A. and Banks, B.E.C. (1971) Phospholipase A from bee venom. Europ. J. Biochem. 20, 453-468.

Strong, P.N., Goerke, J., Oberg, S.G. and Kelly, R.B. (1976) β-bungarotoxin: a presynaptic toxin with enzymatic activity. Proc. Natl. Acad. Sci. USA 73, 178-182.

Tal, N., Goldberg, O., Luini, A. and Teichberg, V.I. (1982) An evaluation of gamma-glutamyl dipeptide derivatives as antagonists of amino acid induced Na$^+$ fluxes in rat striatum slices. J. Neurochem. <u>39</u>, 574-576.

Teichberg, V.I., Goldberg, O., Tal, N. and Luini, A. (1980) The interactions of excitatory amino acids with brain tissue: A study of amino acid-induced ion fluxes in rat striatal slices. in "Neurotransmitters and their receptors" Littauer, U.Z., Dudai, Y., Silman, I., Teichberg, V.I. and Vogel, Z. (eds.) John Wiley, New York, pp. 349-368.

Teichberg, V.I., Goldberg, O. and Luini, A. (1981) The stimulation of ion fluxes in brain slices by glutamate and other excitatory amino acids. Molec. Cellular Biochem. <u>39</u>, 281-295.

Teichberg, V.I., Tal, N., Goldberg, O. and Luini, A. (1984) Barbiturates, alcohols and the CNS excitatory neurotransmission: specific effects on the kainate and quisqualate receptors. Brain Res. <u>291</u>, 285-292.

Watkins, J.C. and Evans, R.H. (1981) Excitatory amino acid transmitters. Ann. Rev. Pharmacol. Toxicol. <u>21</u>, 165-204.

23

Glutamate Receptors and Glutamate Corticofugal Pathways

A.B. Young, J.T. Greenamyre, J.B. Penney and M.B. Bromberg

INTRODUCTION

Although the neurochemical anatomy of many pathways in human brain has been studied, little attention has been paid to the human pyramidal system. This pathway is very important for voluntary motor control in vertebrates, and pyramidal tract pathology leads to hemiparesis or hemiplegia in primates and man (Phillips and Porter, 1977). Identification of the neurotransmitter of the pyramidal tract would allow a more systematic investigation of drugs for the treatment of the neurological consequences of stroke, spinal cord and head trauma, multiple sclerosis, amyotrophic lateral sclerosis and cerebral palsy.

The cells of origin of the pyramidal tract are the triangular shaped neurons in layer V of cerebral cortex (Kuypers, 1981; Phillips and Porter, 1977). The majority of pyramidal tract fibers originate from the sensorimotor cortex in rat, from the pericruciate cortex in cat and from Brodmann's areas 4 and 6 in monkey. In these three species, fibers that originate from the above areas constitute approximately 60% of the pyramidal tract, and the remaining pyramidal fibers originate in other portions of adjacent association cortex. Axons from these neurons project to ipsilateral caudate nucleus (with a small crossed component to contralateral caudate), putamen, motor thalamus, red nucleus and pontine nuclei and to the contralateral cervical and lumbar spinal cord. Electrophysiologic studies suggest that the transmitter of these neurons is an excitatory agent (Stone, 1979). The excitatory action of the endogenous transmitter is mimicked by certain of the excitatory amino acids, particularly glutamate and asparate. Although acetylcholine is also excitatory, the characteristics of its excitatory effects do not match those of the natural substance. Furthermore, other potential neurotransmitters such as the neuropeptides and biogenic amines do not serve as likely candidates for the transmitter of the pyramidal tract fibers since their regional distribution is quite different than that expected for a transmitter of this pathway.

Because electrophysiological and biochemical studies suggested glutamate may be a transmitter of cortical fibers projecting
to the striatum (McGeer et al., 1977; Kim et al., 1977; Divac et
al., 1977; Stone, 1979), we examined the possibility that glutamate might also be a transmitter of other corticofugal pathways.
Specifically, we measured high affinity uptake of glutamate (and
other amino acids) as well as endogenous amino acid levels in several subcortical regions following ablation of the sensorimotor or
motor cortex of rats, cats and monkeys (Bromberg et al., 1981;
Young et al., 1981 and 1983). In addition, we investigated the
regional distribution and pharmacology of glutamate receptors with
respect to various projection areas of the pyramidal tract
(Greenamyre, 1984 and 1985a). Finally, we have extended our examination of glutamate receptors to studies of postmortem human
brains of patients with Alzheimer's disease and Huntington's
disease (Greenamyre et al., 1985b).

STUDIES OF GLUTAMATE UPTAKE AND GLUTAMATE LEVELS AFTER SENSORI-
MOTOR CORTEX ABLATIONS.

Initially, as part of a study of mechanisms of spasticity, we
investigated the effects of cortical ablation on the high affinity
uptake of various neurotransmitter amino acids in subcortical regions and spinal cord in the rat (Bromberg et al., 1981). After
unilateral ablation of sensorimotor cortex, we measured amino acid
levels and sodium-dependent high affinity glutamate uptake in synaptosomes from tissue punches of ventrolateral thalamus, red nucleus and ventral gray of spinal cord by methods described previously.
We found a 48% decrease in high affinity glutamate uptake in ventrolateral thalamus ipsilateral to the lesion and a 67% reduction
in high affinity glutamate uptake in the ipsilateral red nucleus
(Table 1). Glutamate levels were decreased by 33% in ventrolateral
thalamus without any changes in aspartate, glutamine, GABA, glycine
or taurine levels. In rats, we found no significant change in glutamate uptake in spinal cord.

Since electrophysiological studies have indicated that many
individual pyramidal tract neurons send projections to more than
one subcortical region via axon collaterals, we decided to investigate the possibility that glutamate was a neurotransmitter for
corticofugal fibers in the cat which has a large pyramidal tract
relative to the rat (Young et al., 1981). After defining the most
prominant discrete projection areas of pericruciate cortex by the
Fink-Heimer degeneration method (1967), we performed unilateral
ablation of the pericruciate cortex in a series of cats. After one
week, we measured glutamate uptake, choline acetyltransferase activity and amino acid levels in various subcortical regions. Significant decreases in glutamate uptake occurred in the ipsilateral
caudate-putamen, thalamus, red nucleus, pons and contralateral cer-

Table 1. L-Glutamate uptake in cortical projection areas
of three species after unilateral motor or sensorimotor ablations.

Region of Brain	Uptake in regions with cortical fugal tract degeneration as a % of the uptake in the sane region on the opposite side. (Mean ± SEM)		
	Rat	Cat	Monkey
Caudate nucleus	NT	83+9*	73+10*
Ventrolateral nucleus of thalamus	52+7*	56+12*	72+9*
Red nucleus	33+10*	64+17*	88+16
Basis pontis	NT	80+20*	78+5*
Medullary reticular formation	NT	82+22	NT
Dorsal column nuclei	NT	98+38	95+17
Cerebellum	NT	98+3	100+5
Cervical spinal cord	NT†	63+21*	79+9*
Thoracic spinal cord	NT†	90+29	92+8
Lumbar spinal cord	NT†	76+9%*	86+7*

* P<.05 Statistics calculated from the raw data using Student's paired t-test comparing lesioned to control side.
† Preliminary studies indicated no change in these structures.
NT = Not tested

vical and lumbar spinal cord (Table 1).   No changes in glutamate
uptake were seen in the dorsal column nuclei or in the cerebellum.
The uptake of other amino acids and the activity of choline acetyltransferase were unchanged.   Amino acid levels did not change
except in ventrolateral thalamus where a 33% reduction in glutamate levels was observed.

Finally, we had an opportunity to study a series of monkeys
one week after ablation of Brodmann's areas 4 and 6 (Young et al.,
1983).   Initially these lesions produced a contralateral flaccid
hemiparesis but in other animals a classic spastic hemiparesis developed if followed 8 weeks.   As in the cat, glutamate uptake in
ipsilateral caudate-putamen, ventrolateral thalamus, pons and contralateral cervical and lumbar spinal cord was reduced significantly (Table 1). However, no changes in glutamate levels were
seen in any of the regions.

These studies suggested that glutamate (or perhaps an associated dipeptide or excitatory amino acid) is a likely candidate as
the neurotransmitter of pyramidal tract fibers. Recent studies by
other investigators have substantiated this hypothesis by demon-

trating a decrease in calcium-dependent D-aspartate release in the
pons and thalamus after unilateral cortical ablation (Thangnipon
and Storm-Mathisen, 1981; Thangnipon et al., 1983).

THE DISTRIBUTION OF GLUTAMATE RECEPTORS.

    Since studies of presynaptic markers for "glutamatergic" ter-
minals supported a role for glutamate, or a glutamate-like substance, as the transmitter of corticofugal fibers, it was of interest to study postsynaptic markers of glutamate function, i.e.,
glutamate receptors.   Recently, we and others have developed a
technique for measuring glutamate receptors autoradiographically
(Greenamyre et al., 1983, 1984 and 1985a; Monaghan et al., 1983;
Halpain et al., 1983 and 1984).   Frozen sections of rat or human
brain were mounted onto gelatin-coated glass slides and incubated
in various concentrations of $[^3H]$glutamate in the presence or absence of various competitors in 50 mM Tris-HCl buffer, pH 7.4, with
2.5 mM calcium chloride.   The subsequent autoradiograms showed a
marked regional variation in glutamate binding (Figure 1) and the
distribution of binding sites correlated very well with the projection areas of putative glutamatergic pathways (Greenamyre et al.,
1984).   Consistent with the proposal that corticofugal pathways use
glutamate as a transmitter, there was heavy glutamate binding in
cortex, caudate-putamen, motor and sensory thalamus, red nucleus,
pontine nuclei and spinal cord.   In addition, there was a high density of binding sites in the hippocampal formation and cerebellar
cortex.

    Competition studies with various glutamate analogs has shown
that glutamate binds to several receptor subtypes although glutamate itself does not distinguish between these sites (Greenamyre
et al., 1984).   However, the agonist, quisqualate, does differentiate two sites in brain.   One site has a high affinity for quis-
ualate with a K_I of about 10 nM and the other, a low affinity for
quisqualate with a K_I of about 100 μM (Greenamyre et al., 1984).
Detailed cross competition studies indicate that the low affinity
quisqualate site is identical to the N-methyl-D-aspartate site
since the inclusion of 100 μM N-methyl-D-aspartate significantly
decreases the number of low affinity quisqualate sites (Greenamyre
et al., 1985a)(Figures 2 and 3).   In the absence of added calcium,
glutamate binding to kainate sites also occurs.   This is particularly evident in stratum lucidum of hippocampus.   Elsewhere, how-
ver, the proportion of kainate sites is small in comparison to
total specific glutamate binding (about 5-10%).

    At the present time, the responses of the various glutamate
receptor subtypes to lesions of the pyramidal tract are unknown;
however, in preliminary experiments we have found that glutamate
binding does regulate after cortical ablation.   It will be interesting to ascertain whether subtypes of glutamate receptors respond differentially to lesions of motor pathways.   If this proves

to be the case, then more rational drug therapies using selective antagonists or agonists may be devised.

STUDIES OF GLUTAMATE RECEPTORS IN HUMAN BRAIN.

Recent studies of glutamate function in brain (particularly in hippocampus) have lead to the hypothesis by Lynch and Baudry (1984) that the amino acid, glutamate, may be intimately involved in memory formation. Glutamate has also been implicated as a neurotoxic agent and potential endogenous toxin in Huntington's disease (Shoulson, 1984a). It was therefore of interest to measure glutamate receptors in postmortem brain tissue from patients dying with Alzheimer's disease and with Huntington's disease (Greenamyre et al., 1985b). We found striking reductions in glutamate binding in the cerebral cortex of patients with Alzheimer's disease, particularly in the outer layers of cerebral cortex (layers 1 and 2). We have also found large reductions in glutamate binding in hippocampus in Alzheimer's disease (unpublished data). There was no significant change in benzodiazepine, GABA or muscarinic choliner-

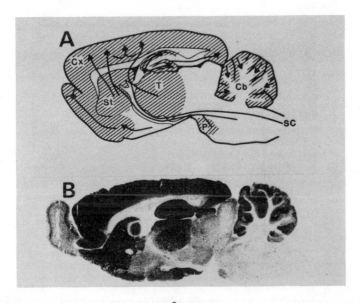

Figure 1. Distribution of ^{3}H-L-glutamate binding in a saggittal section of rat brain. A). Schematic map of glutamatergic neurons (▲) in a saggittal section of rat brain. Cx, cerebral cortex; St, striatum; T, thalamus; Cb, cerebellum; P, pontine nuclei; Sc, spinal cord. B). Autoradiograph of ^{3}H-L-glutamate binding in 200 nM ^{3}H-L-glutamate in 50 mM Tris-HCl buffer, pH 7.4, plus 2.5 mM $CaCl_2$.

gic receptor binding in adjacent or neighboring sections of the same brains (Table 2). Thus, the change in glutamate binding appears to be relatively selective and the possibility is raised that there may be an abnormality in glutamate receptor function on cortical pyramidal neurons in Alzheimer's disease. Cortical neurons not only give rise to the pyramidal tract but also give rise to the cortical association fibers which allow communication between various areas of cerebral cortex; these fibers are also

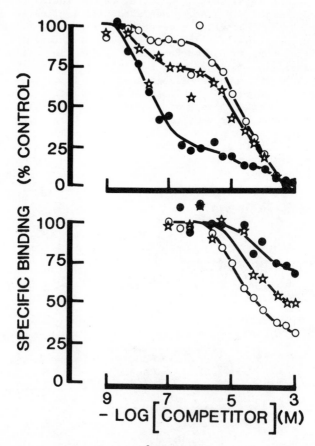

Figure 2. Competition for ^{3}H-L-glutamate binding by NMDA and quisqualate. Glutamate binding was carried out in 200 nM ^{3}H-L-glutamate in 50 mM Tris-HCl, pH 7.4, plus 2.5 mM CaCl$_2$ and increasing amounts of NMDA or quisqualate. There is a reciprocal relationship between the number of low affinity QQA sites (A) and NMDA sites (B). Areas that have high numbers of high affinity quisqualate sites have low numbers of NMDA sites (Greenamyre et al., 1985a). O, stratum moleculare; ☆, striatum; ●, molecular layer of cerebellum.

thought to be glutamatergic. A loss of glutamate receptors on these neurons could conceivably impair these glutamatergic connections and might be expected to lead to cortical disconnection syndromes which are a common finding in Alzheimer's disease (Terry and Katzman, 1983; Foster et al., 1983).

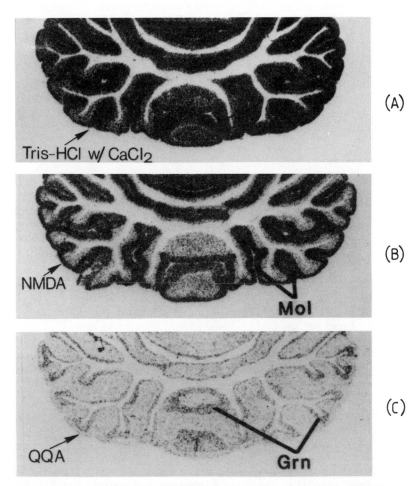

<u>Figure 3.</u> Competition for ^{3}H-L-glutamate binding in cerebellum. Sections were incubated in 200 nM ^{3}H-L-glutamate in 50 mM Tris-HCl buffer, pH 7.4, plus 2.5 mM $CaCl_2$ alone (A) or in the presence of 100 μM NMDA (B) or 2.5 μM quisqualate (C). Glutamate binding in the molecular layer of cerebellum (Mol) represents predominantly binding to high affinity quisqualate sites and binding in granule cell layer (Grn) is selectively sensitive to NMDA.

Table 2.   Receptor Binding in Brains from Normals, Alzheimer's and Huntington's Diseases Patients.

Area and Ligand	Normal	HD	SDAT
CORTEX, LAYERS 1 & 2			
L-[^{3}H]glutamate (Bmax)	6.0+0.6	5.9+0.6	3.6+0.7*
Low affinity quis- qualate site (Bmax)	3.4+0.5	n.d.	1.5+0.2**
[^{3}H]muscimol (Bmax)	4.3+0.1	2.8+0.2	2.7+0.3
[^{3}H]flunitrazepam (Bmax)	1.5+0.2	1.8+0.3	1.2+0.1
[^{3}H]quinuclidinyl benzilate (Bmax)	1.7+0.2	1.0+0.1	1.6+0.2
CAUDATE			
L-[^{3}H]glutamate (at 200 nM)	1.3+0.2	0.4+0.1***	1.3+0.2
PUTAMEN			
L-[^{3}H]glutamate (at 200 nM)	1.3+0.2	0.4+0.1*	1.1+0.2

Values are averages $\pm$ SEM from 5 normal controls (age 63+4
years; 16 $\pm$ 3.2 hours post-mortem delay), 6 SDAT patients
(age 71+4 years; 11.4+2.9 hours postmortem delay) and four
advanced HD patients (age 35 $\pm$ 7 years; 20.1 $\pm$ 11.8 hours
postmortem delay).   ***$p < 0.005$, ** $p < 0.01$, *$p < .05$ compared
to both other groups using univariate analysis of variance
with Bonferroni inequality. (Modified from Greenamyre et
al., 1985b).

     In Huntington's disease, glutamate receptors were normal in
cerebral cortex but severely reduced in caudate and putamen (Table
2).   The loss of receptors in these regions likely reflects the
cell death that occurs in caudate and putamen (Shoulson, 1984b).
Interestingly, the memory loss in Huntington's disease is quite
different than that seen in Alzheimer's disease and does not involve classical cortical deficits.   The fact that glutamate receptors in the cortex of Huntington's patients are normal is consistent with this differentiation between the memory deficits of
these two diseases.

ACKNOWLEDGEMENTS

We would like to thank Michaela Weeks for secretarial assistance. these studies were supported by National Science Foundation Grant BNS-8118765, The Hereditary Disease Foundation, the Alfred C. and Ersa Arbogast Foundation and a National Institute of Mental Health Individual Predoctoral National Research Service Award 1-F31-MH08922 (to JTG).

REFERENCES

Bromberg, M.B., Penney Jr., J.B., Stephenson, B.S. and Young, A.B. (1981). Evidence for glutamate as the neurotransmitter of corticothalamic and corticorubral pathways. Brain Res., 215, 369-374.

Divac, I., Fonnum, F. and Storm-Mathisen, J. (1977). High affinity uptake of glutamate in terminals of corticostriatal axons. Nature, 266, 377-378.

Fink, R.P. and Heimer, L. (1967). Two methods for selective impregnation of degenerating axons and their synaptic endings in the central nervous system. Brain Res., 4, 369-374.

Fonnum, R., Soreide, A., Kvale, I., Walker, J. and Walaas, I. (1981). Glutamate in cortical fibers. Biochem. Psychopharmacol., 27. 29-42.

Foster, N.L., Chase, T.N., Fedio, P., Petronas, N.J., Brooks, R.A., and DiChiro, G. (1983). Alzheimer's disease: Focal cortical changes shown by positron emission tomography. Neurology, 33, 961-965.

Greenamyre, J.T., Olson, J.M.M., Penney, J.B. and Young, A.B. (1985a). Autoradiographic characterization of N-methyl-D-aspartate-, quisqualate- and kainate-sensitive glutamate binding sites. J. Pharmacol. Exp. Ther., 233, 254-263.

Greenamyre, J.T., Penney, J.B., Young, A.B., D'Amato, C.J., Hicks, S.P. and Shoulson, I. (1985b). Alterations in L-glutamate binding in Alzheimer's and Huntington's diseases. Science, 227, 1496-1499.

Greenamyre, J.T., Young, A.B. and Penney, J.B. (1983). Quantitative autoradiography of L-[^{3}H]glutamate binding to rat brain. Neurosci. Lett., 37, 155-160.

Greenamyre, J.T., Young, A.B. and Penney, J.B. (1984). Quantitative autoradiographic distribution of L-[^{3}H]glutamate binding sites in rat central nervous system. J. Neurosci., 4, 2133-2144.

Halpain, S.H., Parsons, B. and Rainbow, T.C. (1983). Tritium-film autoradiography of sodium-independent glutamate binding sites in rat brain. Eur. J. Pharmacol., 86, 313-314.

Halpain, S.H., Wieczorek, C.M. and Rainbow, T.C. (1984). Localization of L-glutamate receptors in rat brain by quantitative autoradiography. J. Neurosci., 4, 2247–2258.

Kim, J.S., Hassler, R., Haug, P. and Paik, K.-S. (1977). Effect of frontal cortex ablation on striatal glutamic acid level in the rat. Brain Res., 132, 370–374.

Kuypers, H.G.J.M. (1981). Anatomy of the descending pathways. In Handbook of Physiology: The Nervous System II. (ed. V.B. Brooks). American Physiology Society, Washington, D.C.

Lynch, G. and Baudry, M. (1984). The biochemistry of memory: A new and specific hypothesis. Science, 224, 1057–1063.

McGeer, P.L., McGeer, E.G., Scherer, U. and Singh, K. (1977). A glutamatergic corticostriatal path? Brain Res., 128, 369–373.

Monaghan, D.T., Holets, V.R., Toy, D.W. and Cotman, C.W. (1983). Anatomical distributions of four pharmacologically distinct ^{3}H-L-glutamate binding sites. Nature, 306, 176–179.

Phillips, C.G. and Porter, R. (1977). Corticospinal Neurones: Their Role in Movement. Academic Press, New York.

Shoulson, I. (1984a). Huntington's disease: Anti-neurotoxic therapeutic strategies. In Excitotoxins. (eds. K. Fuxe, P. Roberts and R. Schwarcz). Plenum Press, New York.

Shoulson, I. (1984b). Huntington's disease: A decade of progress. Neurologic Clin., 2, 515–526.

Stone, T.W. (1979). Amino acids as neurotransmitter of corticofugal neurones in the rat: A comparison of glutamate and aspartate. Br. J. Pharmacol., 67, 545–551.

Terry, R.D. and Katzman, R. (1983). Senile dementia of the Alzheimer type. Ann. Neurol., 14, 497–506.

Thangnipon, W. and Storm-Mathisen, J. (1981). K^+-evoked Ca^{2+}-dependent release of D-[^{3}H]asparate from terminals of the corticopontine pathway. Neurosci. Lett., 23, 181–186.

Thangnipon, W., Taxt, T., Brodal, P. and Storm-Mathisen, J. (1983). The cortico-pontine projection: axotomy-induced loss of high affinity L-glutamate and D-aspartate uptake, but not of GABA uptake, glutamate decarboxylase or choline acetyltransferase, in the pontine nuclei. Neuroscience, 8, 449–458.

Young, A.B., Bromberg, M.B. and Penney Jr., J.B. (1981). Decreased glutamate uptake in subcortical areas deafferented by sensorimotor cortical ablation in the cat. J. Neurosci., 1, 241–249.

Young, A.B., Penney, J.B., Dauth, G.W., Bromberg, M.B. and Gilman, S. (1983). Glutamate or aspartate as a possible neurotransmitter of cerebral corticofugal fibers in the monkey. Neurology, 33, 1513-1516.

24

The Neuropharmacology of Quinolinic Acid and the Kynurenines

T.W. Stone, J.H. Connick, J.I. Addae, D.A.S. Smith and P.A. Brooks

Summary

Quinolinic acid is an endogenous compound with actions on the NMA receptor to produce neuronal excitation, toxicity and convulsions. It does not appear to be taken up by membrane transport processes in the brain, and it may therefore function as a long term modulator of neuronal excitability rather than as a classical neurotransmitter. Alternatively it may accumulate in the brain with age or disease to mediate toxic effects or neurodegeneration. Kynurenic acid is another compound in the same metabolic pathway from tryptophan, and is one of the most effective blockers of (putatively) excitatory amino acid mediated synapses presently available. A balance between these two compounds may have some importance in the regulation of cerebral excitability.

ELECTROPHYSIOLOGY

Quinolinic acid

The discovery that quinolinic acid (2,3-pyridine dicarboxylic acid) had an excitant action on central neurones came about when testing a series of aspartate analogues for excitatory activity. Quinolinic acid, a rigid analogue of aspartate was found to excite single neurones in the rat neocortex when applied by microiontophoresis (Stone and Perkins, 1981). The other ring isomers of quinolinate turned out to be far less active (Birley et al., 1982) with the exception of the extended configuration homologue, homoquinolinic acid, which was considerably more active and has a potency comparable with N-methyl-aspartate (NMA) (Stone, 1984). Quinolinate gave a potency similar to

that of glutamate and could be antagonised in parallel
with responses to NMA by the specific antagonists
2-aminophosphonovalerate. (APV) and heptanoate (APH)
amongst others (Perkins and Stone, 1983c).

These results have been confirmed using peripheral
injections of quinolinate and blocking the neuronal
excitation with iontophoretically applied APH and vice
versa (Perkins and Stone, 1983d).

Whereas glutamate, aspartate and NMA tend to be
universally excitatory, quinolinic acid was only found
to be active in certain areas of the CNS.   The
neocortex, hippocampus and striatum were highly
responsive, whereas the cerebellum and spinal cord
showed little or no sensitivity in comparison with
either glutamate or NMA (Perkins and Stone, 1983a and
c).   These observations have since been confirmed
independently (McLennan 1984).

This variation of responses to quinolinate and NMA
has been interpreted as implying the existence of a
subclass of NMA receptor. For example an NMA1
quinolinate resistant site might exist in the
cerebellum and spinal cord and an NMA2 quinolinate
responsive site in the cortex, striatum and hippocampus
(Perkins and Stone, 1983a).

A number of other studies are consistent with an
action of quinolinate on some form of NMA receptor.
Herrling et al. (1983), for example, have demonstrated
that both NMA and quinolinate produce similar patterns
of burst firing of striatal neurones, while quisqualate
produces only a progressive depolarisation.   Using
biochemical methods to examine acetylcholine release
and sodium fluxes Lehmann et al. (1983) and Luini et
al. (1984) respectively have also shown the blockade of
quinolinate effects by NMA receptor antagonists.

Using topical applications to study the effects of
amino acids on evoked potentials, we have observed a
cross-desensitisation between NMA and quinolinate which
may also be consistent with activity at a common
receptor.

Kynurenic acid

Since quinolinic acid is an important member of a
family of tryptophan metabolites known as the
kynurenines (for review see Stone and Connick, 1985),
several of which have convulsive properties, the
excitatory properties of quinolinate led to an interest

in the possible neuroactive properties of the other
kynurenines. To date, however, none of the other
kynurenine metabolites have shown any excitatory
activity when tested iontophoretically (Perkins and
Stone, 1982) or in the hippocampal slice preparation
(Perkins and Stone, 1984b) in concentrations up to 1mM.

Kynurenic acid however was able to antagonise the
responses of cortical neurones to iontophoretically
applied NMA, quisqualate, quinolinate (Perkins and
Stone, 1982) and kainate (Ganong et al., 1983),
although with little discrimination between the amino
acids. Kynurenate did show selectivity between amino
acid induced responses and those evoked by
acetylcholine. These studies led to the proposal that
quinolinic acid and kynurenic acid act as mutually
antagonistic regulators of CNS activity, and that an
imbalance in their respective concentrations might be
relevant to a variety of disorders such as epilepsy
(Perkins and Stone, 1982). Kynurenic acid has since
been shown to act on a variety of central pathways
(Cochran 1983; Cochran et al., 1984; Robinson et al.,
1984). Ganong et al. (1983) showed an effect of
kynurenic acid on rat hemisected spinal cord and
hippocampal slices. 100μM kynurenate reduced responses
by up to 50% in the spinal cord. These effects appear
to be conserved over a wide phylogenetic range; recent
examinations have indicated that primate hippocampal
slices show a blockade of Schaffer collateral evoked
excitation of CA1 pyramidal neurones by kynurenic acid
at similar doses to those effective in rat (Stone and
Perkins, 1984). In addition NMA and quinolinic acid
were shown to produce excitation in this preparation
which was blocked by kynurenate.

A number of reports also suggest that kynurenic
acid is able to differentiate between NMA and
quisqualate responses (Ganong et al., 1983)(Perkins and
Stone, 1984a), and even between responses to NMA and
quinolinate (Perkins & Stone, 1985), an observation
consistent with the idea of 2 NMA receptor subtypes.

The iontophoretic application of kynurenic acid
(Ganong et al., 1983)(Herrling et al., 1983) produces
little or no change in membrane potential or input
resistance of cells or the ability of cells to generate
action potentials in response to intracellular stimuli.

In our own laboratories, intracellular recordings
have been made from granule cells of the rat dentate
gyrus in vitro. A multi-barrelled iontophoretic
pipette was positioned in the middle one third of the

molecular layer such that a rapid response to glutamate
could be obtained. A saline-filled glass stimulating
electrode was used to activate the medial perforant
path, to produce 5-15mV amplitude e.p.s.p.s. Two
hundred such potentials were recorded before, during
and after the iontophoretic application of kynurenic
acid (70-190nA).

A quantal analysis was performed using a RDP 11/23
computer. While the number of quanta released per
stimulus was clearly unaffected by kynurenate, there
was a significant reduction in the quantal content,
which correlated well with the reduction of e.p.s.p.
amplitude.

These various results suggest that the inhibitory
effects of kynurenic acid on neurotoxicity (see later)
may also be primarily due to interference with the
post-junctional activity of a transmitter, rather than
a direct depressant action on the postsynaptic cell.
In view of the differential sensitivity of pyramidal
and granule cells of the hippocampus to the neurotoxic
action of quinolinic acid, a study has been carried out
to compare the depolarising potency of the agonist on
cells of the hippocampal slice. The most instructive
conclusion was that the ratio of the ED50 values on
granule cells compared with the CA1 pyramidal cells was
greatest for kainate and quinolinate, the same
compounds which most clearly prefer the pyramidal cells
on which to exhibit their neurotoxic potential. The
ED50 ratios were lower for NMA and ibotenic acids
(Stone, 1985b).

CONVULSANT STUDIES

Independently of Stone and Perkins, Lapin was
investigating the convulsive properties of kynurenine
metabolites. It should be noted at this point that
none of the electrophysiological or neurochemical data
obtained to date can explain the convulsant effects of
the kynurenines. Kynurenine itself has no effect
whatsoever on either the electrical activity of single
neurones (Perkins and Stone, 1982) or on the uptake of
tritiated D-aspartate into brain tissue (Connick and
Stone, 1985). In addition, in a study of NMA
antagonists as anticonvulsants it was reported that APH
would protect mice against NMA seizures but not against
those produced by quinolinate (Czuczwar and Meldrum,
1982). This observation was investigated further using
electrophysiological methods, with the findings that
peripherally injected quinolinate would indeed cause
excitation of neurones in the CNS, and that excitation

could be reduced by the microiontophoretic application of APH (Perkins and Stone, 1983d). Equally, i.p. injections of APH proved effective in reducing the excitatory responses of cortical neurones to quinolinic acid.

Lapin demonstrated that intraventricular injections of kynurenines can produce convulsions in mice and frogs, and even intraperitoneal injections to immature rats have the same effect (Lapin, 1981a). The possibility that these endogenous compounds may have a role to play in epilepsy was compounded by the observation that kynurenine induced seizures are often preceded by a phase of locomotor hyperactivity (Lapin 1981a) which is also seen in the convulsive phase of audiogenic seizure prone mice, an animal model of epilepsy.

The most potent of the kynurenines in producing convulsions are quinolinic acid and kynurenine itself. Intracerebro- ventricular injections of quinolinic acid show a threshold dose of approx. 1µg in mice, whereas the majority of the kynurenines require 25µg or greater to produce an effect. 50mg/kg i.p. quinolinate is sufficient to produce seizures in young rats, adults requiring doses of about 1g/kg (Czuczwar and Meldrum, 1982).

Several of the kynurenines, even those which produce convulsions after icv administration do not produce seizures on i.p. injection. Thus it becomes necessary to conclude that under normal physiological conditions the amounts of kynurenines in plasma, even though they may be able to penetrate into the CSF or across the blood brain barrier, cannot do so in amounts sufficient to cause seizure activity. As Lapin has noted however, this conclusion should not be taken to exclude the possibility that under conditions where blood brain barrier permeability is increased, or where peripheral synthesis of kynurenines is increased to extreme physiological or pathological levels, enough kynurenines might be able to cross the barrier to have convulsant effects.

Recently Pinelli et al. (1984) have confirmed the ability of kynurenine to produce seizures when injected icv and in addition demonstrated that L-kynurenine was active, whereas the D-isomer was inactive. They also noted that the formation of the methyl ester enhances kynurenine activity.

In an attempt to clarify the mechanism of
kynurenine induced seizures, Lapin has observed many
interesting phenomena, although the only definite
conclusion that can be drawn is that quinolinate and
kynurenine are not identical in the way they induce
convulsions. One of the most interesting aspects of
these studies has been that some kynurenines can
antagonise the convulsant properties of others. For
example, at doses substantially less than are required
to produce convulsions, picolinic, xanthurenic,
kynurenic and nicotinic acids will all reduce or delay
the incidence of seizures produced by icv
administration of L-kynurenine in mice (Lapin, 1983b).
None of the above had any activity against quinolinate
seizures, other than some increase in latency.

Lapin also performed a number of experiments on
the interactions between kynurenine convulsions and
amino acids. Glycine, for example, administered icv
reduces convulsions in mice (Lapin 1981a). This is
also true of GABA, leading to the suggestion that
inhibitory systems may be involved in kynurenine
convulsions, a suggestion that may be supported by the
ability of D,L-kynurenine to potentiate penicilin
induced epileptiform activity (Gusel and Mikhailov,
1980). Glycine and taurine, however had no effect on
quinolinic acid induced seizures, even though they
successfully prevented L-kynurenine induced seizures.

GABA and its analogues, including muscarine,
4-hydroxy-butyrate and baclofen reduced the severity of
both kynurenine and quinolinate seizures.

Kynurenine has been reported to inhibit the
turnover of GABA in brain (Lapin 1981a), although we
have been unable to find any significant effect of
either kynurenine, kynurenic acid or quinolinic acid on
GAD activity in rat brain homogenates (Connick and
Stone unpublished observations).

Nicotinamide, hypoxanthine and inosine are all
proposed ligands at the benzodiazepine receptor, and
all were able to prevent seizures produced by
D,L-kynurenine, although they produced little or no
protective effect against pentylenetetrazole seizures.
It was therefore considered of interest that diazepam,
an effective anticonvulsant agent against
pentylenetetrazole, had no effect on kynurenine
seizures, even at doses as high as 20mg/kg (Lapin
1981b). In addition it should be noted that
L-kynurenine could prevent the antagonistic action of
diazepam towards caffeine induced seizures (Lapin

1983a).   Neither D-L,kynurenine, kynurenic acid or
quinolinate displace diazepam from rat brain receptors
at reasonable concentrations (Kd>mM).   (V.V. Rozanec in
Lapin 1983a, P.F. Morgan and Stone, Unpub.).   Any link
between kynurenine, GABA and the benzodiazepines
remains obscure.

NEUROTOXICITY

Following the observation by Stone and Perkins
(1981) that quinolinic acid was an endogenous
neuroexcitant, Schwarcz et al. (1983) examined the
possibility that quinolinate might possess toxic
properties similar to those produced by kainic and
ibotenic acids.   They reported that injections of small
amounts (as little as 12nmoles) of quinolinate into the
rat neostriatum would produce neurodegeneration, larger
amounts causing clonic movements of the front
forelimbs.   The localised effects of quinolinate were
similar to those produced by kainate; a loss of
synaptic complexes and a loss or swelling of dendrites
with little evidence of any loss of presynaptic
components or of axons.   Unlike kainate, however, no
distant nerve cell loss was encountered.

In the hippocampus quinolinic acid produces a
selective neuronal loss, with a preferential
degeneration of pyramidal cells or granule cells in the
dentate gyrus (Schwarcz et al., 1984).   This bears
comparison with kainic acid rather than ibotenic acid
and NMA, which do not show such selectivity and is
consistent with the electrophysiological results
presented above, in which a greater excitant potency is
observed on the CAl pyramids compared with the dentate
granule cells (Stone, 1985b).

Schwarcz and Kohler (1983) later examined the
regional sensitivity of neurones in the CNS to
quinolinate toxicity.   In parallel with
electrophysiological data (Perkins and Stone 1983a,
1983c) the striatum, globus pallidus and hippocampus
were particularly vulnerable, whereas the cerebellum,
amygdala, septum, hypothalamus, substantia nigra and
spinal cord were relatively resistant.   A common
feature of the three neurotoxins quinolinate, kainate
and ibotenate is the preferential resistance of large
neurones in the striatum, Purkinje cells in the
cerebellum and lack of effect on the hypothlamus.
However, kainate differs by causing distant cell loss,
whereas quinolinate and ibotenate create well defined
local lesions.   But, whereas quinolinate damage is
selective, ibotenate causes widespread damage to
neurones and nerve terminals, as does NMA.

Electrophysiological data which showed that quinolinic acid acts on NMA type receptors is reinforced by the demonstration that APH was able to block the toxic effects of quinolinate as well as cis and trans 2,3-PDA (Foster et al., 1983). Other quinolinic acid homologues were found to have little neurotoxic activity (Foster et al., 1983) in common with their excitatory activity on single neurones (Birley et al., 1982).

It is interesting that NMA, ibotenate and trans PDA were able to produce toxicity when injected into the striatum of 7 day old rats, as did kainate, whereas quinolinate and cis PDA were non-toxic in 7 day old striatum. Because of this difference, Foster et al. (1983) suggested that different mechanisms were involved in the neurotoxic properties of trans PDA compared with cis PDA and quinolinic acid. In an attempt to explain these differences, Foster et al. (1983) hypothesised the involvement of a novel amino acid receptor with properties common to NMA and kainate receptors. However, with little evidence for the existence of such a receptor, they proposed that quinolinic acid and cis PDA may act presynaptically to modulate the release of some unidentified compounds.

Relatively little work has been performed on the effects of kynurenic acid or other kynurenines on neurotoxic actions in the CNS. However, Foster et al. (1984b) have demonstrated that kynurenic acid will block the neurotoxic effects of kainic acid.

Of great interest was their additional observation that kynurenic acid was 6 to 7 times more active as an antagonist of quinolinate toxicity than of NMA or kainate toxicity in the striatum. Indeed kynurenate was even found to enhance the degenerative effects of ibotenic acid (Foster et al., 1984b). This latter finding does not to date have any counterpart in electrophysiological studies, although carefully controlled microiontophoretic experiments in vivo have revealed some preference of kynurenate for antagonising quinolinic acid rather than NMA on a proportion of hippocampal neurones (Perkins & Stone, 1985).

NEUROCHEMISTRY

Although the kynurenine pathway of tryptophan metabolism is of major importance in the liver and kidney, and as such is well characterised in these tissues, little is known about its importance in the CNS.

The existence of the kynurenine pathway in brain tissue was demonstrated by Gal et al. (1966), when it was shown that kynurenine could be formed by incubating brain homogenates with labelled tryptophan. Further evidence was obtained when Joseph et al. (1978), and Gal and Sherman (1978; 1980) demonstrated the presence of kynurenine and some of its metabolites in rat brain. Kynurenine itself appears to occur at a concentration of approx. 200ng/g tissue i.e. approx. 1μM. With the rise of interest in quinolinic acid it was considered important to confirm its presence and determine its concentration in brain. Wolfensberger et al. (1983) were the first to demonstrate quinolinic acid in brain. Using a sensitive mass fragmentographic method they found quinolinic acid to be present in concentrations equivalent to noradrenaline i.e. μM. In addition they found quinolinate to be unevenly distributed, the regional variation roughly paralleling the excitatory potency of quinolinate (Perkins & Stone, 1983a). Moroni et al. (1984b) also noted that quinolinate concentrations in rat brain increased with age, and in about half the 30 month old rats examined the level of quinolinic acid found approached that seen to cause toxicity.

More recently Schwarcz and Foster (1984) demonstrated the presence of the enzymatic machinery necessary for the synthesis and degradation of quinolinic acid in rat brain.

Few workers have so far investigated the effects of the kynurenines on uptake and release systems. In view of the electrophysiological correlations between quinolinate and excitatory amino acid receptors, a study has recently been made of the effects of a wide range of kynurenines on the uptake of tritiated D-aspartate into rat brain slices (Connick and Stone, 1985). Since none of the kynurenines, except nicotinic acid, had any effect on D-aspartate uptake, it is unlikely that either the excitatory, toxic or convulsive effects of these compounds can be explained by a build up of amino acids caused by a suppression of their removal, whether by neurones or glia.

We have also recently examined the uptake of tritiated quinolinic acid itself, into slices of neocortex and cerebellum (Collins, Connick and Stone, 1985). There did not appear to be any active uptake. Foster et al. (1984a) also reported the absence of

active uptake into slices and synaptosomes prepared
from rat neostriatum or hippocampus. This lack of an
uptake mechanism would explain why after an injection
of tritiated quinolinic acid into the rat striatum, the
label disappeared relatively slowly (half life 22 min)
and all the label remaining 2 hours after injection was
still unmetabolised.

One of the first studies on the neurochemical
effect of quinolinate was by Lehmann et al. (1983)
involving the release of labelled acetylcholine from
slices of rat striatum. Both NMA and quinolinate would
promote the release of acetylcholine with the same
effect but different potencies.

NMA > Ibotenate > Quinolinate > Kainate

NMA and quinolinate were blocked by APV, APH and Mg^{++}.
These results are consistent with quinolinate being a
selective NMA agonist. Lehmann also reported a
regional difference in the ratio of NMA and quinolinate
potencies in producing acetylcholine release. This
supports the idea (Perkins and Stone, 1983a) that two
subpopulations of NMA receptor may exist.

Perkins and Stone (1983b) examined the relese of
pre-loaded purines in the cerebral cortex of
anaesthetized rats. Various excitatory amino acids
were able to produce release of the labelled purines
into a cup on the cortical surface. As in the study on
acetylcholine release NMA and quinolinate were very
much more potent than glutamate, quisqualate and
kainate. Again NMA and quinolinate induced release
were preferentially reduced by APH. Interestingly
kynurenic acid was able to reduce the release produced
by NMA and quinolinate without affecting that by
glutamate.

In the sodium efflux assay of Luini et al. (1984)
quinolinic acid enhanced the efflux of $^{22}Na^{+}$ from
striatial slices and this action was blocked by NMA
antagonists. In this study quinolinic acid was found
to be twice as potent as glutamate.

Binding

Despite the widespread nature of NMA receptors as
determined in electrophysiological studies, it has not
proved possible to demonstrate any specific saturable
binding of an NMA agonist to synaptic membranes
prepared from brain. This is also true of quinolinic
acid. Even data obtained indirectly via displacement

of other ligands e.g. glutamate, kainate etc. (Foster et al., 1983) has not proved encouraging. For example, quinolinic acid shows a comparatively high Kd of 350µM against the binding of 3H APV (Olverman et al., 1984). More recently quinolinic acid and kynurenic acid have been shown to displace the residual glutamate binding in hippocampal membranes after they had been treated in such a way as to diminish binding to Cl^-/Ca^{++} dependent sites. The IC50 was then approx. 180µM.

Acknowledgements
We are grateful to the Wellcome Trust, Medical Research Council and National Fund for Research into Crippling Diseases for supporting the author s work.

References
Birley, S., Collins, J.F., Perkins, M.N. & Stone, T.W. (1982). The effects of cyclic dicarboxylic acids on spontaneous and amino acid evoked activity of rat cortical neurones. Brit. J. Pharmacol. 77, 7-12.

Cochran, S.L. (1983). Kynurenic acid: competitive antagonist of excitatory synaptic transmission? Neurosci. Lett. Suppl. 14, 568.

Cochran, S.L., Kasik, P. & Precht, W. (1984). Evidence for 'glutamate' as the transmitter of VIII nerve afferents in the frog. Neurosci. Lett. Suppl. 18, 5191.

Collins, J.F., Connick, J.H. & Stone, T.W. (1985). Absence of uptake and binding of radiolabelled quinolinic acid in rat brain. Brit. J. Pharmacol. **85**, P.

Connick, J.H. & Stone, T.W. (1985). The effect of quinolinic acid and the kynurenines on the uptake of [3H]-D-aspartic acid in rat brain. Brit. J. Pharmacol. 84, 92P.

Czuczwar, S.J. & Meldrum, B.S. (1982). Protection against chemically induced seizures by 2-amino-7-phosphono-heptanoic acid. Europ. J. Pharmacol. 83, 335-338.

Foster, A.C., Collins, J.F. & Schwarcz, R. (1983). On the excitotoxic properties of quinolinic acid, 2,3-piperidine dicarboxylic acids and structurally related compounds. Neuropharmacol. 22, 1331-1342.

Foster, A.C., Miller, L.P., Oldendorf, W.H. & Schwarcz, R. (1984a). Studies on the disposition of quinolinic acid after intracerebral or systemic administration in the rat. Exp. Neurol. 84, 428-440.

Foster, A.C., Vezzani, A., French, E.D. & Schwarcz, R. (1984b). Kynurenic acid blocks neurotoxicity and

seizures induced in rats by the related brain
metabolite quinolinic acid. Neurosci. Lett. 48,
273-278.

Gal, E.M. & Sherman, A.D. (1978). Synthesis and
metabolism of L-kynurenine in rat brain. J.
Neurochem. 30, 607-613.

Gal, E.M. & Sherman, A.D. (1980). L-kynurenine: its
synthesis and possible regulatory function in
brain. Neurochem. Res. 5, 223-239.

Gal, E.M., Armstrong, J.C. & Ginsberg, B. (1966). The
nature of in vitro hydroxylation of L-tryptophan
by brain tissue. J. Neurochem. 13, 643-654.

Ganong, A.H., Lanthorn, T.H. & Cotman, C.W. (1983).
Kynurenic acid inhibits synaptic and acidic amino
acid-induced responses in the rat hippocampus and
spinal cord. Brain Research 273, 170-174.

Gusel, W.A. & Mikhailov, I.B. (1980). Effect of
tryptophan metabolites on activity of the
epileptogenic focus in the frog hippocampus. J.
Neural Trans. 47, 41-52.

Herrling, P.L. Morris, R. & Salt, T.E. (1983). Effects
of excitatory aminoacids and their antagonists on
membrane and action potentials of cat caudate
neurones. J. Physiol. 339, 207-222.

Joseph, M.H., Baker, H.F. & Lawson, A.M. (1978).
Positive identification of kynurenine in rat and
human brain. Biochem. Soc. Trans. 6, 123-126.

Lapin, I.P. (1981a). Kynurenines and Seizures.
Epilepsia 22, 257-265.

Lapin, I.P. (1981). Nicotinamide, inosine and
hypoxanthine, putative endogenous ligands of the
benzodiazepine receptor, opposite to diazepam are
much more effective against kynurenine-induced
seizures than against leptazol-induced seizures.
Pharmacol. Biochem. Behav. 14, 589-593.

Lapin, I.P. (1983a). Structure-activity relationships
in kynurenine, diazepam and some putative
endogenous ligands of the benzodiazepine
receptors. Neurosci. Biobehav. Revs. 7, 107-118.

Lapin, I.P. (1983b). Antagonism of kynurenine-induced
seizures by picolinic, kynurenic and xanthurenic
acids. J. Neural Trans. 56, 177-185.

Lehmann, J., Schaefer, P., Ferkany, J.W. & Coyle, J.T.
(1983). Quinolinic acid evoked [3H]acetylcholine
release in striatal slices: mediation by NMDA-type
excitatory amino acid receptors. Europ. J.
Pharmacol. 96, 111-115.

Luini, A., Tal, N., Golberg, O. & Teichberg, V.I.
(1984). An evaluation of selected brain
constituents as putative excitatory
neurotransmitters. Brain Research 324, 271-278.

McLennan, H. (1984). A comparison of the effects of NMDA and quinolinate on central neurones in the rat. Neurosci. Lett. 46, 157-160.

Moroni, F., Lombardi, G., Carla, V. & Moneti, G. (1984a). The excitotoxin quinolinic acid is present and unevenly distributed in the rat brain. Brain Research 295, 352-355.

Moroni, F., Lombardi, G., Moneti, G. & Aldinio, C. (1984b). The excitotoxin quinolinic acid is present in the brain of several animal species and its cortical content increases during the aging process. Neurosci. Lett. 47, 51-56.

Olverman, H.J., Jones, A.W. & Watkins, J.C. (1984). L-glutamate has higher affinity than other amino-acids for [3H]-D-AP5 binding sites in rat brain membranes. Nature 307, 460-462.

Perkins, M.N. & Stone, T.W. (1982). An iontophoretic investigation of the actions of convulsant kynurenines and their interaction with the endogenous excitant quinolinic acid. Brain Research 247, 184-187.

Perkins, M.N. & Stone, T.W. (1983a). Quinolinic acid: regional variations in neuronal sensitivity. Brain Research 259, 172-176.

Perkins, M.N. & Stone, T.W. (1983b). In vivo release of [3H]-purines by quinolinic acid and related compounds. Brit. J. Pharmacol. 80, 263-267.

Perkins, M.N. & Stone, T.W. (1983c). Pharmacology and regional variations of quinolinic acid-evoked excitations in the rat central nervous system. J. Pharmacol. Exp. Ther. 226, 551-557.

Perkins, M.N. & Stone, T.W. (1983d). On the interaction of 2-amino-7-phosphono-heptanoic acid and quinolinic acid in mice. Europ. J. Pharmacol. 89, 297-300.

Perkins, M.N. & Stone, T.W. (1984a). A study of kynurenic acid and excitatory amino acids in the rat hippocampus. J. Physiol. 357, 118P.

Perkins, M.N. & Stone, T.W. (1984b). Specificity of kynurenic acid as an antagonist of synaptic transmission in rat hippocampal slices. Neurosci. Lett. Suppl. 18, 432.

Perkins, M.N. & Stone, T.W. (1985). Actions of kynurenic acid and quinolinic acid in the rat hippocampus in vivo. Exp. Neurol. in press.

Pinelli, A., Ossi, C., Colombo, R., Tofanetti, O. & Spazzi, L. (1984). Experimental convulsions in rats produced by intraventricular administration of kynurenine and related compounds. Neuropharmacol. 23, 333-338.

Robinson, M.B., Anderson, K.D. & Koerner, J.F. (1984). Kynurenic acid as an antagonist of hippocampal excitatory transmission. Brain Research <u>309</u>, 119-126.

Schwarcz, R. & Foster, A.C. (1984). Studies of quinolinic acid metabolism in rat brain. Clin. Neuropharmacol. <u>7</u> suppl., 450-449.

Schwarcz, R. & Kohler, C. (1983). Differential vulnerability of central neurons of the rat to quinolinic acid. Neurosci. Lett. <u>38</u>, 85-90.

Schwarcz, R., Brush, G.S., Foster, A.C. & French, A.D. (1984). Seizure activity and lesions after intrahippocampal quinolinic acid injection. Exp. Neurol. <u>84</u>, 1-17.

Schwarcz, R., Whetsel, W.O. & Mangao, R.M. (1983). Quinolinic acid: an endogenous metabolite that produces axon-sparing lesions in rat brain. Science <u>219</u>, 316-318.

Stone, T.W. (1984). Excitant activity of methyl derivatives of quinolinic acid on rat cortical neurones. Brit. J. Pharmacol. <u>81</u>, 175-181.

Stone, T.W. (1985a). Microiontophoresis and Presure Ejection. Wiley, Chichester.

Stone, T.W. (1985b). Relative potencies of excitatory amino acids on pyramidal and granule cells in the rat hippocampal slice. Neurosci. Lett. in press.

Stone, T.W. & Connick, J.H. (1985). Quinolinic acid and other kynurenines in CNS. Neuroscience in press.

Stone, T.W. & Perkins, M.N. (1981). Quinolinic acid: a potent endogenous excitant at amino acid receptors in rat CNS. Europ. J. Pharmacol. <u>72</u>, 411-412.

Stone, T.W. & Perkins, M.N. (1984). Actions of amino acids and kynurenic acid in the primate hippocampus. Neurosci. Lett. <u>52</u>, 335-340.

Wolfensberger, M., Amsler, U., Cuenod, M., Foster, A.C., Whetsell, W.O. & Schwarcz, R. (1983). Identification of quinolinic acid in rat and human brain tissue. Neurosci. Lett. <u>41</u>, 247-457.

25

Endogenous Excitotoxins: Focus on Quinolinic Acid

R. Schwarcz, E. Okuno and C. Köhler

EXCITOTOXINS

The neurotoxic properties of acidic amino acids such as glutamate and aspartate have been recognized for more than 25 years (Lucas and Newhouse, 1957; Potts et al., 1960). However, not much attention was paid to those early reports because of the rapidly accumulating evidence for the neuroexcitatory properties and later, as reviewed in detail in this volume, for a neurotransmitter role of glutamate and its congeners. The work of Olney in the late 1960's and early 1970's describing the particular "axon-sparing" quality of lesions caused by excitatory amino acids (Olney, 1969; Olney et al; 1971), at the time met with some skepticism (Reynolds et al., 1971)). However, Olney's pioneering studies, as well as his admonitions regarding the addition of monosodium glutamate to baby foods (Olney, 1979), led to intensified research efforts which were aimed at a thorough evaluation of the neurodegenerative properties of a spectrum of excitatory amino acids in brain tissue. Important results and hypotheses emerged from those investigations: 1) The neuroexcitatory and neurotoxic properties of acidic amino acids parallel each other; hence, the operational term "excitotoxins" was introduced (Olney, 1974); 2) Several heterocyclic amino acids, isolated from plants or fungi, are powerful excitotoxins; compounds such as kainic, ibotenic and quisqualic acid have found widespread use in many areas of the experimental neurosciences (Coyle et al., 1981); 3) There exist substantial qualitative differences between excitotoxins with regard to their neurodegenerative properties in the brain (see Fuxe et al., 1983 for review); still not fully understood, they may be related to differences in a) the activation of certain receptor populations,

*This work was supported by USPHS grants NS 16102 and 20509, and grants from the Lena Marcus Fund and the American Parkinson's Disease Foundation (to R.S.). The contributions of Dr. Alan Foster are gratefully acknowledged.

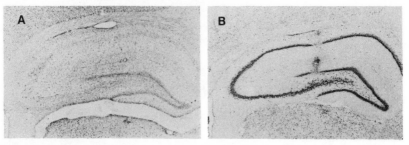

Figure 1: Micrographs of hippocampal morphology 4 days after hippocampal injections of 120 nmol QUIN (A) and 120 nmol QUIN pretreated (24 hours) with a local injection of 444 nmol -APH (B) in the rat. Note the total protection from QUIN's neurotoxic effects in B. Magnification: 13X.

b) the dependency on certain neuronal pathways, c) the convulsive properties, d) the penetration of the blood-brain barrier after systemic administration, e) the developmental characteristics of a given brain area, f) a combination of two or more of the above; 4) Excitotoxic lesions of certain areas of the brain provide animal models for human neurodegenerative disorders (Schwarcz et al., 1984); by inference, an involvement of (an overabundance of) an endogenous excitotoxin in the pathogenesis of those diseases, has been hypothesized.

For several years, the latter issue, a possible participation of an endogenous excitotoxin in disorders such as Huntington's disease and temporal lobe epilepsy, has been examined in our laboratories. While earlier studies focussed on glutamate itself as the putative pathogen, it soon became clear that the normal brain is very well equipped to inactivate even large doses of extracellular glutamate before irreversible damage occurs. Only by artificial interference with these inactivation processes can glutamate be rendered toxic in the adult rodent brain (Köhler and Schwarcz, 1981; McBean and Roberts, 1985). A dysfunctional brain glutamate system may indeed underlie certain neurodegenerative disorders (Plaitakis, 1984) but no unequivocal such link has been established to date. Clearly, studies of neurochemical irregularities of glutamate or related excitatory, and possibly excitotoxic, neurotransmitters in human neuropsychiatric diseases are still in their infancy and will have to be pursued vigorously in the years to come.

QUINOLINIC ACID

In the search for potential endogenous excitotoxins other than glutamate and aspartate, our attention was recently drawn to quinolinic acid (QUIN; pyridine 2,3-dicarboxylic acid). While QUIN has long been known to be present in peripheral tissues and urine (Henderson and Hirsch, 1949) and to constitute a link in the kynurenine pathway converting tryptophan to NAD (Gholson et al., 1964), only sporadic suggestions had appeared in the literature concerning its possible physiological role. This perception changed dramatically in 1981 with the publication of a short report on QUIN's neuroexcitatory properties (Stone and Perkins, 1981). QUIN, when applied iontophoretically to rat cortical neurons, could be demonstrated to depolarize cells by acting specifically on N-methyl-D-aspartate (NMDA) receptors (cf. Watkins, this volume). Interestingly, QUIN appeared to be significantly more potent in the brain than in the spinal cord (Watkins, 1978) and subsequent studies also established pronounced differences in QUIN's neuroexcitatory properties between various regions of the brain (Perkins and Stone, 1983).

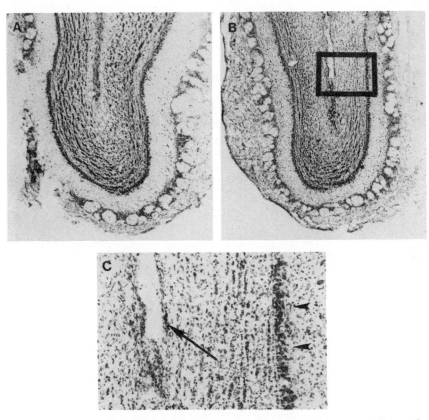

Figure 2: Micrographs depicting the marginal neurotoxic effect of QUIN on olfactory bulb neurons. A: control; B: 4 days following the local injection of 240 nmol QUIN; C: Higher magnification of the enclosed area marked in B; arrow: track of the injection needle; arrowheads point to intact mitral cells close to the injection site. Magnifications: A,B: 30X; C: 150X.

When injected directly into different parts of the rat brain, QUIN showed prominent neurodegenerative qualities (Schwarcz et al., 1983). In fact, QUIN's potency as a neurotoxin is comparable to that of exogenous agents like ibotenic acid and NMDA and greatly exceeds that of any other known endogenous excitatory amino acids. However, as reported for its excitatory effects, there exists a definite variability of neuronal vulnerability to QUIN both between and within areas of the brain (Schwarcz and Köhler, 1983). Examples of this regional heterogeneity are depicted in Figures 1 and 2. 120 nmol QUIN, injected into the hippocampal formation, destroy the entire structure (Figure 1A), while 240 nmol QUIN cause only limited, probably unspecific, damage in the olfactory bulb (Figures 2B and C). Notably, QUIN-toxicity can be prevented by pre-treatment with selective NMDA-antagonists such as D-(-)2-amino-7-phosphonoheptanoic acid (-APH), thus substantiating the parallelism between QUIN's excitatory and toxic properties and establishing QUIN as a excitotoxin. (Note the remarkable potency of -APH, which was administered in a single injection 24 hours before QUIN into the hippocampus pictured in Figure 1B).

Since the neuronal populations which showed particular vulnerability to QUIN in the rat brain correspond to those known to be preferentially affected in human neurodegenerative disorders such as Huntington's disease and temporal lobe epilepsy, QUIN soon came to be considered as a potential pathogen in human diseases - and QUIN-antagonists, following the same reasoning, were concept- ualized as novel therapeutic agents (Schwarcz and Meldrum, 1985).

Using GC-MS methodology, QUIN has been identified in both rat and human brain tissue (Wolfensberger et al., 1983; Moroni et al., 1984). Its concentration was found to be approximately 1 micro- molar, only one order of magnitude lower than the minimal toxic dose established in hippocampal organotypic cultures (Whetsell and Schwarcz, 1983). It also soon became clear that QUIN does not qualify as a neurotransmitter candidate, since one of the main criteria, rapid removal from the extracellular space by reuptake mechanisms or metabolism, clearly is not fulfilled. In fact, QUIN is remarkably stable following introduction in the brain and does not appear to be eliminated from the brain by active transport processes (Foster et al., 1984). Thus, the brain appears ill equipped to effectively handle ("detoxify") extracellular QUIN should an accumulation occur under certain pathological conditions. This is in marked contrast to the situation of several other endogenous excitatory amino acids (glutamate, aspartate, etc.) and gave further impetus to carefully investigate other features of the brain's quinolinergic system.

METABOLISM OF QUINOLINIC ACID IN THE BRAIN

Since QUIN does not cross the blood-brain barrier under normal physiological conditions (Foster et al., 1984), it had to be

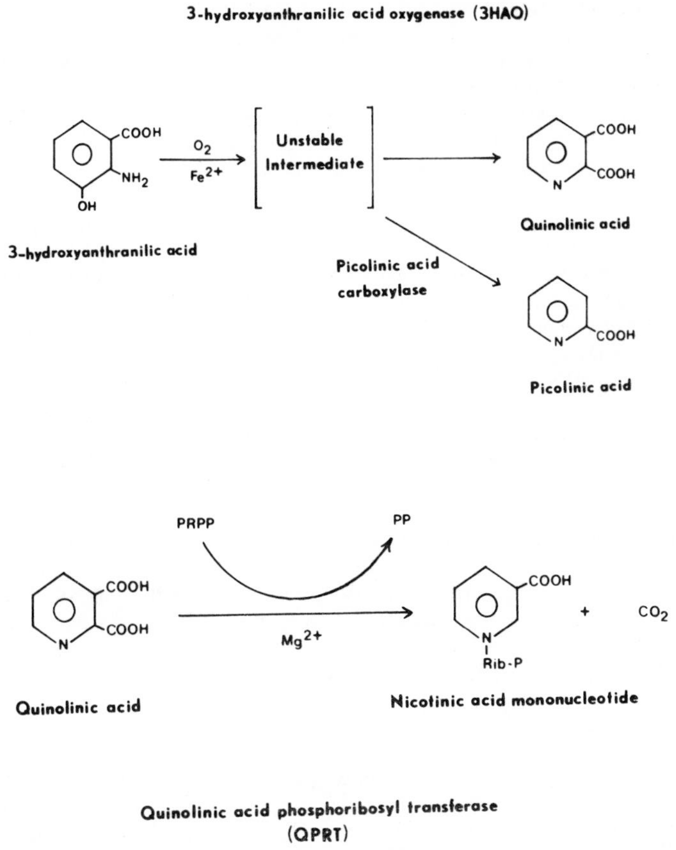

3-hydroxyanthranilic acid oxygenase (3HAO)

Quinolinic acid phosphoribosyl transferase
(QPRT)

Figure 3: Schematic diagrams of the biosynthetic and degradative enzymes of quinolinic acid metabolism.

assumed that the metabolic machinery responsible for its biosynthesis and catabolism is present in the brain. The two enzymes directly involved in the production and breakdown of QUIN, 3-hydroxyanthranilic acid oxygenase (3HAO) and quinolinic acid phosphoribosyltransferase (QPRT), had been examined in peripheral tissues (Bokman and Schweigert, 1951; Gholson et al., 1964). From those studies, it appeared that QPRT constitutes a rate-limiting step in the conversion of tryptophan to NAD via the kynurenine pathway. However, no explicit mention of the presence of 3HAO or QPRT in the brain had been made – probably for two reasons: a) the lack of sufficiently sensitive assay methodologies to detect the relatively low levels of the enzymes in the brain and b) the lack of a rationale to carefully examine brain QUIN metabolism.

We have recently developed novel assays for 3HAO and QPRT which are based on the conversion of the radiolabelled substrates (^{14}C-3-hydroxyanthranilic acid and ^{3}H-quinolinic acid; cf. reaction schemes in Figure 3) to the respective reaction products. Using simple and sensitive procedures, we were able to unequivocally identify both enzymes in rat brain tissue and have begun to examine some of their characteristics (Foster and Schwarcz, 1984; Foster et al., 1985b).

Kinetic analysis of the enzymes in whole rat forebrain preparations revealed that both 3HAO and QPRT possess a K_m for their respective substrates in the low micromolar range (Foster et al., 1985; Foster and Schwarcz, 1984). While the concentration of 3-hydroxyanthranilic acid in brain tissue is unknown at present, this is certainly compatible with the reported concentration of brain QUIN (Wolfensberger et al., 1983; Moroni et al., 1984). V_{max}-data (26.9 pmol QUIN formed/h/mg tissue for 3HAO and 0.9 pmol nicotinic acid mononucleotide formed/h/mg tissue for QPRT, respectively), confirm that QPRT, rather than 3HAO, constitutes the rate-limiting step in QUIN metabolism. In other words, the brain appears to have a more than 80-times greater ability to synthesize than to catabolize the excitotoxin. These values clearly indicate that QUIN production in vivo must be carefully regulated - either by the bioavailability of 3-hydroxyanthranilic

Table 1

Regional distribution of rat brain 3HAO and QPRT activities

	3HAO	QPRT
Olfactory bulb	322 ± 40	47.7 ± 3.2
Frontal cortex	200 ± 12	2.2 ± 0.2
Striatum	193 ± 8	2.5 ± 0.6
Hypothalamus	167 ± 6	16.7 ± 2.4
Hippocampus	154 ± 11	2.9 ± 0.3
Thalamus	120 ± 16	9.0 ± 1.0
Medulla	104 ± 9	8.5 ± 0.8
Spinal cord	88 ± 8	4.7 ± 0.7
Cerebellum	84 ± 12	4.0 ± 0.5
Substantia nigra	82 ± 24	6.6 ± 0.5
Retina	38 ± 19	2.2 ± 0.4

Data represent specific activities (pmol QUIN formed/hr/mg tissue for 3HAO and fmol nicotinic acid mononucleotide formed/hr/mg tissue for QPRT) and are the means ± S.E.M. of 3-6 separate animals.

NOTE: The apparent difference in absolute enzyme activities is in part due to the choice of substrate concentrations ($>K_m$ for 3HAO and $<<K_m$ for QPRT).

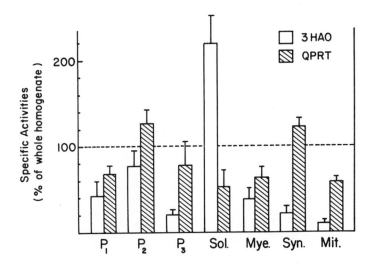

Figure 4: Subcellular distribution of 3HAO and QPRT in rat forebrain fractions. Abbreviations: Sol.: soluble fraction; Mye.: myelin; Syn.: synaptosomes; Mit.: mitochondria.

acid or by a modulatory shift towards the production of picolinic acid (cf. Figure 3).

The subcellular distribution of 3HAO and QPRT was examined in rat forebrain preparations (Mena et al., 1980). As shown in Figure 4, 3HAO is highly concentrated in the soluble fraction while QPRT is apparently largely membrane-bound, as indicated by its presence in non-soluble fractions such as the P_2 and synaptosomes. Assuming that the two enzymes are indeed present in the same cell, this differential intracellular localization suggests that a physical transfer of QUIN, possibly coupled with a specific cellular function, must take place in the course of QUIN metabolism.

The regional distribution of 3HAO and QPRT in rat (Table 1) and human (Foster et al., 1985a; Schwarcz et al., 1985) brain indicates a pronounced heterogeneity of the QUIN-related enzymes. In the rat, both enzyme activities are highest in the olfactory bulb and show substantial, but not parallel, variation in other areas of the brain. It may be of interest that the brain regions with the lowest QPRT activity and the largest 3HAO/QPRT ratio also

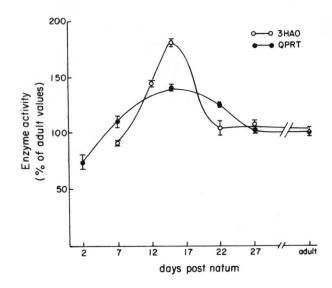

Figure 5: Ontogenetic pattern of 3HAO and QPRT in rat forebrain homogenates.

appear to be the most vulnerable to QUIN's neurotoxic effects. Thus, of the brain regions examined so far, the striatum and hippocampus are the two areas with both the largest probability to accumulate the toxin intracellularly and a high sensitivity to postsynaptically acting QUIN (Perkins and Stone, 1983). It remains to be seen if the two phenomena occur independently and if they, alone or in concert, may be related to the prominent involvement of these two structures in human neurodegenerative disorders.

Figure 6: Photomicrographs showing QPRT-immunoreactivity (i) in horizontal sections of the rat brain. (A) QPRT-i in small glia cells (some cells marked by arrows) in the striatum and the lateral septum. The two structures are situated to the right and left, respectively, of the lateral ventricle (V). (B) A section incubated with QPRT-antiserum preabsorbed with purified QPRT. No specific staining is present in this section. (C) QPRT-i glia cells in the striatum and in the wall of the lateral ventricle (all QPRT-stained cells are marked by arrows). (D) QPRT-i (arrows) in the hilus of the area dentata. (E) QPRT-i glia cells in close association with large neurons (large arrows) of the area dentata. Magnifications: A,B: 210X; C,E:520X, D:325X.

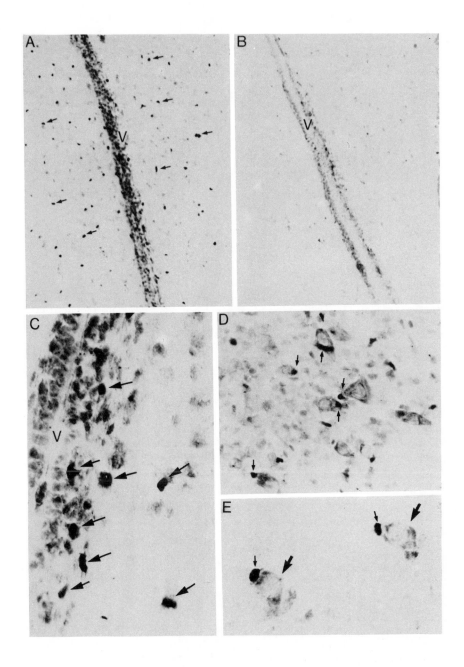

In contrast to the differences between 3HAO and QPRT in terms of subcellular and regional distribution, there exists a remarkable congruence in the ontogenetic pattern of the two enzymes in the rat forebrain. Both the anabolic and the catabolic enzyme reach adult levels during the first postnatal week and are substantially elevated (3HAO > QPRT) in the second and third week post natum (Figure 5). The reason for this transient increase in enzyme activities is unclear at present. However, in view of the above-mentioned tentative parallelism between the activities of 3HAO and QPRT on one hand and the vulnerability of neurons to QUIN on the other, it is interesting to point out the apparent resistance of developing neurons to the toxic effects of QUIN (Foster et al., 1983).

LOCALIZATION OF QPRT IN THE BRAIN

In order to reach a better understanding of the brain's QUIN system, it became imperative to identify the cellular elements responsible for the metabolism and storage of the excitotoxin. In a first approach to assess this issue, we recently purified rat liver QPRT to homogeneity and established the physico-chemical and immunological identity between the liver and the brain enzyme (Okuno and Schwarcz, 1985). Subsequently, anti-rat QPRT antibodies were raised, purified and used for immunohistochemical localization of QPRT in the rat brain. Briefly, rats were perfused transcardially according to the method of McLean and Nakane (1974). The fixed brains were cut on a freezing microtome and 30 μm thick horizontal sections were incubated floating free in rabbit anti-QPRT antiserum diluted 1:6000 in phosphate buffered saline (pH 7.4) containing 0.2% Triton X-100 and 1% normal goat-serum. The antibody-antigen complex was made visible by the avidin-biotin-complex method of Hsu et al. (1981) using a commercially available kit (Vector laboratories, Burlingame, CA, USA). The reacted sections were stained with thionin in order to visualize neuronal cell bodies. In control experiments, sections were incubated with the QPRT-antiserum pre-absorbed with purified QPRT.

Specific QPRT-immunoreactivity (Figure 6) was present in all regions of the rat brain known to contain enzymatic activity (Table 1). In all brain areas tested, for instance the olfactory bulb, the hypothalamus, the striatum and the hippocampus, QPRT-immunoreactivity could be observed in glia cells. QPRT-containing cells were of small size but morphological differences existed. Some cells were round or triangular in shape with few visible processes while others were slightly elongated with one or two distinct and widely branching processes. In the striatum, the QPRT-immunoreactive cells were present throughout the structure with numerous round immunostained glia cells situated along the wall of the ventricle (Figure 6A). In the hippocampal region, small intensely immunoreactive cells were scattered throughout the hilus of the area dentata and the pyramidal cell layer of the

Ammon's horn. In many instances, immunoreactivity occurred in satellite glia cells found to be in intimate association with neuronal cell bodies (Figure 6E).

Preliminary examination of QPRT immunoreactivity in the rat brain not only indicates that the QUIN system may be exclusively localized in glial elements but also raises the possibility of a functional relevance of the sporadically observed close proximity of QPRT-positive glia with neuronal somata. Future careful analyses, conducted at both the light and electron microscopic levels and ideally including experiments using antibodies raised against 3HAO, will be essential in order to fully understand the significance of these anatomical arrangements.

EFFECTS OF LESIONS ON BRAIN 3HAO AND QPRT

Using a simpler approach to examine the cellular localization of QUIN-related enzymes, neuronal degeneration was caused in the rat striatum by the local application of kainic or ibotenic acid. Measurement of both 3HAO and QPRT activities at various timepoints after the lesion revealed a dramatic increase in both enzyme activities in response to excitotoxin-induced nerve cell death (Table 2). The changes were not noted after two days when histological and biochemical analyses clearly indicate the necrosis of all vulnerable neurons. Kinetic analyses of the increases in enzyme activities in the lesioned striata

Table 2

Excitotoxic lesions of the rat striatum: effects on 3HAO and QPRT

Enzyme activities in the lesioned striatum		
Days after the lesion	3HAO[a]	QPRT[b]
2	136 ± 27	108 ± 17
7	439 ± 44**	125 ± 21
21	316 ± 37**	256 ± 15**

[a]Striata were injected with 40 µg ibotenic acid in 2 µl; [b]striata were injected with 2 µg kainic acid in 1 µl. All enzyme activities are expressed as a percentage ± S.E.M. of the contralateral uninjected striatum. N = 5-6 at each timepoint. **$p < 0.01$ as compared to the contralateral striatum (paired t-test).

indicated an accumulation of enzyme protein (data not shown).
Thus, it can be reasonably assumed that both enzymes are contained
in non-neuronal elements, which proliferate following neuronal
injury. This interpretation is compatible with the notion of a
glial localization of 3HAO and QPRT (cf. above).

Increases in the activities of the two QUIN-related enzymes are
clearly not restricted to excitotoxic striatal lesions since
similar phenomena could be observed after hippocampal kainate
lesions and in the substantia nigra following the local injection
of the dopaminergic neurotoxin 6-hydroxydopamine (data not shown).
Detailed studies are currently under way to examine the
mechanism(s) and possible functional consequences of these
effects.

OUTLOOK: EXCITOTOXINS AND NEURODEGENERATION IN MAN

At this point in time, it is somewhat difficult to
realistically assess the possible involvement of an endogenous
excitotoxin, such as QUIN, in the pathogenesis of neurodegenerative phenomena in humans. In order to clarify any such link,
neuropathologists, in particular, must address the cardinal
question if axon-sparing lesions indeed occur under various
pathological conditions. To date, affirmative answers can
probably be given for Huntington's disease and hypoglycemic brain
damage (Wieloch, this volume). Similar data should be forthcoming
with regard to ischemia, epileptic disorders, Alzheimer's disease
and other human conditions in which selective but
ultrastructurally ill-defined neuronal loss has been documented.
Rapidly accumulating information clearly points to a prominent
involvement of endogenous excitatory amino acids, mainly
glutamate and aspartate, in processes associated with (but not
necessarily causally related to) neurodegenerative events
(Benveniste et al., 1984; Auer, 1985). In addition, the efficacy
of recently developed excitatory amino acid receptor antagonists,
which can selectively prevent nerve cell death under a variety of
experimental conditions (see also Figure 1B), suggests that a
pathologic overabundance of an endogenous agonist at those
receptors may have caused the problem in the first place (Schwarcz
and Meldrum, 1985). Thus, several lines of circumstantial
evidence indicate that excitatory amino acids may play a central
role in the etiology of a spectrum of neuropsychiatric disorders
in which neuronal loss can be observed. The substantial increase
in the research efforts geared at the examination of such a
connection can be anticipated to yield relevant new data in the
near future.

REFERENCES

Auer, R.N. (1985). Hypoglycemic brain damage: an experimental neuropathologic study in the rat. Ph.D. Thesis, Lund University, Sweden.

Benveniste, H., Drejer, J., Schousboe, A. and Diemer, N. (1984). Elevation of the extracellular concentrations of glutamate and aspartate in rat hippocampus during transient cerebral ischemia monitored by intracerebral microdialysis. J. Neurochem. 43, 1369-1374.

Bokman, A.H. and Schweigert, B.S. (1951). 3-Hydroxyanthranilic acid metabolism. IV. Spectrophotometric evidence for the formation of an intermediate. Arch. Biochem. Biophys., 33, 270-276.

Coyle, J.T., Bird, S.J., Evans, R.H., Gulley, R.L., Nadler, J.V., Nicklas, W.J. and Olney, J.W. (1981). Excitatory amino acid neurotoxins: selectivity, specificity and mechanisms of action. Neurosci. Res. Progr. Bull., 19, 331-427.

Foster, A.C., Collins, J.F. and Schwarcz, R. (1983). On the excitotoxic properties of quinolinic acid, 2,3-piperidine dicarboxylic acids and structurally related compounds, Neuropharmacology, 22, 1331-1342.

Foster, A.C., Miller, L.P., Oldendorf, W.H. and Schwarcz, R. (1984). Studies on the disposition of quinolinic acid after intracerebral or systemic administration in the rat. Exp. Neurol. 84, 428-440.

Foster, A.C. and Schwarcz, R. (1984). Synthesis of quinolinic acid by 3-hydroxyanthranilic acid oxygenase in rat brain tissue. Soc. Neurosci. Abstr., 10, 11.5.

Foster, A.C., Whetsell, W.O., Jr., Bird, E.D. and Schwarcz, R. (1985a). Quinolinic adic phosphoribosyltransferase in human and rat brain: activity in Huntington's disease and in quinolinate-lesioned rat striatum. Brain Res., in press.

Foster, A.C., Zinkand, W.C. and Schwarcz, R. (1985b). Quinolinic acid phosphoribosyltransferase in rat brain. J. Neurochem., 44, 446-454.

Fuxe, K., Roberts, P. and Schwarcz, R. (1983). Excitotoxins. Macmillan Press, London.

Gholson, R.K., Ueda, I., Ogasawara, N. and Henderson, L.M. (1964). The enzymatic conversion of quinolinate to nicotinic acid mononucleotide in mammalian liver. J. Biol. Chem., 239, 1208-1214.

Henderson, L.M. and Hirsch, H.M. (1949). Quinolinic acid metabolism. I. Urinary excretion by the rat following tryptophan and 3-hydroxyanthranilic acid administration. J. Biol. Chem., 181, 667-675.

Hsu, S.M., Raine, L. and Fanger, H. (1981). Use of avidin-biotin-peroxidase complex (ABC) in immunoperoxidase techniques: a comparison between ABC and unlabelled antibody (PAP) procedures. J. Histochem. Cytochem., 29, 577-580.

Köhler, C. and Schwarcz, R. (1981). Monosodium glutamate: increased neurotoxicity after removal of neuronal re-uptake sites. Brain Res., 211, 485-491.

Lucas, D.R. and Newhouse, J.P. (1957). The toxic effect of L-glutamate on the inner layers of the retina. Arch. Ophthalmol., 58, 193-201.

McBean, G.J. and Roberts, P.J. (1985). Neurotoxicity of L-glutamate and DL-threo-3-hydroxyaspartate in the rat striatum. J. Neurochem. 44, 247-254.

McLean, I. and Nakane, P. (1974). Periodate-lysine-paraform-aldehyde fixative: a new fixative for immunoelectron-microscopy. J. Histochem. Cytochem., 22, 1077-1083.

Mena, E.E., Hoeser, C.A. and Moore, B.W. (1980). Improved method of preparing synaptic plasma membranes: elimination of a contaminating membrane containing 2',3'-cyclic nucleotide 3'-phosphohydrolase activity. Brain Res., 188, 207-231.

Moroni, F., Lombardi, G., Carla, V. and Moneti, G. (1984). The excitotoxin quinolinic acid is present and unevenly distributed in the rat brain. Brain Res., 295, 352-356.

Okuno, E. and Schwarcz, R. (1985). Purification of quinolinic acid phosphoribosyltransferase from rat liver and brain. Biochim. Biophys. Acta, in press.

Olney, J.W. (1969). Brain lesions, obesity and other disturbances in mice treated with monosodium glutamate. Science, 164, 719-721.

Olney, J.W. (1974). Toxic effects of glutamate and related amino acids on the developing central nervous system. In Heritable Disorders of Amino Acid Metabolism. (ed. W.L. Nyhan). Pp. 501-512. Wiley, New York.

Olney, J.W. (1979). Excitotoxic amino acids: research applications and safety implications. In Glutamic Acid: Advances in Biochemistry and Physiology. (ed. L.J. Filer, S. Garattini, M.R. Kare, W.A. Reynolds and R.J. Wurtman). Pp. 287-319. Raven Press, New York.

Olney, J.W., Ho, O.L. and Rhee, V. (1971). Cytotoxic effects of acidic and sulphur containing amino acids on the infant mouse central nervous system. Exp. Brain Res., 14, 61-76.

Perkins, M.N. and Stone, T.W. (1983). Pharmacology and regional variations of quinolinic acid-evoked excitations in the rat central nervous system. J. Pharmacol. Exp. Ther. 226, 551-557.

Plaitakis, A. (1984). Abnormal metabolism of neuroexcitatory amino acids in olivopontocerebellar atrophy. In The Olivopontocerebellar Atrophies. (eds. R.C. Duvoisin and A. Plaitakis). Pp. 225-243. Raven Press, New York.

Potts, A.M., Modrell, R.W. and Kingsbury, C. (1960). Permanent fractionation of the electroretinogram by sodium glutamate. Amer. J. Ophthalmol., 50, 900-907.

Reynolds, W.A., Lemkey-Johnston, N., Filer, L.J., Jr. and Pitkin, R.M. (1971). Monosodium glutamate: absence of hypothalamic lesions after ingestion by newborn primates. Science, 172, 1342-1344.

Schwarcz, R., Foster, A.C., French, E.D., Whetsell, W.O., Jr. and Köhler, C. (1984). Excitotoxic models for neurodegenerative disorders. Life Sci., 35, 19-32.

Schwarcz, R. and Köhler, C. (1983). Differential vulnerability of central neurons of the rat to quinolinic acid. Neurosci. Lett., 38, 85-90.

Schwarcz, R. and Meldrum, B. (1985). Excitatory amino acid antagonists provide a novel therapeutic approach to neurological disorders. Lancet, in press.

Schwarcz, R., Whetsell, W.O., Jr. and Mangano, R.M. (1983). Quinolinic acid: an endogenous metabolite that causes axon-sparing lesions in rat brain. Science, 219, 316-318.

Schwarcz, R., White, R.J. and Whetsell, W.O., Jr. (1985). 3-hydroxyanthranilic acid oxygenase: activity in lesioned rat striatum and in human brain tissue. Soc. Neurosci. Abstr., in press.

Stone, T.W. and Perkins, M.N. (1981). Quinolinic acid: a potent endogenous excitant of amino acid receptors in CNS. Europ. J. Pharmacol., 72, 411-412.

Watkins, J.C. (1978). Excitatory amino acids. In Kainic Acid as a Tool in Neurobiology. (eds. E.G. McGeer, J.W. Olney and P.L. McGeer). Pp. 37-69. Raven Press, New York.

Whetsell, W.O., Jr. and Schwarcz, R. (1983). Mechanisms of excitotoxins examined in organotypic cultures of rat central nervous system. In Excitotoxins. (eds. K. Fuxe, P. Roberts and R. Schwarcz). Pp. 207-219. Raven Press, New York.

Wolfensberger, M., Amsler, U., Cuénod, M., Foster, A.C., Whetsell, W.O. Jr. and Schwarcz, R. (1983). Identification of quinolinic acid in rat and human brain tissue. Neurosci. Lett., 41, 247-252.

26

Interactions of Acidic Peptides: Excitatory Amino Acid Receptors

R. Zaczek, K. Koller, D.O. Carpenter, R. Fisher, J.M. ffrench-Mullen

and J.T. Coyle

INTRODUCTION

Glutamic acid (Glu) and aspartic acid (Asp) are now generally accepted as the major excitatory neurotransmitters in the mammalian brain. Nevertheless, reservations remain about this conclusion because the amino acids are involved in several metabolic pathways including protein synthesis and because they exhibit broad, nearly universal excitatory effects on brain neurones. Furthermore, several reports of inconsistencies between the pharmacology of iontophoretically applied Glu/Asp and that of the endogenous excitatory neurotransmitter released at putative Glu/Asp synapses have appeared. For example, Hori et al. (1981) reported that α-amino-phosphonobutyric acid (APB) antagonized the effects of the excitatory neurotransmitter released by the lateral olfactory tract (LOT) but not the excitatory effects of iontophoretically applied Glu and Asp which have been proposed as the neurotransmitters of the LOT based on their selective uptake and evoked release. Furthermore, the sodium-dependent high affinity uptake process for Glu, considered a presynaptic marker for Glu terminals, efficiently transports Asp; and it has been difficult to effectively isolate calcium dependent evoked release of Glu from Asp. These observations prompted our speculation that small peptides containing Glu and/or Asp might be utilized as the endogenous neurotransmitters at purported Glu/Asp neuronal pathways.

Nearly thirty years ago, Tallon et al. (1956) first reported the existence in brain of high concentration of N-acetyl-aspartic acid (NAA) and several larger acidic peptides containing N-acetyl-aspartate. Although the function of NAA remained obscure, subsequent studies demonstrated that it was restricted primarily to the brain, where its millimolar concentrations are surpassed

only by L-glutamate. Zaczek et al. (1983), in screening brain
extracts for acidic peptides that interact with the receptors for
Glu, demonstrated that N-acetyl-aspartyl-glutamate (NAAG), one of
the acidic peptides first identified by Tallon, exhibited high
affinity for Glu receptors and had potent convulsive effects
when injected in the hippocampus. Accordingly, Zaczek et al.
(1983) proposed that NAAG and perhaps related acidic peptides
might be the endogenous neurotransmitters at putative Glu/Asp
synapses. This chapter will review recent findings relevant to
the excitatory role of NAAG in brain and the effects of related
acidic peptides on Glu receptors.

RESULTS

Neuronal localization and regional distribution of NAAG and NAA in brain.

        The levels of NAA and NAAG were measured in 14 brain regions
and in peripheral tissues of the rat (Koller et al., 1984a) with
an isocratic modification of an anion exchange HPLC method of
Lenda (1981). While extremely low levels of NAA were detected in
the heart (0.3 + 0.1 nmol/mg prot.) and liver (1.7 + 0.6 nmol/mg
prot.), brain levels were at least 20-fold greater except in the
retina and pituitary. Interestingly, the lowest level of NAA in
brain was observed in the pons and the highest level in the frontal cortex, consistent with a modest caudal to rostral gradient.
In contrast, with this sensitive technique, NAAG was undetectable
in heart and liver and exhibited a much more robust regional
variation consistent with a rostral to caudal gradient of distribution (Table 1). Thus, concentration of NAAG in spinal cord (22
+ 4 nmol/ mg prot.) was nearly 10-fold greater than in the
striatum (2.4 + 0.1 nmol/mg prot.). The 10-fold higher concentration of NAA than NAAG and the poor correlation in their
regional distributions does not seem to be consistent with the
hypothesis that NAA is merely a precursor or an inactive metabolite of NAAG.

        To assess the possible neuronal localization of NAA or NAAG,
the effects of selective brain lesions were examined. Unilateral
excitotoxin lesions of the corpus striatum were made by
stereotaxic injection of kainic acid. Three days after lesion,
the levels of NAAG were reduced by 30% bilaterally (p < 0.05)
whereas NAA was decreased by 58% on the injected side but only by
21% on the contralateral side. Unilateral decortication to
induce degeneration of the excitatory cortico-striatal pathway
resulted in a significant 28% decrease of NAAG ipsilateral to the
lesion but a nonsignificant 16% decrease on the contralateral
side. While the reduction in levels of NAA were greater on the

Table 1. Regional Levels of N-Acetyl-Aspartylglutamate in
          Rat Brain.

Area	NAAG nmole/mg prot.
Spinal cord	23 + 3
Medulla	16 + 1
Cerebellum	6 + 0.5
Thalamus	7 + 1
Neocortex	3 + 0.2
Striatum	2 + 0.2
Heart	N.D.
Liver	N.D.

Levels were measured by the isocratic anion exchange HPLC
method of Koller et al. (1984a) and are expressed as the
means of 6 animals + S.E.M.

side of the decortication (-48%), they mirrored the effects on
NAAG on the contralateral side (-16%). Since levels of NAAG are
highest in the spinal cord, it was of interest to determine
whether NAAG might be associated with ascending or descending
pathways. After a midthoracic section, the levels of NAAG were
consistently and significantly reduced from 37 to 51% in three
segments caudal to the lesion and from 25 to 40% in three
segments rostral to the lesion. In contrast, significant reductions in the levels of NAA were restricted to those segments clo-
sest to the transsection and were unaffected in the segments more
distant from the lesion. The results of these studies are consistent with the interpretation that NAAG is associated with
neuronal elements affected by these specific lesions. The findings with regard to NAA are more difficult to interpret in
regards to its neuronal association.

Interaction of NAAG and related acidic dipeptides with glutamate
receptors.

     The observation that initially brought NAAG to our attention
was the fact that it was a potent inhibitor (K_I = 400 nM) of a
subpopulation of the chloride-dependent specific binding sites
for [^{3}H]-Glu in rat forebrain membranes (Zaczek et al., 1984;
Koller and Coyle, 1984b). The percentage of chloride-dependent
binding sites for Glu displaced with high affinity by NAAG varies
among brain regions, and the absolute density of these NAAG
displaceable sites ranges over 10-fold in the subareas of the rat

Table 2. Potency and Efficacy of Acidic Peptides and
Other Excitatory Amino Acid Analogues in
Inhibiting the Binding of [3H]-L-Glutamic Acid
to Rat Brain Membranes.

Compound	K_I (nM)		% Inhibited	
	Site 1	Site 2	Site 1	Site 2
Glutamate	0.75		100	
Quisqualate	0.014	0.78	22	66
Ibotenate	2.0		88	
NAAG	0.23		66	
Asp-Glu	0.03	0.71	18	67
ɣ-Glu-Glu	0.006	0.55	16	78
α-Glu-Glu	0.13		64	
ɣ-Glu-Phe	2.7		78	
ɣ-Glu-Gly	1.4		75	
Glutathione	0.51		78	
Glu-Asp	>100		78	

Well washed membranes were prepared from rat forebrain and
were preincubated for 40 min. at 37°C. After centrifugation, the supernatant was removed and the tissue
resuspended in fresh 50 mM Tris-HCl (pH 7.1).
Displacement curves were run against 14 nM [3H]-L-glutamic
acid. Assay was incubated at 37°C for 40 min. and terminated by centrifugation. Eadee-Hofstee plots, from
which K_I and % inhibition values were estimated, were made
of composites of three separate displacement assays.

brain. In contrast, NAAG is ineffective in inhibiting the specific binding of [3H]-Glu to the chloride-independent sites
measured in the presence of Tris-citrate buffer (K_I > 100 mM).

Recently, we have found that many other acidic dipeptides
compete at the chloride-dependent binding sites labelled by
[3H]-Glu (Zaczek et al., 1985b). For example, aspartyl-glutamate
(Asp-Glu) and ɣ-glutamyl-glutamate inhibit binding of [3H]-Glu in
a biphasic manner (Table 2). These peptides exhibit a high affinity (K_I = 20 nM) to a small subpopulation (20%) of the [3H]-Glu
binding sites and a lower affinity (K_I= 600nM) for the remaining
sites. This pattern of biphasic displacement resembles that
exhibited by quisqualic acid. In contrast, α-glutamyl-glutamate
resembles NAAG in terms of its partial displacement with high

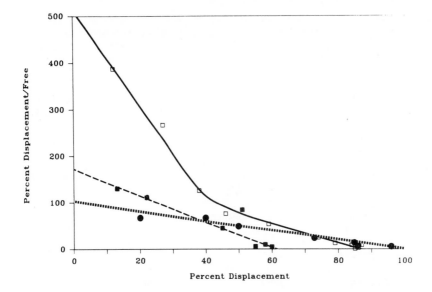

Figure 1. Membranes were prepared as in Table 2.
Displacement curves were performed against 14 nM [3H]-L-
Glutamate using: glutamate (●), NAAG (■) and Asp-Glu (□).

affinity affecting 60% of the [3H]-Glu binding sites (Figure 1).
Other γ-glutamyl peptides such as γ-glutamyl-glycine and glutathione are also effective but less potent inhibitors of specific binding of [3H]-Glu to the chloride-dependent sites.
However, with the exception of x-glutamyl-glutamate and aspartyl-
glutamate, the x-glutamyl peptides like α-glutamyl-aspartate and
x-glutamyl-alanine are rather ineffective inhibitors.

With the recent availability of [3H]-NAAG of high specific
radioactivity, it has been possible to characterize directly the
chloride-dependent subpopulation of Glu receptors with which this
ligand interacts (Koller and Coyle, 1984a,b). [3H]-NAAG binds in
a specific, saturable and reversible fashion. This ligand labels
an apparent single population of noninteracting sites with a Hill
coefficient of 1.0 and a K_D of 300 nM. Notably, the K_D for
[3H]-NAAG is similar to its K_I for the subpopulation of sites
labelled by [3H]-Glu, and the K_I for L-Glu at the binding site
for [3H]-NAAG is similar to its K_D at the [3H]-Glu sites. The
kinetic, pharmacologic and brain regional studies indicate that
[3H]-NAAG is labelling the subpopulation of receptor recognition
sites which are revealed by the displacement of the specific
binding of [3H]-Glu by endogeous or synthetic NAAG. The specific

binding site for [^{3}H]-NAAG exhibits a strict requirement for
chloride, is increased in density by preincubation with Ca^{++} and
is inhibited by Na$^+$. Whereas quisqualate is a potent displacer
at the [^{3}H]-NAAG receptor, kainic acid and N-methyl-D-aspartic
acid (NMDA) exhibit negligible affinities (K$_I$ > 100 uM). Of the
neurophysiologically effective inhibitors of excitatory amino
acid receptors, APB exhibits the highest affinity for the sites
labelled with [^{3}H]-NAAG, consistent with the specific effects of
APB in inhibiting the excitatory effects of the neurotransmitter
released by the LOT and iontophoretically applied NAAG on pyriform cortex pyramidal cells. The ionic and pharmacologic characteristics of the receptor labelled with [^{3}H]-NAAG is consistent
with the site designated as "A4" by Foster and Fagg (1984) and
recognized by others as the APB site.

Neurophysiologic effects of NAAG and Asp-Glu

The studies of Hori et al. (1981) suggested that neither Glu
nor Asp exhibited pharmacologic characteristics consistent with
the excitatory neurotransmitter released by the LOT.
french-Mullen et al. (1985) observed that NAAG, iontophretically
applied to the pyramidal cell neurons in the pyriform cortex
receiving LOT input, demonstrated excitatory effects that were
more potent than Glu, Asp or NMDA, equivalent to quisqualate and
less potent than kainic acid (ffrench-Mullen et al., 1985). The
onset and offset of the excitatory effects of NAAG were rapid,
which was consistent with a direct action of the peptide and
inconsistent with its proteolysis to an active metabolite.
Similar to the effects on the endogenous neurotransmitter
released by the LOT, the excitatory effects of NAAG iontophoretically applied to pyriform pyramidal cells were antagonized by bath applied APB which was ineffective against either
Glu or Asp. The possible localization of NAAG to the LOT pathway
was supported by the observation that olfactory bulbectomy, which
causes degeneration of the LOT, resulted in a significant reduction in the levels of NAAG in the pyriform cortex.

In hippocampal slice studies, NAAG was observed to be excitatory when pressure applied to 15 of 20 CA-1 pyramidal cells
(Bernstein et al., 1985). ɣ-Asp-Glu also excited 14 of 14 cells
when examined in this paradigm. While the excitatory potency of
Glu was similar in the stratum radiatum and the stratum pyramidale, both Asp-Glu and NAAG were more potent in the dendritic
tree of the pyramids (stratum radiatum), the area of termination
of the Schaffer collaterals and commiseral afferents than in the
stratum pyramidale. This differential distribution of sensitivity for the dipeptides on dendrites as compared to the uniform effects of Glu itself is consistent with a specific,
synaptic role of the dipeptides.

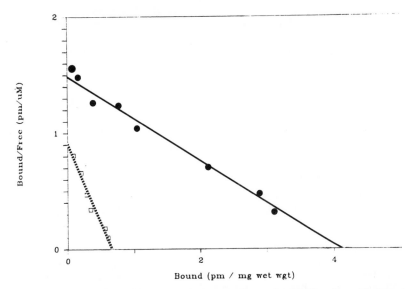

Figure 2. Membranes were prepared as in Table 2. Glutamate saturation isotherms were performed at 37°C for 40 min. in the presence (●) and absence (□) of 10 uM Phe-Glu.

Acidic dipeptides which enhance [3H]-Glu binding

/   Several dipeptides, of which phenylalanyl-glutamate (Phe-Glu) is a prototype, have been demonstrated to enhance the specific binding of [3H]-2-amino-7-phosphono-heptanoic acid, a potent inhibitor of excitation induced by NMDA, and to decrease the specific binding of [3H]-kainic acid to rat brain membranes (Ferkany et al., 1984). The ability of these dipeptides to increase the binding of [3H]-Glu has also been demonstrated (Zaczek et al., 1985a). Phe-Glu increases the binding of [3H]-Glu in a saturable (EC_{50} = 50 uM) and reversible manner. Figure 2 presents Scatchard transformation of saturation isotherms for [3H]-Glu in the presence and absence of Phe-Glu. The major effect of Phe-Glu on the specific binding of [3H]-Glu to chloride-dependent sites is a large increase in the B_{max} although a 3-fold decrease in the K_D (700 uM to 2.0 uM) was also observed. This augmentation of the specific binding of [3H]-Glu is temperature dependent with an optimum of 37°C. Phe-Glu also augments the binding of [3H]-Glu to the chloride independent sites as well as chloride dependent sites.

Table 3.   Stimulation of [^{3}H]-Glutamate Binding by
           Dipeptides

Compound	% Stimulation
Arg-Glu	329 $\pm$ 62
Phe-Glu	303 $\pm$ 59
Tryp-Glu	191 $\pm$ 42
Phe-Arg	90 $\pm$ 6
Pro-Glu	68 $\pm$ 6

Membranes were prepared from rat forebrain and pre-incubated for 40 min. at 37°C. After centrifugation, the supernatant was discarded and the membranes resuspended in fresh Tris-HCl buffer (pH 7.1). The membranes were then incubated in the presence of 1.0 uM [^{3}H]-L-Glutamate and 100 uM peptide at 37°C for 40 min. The assay was terminated by centrifugation. Values are expressed as the means of four experiments $\pm$ S.E.M.

Incubation with Phe-Glu discloses a distinctily different set of binding sites for [^{3}H]-Glu that are virtually undetectable in its absence. Several lines of evidence support this conclusion. First, the affinity of these sites is substantially lower in well washed and pre-incubated membranes in the presence of Phe-Glu as compared to its absence. Secondly, the pharmacology of the two sites also differ. While NAAG is a potent inhibitor of the specific binding of [^{3}H]-Glu to the chloride-dependent sites in the absence of Phe-Glu, NAAG acts as a very weak inhibitor ($K_I > 100$ mM) at the sites uncovered by Phe-Glu. Finally, the chloride-dependent, Phe-Glu-independent binding sites are present in the rat forebrain at birth and increase approximately 2-fold in density from birth to adulthood. In contrast, augmentation in the specific binding of [^{3}H]-Glu by Phe-Glu is not detectable before postnatal day 7; and between day 7 after birth and adulthood, the density of chloride-dependent, Phe-Glu stimulated binding sites for [^{3}H]-Glu increases 6-fold.

Other peptides having L-Glu at the carboxy terminus also stimulate the binding of [^{3}H]-Glu (Table 3). These include arginyl-glutamate, tyrosyl-glutamate and tryptophanyl-glutamate but not prolyl-glutamate or aspartyl-glutamate. The existence of an endogenous substance in brain with Phe-Glu-like characteristics has been suggested by pre-incubation experiments. Under standard conditions in which rat forebrain membranes are washed extensively, subjected to a 40 min. pre-incubation at 37°C, then pelleted and washed again, saturation isotherms of [^{3}H]-Glu under chloride conditions reveal a single class of sites (Hill coefficient = 1.0) with an apparent K_D of 520 nM. With forebrain membranes not subjected to the pre-incubation at 37°C, saturation isotherms reveal biphasic binding on Scatchard transformation (Hill coefficient = 0.5) with two apparent sites for [^{3}H]-Glu with K_D's respectively of 400 nM and 5.2 uM. In addition to eliminating the lower affinity site, membrane pre-incubation at 37°C reduces the B_{max} for [^{3}H]-Glu and increases the percentage of NAAG displaceable sites. The most parsimonious explanation for these findings is that pre-incubation at 37°C results in the dissociation or degradation of a functional analog of Phe-Glu, which is essential for the binding of [^{3}H]-Glu to the lower affinity site.

DISCUSSION

Considerable circumstantial evidence supports the role of Glu and Asp in excitatory neurotransmission in a number of pathways within the brain. These amino acids are found in high concentration in brain; there is a neuronally localized sodium-dependent, high affinity process for Glu/Asp; and Glu/Asp are released from specific pathways by a calcium-dependent process upon depolarization. Finally, Glu and Asp are agonists at specific receptors that have been defined by neurophysiologic and ligand binding techniques. Nevertheless, the conclusion that Glu/Asp are the excitatory neurotransmitters must be viewed with caution since the same criteria would support the conclusion that choline and not acetylcholine is the neurotransmitter in brain. Thus, neuronal specific, sodium-dependent high affinity uptake process exists for choline; in the absence of an acetylcholinesterase inhibitor, a calcium-dependent depolarization-induced release of [^{3}H]-choline occurs; and choline is an agonist at both muscarinic and nicotinic cholinergic receptors. Accordingly, there is reason to speculate that Glu and Asp might serve as precursors analogous to choline for rapidly degraded acidic peptides, which serve as the neurotransmitters in various excitatory pathways.

Data accumulated thus far support the hypothesis that NAAG may be one of a larger family of small acidic peptides that serve as excitatory neurotransmitters in brain. NAAG is restriced to brain, appears to be concentrated in neurons and exhibits a high affinity for a subpopulation of Glu receptors. Furthermore, NAAG has excitatory effects when iontophoretically applied to dendritic fields of pyramidal cells in the hippocampus and in the pyriform cortex. In the case of the pyriform cortex, the pharmacology of NAAG more closely resembles that of the excitatory neurotransmitter released by the LOT than do either Glu or Asp. The markedly uneven regional brain distribution of NAAG suggest that it is restricted to a subpopulation of excitatory systems, primarily localized to hindbrain and spinal cord, and that other substances are more likely candidates for forebrain excitatory systems. Indeed, the brain is rich in small acidic peptides containing Glu and/or Asp. For example, Kanazawa and Sutoo (1981) have reported that β-hydroxybutyrylaspartyl-aspartylglutamate, a tripeptide localized in spinal pathways, exerts excitatory effects on spinal neurons.

The effects of Phe-Glu and related dipeptides on the specific binding sites of [^{3}H]-Glu suggests that both chloride-dependent as well as chloride-independent Glu receptors may be subject to allosteric regulation. The appearance of a pharmacologically and kinetically distinct binding sites for [^{3}H]-Glu in the presence of Phe-Glu suggests that Phe-Glu may transform a population of receptors whose affinity is too low to be detected in ligand binding assays to a higher affinity state (e.g. K_D = 2-5 uM) that are measurable under these assay conditions. Since this population of Glu receptors can be eliminated by pre-incubation of forebrain membranes at 37°C, it appears that there may be endogenous substances in brain that allosterically regulate Glu receptors in a fashion analogous to Phe-Glu. However, it is not possible, on the basis of the ligand binding results, to determine whether Phe-Glu would enhance or inhibit the excitatory action of Glu. This issue is currently under investigation. Nevertheless, these findings suggest that excitatory receptors may be allosterically regulated similar to the interaction of benzodiazepines with the inhibitory GABA receptor. *

REFERENCES

Bernstein, J., Fisher, R.S., Zaczek, R. and Coyle, J.T. (1985). Dipeptides of Glutamate and Aspartate May Be Endogenous Neuroexcitatnts in the Rat Hippocampal Slice. J. Neurosci., in press.

Ferkany, J.W. and Coyle, J.T. (1983). Kainic Acid Selectively Stimulates the Release of Endogenous Excitatory Acidic Amino Acids. J. Pharmacol. Exp. Ther., 225, 399-406.

Ferkany, J.W., Zaczek, R., Markl, A. and Coyle, J.T. (1984). Glutamate-Containing Dipeptides Enhance the Specific Binding at Glutamate Receptors and Inhibit Specific Binding at Kainate Receptors in Rat Brain. Neurosci. Letts., 44, 281-286.

*   Acknowledgement. This research was supported by USPHS grants RSDA Type II MH00125 and NS13584.

Foster, A.C. and Fagg, G.E. (1984). Acidic Amino Acid Binding Sites in Mammalian Neuronal Membranes: Their Characteristics and Relationship to Synaptic Receptors. Brain Res. Rev., 7, 103-164.

ffrench-Mullen, J.M.H., Koller, K., Zaczek, R., Coyle, J.T., Hori, N. and Carpenter, D.O. (1985). N-Acetylaspartylglutamate: Possible Role as the Neurotransmitter of the Lateral Olfactory Tract. Proc. Natl. Acad. Sci. USA, in press.

Hori, N., Auter, C.R., Braitman, D.J. and Carpenter, D.O. (1981). Lateral Olfactory Tract Transmitter: Glutamate, Aspartate or Neither? Cell. Mol. Neurobiol. 1, 115-120.

Kanazawa, I. and Sutoo, D. (1981). Decrease of a Peptide in the Cat Spinal Cord After Upper Cervical Transsection. Neurosci. Letts. 26, 113-117.

Koller, K.J., Zaczek, R. and Coyle, J.T. (1984a). N-Acetyl-Aspartyl-Glutamate: Regional Levels in Rat Brain and the Effects of Brain Lesions as Determined by a New HPLC Method. J. Neurochem. 43, 1136-1142.

Koller, K.J. and Coyle, J.T. (1984b). Characterization of the Interactions of N-Acetyl-Aspartyl-Glutamate with [^{3}H]L-Glutamate Receptors. Eur. J. Pharmacol., 98, 193-199.

Koller, K.J. and Coyle, J.T. (1985). The Characterization of the Specific Binding of [^{3}H]N-Acetyl-Aspartyl-Glutamate to Rat Brain Membranes. J. Neurosci., in press.

Lenda, K. (1981). Ion Exhanger Liquid Chromatography of N-Acetyl-Aspartic Acid and Some N-Acetyl-Aspartyl Peptides. J. Liq. Chromatog. 4, 863-869.

Miyake, M., Kakimoto, Y. and Sorimachi, M. (1981). A Gas Chromatographic Method for the Determination of N-Acetyl-Aspartic Acid, N-Acetyl-Aspartylglutamic Acid and Beta-Citryl-L-Glutamic Acid and Their Distributions in the Brain and Other Organs of Various Species of Animals. J. Neuochem. 36, 804-810.

Reichert, K.L. and Fonnum, F. (1969). Subcellular Localization of N-Acetyl-Aspartyl-Glutamate, N-Acetyl-Glutamate and Glutathione in Brain. J. Neurochem. 16, 1409-1416.

Tallon, H.H., Moore, S. and Stein, W.H. (1956). N-Acetyl-L-Aspartic Acid in Brain. J. Biol. Chem., 224, 257-264.

Zaczek, R., Koller, K.J., Cotter, R., Heller, D. and Coyle, J.T. (1983). N-Acetyl-Aspartyl-Glutamte: An Endogenous Peptide With

High Affinity for a Brain "Glutamate" Receptor. Proc. Natl.
Acad. Sci. U.S.A., 80, 1116-119.

Zaczek, R., Arlis, S., Markl, A., Murphy, T. and Coyle, J.T.
(1985a). Effects of Glutamate Containing Dipeptides on the
Binding of [^{3}H]-L-Glutamate Binding to Rat Brain Membranes. Abs.
Soc. Neurosci., 11.

Zaczek, R., Arlis, S. Murphy, T. and Coyle, J.T. (1985b).
Peptide Interactions with Brain Excitatory Amino Acid Receptor
Binding Sites. In preparation.

27

Extracellular Concentration of Excitatory Amino Acids: Effects of Hyperexcitation, Hypoglycemia and Ischemia

A. Hamberger, S.P. Butcher, H. Hagberg, I. Jacobson, A. Lehmann

and M. Sandberg

Introduction

Although acidic amino acids are widely believed to act as chemical transmitters at a number of synapses in the brain, the precise nature of the transmitter substance remains unclear. Several researchers have attempted to evaluate the identity of the neurotransmitter utilised in these pathways by examining the in vitro release of transmitter candidates such as glutamate and aspartate (Fagg and Lane, 1979; Fagg and Foster, 1983; Fonnum et al., 1983). This approach has indeed provided some relevant information especially when used in combination with lesion experiments. However, any extrapolation of in vitro data to the in vivo situation is fraught with difficulties, and in vivo neurochemical studies will clearly be necessary before a physiological role for any putative transmitter can be confirmed.

There is also evidence to suggest that rapid flooding of the extracellular compartment with excitatory transmitters is intimately linked with the induction of cell injury found in neurological disorders, such as epilepsy, hypoglycemia and ischemia . In this respect, the neurotoxic properties of certain acidic amino acids are well documented and these endogenous compounds may therefore play a central role in the complex sequence of events, involving Ca^{2+} ions, proteolytic activity and changes in high energy compounds, responsible for cell death in these pathophysiological situations.

Recent advances in methods for sampling the extracellular fluid surrounding neurones have made the investigation of these two aspects of excitatory amino acid function possible under in vivo conditions. This chapter deals mainly with our studies using the in vivo brain dialysis technique for monitoring extracellular amino acid levels. Methodological aspects of this technique have been described in two earlier reviews (Hamberger et al., 1983; 1985).

Stimulation of specific pathways

Depolarising stimuli, such as veratridine and high potassium concentration, increase the in vivo extracellular content of glutamate and aspartate in several brain areas, including the hippocampus and the olfactory bulb (Lehmann et al., 1983; Jacobson and Hamberger, 1984). These findings are wholly consistent with the proposed transmitter role of glutamate and/or aspartate in these structures (Cotman et al., 1981; Halasz and Shepherd, 1983). However, further refinements are required to permit the study of transmitter release in response to physiological stimuli. In this respect, we have recently monitored the extracellular content of amino acids in the terminal regions of several presumably glutamatergic/aspartergic pathways following electrical stimulation of the relevant neuronal input:

a) Olfactory bulb neurons activated by stimulation of the lateral olfactory tract.
b) The descending cortical input into the lateral geniculate nucleus.
c) The perforant path input into the dentate gyrus of the hippocampal formation.

In each case, recording electrodes placed alongside a dialysis probe were used to confirm the correct positioning of the device within the terminal field (Butcher and Jacobson, unpublished data; Sandberg and Lindström, 1983; Jacobson and Hamberger, 1984). This approach permits the direct correlation of discrete electrophysiological and neurochemical data.

Upon stimulation of these pathways, using stimulus parameters sufficient to induce electrophysiological events, no consistent pattern of amino acid release could be demonstrated. In some cases, relatively small changes in extracellular amino acid contents were detected and these appeared to be restricted to neuroactive amino acids such as glutamate, aspartate, GABA and taurine. However, the inconsistent nature of these responses suggest that, without resorting to the use of chemical or metabolic manipulations of brain tissue surrounding the sampling probe, neurochemical methods based on dialysis techniques are not suitable for monitoring transmitter release in vivo.

It is only possible to speculate about the factor(s) responsible for the inability of dialysis probes to detect acidic amino acids transmitter release in "pseudo-physiological" circumstances. The possibility that neither glutamate nor aspartate is the transmitter in these pathways cannot be discounted. However, a more likely explanation would appear to be the presence of avid, high affinity uptake mechanisms which restrict the diffusion of neuroactive amino acids beyond the synaptic cleft region, a prerequisite for the presently used technique. Chemical stimuli, such as veratridine and potassium, most probably "overload" the uptake processes since they cause a massive, generalised depolarisation. In order to deal with this problem we are at present examining a number of approaches to decrease, but not abolish, the capacity of the uptake system.

Chemically induced seizures

A majority of studies on the effect of sustained seizures on the content of amino acids in the CNS have been concerned with total tissue content. However, there appears to be no standardized pattern of changes during status epilepticus. Instead, alterations in amino acid content seem to be dependent on the convulsant drug employed (Chapman et al., 1977; 1984; Nitsch et al., 1983; van Gelder et al., 1983). It is obvious that extracellular rather than intracellular amino acid concentrations represent the critical factor. Although there is a paucity of information in this respect, a few reports on the liberation of amino acids from the electrically kindled amygdala (Peterson et al., 1983) and from epileptic foci (Dodd and Bradford, 1976; Dodd et al., 1980; Koyama, 1972) have appeared in the literature.

In an attempt to relate extracellular neurochemical events to status epilepticus we have monitered extracellular amino acids in the hippocampus during sustained seizures induced acutely with systemic administration of kainic acid (KA) or bicuculline (BC). KA-induced epilepsy was associated with marked increases in phosphoethanolamine (PEA) and taurine levels (Lehmann et al., 1985). Alanine and ethanolamine were moderately raised whereas other amino acids were unaffected. Flooding of the interstitial space with taurine may be an autoprotective mechanism against KA toxicity since taurine prevents KA induced neuronal degeneration in the hippocampus (Fariello et al., 1982) and in the striatum (Sandberg et al., 1979). The reduction in the total tissue content of taurine in the rat hippocampus found after i.v. administration of KA indicates that the increase in extracellular levels is derived from intracellular compartments. BC seizures, in contrast to KA seizures, had no effect on taurine whereas significant increments were observed for PEA and alanine. This finding is noteworthy because BC has the opposite effect as KA on total hippocampal taurine (Chapman et al., 1984).

The unaltered levels of extracellular glutamate and GABA during either BC or KA seizures is rather surprising in view of the repeated demonstrations of increased glutamate release from epileptic tissue (Dodd and Bradford, 1976; Dodd et al., 1980; Koyama, 1972; Peterson et al., 1983) and the established antiepileptic effect of antagonists to excitatory amino acid receptors (Meldrum 1984). However, our findings do not preclude the possibility that glutamate and/or aspartate are released from presynaptic pools during epileptic discharges because the negative result obtained with the dialysis system may again be due to efficient uptake mechanisms.

Effects of locally applied kainic acid

Powerful excitatory and toxic actions of KA have been described in several brain areas. This effect may be associated with amino acid efflux because KA is reported to enhance the release of both aspartate and glutamate from brain slices in a Ca^{2+}-dependent (Ferkany and Coyle, 1983) and TTX insensitive fashion (Ferkany and Coyle, 1983; Krespan et al., 1982) This finding has led to the

Table I.

Effects of kainic acid on extracellular amino acid concentrations
in the rabbit hippocampus and olfactory bulb.

amino acid	percent increase	
	hippocampus	olfactory bulb
Asp	n.m	80 ± 33
Glu	45 ± 7	47 ± 14
PEA	242 ± 150	84 ± 40
Tau	400 ± 126	86 ± 49
GABA	n.m	79 ± 65

Effects of KA on extracellular amino acids in vivo. The results
are expressed as percent change compared to basal levels. In the
olfactory bulb the basal level was determined as the mean of the
two consecutive fractions immediately before superfusion with KA.
Five min fractions were collected and the flow rate was 6.5 μl/
min. Total superfusion time was 130 min including a 10 min period
with 1 mM KA, n=4, Mean + S.E.M. (Anesthetized animals).
In the hippocampus baseline values were obtained by perfusion with
control medium for 2 h, during which time three 10-min fractions
were collected. The animals were awake and freely moving when the
perfusates were collected (2.5 μl/min). The data refer to the
first ten min of 1 mM KA perfusion. n=5, Mean ± S.E.M.

suggestion that KA may act at two distinct loci (Ferkany and
Coyle, 1983); one postsynaptic site which accounts for the direct
excitatory action of KA and one presynaptic site which is responsible for the release of excitatory amino acids. The existence of
presynaptic KA receptors would explain the finding that the neurotoxic effect of KA is dependent on glutamatergic synaptic inputs
in some brain areas (Biziere and Coyle, 1978; McGeer et al., 1978;
Streit et al., 1980). We have found that KA selectively enhances
the efflux of aspartate and glutamate from olfactory bulb slices
(Jacobson and Hamberger, 1985). However, in vivo studies with the
brain dialysis technique suggest that the acute (within 10 min)
response to KA involves not only the excitatory amino acids glutamate and aspartate, but also GABA, PEA and taurine. These effects
are similar in the hippocampus and in the olfactory bulb (Table
I). The relatively small increases in aspartate, glutamate and
GABA found in these studies may be due to efficient in vivo cellular uptake mechanisms for these amino acids, in contrast to the

case for taurine (Richelson and Thompson, 1973; Schrier and Thompson, 1977; Kontro and Oja, 1978).

However, the suggestion that KA evokes a significant release of the excitatory amino acids exclusively from nerve terminals remains controversial (Krespan et al., 1982; Pastuszko et al., 1984; Poli et al., 1984) and it has been suggested that KA released glutamate is of glial origin (Krespan et al., 1982).

The saturated derivative of KA, dihydrokainic acid (DKA), exhibits a different profile of biological actions. The excitatory (Johnston et al., 1978) and toxic (Nadler et al., 1981) properties of KA are less pronounced, whereas DKA is a more potent inhibitor of high-affinity uptake of acidic amino (Johnston et al., 1979). This compound increases extracellular aspartate and glutamate nearly as efficiently as KA. In fact, prolonged perfusion with DKA has a similar effect on extracellular amino acids as KA (Lehmann and Hamberger, 1983). The effect of DKA on extracellular aspartate and glutamate clearly indicates that a block of amino acid reuptake might explain part of the response to KA in vivo.

Hypoglycemia and extracellular amino acids

Severe hypoglycemia (blood glucose below 1 mM) leads to irreversible cell-loss in brain areas such as the cortex, corpus striatum and hippocampus (Auer et al., 1984). Although the normal adult brain almost exclusively uses glucose as the source of energy, alternative substrates appear to be utilized when blood glucose falls below a certain level (Agardh et al., 1981; Siesjö and Agardh, 1983). For example, tissue levels of amino acids are altered probably due to their use as substrates for the citric acid cycle (Dawson, 1950; Tews et al., 1965; Norberg and Siesjö, 1976; Butterworth et al., 1982; Behar et al., 1985). Aspartate is greatly enhanced during severe hypoglycemia whereas glutamine, glutamate and GABA are reduced. These changes may be involved in the mechanism underlying the disturbed EEG pattern observed during progressive hypoglycemia (Butterworth et al., 1982; Norberg and Siesjö, 1976). Since direct injection of aspartate (or glutamate) results in neuronal loss in the hippocampus (Nadler et al., 1981; Mangano and Schwarcz, 1983), an overflow of aspartate from the increased intracellular pool may be involved in the aethiology of hypoglycemic cell death (Auer et al., 1984). In order to test this hypothesis, changes in the extracellular amino acids were monitored in the hippocampus during the following conditions:

a) Local blockage of the glycolysis by perfusion of the hippocampus with iodoacetic acid in vivo (IAA, Sandberg et al.,1985a). At low concentrations IAA is an inhibitor of glycolysis (Webb, 1966).

b) Insulin-induced hypoglycemia (Sandberg et al., 1985b).

Our results demonstrate that the extracellular concentration of neuroactive amino acids, particularly aspartate, is increased in both situations (Table II). The increase in the neuroactive amino acids may reflect reduced uptake capacity due to lowered energy

Table II

Effects of iodoacetate and severe insulin-induced hypoglycemia on
extracellular amino acids in the rabbit and rat hippocampus.

Amino acid	Iodoacetate	Severe hypoglycemia
Asp	805 + 220	1550 + 395
Glu	95 + 30	265 + 55
PEA	350 + 40	305 + 85
TAU	220 + 45	450 + 100
GABA	n.m.	260 + 50

The results are expressed as mean percent change + S.E.M. compared to basal levels after 90 min of perfusion with iodoacetate
(0.1 mM, rabbit hippocampus) and after 30 min of isoelectric EEG
induced by severe hypoglycemia (rat hippocampus) n=5-6.

reserves in combination with the depolarisation which occurs during severe hypoglycemia (Wieloch et al., 1984).
Several queries arise from these data:

1) Is the increase in extracellular aspartate (and glutamate)
   related to toxicity?
2) If an excitotoxic effect of extracellular aspartate (and glutamate) is involved in the neuronal loss observed in ischemia
   and hypoglycemia, why does the distribution pattern of injury
   differ (Auer et al., 1984; Smith et al., 1984) ?
3) What is the mechanism behind the possible excitotoxic effect?

Auer and coworkers (1984) have proposed previously that Ca^{2+},
an endogenous amino acid or some yet unknown substance in the CSF
and tissue fluid might be involved in hypoglycemic cell-loss.
Assuming that extracellular aspartate (and glutamate) is excitotoxic under these conditions, important factors concerning neuronal susceptibility may be the ability to use low concentrations
of glucose (La Manna and Harik, 1985), the relative activity of
enzymes involved in the metabolism of amino acids and regional
differences in the efflux of excitotoxic amino acids and receptor
populations. Aspartate amino transferase (AspAT) activity in the
hippocampus is high in the dentate layer (Schmidt and Wolf,
1984), an area which is more susceptible in hypoglycemia compared
with ischemia (Auer et al., 1984; Smith et al., 1984). Thus, in
some areas there may be an interrelation between toxicity and
AspAT activity. The mechanism of aspartate (and glutamate) toxicity is not known. Since application of acidic amino acids to

hippocampal slices induces a drop in extracellular Ca^{2+} (Zanotto and Heinemann, 1983), an increase in the intracellular levels of Ca^{2+}, caused by the overflow of aspartate observed during hypoglycemia, may be one factor involved in hypoglycemic cell injury. It is interesting to note that the effects of aspartate and glutamate on cation fluxes are reduced by GABA (Zanotto and Heinemann, 1983), another neuroactive amino acid released during hypoglycemia. The pattern of cell loss induced by hypoglycemia may therefore, in part, be determined by the relative density of excitatory and inhibitory receptors on affected neurons.

Ischemia and extracellular amino acids

Attention has recently focused on the possible role of excitatory amino acids in the pathophysiology of ischemic brain damage (Jörgensen and Diemer, 1982; Rothman, 1984; Simon et al., 1985; Siesjö and Wieloch, 1985). Ischemia-induced loss of neurons appears to occur selectively in brain regions recieving afferent projections in which glutamate or aspartate are the putative neurotransmitters (Jörgensen and Diemer, 1982). Cell damage is primarly localized in post-synaptic elements, and thus exhibits the same distribution as that seen after administration of the glutamate analogues, NMDA or KA (Diemer et al., 1983; Fryd Johansen et al., 1983; Olney 1983).

Recent evidence suggests that excitatory amino acids are released in response to ischemia, as recorded with in vivo brain dialysis (Benveniste et al., 1984, Hagberg et al., 1985). Utilizing the 4-vessel occlusion model in rabbits, 10 min of ischemia evoked a 4-fold elevation in glutamate and a 2-fold increase in aspartate (Fig. 1). Circulatory arrest for 10 min was followed by a complete recovery of high-energy phosphates and EEG activity. The outflow of excitatory amino acids increased progressively when ischemia was prolonged further. Glutamate and aspartate increased by 160 and 30 times respectively after 30 min of circulatory occlusion (Fig. 1). Ischemia of 30 min duration was only partially reversible with regard to high energy compounds and EEG activity (Hagberg et al., 1985).

The assumption that extracellular overflow of excitotoxic amino acids could be relevant in ischemic cell death receives further support from recent studies with different glu/asp-receptor antagonists: Preischemic blockage of the NMDA receptors by in vivo focal injections of 2-amino-7-phosphoheptanoic acid (APH) attenuates ischemic cell injury in all hippocampal regions (Simon et al. 1984). Furthermore, administration of gamma-D-glutamylglycine, a blocker of NMDA and KA receptors, protects against anoxia-induced damage in neuronal cultures (Rothman 1984).
It is not known whether outflow of excitatory amino acids during ischemia (Benveniste et al., 1984) or burst firing after ischemia constitute the critical event (Suzuki et al. 1983). The extracellular content of excitatory amino acids returns to normal within 30 min of reflow and remains unaltered during 4 h of reperfusion (Fig. 1). However, receptor activation with influx of Ca^{2+} during ischemia may trigger alterations in protein

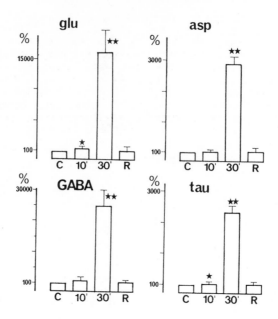

<u>Fig. 1. Extracellular amino acids during ischemia in rabbit hippocampus.</u>
Global ischemia was induced by carotid clamping after electro-cauterization of the vertebral arteries. The extracellular amino acid levels were recorded with the <u>in vivo</u> dialysis technique. Extracellular glutamate (glu), aspart<u>ate (</u>asp), GABA and taurine (tau) levels after 10 min ischemia (10'), 30 min-ischemia (30') and 30 min of reperfusion (R) given as percent of control + SEM. Levels of significance (vs. control): *p ≤ 0.05, **p $\leq 0.0\overline{1}$;(n = 5-7).

phosphorylation or initiate proteolytic processes which could lead to cell death several days after the ischemic period (Siesjö and Wieloch, 1985). Post-synaptic alterations triggered by ische-mia might, for example, cause an increase in the glutamate recep-tor population with a resulting overactivation upon resumption of electrical activity (Bodsch et al., 1984).

The ischemia-evoked amino acid overflow was not limited to excitatory compounds (Fig. 1). Taurine and GABA increased by 20 and 250 times the basal level, respectively. Outflow of inhibit-ory amino acids may to some extent counterbalance the proposed excitotoxic effects of glutamate/aspartate (Zanotto and Heine-mann, 1983; Berdichevsky et al., 1983). Further studies are re-quired to establish whether an interaction between excitatory and inhibitory amino acids during and/or after ischemia is connected with the induction of ischemic injury.

Conclusion

Our studies demonstrate that neuroactive amino acids are released into the extracellular compartment during KA perfusion, hypoglycemia and ischemia. This finding is of particular interest because the increase in extracellular levels of excitotoxic amino acids, such as glutamate and aspartate, may be associated with subsequent neuropathological damage. In contrast, acidic amino acid efflux could not be detected in two situations (electrical stimulation of glutamatergic/aspartergic pathway and status epilepticus) where it might be expected. This may be due, at least in part, to the presence of acid uptake mechansims which prevent any overflow of neuroactive amino acids into the extracellular compartment. Despite this potential limitation, the in vivo dilaysis fibre technique is proving to be a powerful tool for the study of many aspects of amino acid function in the brain.

References

Auer, R.N., Wieloch, T., Olsson, Y. and Siesjö, B.K. (1984). The distribution of hypoglycemic brain damage. Acta Neuropathol. (Berl), 64, 177-191.

Agardh C.-D., Chapman A.G., Nilsson B. and Siesjö B.K. (1981). Endogenous substrates utilized by rat brain in severe insulin-induced hypoglycemia. J.Neurochem., 36, 490-500.

Behar, K.L., den Hollander, J.A., Petroff, O.A.C., Hetherington, H.P., Prichard, J.W. and Shulman, R.G. (1985). Effect of hypoglycemic encephalopathy upon amino acids, high-energy phosphates, and pHi in the rat brain in vivo: Detection by sequential 1H and 32P NMR spectroscopy. J. Neurochem., 44, 1045-1055.

Benveniste, H., Drejer, J., Schousboe, A. and Diemer, N.H. (1984). Elevation of the extracellular concentrations of glutamate and aspartate in rat hippocampus during transient cerebral ischemia monitored by intracerebral microdialysis. J. Neurochem. 43, 1369-1374.

Berdichevsky, E ., Riveros, N., Sánchez-Armáss, S. and Orrego, F. (1983). Kainate, N-methylaspartate and other excitatory amino acids increase calcium influx into rat brain cortex cells in vitro. Neurosci.Lett., 36, 75-80.

Biziere, K. and Coyle, J.T. (1978). Influence of cortico-striatal afferents on kainic acid neurotoxicity. Neurosci.Lett., 8, 303-310.

Bodsch, W., Takahashi,K. and Hossmann, K.-A. (1984). Functional implications of glutamate receptor and protein phosphorylation in protein synthesis regulation after transient forebrain ischemia. Abstract. Proc. of the 5th Meeting of Eur.Soc.for Neurochem., Budapest, Hungary,21-26 August,1984, p 35.

Butterwoth, R.F., Merkel, A.D. and Landreville, F. (1982). Regional amino acid distribution in relation to function in insulin hypoglycemia. J. Neurochem., 38, 1483-1489.

Chapman, A.G., Meldrum, B.S.and Siesjö, B.K. (1977). Cerebral metabolic changes during prolonged epileptic seizures in rats. J. Neurochem., 28, 1025-1035.

Chapman, A.G., Westerberg, E., Premachandra, M. and Meldrum, B.S. (1984). Changes in regional neurotransmitter amino acid levels in rat brain during seizures induced by L-allylglycine, bicuculline and kainic acid. J. Neurochem., 43, 62-70.

Cotman, C.W., Foster, A. and Lanthorn, T. (1981). An overview of glutamate as a neurotransmitter. In:Glutamate as a Neurotransmitter (eds. G. Di Chiara and G.L. Gessa) Raven Press, New York, pp 1-28.

Dawson, R.M.C. (1950) Studies on the glutamine and glutamic acid content of rat brain during insulin hypoglycemia. Biochem. J., 47, 386-391.

Diemer, N.H., von Lubitz, D.E., Fryd Johansen, F., Balslev Jörgensen, M., Benveniste, H., Drejer, J. and Schousboe, A. (1983) Ischemic damage of hippocampal CA1 neurons. Possible neurotoxicity of glutamate released during ischemia. Acta Neurol. Scand. 61-69.

Dodd, P.R. and Bradford, H.F. (1976). Release of amino acids from the maturing cobalt-induced epileptic focus. Brain Res., 111, 377-388.

Dodd, P.R., Bradford, H.F., Abdul-Ghani, A.S., Cox, D.W.G. and Continho-Netto, J. (1980). Release of amino acids from chronic epileptic and subepileptic foci in vivo. Brain Res., 193, 505-517.

Fagg, G.E. and Foster, A.C. (1983). Amino acid neurotransmitters and their pathways in the mammalian central nervous system. Neurosci., 9, 701-709.

Fagg, G.E. and Lane, J.D. (1979). The uptake and release of putative amino acid neurotransmitters. Neurosci., 4, 1015-1036.

Fariello, R.G., Golden, G.T. and Pisa, M. (1982). Homotaurine (3 aminopropanesulfonic acid; 3APS) protects from the convulsant and cytotoxic effect of systemically administered kainic acid. Neurology(Ny), 32, 241-245.

Kontro, R. and Oja, S.(1978). Taurine uptake by rat brain synapto-somes. J. Neurochem. 30, 1297-1304.

Koyama, I. (1972). Amino acids in the cobalt-induced epileptogenic and nonepileptogenic cat's cortex. Can. J. Physiol. Pharmacol., 50, 740-752.

Krespan, B., Berl, S. and Nicklas, W.J. (1982). Alteration in neuronal-glial metabolism of glutamate by the neurotoxin kainic acid. J. Neurochem., 38, 509-518.

La Manna, J.C. and Harik, S.I. (1985). Regional comparisons of brain glucose influx. Brain Res., 326, 299-305.

Lehmann, A. and Hamberger, A. (1983). Dihydrokainic acid affects extracellular taurine and phosphoethanolamine levels in the hippo-campus. Neurosci.Lett., 38, 67-72.

Lehmann, A., Isacsson, H. and Hamberger, A. (1983). Effects of in vivo administration of kainic acid on the extracellular amino acid pool in the rabbit hippocampus. J. Neurochem., 40, 1314-1320.

Lehmann, A, Hagberg, H., Jacobson I. and Hamberger A. (1985). Effects of status epilepticus on extracellular amino acids in the hippocampus. Brain Res.,In Press.

Mangano, R.M. and Schwarcz, R.M. (1983). Chronic infusions of endogenous excitatory amino acids into rat striatum and hippocam-pus. Brain Res. Bull., 10, 47-51.

McGeer, E.G., McGeer, P.L. and Singh, K. (1978). Kainate-induced degeneration of neostriatal neurons: dependency upon cortico-striatal tract. Brain Res., 139, 381-383.

Meldrum, B.S. (1984). Amino acid neurotransmitters and new approa-ches to anticonvulsant drug action. Epilepsia, 25 (suppl. 2), 5140-5149.

Nadler, J.V., Evenson, D.A. and Cuthbertson, G.J. (1981). Compara-tive toxicity of kainic acid and other acidic amino acids toward rat hippocampal neurons. Neurosci., 6, 2505-2517.

Nitsch, C., Schmude, B. and Haug, P. (1983). Alterations in the content of amino acid neurotransmitters before the onset and during the course of methoxypyridoxine-induced seizures in indivi-dual brain regions. J. Neurochem., 40, 1571-1581.

Norberg, K. and Siesjö B.K. (1976). Oxidative metabolism of the cerebral cortex of the rat in severe insulin-induced hypoglycemia. J. Neurochem., 26, 345-352.

Olney, J.W. (1983) Excitotoxins: An overview. In: Excitotoxins. (Eds. K. Fuxe, P. Roberts and R. Schwarcz) The MacMillan Press Ltd.,London.

Ferkany, J.W. and Coyle, J.T. (1983). Kainic acid selectively stimulates the release of endogenous excitatory acidic amino acids. J. Pharmacol. Exper. Ther., 225, 399-406.

Fonnum, F., Fosse, V.M. and Allen, C.N. (1983). Identification of excitatory amino acid pathways in the mammalian nervous system. In: Excitotoxins. (eds. K. Fuxe, P. Roberts and R. Schwarcz) MacMillan Press, London, pp 1-18.

Fryd Johansen., F., Balslev Jörgensen, M. and Diemer, N.H. (1983) Resistance of hippocampal CA-1 interneurons to 20 min of transient cerebral ischemia in the rat. Acta Neuropathol. 61, 135-140.

Hagberg, H., Lehmann, A., Sandberg, M., Nyström, B., Jacobson, I. and Hamberger, A. (1985). Ischemia-induced shift of inhibitory and excitatory amino acids from intrato extracellular compartments. J Cereb Blood Flow Metabol, In Press.

Halász, N. and Shepherd, G.M. (1983). Neurochemistry of the vertebrate olfactory bulb. Neurosci., 10, 579-619.

Hamberger, A., Berthold, C.-H., Karlsson, B., Lehmann, A. and Nyström, B. (1983). Extracellular GABA, glutamate and glutamine in vivo - perfusion-dialysis of the rabbit hippocampus. In:Glutamine, glutamate and GABA in the central nervous system. (Eds. L. Hertz, E. Kvamme, E.G. McGeer and A. Schousboe). Alan R. Liss, New York, pp 473-492.

Hamberger, A., Berthold, C.-H., Jacobson, I., Karlsson, B., Lehmann, A., Nyström, B. and Sandberg, M. (1985). In vivo brain dialysis of extracellular nontransmitter and putative transmitter amino acids. In:In vivo perfusion and release of neuroactive substances. (eds. A. Bayon and R. Drucker-Colin). Academic Press Inc., New York,pp 119.

Jacobson, I. and Hamberger, A. (1984). Veratridine induced release in vivo and in vitro of amino acids in the rabbit olfactory bulb. Brain Res., 299, 103-112.

Jacobson, I. and Hamberger, A. (1985). Kainic acid-induced changes of extracellular amino acid levels, evoked potentials and EEG activity in the rabbit olfactory bulb. Brain Research in press.

Johnston, G.A.R., Curtis, D.R., Davies, I. and McCulloch, R.M. (1978). Spinal interneurone excitation by conformationally restricted analogues of L-glutamic acid. Nature (Lond.), 248, 804.

Johnston, G.A.R., Kennedy, S.M.E. and Twitchin, B. (1979). Action of the neurotoxin kainic acid on high-affinity uptake of L-glutamic acid in rat brain slices. J. neurochem., 32, 121-127.

Jörgensen, B.M. and Diemer, N.H. (1982). Selective neuron loss after cerebral ischemia in the rat: Possible role of trans- mitter glutamate. Acta Neurol. Scand. 66, 536-546.

Pastuszko, A., Wilson, D.F. and Erecinska, M. (1984). Effects of kainic acid in rat brain synaptosomes: The involvement of calcium. J. Neurochem., 43, 747-754.

Peterson, D.W., Collins, J.F. and Bradford, H.F. (1983). The kindled amygdala model of epilepsy: Anticonvulsant action of amino acid antagonists. Brain Res., 275, 169-172.

Poli, A., Contestabile, A., Migani, P., Rondelli, C., Rossi, L. and Barnabei, O. (1984). Effects of kainic acid on the release of endogenous and exogenous amino acids from synaptosome preparations. Neurosci.Lett.,Suppl 18,pp 191.

Richelson, E. and Thompson, E.J. (1973). Transport of neurotransmitter precursors into cultured cells. Nature, 241, 201-204.

Rothman, S. (1984). Synaptic release of excitatory amino acid neurotransmitter mediates anoxic neuronal death. J. Neurosci., 4, 1884-1891.

Sanberg, P.R., Staines, W. and McGeer, E.G. (1979). Chronic taurine effects on various neurochemical indices in control and kainic acid-lesioned neostriatum. Brain Res., 161, 367-370.

Sandberg, M. and Lindström, S. (1983). Amino acids in dorsal lateral geniculate nucleus of the cat-collection in vivo. J. Neurosci. Lett., 9, 65-72.

Sandberg, M., Nyström, B. and Hamberger, A. (1985a). Metabolically derived aspartate-elevated extracellular levels in vivo in iodoacetate poisoning. J. Neurosci. Res., In press.

Sandberg, M., Butcher, S.P. and Hagberg, H. (1985b). Extracellular overflow of neuroactive amino acids during severe insulin-induced hypoglycemia: in vivo dialysis of the rat hippocampus. Manuscript in preparation.

Schmidt, W. and Wolf, G. (1984). Histochemical localization of aspartate aminotransferase activity in the hippocampal formation and in peripheral ganglia of the rat with special reference to the glutamate transmitter metabolism. J. Hirnforsch. 25, 505-510.

Schrier, B.K. and Thompson, E.J. (1977). On the role of glial cells in the mammalian nervous system. J. Biol. Chem., 249, 1769-1780.

Siesjö, B.K. and Agardh, C.-D. (1983) Hypoglycemia. In:"Handbook of Neurochemistry", Ed. A. Lajtha. Plenum Press, N.Y.

Siesjö, B.K. and Wieloch, T. (1985). Cerebral metabolism in ischaemia: Neurochemical basis for therapy. Br. J. Anaesth. 57, 47-62.

Simon, R.P., Swan, J.H., Griffiths, T. and Meldrum, B.S. (1984). Blockade of N-methyl-D-aspartate receptors may protect against ischemic damage in the brain. Science, 226, 850-852.

Smith, M.-L., Auer, R.N. and Siesjö, B.K. (1984). The density and distribution of ischemic brain injury in the rat following 2-10 min of forebrain ischemia. Acta Neuropathol.(Berl), 64, 319-332.

Streit, P., Stella, M. and Cuénod, M. (1980). Kainate-induced lesion in the optic tectum: Dependency upon optic nerve afferents or glutamate. Brain Res., 187, 47-57.

Suzuki, R., Yamaguchi, T., Choh-Luh, L. and Klatzo, I. (1983) The effects of a 5-minute ischemia in mongolian gerbils: II. Changes of spontaneous neuronal activity in cerebral cortex and CA1 sector of hippocampus. Acta Neuropathol. (Berl.) 60, 217.

Tews, J.K., Carter, S.H. and Stone, W.E. (1965). Chemical changes in the brain during insulin hypoglycemia and recovery. J. Neurochem., 12, 679-693.

Webb, J.L. (1966). Iodoacetate and iodoacetamide. Enzyme and metabolic inhibitors, 3, 1-283.

Wieloch, T., Harris, R.J., Symon, L. and Siesjö, B.K. (1984). Influence of severe hypoglycemia on brain extracellular calcium and potassium activities, energy and phospholipid metabolism. J. Neurochem., 43, 160-168.

Zanotto, L. and Heinemann, U. (1983). Aspartate and glutamate induced reductions in extracellular free calcium and sodium concentration in area CA1 of in vitro hippocampal slices of rats. Neurosci. Lett., 35, 79-84.

van Gelder, N.M., Siatitsas, I., Mébini, C. and Gloor, P. (1983). Feline generalized penicillin epilepsy: Changes of glutamic acid and taurine parallel the progressive increase in excitability of the cortex. Epilepsia, 24, 200-213.

28

Excitatory Amino Acids: Approaches to Rational Anticonvulsant Therapy

B.S. Meldrum, A.G. Chapman, L.M. Mello, M.H. Millan, S. Patel and L. Turski

INTRODUCTION

In recent years, enhancement of inhibitory transmission has come to be considered the most important rational approach to the design of novel anticonvulsant drugs (Meldrum, 1983). This has led to clinical trials involving several novel compounds, designed to enhance GABA-mediated inhibition such as Progabide and Vigabatrin. The alternative approach of seeking compounds that impair excitatory transmission has only recently attracted attention. It might be thought that impairing excitatory transmission is unlikely to suppress epileptic activity selectively while leaving normal function intact. However, as will be explained below, there are reasons why it may be possible to act selectively on the sustained abnormal discharge of epilepsy.

Among possible mechanisms by which drugs could diminish excitatory transmission, we shall consider, firstly compounds acting as enzyme inhibitors to reduce the maximal rate of synthesis of excitatory amino acids, secondly, compounds that decrease the synaptic release of excitatory amino acids, and thirdly, compounds that act on post-synaptic sites to decrease the excitatory effect of these amino acids.

INHIBITION OF SYNTHESIS

On theoretical grounds, this could be a valid approach to anticonvulsant action as it might be possible to decrease the maximal rate of synthesis of neurotransmitter glutamate or aspartate so that the sustained release associated with transmission of burst firing to other brain regions would be prevented, but normal levels of activity would not be modified. However at present there are great uncertainties about the metabolic pathways involved, and the optimal enzyme inhibitors. Glutamate can be synthesized by many separate routes involving, for example, glutamate dehydrogenase (EC 1.4.1.2) and aspartate transaminase (EC 2.6.1.1.) and glutaminase (EC 3.5.1.2). There is evidence, particularly, from

in vitro studies for a significant role for glutaminase in the
synthesis of neurotransmitter glutamate. It has been proposed that
asparaginase may be involved in aspartate synthesis (Reubi et al.,
1980). However, transamination of oxaloacetate appears to be the
main route for synthesis. Studies relating to these pathways have
been reviewed in an earlier chapter in this volume (Shank, pp.
Azetazolamide which has anticonvulsant action in absence seizures
inhibits phosphate-activated glutaminase (Beaton, 1961). It is not
clear if this is relevant to its anticonvulsant action, which is
more usually attributed to carbonic anhydrase inhibition. Two
potent glutaminase inhibitors 6-diazo-5-oxo-L-norleucine (Don) and
azaserine, when tested for anticonvulsant action against audiogenic
seizures in mice show a weak protective action (Chung and Johnson,
1984).

In experimental animals, sodium valproate administration
induces a decrease in brain aspartate concentration (Chapman et al,
1982; Schechter et al, 1978). Analogues of valproate with anticonvulsant action also show this effect on brain aspartate
concentration (Chapman et al, 1983a). It is not known whether
there is impairment of aspartergic transmission following valproate
administration. Ethosuximide also decreases aspartate concentration
in rat brain (Nahorski, 1972) suggesting a link with antiabsence
activity.

INHIBITION OF RELEASE OF EXCITATORY AMINO ACIDS

Release of excitatory amino acids can be modified by compounds
acting directly on receptors or other membrane components in presynaptic terminals. These receptors are very diverse ranging from
"autoreceptors" responding to glutamate or its analogues, to
"heteroreceptors" found presynaptically on a wide range of neurons,
such as adenosine A_1, $GABA_B$ (baclofen), $GABA_A$/benzodiazepine
receptors, and α and β-adrenergic receptors. There are also various
mechanisms involving Ca^{2++} transport or entry, and protein
phosphorylation.

'Glutamate autoreceptors'. In the rat hippocampus L-glutamate,
L-cysteate, and L-homocysteate decrease the K^+ stimulated release
of exogenous aspartate (apparently through an action on a presynaptic receptor) whereas kainate and N-methyl-D-aspartate do not
do this suggesting that only the glutamate/quisqualate receptor is
involved (McBean and Roberts, 1981). However in the olfactory
cortex release of endogenous aspartate is decreased selectively by
N-methyl-D-aspartate (Collins et al., 1983). L(+)2 amino-4-
phosphonobutyric acid (2APB) apparently acts presynaptically to
block glutamatergic transmission in the hippocampus. 2APB is not
generally active as an anticonvulsant agent in animal models but
it appears to have a particular protective action against the
kindling process (Peterson et al, 1983).

Adenosine presynaptic receptors. Purinergic agonists act presynaptically to inhibit the release of acetylcholine and noradrenaline. In slices of rat dentate gyrus 2-chloroadenosine inhibits

the stimulated release of endogenous glutamate (Dolphin and Archer, 1983). There is also autoradiographic evidence suggesting a pre-synaptic location of adenosine (A_1) receptors on excitatory afferents (Goodman et al, 1983). Adenosine and 2-chloroadenosine and other analogues show anticonvulsant activity in some animal test systems (Dunwiddle and Worth, 1982; Barraco et al, 1984). However it is not proven that the anticonvulsant action relates to an effect on excitatory amino acid release as there is evidence for a post-synaptic action of adenosine that correlates with an in vitro anticonvulsant action (Lee et al, 1984).

GABA_B receptors. Baclofen selectively decreases the stimulated release of glutamate and aspartate from brain slices (Potashner, 1979). This may be due to an action at a presynaptic GABA receptor that is bicuculline-insensitive (the $GABA_B$ receptor) (Bowery et al, 1980).

In the hippocampal slice baclofen attenuates electrically induced after-discharges but also abolishes recurrent inhibition, apparently by decreasing excitatory inputs to inhibitory inter-neurons (Ault & Nadler, 1983).

GABA/Benzodiazepine receptors. Stimulated release of exogenous glutamate or cysteine sulfinate is inhibited by diazepam or GABA (Baba et al, 1983). Chlordiazepoxide decreases the release of aspartate from olfactory cortex (Collins, 1981).

A convulsant β-carboline acting at the benzodiazepine receptor site enhances the stimulated release of exogenous aspartate (Kerwin & Meldrum, 1983).

Physiological experiments indicate that benzodiazepines can decrease excitatory transmission by both pre- and post-synaptic actions but the contribution this makes to their anticonvulsant effect is not yet evaluated (Meldrum & Braestrup, 1984).

Noradrenaline. In cerebellar slices potentiation of glutamate release is produced by β-adrenergic agonists (such as salbutamol) and inhibition of release by $α_2$ adrenergic agonists such as clonidine (Dolphin, 1982). Clonidine is anticonvulsant in DBA/2 mice but not in most other test systems (Horton et al, 1980).

Anticonvulsant compounds. Anticonvulsant and anaesthetic barbiturates inhibit the potassium stimulated release of aspartate from brain slices, and a convulsant barbiturate (CHEB) facilitates this release (Minchin, 1981; Skerritt et al, 1983; Holtman and Richter, 1983).

Phenytoin (100 μM) or carbomazepine (100 μM) inhibit the proto-veratrine-stimulated release of exogenous ^{3}H-D-aspartate from cortical minislices (Skerritt & Johnston 1984). These concentra-tions are more than ten times the plasma concentration occurring therapeutically whereas phenytoin or carbamazepine can prevent sustained repetitive firing of isolated neurons at therapeutic concentrations (MacDonald, 1983).

Valproate and structural analogues with similar anticonvulsant activity decrease brain aspartate content (Chapman et al, 1983a). It has not been shown whether there is an associated decrease in synaptic release of aspartate.

POST-SYNAPTIC BLOCKADE OF EXCITATORY TRANSMISSION

Many analogues of glutamate and aspartate act at the post-synaptic receptor sites to anatagonise their excitatory effects (McLennan, 1984). These antagonists vary markedly in their potency and specificity against the three preferred agonists, N-methyl-D-aspartate, quisqualate and kainate. We have tested these antagonists for anticonvulsant activity using intracerebro-ventricular injection in DBA/2 mice, an inbred strain that show wild running, clonus and tonic seizures in response to a loud sound (Croucher et al, 1982; Meldrum et al, 1983b; Chapman et al, 1984).

Table 1.   Excitatory amino acid antagonists: anticonvulsant potency in comparison with receptor subtype specificity.

	Anticonvulsant potency[1] i.c.v.	Relative antagonist potency at receptor subtypes[2]		
	μmol	NMDA	KA	QA
β-D-aspartylaminomethyl phosphonate	0.0006	+++++	+	+(+)
2-amino-7-phosphonoheptanoate	0.0015	+++++	0	0
γ-D-glutamylaminomethyl phosphonate	0.0018	++++	(+)	+
1-(p-bromobenzoyl) 2,3-piperazine dicarboxylate	0.010	++	+++	++
cis-2,3-piperidine dicarboxylate	0.017	++(+)	++(+)	+(+)
2-amino-5-phosphonovalerate	0.020	+++++	0	0
γ-D-glutamylglycine	0.046	+++	++	+(+)
γ-D-glutamylaminomethyl sulphonate	0.074	+	++	++
β-kainate	0.090			
2-amino-6-phosphonohexanoate	0.14	++	0	0

[1] Values are ED_{50} for suppression of clonic seizures in DBA/2 mice (from Croucher et al 1982, 1984; Meldrum et al, 1983b; Jones et al, 1984; Turski et al, 1985).

[2] Antagonist potency is represented on a five point scale [+ (weak) to +++++ (very strong)] derived from studies of electrical response to selective agonists (NMDA, N-methyl-D-aspartate; KA, kainate; QA, quisqualate) in the spinal cord (Davies et al, 1982; Jones et al, 1984a; Peet et al, 1983).

NMDA antagonists. All compounds that are potent and specific antagonists at the N-methyl-D-aspartate preferring receptor site show significant anticonvulsant activity (see Table 1). There is an appropriate correlation in terms of rank order between anti-convulsant potency and NMDA antagonism. The ω-phosphono derivatives are of particular interest as they are highly selective for the NMDA receptor. 2-amino-7-phosphonoheptanoic acid (2APH) is significantly more potent than 2-amino-5-phosphonovaleric acid (2APV) as an anticonvulsant, but in most studies employing in vitro spinal cord preparations or in vivo electrophysiological measurements of single cell responses. 2APH appears to be either as potent or weaker than 2APV (Evans et al, 1982; Peet et al, 1983). One study employing iontophoresis on cortical neurons found rather weak evidence for greater potency of 2APH (Meldrum et al, 1983b). It may be that there are two types of NMDA receptor, with one subtype that is particularly responsive to quinolinic acid found predominantly in cortex and hippocampus (see McLennan this volume) and another subtype relatively unresponsive to quinolinic and which predominates in the spinal cord. The NMDA receptor on GABAergic interneurons (in hippocampus and cerebellum) may be pharmacologically different from the NMDA receptor on apical dendrites of hippocampal pyramidal neurons.

Systemic administration. Although the NMDA antagonists are all highly polar compounds that would not be expected to cross the blood brain barrier, nevertheless they are active following systemic administration. Their potency relative to diazepam or valproate, is reduced when i.p. administration is used rather than i.c.v. administration (cf Tables I and 2). A study using ^{3}H 2APH (Chapman et al, 1983b) has shown that following systemic injection of an anticonvulsant dose unmetabolised 2APH accumulates in rat brain, at a concentration (0.1-1.0 μmol) that produces an NMDA antagonist effect in vitro.

In DBA/2 mice the relative anticonvulsant potencies of the different antagonists tend to be similar to their relative potencies following i.c.v. injection, except that the dipeptides (such as β-D-aspartyl aminomethylphosphonate) are less active systemically.

An anticonvulsant action of systemically administered 2APH has been observed in a wide range of seizure models. These include several types of chemically induced seizure in mice (Czuczwar & Meldrum, 1982), the high pressure neurological syndrome in rats (Meldrum et al, 1983c), and photically-induced myoclonus in the Senegalese baboon, Papio papio (Meldrum et al, 1983a). The latter model is of particular significance for future clinical studies. In the case of known anticonvulsant drugs it provides a clear indication of neurological side effects. An antiepileptic effect is observed following 2APH in the absence of any acute neurological toxicity of the kind normally associated with anticonvulsants (i.e. ataxia, nystagmus, muscle weakness, sedation).

Table 2.   Excitatory amino acid antagonists: anticonvulsant activity
following systemic administration in reflex epilepsy in
DBA/2 mice and in the photosensitive baboon, Papio papio.

	DBA/2 mice mmol/kg i.p.	Papio papio mmol/kg, i.v.
2-amino-5-phosphonovalerate	0.40	3.3
2-amino-7-phosphonoheptanoate	0.04	1.0
cis-2,3-piperidine dicarboxylate	0.52	3.3
γ-D-glutamyl aminomethyl phosphonate	0.28	—
β-D-aspartyl aminomethylphosphonate	0.18	—
L-glutamic acid diethyl ester	—	>3.3
kynurenate	4.4	>1.5
sodium valproate	1.25	1.4
diazepam	0.0004	0.004
phenobarbital	0.01	0.065

Values in mice are ED_{50} (following intraperitoneal administration)
for suppression of clonic response to a loud sound (from Croucher
et al, 1982, 1984b; Chapman et al, 1984; Jones et al, 1984).
Values in baboons are approximate ED_{100} (following i.v. injection)
for suppression of myoclonic response to intermittent photic
stimulation (from Meldrum et al 1983; Meldrum 1978 and Mello &
Meldrum, unpublished).

        Kainate antagonists.   No potent specific antagonist for the
kainate receptor is known.

        Considering the anticonvulsant potency, following i.c.v.
administration of compounds which are antagonists at all three
receptor sites, in relation to their NMDA antagonist activity (see
Table 1) it appears that antagonism at the kainate receptor can
contribute to the total anticonvulsant action.   Thus 1-(p-bromo-
benzoyl)2,3-piperazine dicarboxylate, cis-2,3-piperidine
dicarboxylate and particularly γ-D-glutamyl aminomethylsulphonate
appear more potent as anticonvulsants than would be expected on the
basis of their NMDA antagonist activity.   However relative to 2APH
their therapeutic ratio is less satisfactory i.e. toxic side
effects, such as impaired motor activity, are seen with anticonvulsant doses.   Studies in isolated neurons (Mayer and Westbrook,
1984) have shown an interaction between kainate and NMDA responses,
whereby a threshold concentration of NMDA enhances responses to
kainate.   Thus simultaneous antagonism at the NMDA and kainate
receptors could have a potentiating effect on anticonvulsant
activity.   It is not yet possible to say if antagonism at the
kainate receptor alone has an anticonvulsant action.   Anticonvulsant
effects of a series of substituted derivatives of kainic acid have
been demonstrated (Collins et al, 1984 and Chapman et al this
volume).   However these effects probably do not depend on kainate
antagonism (see below).

Quisqualate antagonists. The most specific antagonist at the quisqualate receptor is glutamic acid diethyl ester, GDEE, although it is relatively weak and also acts on cholinergic systems. It is inactive against sound-induced seizures when given i.c.v. to DBA/2 mice. However a number of anticonvulsant effects have been described and suggest that blockade at this receptor has a rather different impact on epilepsy compared with blockade at the NMDA receptor.

Given systemically in high doses (2-4 mmol/kg) protection against several epilepsy models is observed. This includes those due to alcohol withdrawal or the administration of homocysteine in mice (Freed and Michaelis, 1978; Freed, 1985). Protection is also seen against the tonic seizures occurring in the high pressure neurological syndrome (HPNS) in rats (Wardley-Smith et al, 1985). This is clearly different from the effect of 2APH in the same syndrome, as 2APH provides protection against the tremor and myoclonus of HPNS whereas GDEE does not.

Intraventricular coinjection of agonists and antagonists. We have obtained further information about the possible interactions between the three receptor systems and epileptic phenomena by studying clonic seizure responses to intracerebroventricular injection of excitatory amino acids, and modification of these responses by co-injection of antagonists (Turski et al, 1985).

Figure 1 shows the relative convulsant potency of 6 agonists. In contrast to the results found in single cell studies, or studies of spinal cord electrical responses in vitro, quisqualate is relatively weak and NMDA is the most potent excitant/convulsant. This suggests that the burst firing pattern associated with NMDA and quinolinate is particularly important in inducing clonic activity, whereas the normal depolarisation/excitation induced by quisqualate is not.

The protective effect of 2APH is most potent against NMDA and quinolinate (ED_{50} increased 153 fold and 70 fold respectively). Nevertheless the ED_{50} for kainate is increased 6 fold. In contrast the only significant effects of GDEE are on responses to quisqualate and D-homocysteinesulphinic acid (ED_{50} increased 3.9 and 3.3 fold respectively).

In the spinal cord γ-D-glutamylaminomethylsulphonate preferentially antagonises excitation induced by kainate (Jones et al, 1984a). With coinjection there is a 90 fold increase in the ED_{50} for α-kainate, compared with a 27 fold increase for NMDA.

β-kainate shows a spectrum of activity closely similar to that of APH, increasing the ED_{50} for NMDA 100 fold (compared with 68 fold for quinolinate, 16 fold for α-kainate and 3 fold for quisqualate). Data at the receptor level that can explain this apparent selective NMDA antagonism are not yet available.

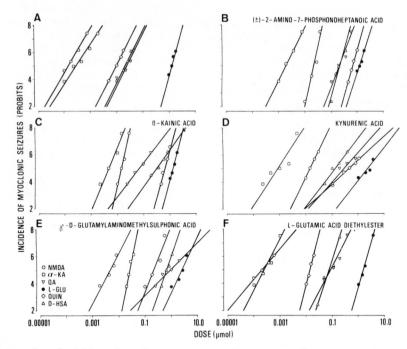

<u>Figure 1</u>. Probit - log dosage regression curves for myoclonic seizures induced by excitatory amino acid agonists in mice, showing the selective protective action of coadministration of 5 antagonists. Drugs (as sodium salts, in phosphate buffer, pH 7.35) were administered intracerebroventricularly in Swiss S mice (as described by Herman, 1975) in a volume of 5 μl at a rate of 1 μl/5s. Mice were observed for the occurrence of clonic and clonic/tonic seizures within 30 min. Graphs show the incidence of myoclonic seizures (as probit-transformed percentages for groups of 10 mice) against the log doses of excitatory amino acids, with linear regression analysis.

A. Administration of N-methyl-D-aspartate (NMDA), α-kainate (α-KA), quisqualate (QA), L-glutamate (L-GLU), quinolinate (QUIN) or D-homocysteinesulphinate (D-HSA).
B. As in A, but with co-injection of (±)2-amino-7-phosphonoheptanoate 0.1 μmol.
C. As in A, but with co-injection of β-kainate, 1.0 μmol.
D. As in A, but with co-injection of Kynurenate, 0.1 μmol.
E. As in A, but with coadministration of γ-D-glutamyl aminomethylsulphonate, 1.0 μmol.
F. As in A but with co-injection of L-glutamic acid diethylester. 3.0 μmol.

NMDA antagonists as potential antiepileptic drugs. Of all
the classes of compounds that may impair excitatory amino acid
transmission in the brain the NMDA antagonists have so far shown
the greatest promise as potential clinically useful anticonvulsant
agents. The discovery of this property of selective NMDA
antagonists preceded our understanding of the particular features
of the NMDA receptor system that link it to epilepsy. However
single unit studies with intracellular recording in striatum and
hippocampus (Herrling et al, 1983; Collingridge et al, 1983) show
that NMDA or quinolinate induce a paroxysmal depolarisation shift
and burst firing in normal neurons, comparable to the abnormal
neuronal activity that has long been recognised in epileptogenic
foci (Schwartzkroin & Wyler, 1980). This pattern of activity
arises from the opening of ionic channels with a non-linear
voltage current relationship (which can be gated by Mg^{2+})
(MacDonald et al, 1982; Nowak et al, 1984).

Our observations in rodents and baboons show that a selective
NMDA antagonist such as 2APH, can produce a powerful anticonvulsant
effect without modifying normal motor activity. Higher doses
reduce muscle tone and eventually produce hind limb weakness.
Possible effects of NMDA antagonists on cognitive functions remain
to be explored. It is known that 2APV and 2APH can block long term
potentiation in hippocampal slices (Harris et al, 1984). Thus it
is possible that NMDA antagonists will impair spatial learning
(Anderson et al, 1985).

The decrease in potency relative to other anticonvulsants
associated with systemic administration (instead of intracerebral
injection) of 2APH suggests that compounds with similar antagonist
activity that penetrated the blood-brain barrier to a greater
extent are required for clinical testing.

SUMMARY

In the search for novel anticonvulsant drugs we are studying
the effect on seizures in animals of selective impairment of
excitatory transmission. There are three principal approaches,
namely 1) impairment of the maximal rate of synthesis of the
excitatory neurotransmitters, 2) selective inhibition of the
synaptic release of excitatory aminoacids, and 3) antagonism of the
post-synaptic excitatory action of amino acids. The latter approach
has to date provided the most potent anticonvulsant compounds. In
particular selective antagonists at the N-methyl-D-aspartate
preferring receptor show anticonvulsant activity in a wide range of
animal models of epilepsy. This is consistent with electrophysiological studies showing that activation of the NMDA receptor in
normal neurons induces paroxysmal depolarisation and burst firing
of the kind recorded in an epileptic focus or during seizure
activity.

The most effective NMDA antagonist, D(-) 2-amino-7-phosphono-
heptanoic acid, is a more potent anticonvulsant than diazepam when
given i.c.v. to mice. It is relatively less potent when given
systemically, apparently because of limited penetration of the
blood-brain barrier. In several test systems complete suppression
of epileptic responses occurs after doses of 2APH that do not
impair normal motor activity.

The validity of this approach to anticonvulsant drug design
is clearly established. However several problems remain to be
resolved before the optimal compound for clinical trial is
identified.

ACKNOWLEDGEMENT

We thank the Wellcome Trust and the Medical Research Council
for financial support.

REFERENCES

Anderson, E., Baudry, M., Lynch, G. and Morris R.G.M. (in Press).
Selective impairment of learning and blockade of long-term
potentiation by an N-methyl-D-Aspartate receptor antagonist, APV-5.
Physiol. Soc. __, 38P.

Ault, B. and Nadler, J.V. (1983). Anticonvulsant like actions of
baclofen in the rat hippocampal slice. Brit. J. Pharmacol. 78,
701-708.

Baba, A., Okumura, S., Mizuo, H. and Iwata, H.(1983). Inhibition by
diazepam and γ-aminobutyric acid of depolarization-induced release
of [^{14}C]cysteine sulfinate and [^{3}H]glutamate in rat hippocampal
slices. J. Neurochem. 40, 280-4.

Barraco, R.A., Swanson, T.H., Phillis, J.W. and Berman, R.F. (1984).
Anticonvulsant effects of adenosine analogues on amygdaloid-kindled
seizures in rats. Neurosci. Lett., 46, 317-322.

Beaton, T. (1961). The inhibition by acetazoleamide of renal
phosphate activated glutaminase in rats. Can. J. Biochem. Physiol.
39, 663-6.

Bowery, N.G., Hill, D.R., Hudson, A.L., Doble, A., Middlemiss, D.N.,
Shaw, J. and Turnbull, M. (1980). (-)Baclofen decreases neurotransmitter release in the mammalian CNS by an action of a novel GABA
receptor. Nature 362, 92-4.

Chapman, A.G., Riley, K., Evans, M.C. and Meldrum, B.S. (1982).
Acute effects of sodium valproate and γ-vinyl GABA on regional
amino acid metabolism in the rat brain. Neurochem. Res. 7, 1089.

Chapman, A.G., Meldrum, B.S. and Mendes, E. (1983a). Acute anticonvulsant activity of structural analogues of valproic acid and
changes in brain GABA and aspartate content. Life Sci. 32, 2023-31.

Chapman, A.G., Collins, J.F., Meldrum, B.S. Westerberg, E.
(1983b). Uptake of a novel anticonvulsant compound, 2-amino-7-
phosphono-(4,5^{3}H)-heptanoic acid, into mouse brain. Neurosci.
Lett. 37, 75-80.

Chapman, A.G., Croucher, M.J. and Meldrum, B.S. (1984).
Evaluation of anticonvulsant drugs in DBA/2 mice with sound-
induced seizures. Arzneimitt. Forsch. 34, 1261-1264.

Chapman, A.G., Hart, G.P., Meldrum, B.S., Turski, L. and Watkins,
J.C. (1985). anticonvulsant activity of two novel piperazine
derivatives with potent kainate antagonist activity. Neurosci.
Lett., 55, 325-330.

Chung, S.H., Johnson, M.S. (1984). Studies on sound-induced
epilepsy in mice. Proc. Roy. Soc B., 221, 145-168.

Collingridge, G.L., Kehl, S.J. and McLennan, H. (1983). The
antagonism of amino acid-induced excitations of rat hippocampal
CA$_1$ neurones in vitro. J. Physiol. (Lond)., 334, 19-31.

Collins, G.G.S. (1981). The effects of chlordiazepox on synaptic
transmission and amino acid neurotransmitter release in slices of
rat olfactory cortex. Brain Res., 224, 389.

Collins, G.G.S., Anson, J. and Surtees, L. (1983). Presynaptic
kainate and N-methyl-D-aspartate receptors regulate excitatory amino
acid release in the olfactory cortex. Brain Res., 265, 157-159.

Collins, J.F., Dixon, A.J., Badman, G., De Sarro, D., Chapman, A.G.
Hart, G.P. and Meldrum, B.S. (1984). Kainic acid derivatives with
anticonvulsant activity. Neurosci. Lett., 51, 371-376.

Croucher, M.J., Collins, J.F. and Meldrum, B.S. (1982). Anticonvulsant action of excitatory amino acid antagonists. Science
216, 899-901.

Croucher, M.J., Meldrum, B.S., Jones, A.W. and Watkins, J.C. (1984a).
γ-D-Glutamylaminomethyl-sulphonic acid (GAMS), a kainate and
quisqualate antagonist, prevents sound-induced seizures in DBA/2
mice. Brain Res. 322, 111-114.

Croucher, M.J., Meldrum, B.S. and Collins, J.F. (1984b). Anticonvulsant and proconvulsant properties of a series of structural
isomers of piperidine decarboxylic acid. Neuropharmacol., 23, 467-
472.

Czuczwar, S.J., Meldrum, B. (1982). Protection against chemically
induced seizures by 2-amino-7-phosphonoheptanoic acid. Eur. J.
Pharmacol. 83, 335-338.

Davies, J., Evans, R.H., Jones, A.W., Smith, D.A.S. and Watkins,
J.C. (1982). Differential activation and blockade of excitatory
amino acid receptors in the mammalian and amphibian central
nervous system. Comp. Biochem. Physiol. 72C, 211-224.

Dolphin, A.C. (1982). Noradrenergic modulation of glutamate release
in the cerebellum. Brain Res., 252, 111-116.

Dolphin, A.C. and Archer, E.R. (1983). An adenosine agonist inhibits and a cyclic AMP analogue enhances the release of glutamate but not GABA from slices of rat dentate gyrus. Neurosci. Lett., $\underline{43}$, 49-54.

Dunwiddle, T.V. and Worth, T. (1982). Sedative and anticonvulsant effects of adenosine analogs in mouse and rat. J. Pharmacol. exp. Therap., $\underline{220}$, 70-76.

Evans, R.H., Francis, A.A., Jones, A.W., Smith, D.A.S., Watkins, J.C. (1982). The effects of a series of ω-phosphonic α-carboxylic amino acids on electrically evoked and excitant amino acid-induced responses in isolated spinal cord preparations. Brit. J. Pharmacol., $\underline{75}$, 65-75.

Freed, W.J. (1985). Selective inhibition of homocysteine-induced seizures by glutamic acid diethyl ester and other glutamate esters. Epilepsia, $\underline{26}$, 10-36.

Freed, W.J. and Michaelis, E.K. (1978). Glutamic acid and ethanol dependence. Pharmacol. Biochem. Behav., $\underline{8}$, 509-514.

Goodman, R.R., Kuhar, M.J., Hester, L. and Snyder, S.H. (1983). Adenosine receptors: Autoradiographic evidence for their location on axon terminals of excitatory neurons. Science, $\underline{220}$, 967-969.

Harris, E.W., Ganong, A.H. and Cotman, C.W. (1984). Long-term potentiation in the hippocampus involves activation of N-methyl-D-aspartate receptors. Brain Res., $\underline{323}$, 132-137.

Herman, Z.S. (1975). Behavioural changes induced in conscious mice by intracerebroventricular injection of catecholamines, acetylcholine and 5-hydroxytryptamine. Br. J. Pharmacol., $\underline{55}$, 351-358.

Herrling, P.L., Morris, R. and Salt, T.E. (1983). Effects of excitatory amino acids and their antagonists on membrane and action potentials of cat caudate neurones. J. Physiol. $\underline{339}$, 207.

Holtman, J.R. and Richter, J.A. (1983). Comparison of the effects of a convulsant barbiturate on the release of endogenous and radio-labelled amino acids from slices of mouse hippocampus. J. Neurochem. $\underline{41}$, 723-728.

Horton, R., anlezark, G. and Meldrum, B. (1980). Noradrenergic influences on sound-induced seizures. J. Pharmacol. Exp. Ther., $\underline{214}$, 437-442.

Jones, A.W., Smith, D.A.S. and Watkins, J.C. (1984a). Structure-activity relations of dipeptide antagonists of excitatory amino acids. Neuroscience, $\underline{13}$, 573-581.

Jones, A.W., Croucher, M.J., Meldrum, B.S. and Watkins, J.C. (1984b). Suppression of audiogenic seizures in DBA/2 mice by two new dipeptide NMDA receptor antagonists. Neurosci. Lett., $\underline{45}$, 157-161.

Kerwin, R.W. and Meldrum, B.S. (1983). Effect on cerebral ^{3}H-D-aspartate release of 3-mercaptopropionic acid and methyl 5,7-dimethoxy-4-ethyl-β-carboline-3-carboxylate. Eur. J. Pharmacol., $\underline{89}$, 265.

Lee, K.S., Schubert, P. and Heinemann, U. (1984). The anti-convulsive action of adenosine: a post-synaptic, dendritic action by a possible endogenous anticonvulsant. Brain Res. 321, 160-164.

MacDonald, J.F., Porietis, A.V. and Wojtowicz, J.M. (1982). Aspartic acid induces a region of negative slope conductance in the current-voltage relationship of cultured spinal cord neurons. Brain Res., 237, 248-253.

MacDonald, R.L. (1983). Mechanisms of anticonvulsant drug action. In Recent Advances in Epilepsy. (eds. T.A. Pedley and B.S. Meldrum). Churchill Livingstone Edinburgh.

McBean, G.J. and Roberts, P.J. (1981). Glutamate-preferring receptors regulate the release of D-^{3}H aspartate from rat hippocampal slices. Nature 291, 593-594.

McLennan, H. (1984). Receptors for the excitatory amino acids in the mammalian central nervous system. Prog. Neurobiol., 20, 251-271.

Mayer, M.L. and Westbrook, G.L. (1984). Mixed-agonist action of excitatory amino acids on mouse spinal cord neurones under voltage clamp. J. Physiol. (Lond.), 354, 29-53.

Meldrum, B.S. (1978). Photosensitive epilepsy in Papio papio as a model for drug studies. In Contemporary Clinical Neurophysiology, Suppl. 34: Electroencephalography Clinical Neurophysiology. (eds. E.A. Cobb and H. van Duijn). Elsevier, Amsterdam.

Meldrum, B.S. (1983). Pharmacological considerations in the search for new anticonvulsant drugs. In Recent Advances in Epilepsy. (eds. T.A. Pedley and B.S. Meldrum). Churchill Livingstone, London.

Meldrum, B. and Braestrup, C. (1984). GABA and the Anticonvulsant Action of Benzodiazepines and Related Drugs. In Actions and Interactions of GABA and Benzodiazepines. (eds. N.G. Bowery). Raven Press, New York.

Meldrum, B.S., Croucher, M.J., Badman, G. and Collins, J.F. (1983a). Antiepileptic action of excitatory amino acid antagonists in the photosensitive baboon, Papio papio. Neurosci. Lett., 39, 101-104.

Meldrum, B.S., Croucher, M.J., Czuczwar, S.J., Collins, J.F., Curry, K., Joseph, M. and Stone, T.W. (1983b). A comparison of the anticonvulsant potency of (±)2-amino-5-phosphonopentanoic acid and (±)2-amino-7-phosphonoheptanoic acid. Neuroscience 9, 925-930.

Meldrum, B., Wardley-Smith, B., Halsey, M., Rostain, J.C. (1983c). Phosphonoheptanoic acid protects against the high pressure neurological syndrome. Eur. J. Pharmacol., 87, 501-502.

Minchin, M.C.W. (1981). The effect of anaesthetics on the uptake and release of γ-aminobutyrate and D-aspartate in rat brain slices. Brit. J. Pharmac., 73, 681-689.

Nahorski, S.R. (1972). Biochemical effects of the anticonvulsants trimethadione, ethosuximide and chlordiazepoxide in rat brain. J. Neurochem. 19, 1937-1946.

Nowak, L., Bregestovski, P., Ascher, P., Herbet, A. and Prochiantz, A. (1984). Magnesium gates glutamate-activated channels in mouse central neurones. Nature (Lond.), 307, 462-465.

Peet, M.J., Leah, J.D. and Curtis, D.R. (1983). Antagonists of synaptic and amino acid excitation of neurons in the cat spinal cord. Brain Res., 266, 83-95.

Peterson, D.W., Collins, J.F. and Bradford, H.F. (1983). The kindled amygdala model of epilepsy: anticonvulsant action of amino acid antagonists. Brain Res., 275, 169-172.

Potashner, S.J. (1979). Baclofen: Effects on amino acid release in slices of guinea pig cerebral cortex. J. Neurochem., 32, 103.

Reubi, J.C., Toggenburger, G. and Guenod, M. (1980). Asparagine as precursor for transmitter aspartate in corticostriatal fibres. J. Neurochem., 35, 1015.

Schechter, P.J., Tranier, Y., Grove, J. (1978). Effect of n-dipropylacetate on amino acid concentrations in mouse brain: correlations with anticonvulsant activity. J. Neurochem., 31, 1325-1327.

Schwartzkroin, P.A., Wyler, A.R. (1980). Mechanisms underlying epileptiform burst discharge. Ann. Neurol. 7, 95.

Skerritt, J.H., Johnston, G.A.R. (1984). Modulation of excitant amino acid release by convulsant and anticonvulsant drugs. In Neurotransmitters, Seizures and Epilepsy II. (eds. R.G. Fariello). pp.215-226, Raven Press, New York.

Skerritt, J.H., Willow, M. and Johnston, G.A.R. (1983). Contrasting effects of a convulsant (CHEB) and an anticonvulsant barbiturate (Phenobarbitone) on amino acid release from rat brain slices. Brain Res., 258, 271-276.

Turski, L., Collins, J.F. and Meldrum, B.S. (1985). Is β-kainic acid an N-methyl-D-aspartate antagonist? Brain Res., 336, 162-166.

Wardley-Smith, B. & Meldrum, B.S. (1984). Effect of excitatory amino acid antagonists on the high pressure neurological syndrome in rats. Eur. J. Pharmacol., 105, 351-354.

Short
Communications

Methionine as a Probe for Studying the Neuron–Glia Glutamate–Glutamine Cycle

J.D. Wood, J.W. Geddes and E. Kurylo

The presence of a glutamate-glutamine cycle between nerve endings and adjacent glial cells was proposed by Benjamin and Quastel (1972). They suggested that glutamate released from nerve endings was taken up by glial cells where it was metabolized to glutamine, the latter compound being in turn released by the glia and taken up by nerve endings where glutamate was reformed by the action of glutaminase. A similar concept, but expanded to include GABA metabolism has also been proposed. Since methionine is known to inhibit glutamine uptake into synaptosomes, a study was initiated to determine whether the administration of methionine to mice brought about a redistribution of glutamine and related amino acids between nerve endings and other cellular structures, and if so, whether the data supported the existence of a neuron-glia glutamate-glutamine cycle. Methods used in the study were as described by Wood et al. (1985).

The intramuscular administration of methionine (3.33 mmol/kg) to mice resulted in changes in the levels of aspartate, glutamate, glutamine and GABA both in nerve endings (synaptosomes) and in other brain cellular structures (as reflected by whole tissue levels) (Fig. 1). However, the amino acid changes in the two locations differed considerably, not only in the time to onset of the changes, but also in the direction of the changes, and in their duration. Since the methionine-induced decreases in the glutamate, GABA and aspartate contents of synaptosomes paralleled the decrease in glutamine content of whole tissue, the data support the Benjamin-Quastel model wherein glial glutamine is a major source of the amino acids in nerve endings. The inhibition by methionine of glutamine uptake by synaptosomes was confirmed, but the observed magnitude of the inhibition made it doubtful whether the effect would be significant in vivo.

Although methionine induced significant changes in the levels of glutamate, GABA and aspartate in whole tissue, the combined concentration of the amino acids remained constant. Since these

438

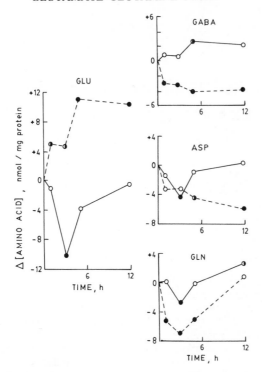

Fig. 1. Effect of methionine (3.33 mmol/kg) on amino acid levels in brain tissue. Broken lines indicate whole tissue, solid lines indicate synaptosomes. Solid and hemisolid circles indicate that the treated group was significantly different from controls at p<0.01 and p<0.05 respectively (from Wood et al., 1985).

amino acids are closely related metabolically by aminotransferase enzymes, it appears that the methionine treatment had shifted the equilibrium away from aspartate and GABA, and towards glutamate. A consequence of this shift would be a concomitant decrease in α-ketoglutarate levels with a possible shortage in the supply of the keto acid from glia to the nerve endings.

REFERENCES

Benjamin, A.M. and Quastel, J.H. (1972). Locations of amino acids in brain slices from the rat. Tetrodotoxin-sensitive release of amino acids. Biochem. j., 128, 631–646.

Wood, J.D., Kurylo, E. and Geddes, J.W. (1985). Methionine induced changes in glutamate, aspartate, glutamine and GABA levels in brain tissue. J. Neurochem., (in press).

Biochemical Effects of Kainic Acid (KA) and Other Excitotoxins in the Rat Superior Colliculus (SC)

V.M. Fosse and F. Fonnum

The superficial laminae of the superior colliculus (SC) receive afferents from retinal ganglion cells and the visual cortex (VC). Previous evidence from our laboratory have suggested that glutamate may be the transmitter in visual corticofugal fibres to SC (Lund Karlsen and Fonnum, 1978). However, the lack of retrograde transport of D-[^{3}H]Asp from SC to visual cortex in cats has cast some doubt on that suggestion (Baughman and Gilbert, 1981). We have therefore performed surgical, chemical (kainic acid) and combined surgical/chemical lesions in rats in order to substantiate the notion of a glutamatergic projection from the VC to SC.

VC ablation reduced high affinity (HA) uptake of D-[^{3}H]Asp by 32% in homogenates of the deafferented SC. Local injection of less than 1 nmol KA into SC reduced HA uptake of D-[^{3}H]Asp, selectively, by 50-60% as compared with the uninjected (control) SC. The GABAergic marker glutamate decarboxylase (GAD) was decreased only at doses exceeding 2 nmol, and the maximal decrease was 60%. Choline acetyltransferase (ChAT) was not affected at any of the doses administered.

Rats subjected to unilateral VC ablation were subsequently injected with 1 nmol KA after either 1, 5 or 12 days into the deafferented SC. Already 1 day after surgery did VC ablation provide some protection and after 5 days the protection was nearly complete. Thus, HA D-Asp did not decrease more than after VC ablation alone. In another group of animals KA was injected first, and 2 days later the ipsilateral VC was ablated. This temporal sequence brought fourth an additive effect of the two lesions. Hence, we infer that visual corticofugal fibres most likely employ an acidic amino acid (Glu/Asp) as their transmitter. Local Glu-neurons, which account for more than 60% of the HA D-Asp uptake in SC, probably contribute to attenuate retrograde transport of D[^{3}H]-Asp from SC to VC (Baughman and Gilbert, 1981; Matute et al, 1984). The notion that visual corticofugal fibres are glutamatergic are supported by our recent observation that the KCl-evoked release of endogenous Glu from SC-slices are decreased by 28% after VC ablation (Fosse et al, in preparation).

Injection of 5 and 10 nmol N-methyl tetrahydrofolic acid, N-MeTHF into SC reduced HA D-Asp by about 50%, only when coinjected with ascorbate (10 nmol), whereas both GAD and ChAT were unaffected. The same amount of other folates (also coinjected with ascorbate) had no significant effects. Ascorbate potentiates the neurotoxicity of 6-hydroxydopamine by enhancing superoxide (O_2 -) formation. Hence, it is tempting to speculate that formation of such radicals were involved after ascorbate/ N-MeTHF injections. O_2 - radicals provoke lipid peroxidation which may inhibit re-uptake of neurotransmitters (Braughler, 1985) and may account for the apparent specificity towards HA D-Asp uptake.

Presumably, quinolinic acid (QA) interacts with the N-methyl-D-aspartate (NMDA) receptor. On a molar basis QA was about 100-fold less potent than KA. Qualitatively, their effects were similar. Thus, HA D-Asp was selectively affected at the lower doses of QA, but the maximal reduction in uptake was only 40% (vs. 60% with KA). In contrast, the maximal reduction of GAD was similar, i.e. 60%, after QA and KA injections. ChAT was not affected by QA. Our findings may indicate that only two thirds of the local Glu neurons, but all the GABA neurons, in SC are endowed with NMDA receptors.

In conclusion, our observations support the notion that visual corticofugal fibres to SC employ Glu (or Asp) as their transmitter and account for about 30% of the HA D-Asp uptake in SC. Most of the remaining uptake (approx 60%) can be accounted for by local Glu neurons in superficial and intermediate layers which most likely are impinged upon by the corticofugal fibres. About 60% of the GAD activity is confined to local GABA neurons in the superficial layers and 40% in terminals of the nigrotectal and reticulotectal projection in deeper layers. The presence of local cholinergic neurons in SC is very unlikely.

REFERENCES

Baughman, R.W. and Gilbert, C.D. (1981). Aspartate and glutamate as possible neurotransmitters in the visual cortex. J. Neurosci., 1, 427-439.

Braughler, J.M. (1985). Lipid peroxidation-induced inhibition of γ-aminobutyric acid uptake in rat brain synaptosomes: Protection by glucocorticoids. J. Neurochem., 44, 1281-1288.

Lund Karlsen, R. and Fonnum, F. (1978). Evidence for glutamate as a neurotransmitter in the corticofugal fibers to the dorsal lateral geniculate body and the superior colliculus in rats. Brain Res., 151, 457-467.

Matute, C., Waldvogel, H.J., Streit, P. and Cuenod, M. (1984). Selective retrograde labeling following D-[^{3}H]-aspartate and [^{3}H]-GABA injections in the albino rat superior colliculus. Neurosci. Lett. Suppl., 18, 190.

Hippocampal Excitatory Neurons: Anterograde and Retrograde Axonal Transport of D–[^{3}H]Aspartate*

B.O. Fischer, J. Storm-Mathisen and O.P. Ottersen

The present method, using a tentatively transmitter spesific technique for labelling neurons and pathways (Streit, P. 1980), is applied on the hippocampal formation.

Tritiated D-aspartate (D-[^{3}H]Asp), a metabolically inert substrate for the high affinity membrane transport of glutamate and aspartate, was pressure injected in vivo by glass micropipettes (o.d. at tip 50 um) into dorsal hippocampus in 5 rats under pentobarbital anesthesia (16 Ci/mmol, about 25 uCi in 100 nl sodium phosphate buffer, pH 7.4). In rats R3, R8 and R13 mainly the fascia dentata was affected, in R35 and R36 the injection centre was located at the CA1/CA3 border. The rats were perfused with 5% glutaraldehyde after 12 or 24 hours and the brains sectioned and processed for autoradiography (Ilford L4, exposure time 5 or 15 weeks).

Labelling was found in:
1) ipsilateral granular cell perikarya and mossy fibres in stratum lucidum (R3 and R13).
2) CA3 pyramidal cell perikarya and neuropil of str. oriens and str. radiatum of CA1 and CA3 both ipsi- and contralateral to the injection (in all rats), and neuropil in lateral septum bilaterally (R13, R35 and R36; sections of septum missing in R3).

---

*Supported by the Norwegian Research Council for Science and the Humanities.

3) ipsi- and contralateral deep hilar neurons and neuropil in inner third of molecular layer of fascia dentata (in all rats).
4) perikarya in ipsilateral entorhinal cortex, layer 2 (R3, R35 and R36).

The neuropil labelling of CA1/CA3 was heavier ipsilateral to the injection, whereas the septal labelling was almost symmetrical. This is consistent with the anatomical data on these projections (Swanson & Cowan, 1979, Taxt & Storm-Mathisen, 1984 and Goldowitz et al. 1979). No labelling was observed in the medial septal nucleus, nucleus of the diagonal band, supramammillary nucleus, raphe nuclei or locus coeruleus. These nuclei are known to project fibres likely to use other transmitters than aspartate or glutamate, to the hippocampal formation (Azmitia, 1981).

These data suggest that D-[^{3}H]Asp is taken up (probably by the terminals) and transported bidirectionally in the axons of 4 types of alleged excitatory neurons in the hippocampal formation. The results of the present study give further evidence that glutamate or aspartate may be transmitters in hippocampal excitatory neurons.

REFERENCES

Azmitia, E.C. (1981). Bilateral serotonergic projections to the dorsal hippocampus of the rat: Simoultaneus localisation of ^{3}H-5HT and HRP after retrograde transport. J. Comp. Neurol., 203, 737-743.

Goldowitz, D., Scheff, S.W. and Cotman, C.W. (1979). The specificity of reactive synaptogenesis: A comparative study in the adult rat hippocampal formation. Brain Res., 170, 427-441.

Streit, P. (1980). Selective retrograde labelling indicating the transmitter of neuronal pathways. J. Comp. Neurol., 191, 429-463.

Swanson, L.W. and Cowan, W.M. (1979). The connections of the septal region in the rat. J. Comp. Neurol., 186, 621-656).

Taxt, T. and Storm-Mathisen, J. (1984). Uptake of D-aspartate and D-glutamate in excitatory axon terminals in hippocampus: Autoradiographic and biochemical comparison with y-aminobutyrate and other amino acids in normal rats and in rats with lesions. Neuroscience, 11, 79-100.

N-Methyl-D-Aspartate (NMDA) Induced Redistribution of Amino Acids in the Hippocampal Formation Visualized by Immunocytochemistry[1]

S. Madsen, A. Lehmann, O.P. Ottersen and J. Storm-Mathisen

The microscopic distribution of amino acids in the brain is now accessible to immunocytochemical analysis. We have employed this technique to study the effect of NMDA, an exogenous excitotoxin, on amino acid localization in the hippocampus.

Rats and rabbits were used in the experiments. In the rats, 1.3 or 2.6 ug NMDA in 0.8 ul 0.1 M sodium phosphate buffer, pH 7.4, was stereotactically injected in CA1 on the left side by means of a Hamilton syringe. Contralaterally we injected 0.8 ul buffer. The rabbit hippocampi were exposed to 5 mM NMDA or buffer via implanted dialysis tubes (Lehmann et al. 1985). The animals were perfused transcardially with 5% glutaraldehyde in sodium phosphate buffer and the brains were removed. Sections were processed (Ottersen and Storm-Mathisen 1984) with sera against conjugated taurine (Tau), glutamate (Glu) and aspartate (Asp).

Both groups of NMDA-treated animals showed a central zone of tissue vacuolization suggestive of cytotoxic changes. Within this zone and extending slightly beyond it there was a pronounced loss of immunoreactivity for Glu, Asp and Tau in pyramidal cell bodies and major dendrites (Fig. 1). Glu-like immunoreactivity was reduced in granular cells. An increase in glial staining, which was most prominent for Tau, was observed at the periphery of the cytotoxicity zone. In the controls only a small zone of tissue damage was observed.

In brain dialysis experiments Lehmann et al. (1985) observed a large increase in Tau efflux from the hippocampus during NMDA exposure, but only a moderately increased efflux of most other amino acids. Our results indicate that Tau release occurs from pyramidal cells. The disappearance of Glu and Asp from these cells may partly be caused by other factors than release, e. g.,

[1]Supported by the Norwegian Research Council for Science and the Humanities.

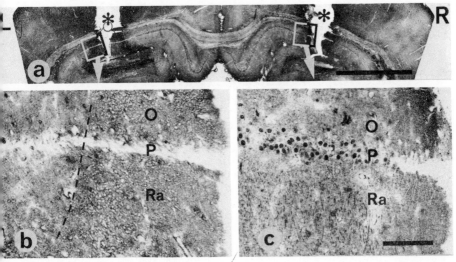

Figure 1. a. Transverse section of the rat brain (Case NMDA 4) showing the two hippocampi with left (L) and right (R) side indicated. The cannula traces (asteriscs) are seen. The frames indicate the parts enlarged in b. and c. Bar: 2 mm.
b. NMDA-injected left side, showing reduced staining for Tau in the pyramidal cells. Broken line indicates extent of cytotoxic changes. O, P, Ra; stratum oriens, pyramidale and radiatum.
c. Buffer injection, with minimal cytotoxic changes. Bar: 100 um, valid for b. and c.

decreased synthesis or increased breakdown. The findings reported here are relevant to brain ischemia, which may be associated with an increased release of endogenous excitotoxins (Benveniste et al. 1984). We will now explore whether experimentally induced ischemia causes changes in amino acid localization similar to those caused by NMDA.

REFERENCES

Benveniste, H., Drejer, J., Schousboe, A., and Diemer, N. H. (1984) Elevation of the extracellular concentrations of glutamate and aspartate in rat hippocampus during transient cerebral ischemia monitored by intracerebral microdialysis. J. Nerurochem. 43, 1369-1374.
Lehmann, A., Lazarewicz, J. W. and Zeise, (1985) N-methyl-aspartate evoked liberation of taurine and phosphoetanolamine in vivo: site of release. J. Neurochem. (in press).
Ottersen, O. P. and Storm-Mathisen, J. (1984) Glutamate- and GABA-containing neurons in the mouse and rat brain, as demonstrated with a new immunocytochemical technique, J. Comp. Neurol. 229, 374-392.

Evidence for Excitatory Amino Acid Afferents to Brain Stem Raphe Nuclei

L. Wiklund, P. Kalén, A. Nieoullon and M. Pritzel

The brain stem raphe nuclei are origins of serotonergic projections to large parts of the central nervous system. We identified afferent inputs to raphe nuclei in Sprague-Dawley rats with retrograde neuroanatomical tracers, and used the method of selective retrograde labelling with D-(^{3}H)asp (Streit, 1980) to indicate which of these may use excitatory amino acids as transmitter.

Nucleus raphe dorsalis (NRD) holds the largest aggregation of serotonergic cells, and projects to cerebral cortex and basal ganglia. For neuroanatomical tracing, wheat germ agglutinin conjugated horseradish peroxidase (WGA-HRP) was implanted into NRD, and after 48 h survival the transported tracer was visualized by the tetramethylbenzidine method (Mesulam, 1978). Retrogradely labelled cells were identified in circa 20 different regions; including band of Broca, dorsomedial and lateral hypothalamus, lateral habenula, ventral tegmental area, substantia nigra, central gray, parabrachial nuclei, raphe magnus and perihypoglossal nuclei. For transmitter selective retrograde labelling microinjections of D-(^{3}H)asp were placed in NRD (25-50 nl, 10^{-2} to 10^{-4}M), rats survived 6-24 h, and brain sections were processed for autoradiography of 4-8 weeks exposure. Large numbers of D-(^{3}H)-asp retrogradely labelled cells were observed in the lateral habenular nuclei, while smaller numbers of labelled cells were found in central gray and substantia nigra.

Nucleus raphe centralis superior (RCS) is the second largest source of ascending serotonergic projections, but contains also many non-serotonergic cells. Retrograde tracing with WGA-HRP demonstrated afferent sources including band of Broca, hypothalamic nuclei, lateral habenula and many brain stem nuclei. As in the NRD experiments, microinjections of D-(^{3}H)asp into RCS resulted in strong selective labelling of cell bodies in lateral habenula.

These autoradiographic results suggested neurochemical investigation. Lateral habenular nuclei were lesioned electrolytically, rats survived 10 days, and NRD and RCS were microdissected from Vibratome^R sections. Habenular lesions induced decrease of high affinity glutamate uptake in NRD to 82$\pm$3.4% (mean $\pm$ S.D., P<0.001,

N=5) and in RCS to $76\pm12.9\%$ (P<0.05, N=5). Taken together, these results strongly suggest excitatory amino acid transmission in the important habenular projection to NRD and RCS, which is believed to mediate limbic influence over the ascending sertonergic systems.

Nucleus raphe magnus (NRM) is the origin of serotonergic and peptidergic projections to the spinal cord, and constitutes an important component of descending nociceptive control. Neuroanatomical tracing demonstrated inputs from frontal cortex, basal forebrain, zona incerta, dorsomedial hypothalamus, nc parafasicularis prerubralis, periaqueductal central gray (PAG), cuneiform nucleus, etc. Selective D-(^{3}H)asp retrograde tracing yielded a relatively complex picture where several afferent sources demonstrated many labelled cells: frontal cortex, dorsomedial hypothalamus and the ventrolateral PAG cell group. The cortical labelling is consistent with suggested glutamate transmission in corticofugal neurons (Fonnum et al, 1981). Labelling of the ventrolateral PAG projection to NRM is of particular interest since this connection is especially important in eliciting analgesia (Fardin et al, 1984). Testing the possible excitatory amino acid transmission, Headley et al (in progress) found that PGA evoked activity in NRM could be suppressed by kynurenate, which is an efficient glutamate receptor antagonist (Perkins and Stone, 1982).

In summary, D-(^{3}H)asp retrograde labelling indicated excitatory amino acid transmission in important afferents to raphe nuclei: habenular afferents to NRD and RCS, and PAG input to NRM. These autoradiographic indications were supported by neurochemical and electrophysiological experiments.

References

Fardin, V., Oliveras, J.-L. and Besson, J.-M. (1984) A reinvestigation of the analgesic effects induced by stimulation of the periaqueductal gray matter in rat. I. The production of behavioral side effects together with analgesia. Brain Res. 306, 105-123.

Fonnum, F., Storm-Mathisen, J. and Divac, I. (1981) Biochemical evidence for glutamate as neurotransmitter in the corticostriatal and corticothalamic fibers in rat brain. Neurosci. 6, 863-873.

Mesulam, M.-M. (1978) Tetramethyl benzidine for horseradish peroxidase neurohistochemistry: A non-carcinogenic blue reaction product with superior sensitivity for visualizing neuronal afferents and efferents. J. Histochem. Cytochem. 26, 106-117.

Perkins, M.N. and Stone, T. W. (1982) An iontophoretic investigation of the actions of convulsant kynurenines and their interactions with the endogenous excitant quinolinic acid. Brain Res. 247, 184-187.

Streit, P. (1980) Selective retrograde labelling indicating the transmitter of neuronal pathways. J. Comp. Neurol. 191, 429-463.

An Antiserum Against Glutamine*

J.H. Laake, V. Gundersen, G. Nordbø, O.P. Ottersen and J. Storm-Mathisen

Glutamine (GlN) serves as a precursor of transmitter glutamate (Glu) and GABA, but has also important roles in intermediary metabolism (e.g., Hertz et al., 1983). Biochemical data and the localization of glutamine synthetase suggest that the production of GlN occurs in glial cells (ibid.). However, the distribution of GlN in the various tissue compartments has so far not been open to direct investigation. Tissue amino acids can now be demonstrated by immunocytochemistry (Storm-Mathisen et al., 1983, and this volume). In the present paper we report preliminary results on specific antibodies directed against GlN.

Rabbits were injected with GlN bound to protein by glutaraldehyde (Storm-Mathisen et al., 1983). The purity of the GlN used (Sigma) was in excess of 99.9% according to amino acid analysis (Rank-Hilger Chromospec, fluorescence detection with o-phthalaldehyde postcolumn, courtesy of Paul Edminson). The protein was either bovine serum albumin for all injections or was changed for every injection, as suggested by Seguela et al. (1984), to augment the immune response to GlN relative to the protein carrier. Both methods gave a specific antibody response. The results presented are from a serum obtained after intracutaneous immunizations with GlN conjugates of a series of proteins (keyhole limpet hemocyanin, casein, human hemoglobin, lysozym; 125 ug in Freund's complete adjuvant) at 6 weeks intervals, and bleeding 2 weeks after the last injection. The serum was purified through Sepharose-bovine serum albumin-glutaraldehyde-Glu on a microcolumn (Ottersen and Storm-Mathisen, 1984). The purified serum showed good specificity (Fig. 1) when tested against a variety of compounds fixed to brain macromolecules by glutaraldehyde (Ottersen and Storm-Mathisen, 1984). When added to the purified serum at a relatively high concentration, free GlN, but not Glu, GABA or NaCl, inhibited the staining of the fixation products of GlN (Fig. 1). This suggests that antibodies in

*Supported by the Norwegian Research Council for Science and the Humanities.

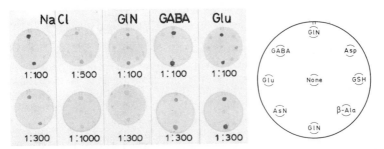

Fig. 1. Characteristics of the GlN antiserum 34 (diluted as stated).
Glutaraldehyde fixation products of amino compounds (GSH = reduced
glutathione) and brain macromolecules were spotted on 13 mm Millipore filters, reacted with the purified serum and developed by the
PAP method (Sternberger, 1979). Products of other compounds tested
without significant crossreactivity included alanine, valine, leucine, proline, tryptophan, methionine, glycine, serine, cysteine,
tyrosine, lysine, arginine, ornithine, histidine, purtrescine, cadaverine, spermine, spermidine, carnosine and homocarnosine. The
diluted sera contained 800 mM of NaCl or of a free amino acid.

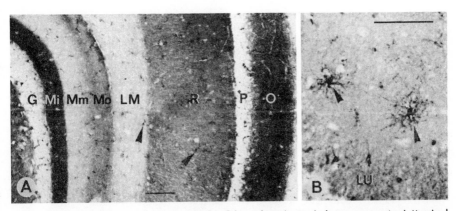

Fig. 2. GlN-LI in a hippocampal slice incubated in oxygenated Krebs'
solution (60 min, 30 °C), fixed in 5% glutaraldehyde and resectioned
at 20 um. A, CA1 and area dentata. B, CA3. Symbols: G, stratum
granulosum; LM, stratum lacunosum-moleculare; LU, stratum lucidum;
Mi, Mm, Mo, sublaminae of stratum moleculare; O, stratum oriens; P,
stratum pyramidale; R, stratum radiatum; ▼ , astroglial cells; bars
100 um.

the absorption purified serum are able to recognize Gln even when it
is not conjugated with glutaraldehyde.

The tissue distribution of Gln-like immunoreactivity (Gln-LI)
was initially studied in immersion fixed hippocampal slices, where
Glu-LI has a very striking laminar distribution corresponding to
that of excitatory nerve terminals (Storm-Mathisen et al., this
volume). Gln-LI also showed a distinct zonal pattern (Fig. 2). This
differed from the distribution of Glu-LI. Notably, Gln-LI was relatively higher than Glu-LI in the inner zone of the dentate molecular
layer and generally lower than Glu-LI in CA3. Nevertheless, the
pattern suggested that Gln-LI was present in excitatory axon terminals. In addition, numerous strongly stained astroglia-like cells
were observed (Fig. 2). In sections of perfusion-fixed brains the
distribution of Gln-LI was more diffuse (Laake et al., in prep.).

The present results suggest that the tissue localization of Gln
can be selectively demonstrated by immunocytochemistry. In addition
to confirming a glial localization of Gln, they also indicate that
Gln can be highly concentrated in certain populations of nerve
endings.

REFERENCES

Hertz, L., Kvamme, E., McGeer, E.G. and Schousboe, A. (eds.) (1983).
    Glutamine, Glutamate and GABA in the Central Nervous System.
    Alan R. Liss, New York.
Ottersen, O.P. and Storm-Mathisen, J. (1984). Glutamate- and GABA-
    containing neurons in the mouse and rat brain, as demonstrated
    with a new immunocytochemical technique. J. Comp. Neurol.,
    229, 374-392.
Seguela, P., Geffard, M., Buijs, R.M. and LeMoal, M. (1984). Anti-
    bodies against y-aminobutyric acid: Specificity studies and
    immunocytochemical results. Proc. Nat. Acad. Sci. USA, 81,
    3888-3892.
Sternberger, L.A. (1979). Immunocytochemistry. Second Ed. Wiley,
    Chichester.
Storm-Mathisen, J., Leknes, A.K., Bore, A.T., Vaaland, J.L.,
    Edminson, P., Haug, F.M. and Ottersen, O.P. (1983). First
    visualization of glutamate and GABA in neurones by immunocyto-
    cehmistry. Nature, 301, 517-520.

The Receptors of the Optic Nerve EPSP
in the Rat LGN

V. Crunelli, N. Leresche, M. Pirchio and J.S. Kelly

Anatomical, biochemical and electrophysiological evidence has
suggested that an excitatory amino acid could be the transmitter of
the optic nerve in the dorsal lateral geniculate nucleus (LGN)
(Kemp and Sillito, 1982). We have now studied the effect of some
of the new excitatory amino acid antagonist on the epsp evoked by
electrical stimulation of the optic nerve using an in vitro slice
preparation of the rat LGN and standard intracellular recording
techniques. LGN slices (250-400µm thick) were prepared and
maintained in vitro as previously described (Godfraind and Kelly
1981; Crunelli et al., 1983). Excitatory amino acids were applied
by iontophoresis while the antagonists were dissolved in the
perfusion medium. Bicuculline (10^{-3}-10^{-4}M) was present throughout
the course of these experiments to block the GABA-mediated ipsp
that follows the optic nerve epsp. Intrasomatic injection of
horseradish peroxidase and subsequent camera lucida reconstruction
was used to characterize impaled cells. They all had morphological
features similar to those described in Golgi preparations for
principal neurones of the rat LGN.

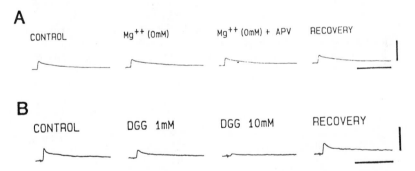

Figure 1. Antagonism by DGG on the optic nerve epsp and lack of
action of APV in the presence and absence of Mg^{2+} ions. Records
are epsps evoked in two principal neurones of the LGN by
stimulation of the optic nerve in the various experimental
conditions indicated above each trace. In A, removal of Mg^{2+} ions
had no effect on the epsp and still in the absence of Mg^{2+} the
addition of APV left the epsp unchanged. In B, DGG at 1mM

decreased the epsp of about 30% of the control amplitude and at 10mM by almost 90%. Recovery was obtained 20 min after switching back to the control solution. Calibration bars equal 15mV and 30msec in A, 20mV and 30msec in B.

D_{5}2-amino-5-phosphonovalerate (APV) applied at a concentration of $10^{-5}M$ that completely blocked the response to N-methyl-D-aspartate (NMDA) had no effect on the epsp (n=10). Since NMDA receptors are also blocked by micromolar concentrations of Mg^{2+}, experiments were also performed in the absence of these ions to assess if a NMDA-sensitive component of the optic nerve epsp is normally blocked in the standard Mg^{2+} concentration (2mM). However perfusion of the slice with a Mg^{2+}-free solution did not produce any change in the amplitude of the epsp nor did APV affect the epsp recorded in OmM Mg^{2+} (n=6) (Fig. 1).

In contrast, γ-D-glutamylglycine (DGG) reversibly inhibit the optic nerve epsp (n=15) with an IC_{50} of 4.7mM and without any effect on resting membrane potential, input resistance and excitability of the cell (Fig. 1). We could also shown that DGG produced a parallel shift towards the right of the stimulus-response curve, indicating that it was acting as a competitive antagonist. The IC_{50} of DGG in the millimolar range was somewhat unexpected, but depolarizations of LGN neurones produced by quisqualate and glutamate were also decreased of about 50% by DGG at a concentration of 4.7mM.

The D-isomer of 2-amino-4-phosphonobutyrate (APB) was completely inactive even at $10^{-2}M$ and the L-isomer had practically no effect on 16 cells. The latter, however, was able to reduce the epsp (30-50%) in another four cells but this effect was always associated with a marked decrease in input resistance and no change in resting potential.

These results indicate that: 1) APB has a non specific effect on the postsynaptic receptors in the LGN, 2) NMDA receptors do not mediate the optic nerve epsp, 3) there is no NMDA-sensitive component of this epsp that is blocked at the physiological Mg^{2+} concentration and 4) the synaptic receptors of the optic nerve epsp in the rat LGN are of the quisqualate/kainate type.

Reference

CRUNELLI, V., FORDA, S. and KELLY, J.S. (1983). Blockade of amino acid-induced depolarizations and inhibition of excitatory post-synaptic potentials in rat dentate gyrus. J. Physiol. **341**, 627-640.

GODFRAIND, J.M. and KELLY, J.S. (1981). Intracellular recording from thin slices of the lateral geniculate nucleus of rats and cats. In Electrophysiology of Isolated Mammalian CNS Preparations (Kerkut, G.A. and Wheal, H., eds) 257-283, Academic Press, New York.

KEMP, J.A. AND SILLITO, A.M. (1982). The nature of the excitatory transmitter mediating X and Y cell inputs to the cat dorsal lateral geniculate nucleus. J. Physiol. **323**, 377-391.

Excitatory Amino Acid Receptors and Synaptic Excitation in the Rat Ventrobasal Thalamus

T.E. Salt

There is evidence that excitatory amino acids may be involved in the excitation by sensory afferents of neurones in the ventrobasal thalamus (Haldeman et.al., 1972; Ben-Ari & Kelly, 1975). Little is however known of the synaptic receptors which may be concerned with these pathways, or of the sensory modalities which may utilise excitatory amino acids as transmitters. The purpose of this investigation was to pharmacologically characterise the synaptic receptors of a physiologically identified afferent pathway to the ventrobasal thalamus using some of the more specific excitatory amino acid antagonists which have recently become available (Davies et.al., 1982).

Extracellular single neurone recordings were made with five-barreled iontophoretic electrodes in the ventrobasal thalamus of urethane-anaesthetised rats. Each drug barrel contained one of the following substances: D,L-homocysteic acid (DLH), N-methylaspartic acid (NMA), kainic acid, quisqualic acid, carbachol, D-alpha-aminoadipic acid (DAA), D-2-amino-5-phosphono valeric acid (APV). In some experiments one barrel contained 1M NaCl for the purpose of automatic current balancing. Electrode penetrations were marked for subsequent histological verification by ejection of pontamine sky blue dye from the recording barrel of the microelectrode. Physiological stimulation of facial hair and vibrissa follicle afferents was performed using an air jet which could be directed at a single vibrissa or small area of hairy skin and whose duration of application was electronically gated. Responses of single neurones to sensory stimuli and exogenously applied agonists, quantified in the form of peri-stimulus time histograms, were challenged with iontophoretically ejected excitatory amino acid antagonists.

In agreement with previous findings (Davies et.al., 1982), DAA and APV were found to antagonise neuronal responses to NMA and DLH rather than responses to kainate, quisqualate or carbachol. Using ejection currents of DAA (n=5) and APV (n=14) which produced such selective antagonism, it was possible to antagonise sensory responses of the same thalamic neurones by between 42 and 100% (mean reduction = 70%) on sixteen of the nineteen neurones studied. Similar results have been obtained in the visual system (Kemp & Sillito, 1982). In the remaining three cases, reduction of the sensory response evoked by stimulation of the hair or vibrissa afferents was observed only when antagonist ejection currents were increased to a degree which also reduced responses to the control excitant (either kainate or quisqualate). These three neurones did not appear to differ in terms of their sensory responses in any obvious way from the majority of the population.

These results indicate that the sensory input to ventrobasal thalamic neurones from non-nociceptive hair and vibrissa follicle afferents is mediated largely by an excitatory amino acid, or similar substance, acting on a synaptic receptor of the NMA type. In view of the finding that it was possible to antagonise such sensory responses completely or almost completely with DAA or APV in many neurones, with little effect on responses to exogenously applied kainate or quisqualate, it appears unlikely that such non-NMA receptors play a significant part in this synaptic response. Preliminary experiments carried out with the selective kainate/quisqualate antagonist, glutamylaminomethylsulphonate (GAMS), also lend support to this hypothesis. For the present, the possibility does however remain that the apparent resistance of the sensory response of some neurones to DAA and APV reflects a non-NMA synaptic receptor, although it is also possible, if not likely, that this resistance is due to the inability of the iontophoretically ejected antagonist to penetrate the synaptic cleft effectively in all cases.

REFERENCES

Ben-Ari, Y. & Kelly, J.S. (1975). Specificity of nuciferine as an antagonist of amino acid and synaptically evoked activity in cells of the feline thalamus. J. Physiol. (Lond.), 251, 25P-27P.

Davies, J., Evans, R.H., Jones, A.W., Smith, D.A.S. & Watkins, J.C. (1982). Differential activation and blockade of excitatory amino acid receptors in the mammalian and amphibian central nervous systems. Comp. Biochem. Physiol., 72C, 211-224.

Haldeman, S., Huffman, R.D., Marshall, K.C. & McLennan, H. (1972). The antagonism of the glutamate-induced and synaptic excitations of thalamic neurones. Brain Res., 39, 419-425.

Kemp, J.A. & Sillito, A.M. (1982). The nature of the excitatory transmitter mediating X and Y cell inputs to the cat dorsal lateral geniculate nucleus. J. Physiol. (Lond.), 323, 377-391.

This work was supported by the Medical Research Council (U.K.)

Interactions of Sulphur-containing Excitatory Amino Acids with Membrane and Synaptic Potentials of Cat Caudate Neurons

P.L. Herrling and W.A. Turski

Neurons in the caudate nucleus of halothane anaesthetized cats were recorded with intracellular electrodes during simultaneous iontophoretic application of drugs and stimulation of the cortico-caudate pathway as described previously (Herrling et al., 1983). The effects of N-methyl-D-aspartate (NMDA) and quisqualate (QUIS) were compared to those of L-cysteine sulfinic acid (CSA) and L-homocysteic acid (HCA) as these two agents were shown to be endogenously released from striatal slices following potassium depolarization (Cuénod et al., this volume).

Cysteine sulfinic acid elicited a firing pattern consisting at the beginning of the iontophoretic application of repetitive bursts similar to the pattern elicited by NMDA (Herrling et al., 1983). This initial pattern then changed after continuing application into the regular pattern usually seen after QUIS applications (ibid.). On cells where CSA and NMDA were cycled, the selective NMDA antagonist 2-amino-7-phosphonoheptanoic acid (AP-7) completely inhibited the effects of NMDA, but only partly those of CSA.

Homocysteic acid, however, elicited a firing pattern much more similar to the pure NMDA pattern (Fig. 1), i.e. the bursts persisted during prolonged applications. These excitations could be completely abolished by co-application of AP-7, while the effects of QUIS on the same cells remained unaffected by this antagonist.

Results from patch- and whole cell clamp studies with excitatory amino acids (Nowak et al., 1984; Mayer & Westbrook, 1984) led to the proposal that NMDA agonists might be capable of modulating excitatory postsynaptic potentials (EPSPs). We therefore applied NMDA, HCA and QUIS to cells on which EPSPs were evoked by stimulation of the precruciate cortex, the site of origin of the cortico-caudate pathway. During equal depolarization of the membrane by

these agents it could be seen that EPSP amplitude and the
number of evoked action potentials were greatly augmented
by NMDA and HCA, but not by QUIS.

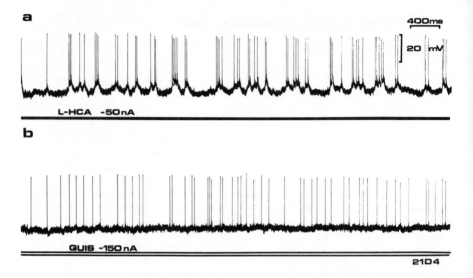

Figure 1. Effects of iontophoretic application of L-homocysteic acid (L-HCA) and quisqualate (QUIS) on the membrane potential of a caudate neuron. a) L-HCA evokes a
bursty firing pattern very similar to the pattern evoked
by NMDA. b) On the same cell QUIS evokes a much more regular pattern.

The present results indicate that CSA is a mixed agonist interacting predominantly with quisqualate or kainate
receptors, while HCA has a predominant action at NMDA-
receptors. Furthermore, they confirm the notion that NMDA
agonists are capable of increasing the effectiveness of
excitatory postsynaptic potentials as predicted by in
vitro experiments.

Herrling, P.L., Morris, R. & Salt, T.E. (1983). Effects
of excitatory amino acids and their antagonists on membrane and action potentials of cat caudate neurones.
J. Physiol. 339, 207-222.

Mayer, M.L. & Westbrook, G.L. (1984). Mixed-agonist
action of excitatory amino acids on mouse spinal cord
neurones under voltage clamp. J. Physiol. 354, 29-53.

Nowak, L., Bregestovski, P., Ascher, P., Herbet, A. & Prochiantz, A. (1984). Magnesium gates glutamate-activated
channels in mouse central neurones. Nature 307, 462-465.

Effects of Magnesium, Temperature and N-Methylaspartate Antagonists on Spontaneous and Evoked Field Potentials in Rat Cortical Slices

J.A. Aram and D. Lodge

Following the recent intracellular studies of cerebral cortical neurones (Thomson, West & Lodge, 1985) demonstrating an excitatory postsynaptic potential with appropriate voltage-dependency and pharmacological sensitivity to implicate receptors of the N-methylaspartate type, we are using in vitro extracellular recording to study the pharmacology of cortical excitation.

Using similar techniques to Thomson et al. (1985), 500 μm slices of rat cerebral cortex were maintained at the interface between an artificial CSF and a humidified gas mixture of 5% carbon dioxide in oxygen. A bipolar nichrome stimulating electrode was placed in the corpus callosum near the cingulum and recordings of spontaneous and evoked extracellular field potentials were made using a glass coated tungsten microelectrode (R = 5Mohm).

The slices were first perfused with medium containing 1 mM magnesium for one hour before changing to one with no added magnesium for a further hour at which stage recording commenced. Field potentials, both spontaneous and evoked by callosal stimulation, could be found in much of the cerebral grey matter. The electrode position was adjusted so that the amplitude of the negative potentials was greatest. Characteristically, in magnesium-free medium at 26-28°C, a large negative potential of 0.5-2 mV was followed by a burst of negative after potentials. Such bursts of up to 12 or more individual after potentials lasted 1-3 sec. Spontaneous and evoked potentials were almost indistinguishable. Reintroduction of medium containing 1 mM magnesium rapidly blocked the late potentials. This block was usually accompanied by an increase in amplitude of the early negative potential which was followed by a positive potential. It thus appears that there are at least three components to the field potentials that follow callosal stimulation. The early negative and positive potentials were apparently not reduced by magnesium and may represent excitatory and inhibitory synaptic potentials seen with intracellular recording (Thomson et al., 1985).

Because low concentrations of magnesium selectively antagonise the action of N-methylaspartate, we investigated the effect of increasing the magnesium from zero to 2 mM. This resulted in a dose-dependent reduction in the number of the after potentials in both the spontaneous and the evoked responses, an effect which could be observed with 50 µM magnesium and reached a maximum at about 400 µM. The amplitude and the time course of individual after potentials were not much affected by changing levels of magnesium.

During these initial studies, we observed that the number of after potentials in magnesium-free medium was reduced if the rate of callosal stimulation was increased. More detailed examination showed that, at a stimulus rate of 1 Hz, no after potentials were evoked and there was no obvious spontaneous activity. The number of after potentials increased as the rate of stimulation was slowed to 0.1 Hz and this was accompanied by the advent of spontaneous potentials between stimuli. Such spontaneous activity became more frequent at lower rates of stimulation. Consequently evoked activity was reduced if the callosal stimulus immediately followed a spontaneous burst. A rate of 0.1 Hz was adopted for later tests.

As the temperature of the slice was changed from 23°C to 30°C, the number of after potentials in evoked and spontaneous bursts increased. Above 30°C the amplitude of the discharges continued to increase but the number of after potentials decreased. At these higher temperatures there was an increase in background random activity and the epileptiform discharges were less easily distinguishable. We subsequently maintained slices near 28°C.

Addition of 5–20 µM 2-amino-5-phosphonovalerate, ketamine or cyclazocine produced a reversible block of the after potentials.

Since like magnesium such substances are selective antagonists of N-methylaspartate (Lodge et al., 1984, Thomson et al., 1985 and cited references), activation of N-methylaspartate receptors is likely to initiate the observed burst of cortical potentials. Such bursts are the presumed neurophysiological substrate of some forms of epilepsy and hence it would appear that this cortical slice preparation may be used for screening putative therapeutic agents. Our results support the accumulating evidence that a wide variety of N-methylaspartate antagonists have anticonvulsant properties.

REFERENCES

Lodge, D., Berry, S.C., Church, J., Martin, D., McGhee, A., Lai, H-M. and Thomson, A.M. (1984). Isomers of cyclazocine as excitatory amino acid antagonists. Neuropeptides, 5, 245–248.

Thomson, A.M., West, D.C. and Lodge, D. (1985). An N-methylaspartate receptor-mediated synapse in rat cerebral cortex: a site of action for ketamine? Nature, 313, 479–481.

Spinal Cord: A Site of Muscle Relaxant Action of Excitatory Amino Acid Antagonists

L. Turski*, T. Klockgether, M. Schwarz, W.A. Turski* and K.-H. Sontag

INTRODUCTION

A growing body of evidence indicates that excitatory amino acids act as neurotransmitters in the mammalian CNS (Watkins and Evans, 1981). The receptors for excitatory amino acids have been designated in terms of their preferred agonists, as N-methyl-D-aspartate (NMDA), kainate (KA) and quisqualate (QA) receptors. The most potent and selective antagonists at the NMDA receptor are the D-isomers of the ω phosphono derivatives of α-amino carboxylic acids, D-(-)-2-amino-5-phosphonopentanoate (D-AP5) and D-(-)-2-amino-7-phosphonoheptanoate (D-AP7)(Watkins and Evans, 1981). γ-D-Glutamyl-aminomethylsulphonic acid (γ-D-GAMS) is an antagonist with a preference for KA and QA mediated excitation (Watkins and Evans, 1981). To investigate the possible role of excitatory amino acid neurotransmission in the regulation of muscle tone, we examined the effects of both agonists, NMDA, KA and QA, and antagonists, AP5, AP7, γ-D-GAMS and L-glutamic acid diethylester (GDEE), of excitatory amino acid receptors in a genetically determined strain of Wistar rats, in which a spontaneous tonic activity can be recorded in the electromyogram (EMG) from the gastrocnemius (GS) muscle.

METHODS

Subjects were genetically spastic Wistar rats (Han:WIST, spa/spa) of mixed sexes at the age of 10-12 weeks, 100-170 g in weight. The spontaneous activity in the EMG was recorded from the GS muscle of unanesthetised rats by means of techniques described elsewhere (Turski et al., 1985). For intrathecal (i.th.) injections rats were chronically implanted with polyethylene catheters under pentobarbitone anesthesia (30 mg/kg i.p.). The catheter was inserted into the subdural space and advanced 70 mm into the lumbar region. The drugs for i.th. administration were dissolved in saline and delivered in a

---
*On leave from the Department of Pharmacology, Institute of Clinical Pathology, Medical School, Jaczewskiego 8, PL-20-090 Lublin, Poland.

volume of 5 µl. KA, QA, kynurenic acid (KYN) and GDEE were obtained
from Sigma (St. Louis, MO, USA). NMDA, AP5, AP7 and γ-D-GAMS were
purchased from Tocris (Buckhurst Hill, Essex, UK). Experimental
groups consisted of four to eight animals. The data from EMG monitoring were statistically analysed by means of Student's t-test.

RESULTS

I.th. administration of NMDA, 0.00001–0.0001 µmol, results in
immediate enhancement of the EMG activity (Fig. 1A). The effect of
NMDA reaches its maximum by 10–15 min after the i.th. injection and
lasts for up to 100–120 min (Fig. 1A). KA have an almost identical
profile and potency of action on the EMG as does NMDA (Fig. 1B). By
contrast, both compounds are consistently about 50 times more potent
as QA, 0.0005–0.005 µmol. Fig. 1C shows that AP7, 0.01–0.1 µmol,
reduces the EMG activity in a dose- and time-dependent manner. AP5
displays comparable potency of muscle relaxant action, while both
KYN, 0.02–0.5 µmol, and γ-D-GAMS, 0.02–0.5 µmol, are much less potent (Fig. 1D). GDEE, up to 1.0 µmol, has no effect on the muscle
tone.

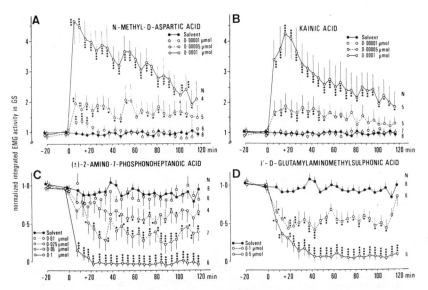

Figure 1. Effect of N-methyl-D-aspartic acid (A), kainic
acid (B), (±)-2-amino-7-phosphonoheptanoic acid (C) and γ-
D-glutamylaminomethylsulphonic acid (D) on spontaneous EMG
activity of the GS muscle in genetically spastic rats. Abscissa: time (min). Ordinate: normalized integrated EMG
activity in the GS. Results are presented as mean ± S.E.M.
in 4-8 animals. *P<0.05, **P<0.01, ***P<0.001 versus sol-
vent, Student's t-test.

DISCUSSION

The most significant novel finding of this study is the demonstration of muscle relaxant action of antagonists of excitation mediated by acidic amino acids at the spinal level in genetically spastic rats. A decrease of the NMDA mediated synaptic excitation within the spinal cord achieved with i.th. infusion of selective antagonists, AP5 and AP7, results in a profound reduction of pathologically increased muscle tone. The specificity of this action is substantiated by demonstrating increases in the muscle tone elicited by i.th. given NMDA. γ-D-GAMS, a preferential KA/QA antagonist, which is not devoid of antagonistic action at NMDA receptors, displays much less potent muscle relaxant effect. By contrast, KA is equiactive with NMDA in its action on the muscle tone. The extent to which muscle relaxant action of γ-D-GAMS depends on the blockade of excitation mediated via NMDA or non-NMDA receptors remains to be established.

CONCLUSION

An effect on the muscle tone of excitatory amino acids and antagonists of amino acids mediated excitation was tested in genetically spastic rats. NMDA, KA and QA, amino acids considered to be selective receptor agonists, increase the muscle tone recorded in the EMG of the GS muscle when given i.th. to rats. NMDA and KA were 50 times more potent than QA in increasing the muscle tone. Antagonists of excitation mediated by dicarboxylic amino acids reduce the muscle tone in genetically spastic rats. In order of decreasing relative potency, AP7, AP5, KYN (preferential NMDA antagonists), and γ-D-GAMS (a preferential KA/QA antagonist) reduce the tonic EMG activity in spastic rats following i.th. administration. GDEE is devoid of muscle relaxant activity. The present data demonstrate that the muscle tone in rats is subjected to the regulation by the excitatory neurotransmission in the spinal cord. Specific antagonists of excitation induced by dicarboxylic amino acids may provide a new class of muscle relaxants offering a new approach to the therapy of states with pathologically increased muscle tone.

REFERENCES

Turski, L., Schwarz, M., Turski, W.A., Klockgether, T., Sontag, K.-H. and Collins, J.F. (1985). Muscle relaxant action of excitatory amino acid antagonists. Neurosci. Lett., 53, 321–326.

Watkins, J.C. and Evans, R.H. (1981). Excitatory amino acid transmitters. Ann. Rev. Pharmacol. Toxicol., 21, 165–204.

Pharmacological Characterisation of Excitatory Amino Acid Receptors on Two–Neuronal Primary Cultures

J. Drejer, T. Honoré and A. Schousboe

Postsynaptic effects of excitatory amino acids may be studied in vitro using electrophysiological techniques or by various neuro-chemical methods such as stimulated Na^+-flux or cGMP formation in brain slices. Here we report on a simple and sensitive method for pharmacological studies on the effects of excitatory amino acids and their antagonists measuring induced transmitter release from cultured neurons mediated via excitatory amino acid receptors. Neuronal cultures prepared from 7–days-old mouse cerebella have been shown to be an almost pure culture of glutamatergic granule cells (Drejer et al. (1985) Neurochem. Res. 10, 49). Cultured cerebral cortex neurons obtained from 15–days-embryo mice contain mainly GABA-ergic inter-neurons (Larsson et al. (1985) Int. J. Devl. Neurosci. 3, 177). For release studies cerebellar granule cells were preloaded with ^{3}H-D-aspartate as a tracer for the proposed transmitter L-glutamate and cerebral cortex neurons were preloaded with ^{3}H-GABA. The release of radioactivity from the cultures was studied in a continuous super-fusion system and neurons were intermittently stimulated by the addition of excitatory amino acid agonists to the superfusion medium in the presence or absence of antagonists.

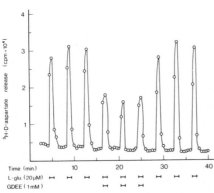

Fig. 1. Release of preloaded ^{3}H-D-aspartate from cultured cere-bellar granule cells as a func-tion of superfusion time. Every 4 min cells were stimulated for 30 sec by changing the superfusion medium from PBS to PBS containing 20 μM L-glutamate. The stimula-tion periods are indicated in the figure by horizontal bars. During the stimulation periods 4-6 the superfusion medium contained 1 mM GDEE in addition to 20 μM L-glutamate.

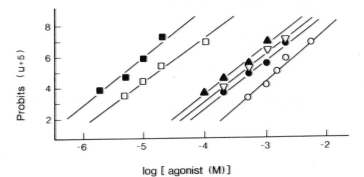

Fig. 2. Dose-response curves (log-probit transformed) from release experiments performed on cortex neurons stimulated with ■ Quisqualate, □ AMPA, ▲ Kainate, ▽ NMDA, ● L-glutamate, o L-aspartate.

It is seen from Fig. 2 that all excitatory amino acids tested could induce calcium-dependent ^{3}H-GABA release from cerebral cortex neurons with quisqualate being the most potent agonist. In granule cells (results not shown) only L-glutamate and L- and D-aspartate induced calcium-dependent release of ^{3}H-D-aspartate from the neurons at micromolar concentrations. Quisqualate, kainate, NMDA, L-α-aminoadipate and D-glutamate were all very weak stimulators of ^{3}H-D-aspartate release (EC_{50} > 1 mM). All stimulations by agonists could be blocked by one or more of the antagonists D-APV, D-APH, GDEE, D-α-AA, γ-DGG and PK 26124.

Binding studies. ^{3}H-L-glutamate binding (2 nM) was studied on a membrane preparation from the cultured neurons in the presence of Ca^{2+} (2.5 mM) and Cl^- (35 mM). It was found that quisqualate was a potent displacer of ^{3}H-L-glutamate binding in cortex neurons whereas in granule cells quisqualate was only a poor displacer. This is in agreement with the findings from the release experiments. Thus, quisqualate could only induce transmitter release in cortex neurons. Also L-glutamate and L-aspartate showed good correlations between binding data and effect on transmitter release (not shown).

Conclusion. The results of this study demonstrate that pharmacological effects (such as evoked transmitter release) of excitatory amino acids may be studied in a simple and sensitive way in neuronal cultures. Such models may represent a useful alternative or supplement to electrophysiological approaches. Using this approach we have demonstrated the presence of functionally active glutamate receptors on cerebellar granule cells and on cerebral cortex interneurons. On the latter cell type all three classical excitatory amino acid receptors (the NMDA-, quisqualate- and kainate receptors) were expressed whereas granule cells express a unique type of receptors recognizing only L-glutamate and L- and D-aspartate, but neither NMDA, quisqualate nor kainate.

Molecular Characterisation of Excitatory Amino Acid Receptors in Rat Cortex

T. Honoré, M. Nielsen and J. Drejer

Different localisation of the quisqualate, kainate and NMDA subtypes of excitatory amino acid receptors have been demonstrated by autoradiographic studies (Cotman et al. this volume). Although the receptor subtypes seem different there is some overlap in the pharmacological specificities of the subtypes and it is therefore important to know if the binding sites are completely different chemical entities or if the differences in binding site specificity are due to the same macromolecule forced into different conformations by the surrounding cell membranes.

Determination of molecular target size

Determination of functional size of a macromolecule by the high energy irradiation technique is based on the fact that the function of biological macromolecules are destroyed by high energy particles according to their size following classical target theory (For review Jung, 1984).

Table 1 summarize the molecular weight of binding sites for excitatory amino acid neurotransmitters.

Table 1. Determination of molecular weights of receptors for excitatory neurotransmitters.

Ligand	Receptor type	Molecular weight (daltons) mean ± SEM (n determination)
^{3}H–AMPA	quisqualate	51,600 ± 3,800 (5)
^{3}H–kainate	kainate	76,600 ± 5,500 (3)
^{3}H–kainate (+ Ca^{2+})	quisqualate?	52,400 ± 5,000 (3)
^{3}H–glutamate (+ Cl^-/Ca^{2+})	APB?	approx. 10,000; >1,000,000 (3)
^{3}H–glutamate (+ Na^+)	uptake site	670,000 ± 50,000 (4)

The quisqualate subtype of receptors as determined by ^{3}H-AMPA has a target size of 51,600 daltons.

The kainate binding site assayed in tris-citrate buffer has a molecular target size of 76,600 daltons. When ^{3}H-kainate binding is assayed in the presence of calcium ions only a low affinity binding site is detected. The molecular weight of this site correspond to the quisqualate site. The fact that kainic acid is a weak inhibitor of ^{3}H-AMPA binding support the suggestion that the low affinity kainate binding site in the presence of calcium is equivalent to the quisqualate site.

Binding of ^{3}H-glutamate in the presence of Cl^- and Ca^{2+} to homogenates prepared from frozen whole cortex is similar to glutamate binding to homogenates from fresh cortex. This type of binding might correspond to a recently described APB-site. Radiation inactivation analysis of this binding site reveal two molecular weights; one of approximately 10,000 daltons and one above 1,000,000 daltons. The low molecular weight could correspond to the molecular weight of a recently isolated glutamate binding protein from bovine brain (Michaelis et al. this volume).

[3]H-glutamate binding in the presence of sodium ions is believed to represent binding to the high affinity glutamate uptake recognition sites. The molecular weight of the uptake carrier recognition site is 670,000 daltons.

Preliminary ^{3}H-glutamate binding experiments using frozen homogenates from rat cortex showed only 10 percent of binding compared to the homogenate before the freezing/thawing procedure. Apparently, this binding represent sixty percent NMDA sites and forty percent quisqualate sites (see also Fagg et al. this volume). Preliminary irradiation inactivation analysis of ^{3}H-glutamate binding to previously frozen homogenates in the presence of quisqualate (10 µM), which should account only for the NMDA site revealed a molecular weight equal to or greater than 140,000 daltons.

Influence of Ca^{2+} and SCN^- ions

Another way to differentiate between different subtypes of excitatory amino acid receptors is by investigating the influence of different ions on the binding of different ligands. Ca^{2+} ions enhanced two-fold ^{3}H-AMPA and ^{3}H-glutamate binding to previous frozen homogenates, whereas ^{3}H-kainate binding was reduced three-fold. The chaotropic ion SCN^- enhanced ^{3}H-AMPA binding eight-fold, whereas ^{3}H-glutamate and ^{3}H-kainate binding to previous frozen homogenate was almost unaltered. ^{3}H-glutamate binding to fresh homogenates was reduced five-fold.

The present results, molecular weights of the binding sites as well as the ionic dependency of the binding sites, are in favour of the explanation that completely different structures of receptors for excitatory amino acids exist.

Reference. Jung, C.Y. (1984) Molecular Weight Determination by Radiation Inactivation. In Molecular and Chemical Characterization of Membrane Receptors (eds. J.C. Venter and L.C. Harrison) Alan R. Liss, Inc., New York.

Striatal Glutamate Uptake and Binding are Activated *in vivo* by Cortical Stimulation: Modulatory Influence of the Nigrostriatal Dopaminergic System on Corticostriatal Transmission

A. Nieoullon, L. Kerkerian and M. Errami

Reciprocal interactions have been shown to occur at presynaptic level in the striatum between the nigrostriatal dopaminergic and the corticostriatal glutamatergic pathways. Glutamate (Glu) was evidenced to exert a facilitatory action on dopamine (DA) release whereas DA has an inhibitory effect on Glu release.

High affinity glutamate utpake (HAGU) and ^{3}H-Glu binding were measured from rat striatal samples. Glu transport was shown to be related for more than 60% to corticostriatal afferent fibres. A single binding site for Glu was detected ($Kd = 1.75 \times 10^{-6}$ M), located for 37% on striatal cells sensitive to kainic acid. A close correlation between this binding site and the cortical input was evidenced by the hypersensitivity following decortication (Bmax : +23%).

HAGU and Glu binding were activated 20 min after cortical stimulation in vivo. Increased Glu uptake was concomitant with an increase in the affinity of the transport system for Glu, while activation of Glu binding correlated with a selective increase of Bmax (+26.4%). Since cortical stimulation has been previously shown to enhance striatal Glu release, these results suggest that an increase in corticostriatal neurons firing rate induces an activation of Glu release and uptake processes at nerve terminals, together with an increase in the number of the binding sites at postsynaptic level.

HAGU was also shown to be inhibited in vitro by DA. DA influence was found to be probably exerted by means of presynaptic D_2 type receptors activation which was previously evidenced to inhibit the Glu K^+-evoked release. These data suggest the existence of a presynaptic dopaminergic control on Glu turnover at corticostriatal nerve endings. Pharmacological manipulations in vivo and in vitro of the dopaminergic nigrostriatal system combined with cortical stimulations were used to further analyze in which conditions the dopaminergic influence on HAGU is exerted. Haloperidol was shown to potentiate the activatory effect of cortical stimulation on HAGU, while it had no effect alone on basal HAGU. A low concentration of DA in vitro, which had no effect on basal HAGU, was shown to be potent to inhibit the stimulation-increased HAGU. The inhibitory effect of DA appeared to be directly proportional to the level of HAGU activation from basal value. These results suggest that DA does not exert a tonic inhibitory control on Glu uptake at corticostriatal nerve terminals but that it acts by counteracting the increase in Glu uptake related to an hyperactivity of the corticostriatal neurons. Thus, the hypothesis of a modulatory presynaptic inhibitory influence of the nigrostriatal dopaminergic input on corticostriatal glutamatergic transmission was developed.

This hypothesis was further investigated by measuring HAGU in rats submitted to 6-OHDA lesions of the nigrostriatal dopaminergic neurons. The results showed that if basal HAGU level is not affected by the lesion, there is on the contrary a slight but significant increase in the activatory effect of cortical stimulation on HAGU in the animals with the lesions when compared to shame operated animals. These results, in agreement with those obtained with haloperidol pretreatment, reinforce the idea that the nigrostriatal dopaminergic system acts as a neuromodulator of the corticostriatal glutamatergic transmission.

NIEOULLON A., KERKERIAN L., DUSTICIER N. Inhibitory effects of dopamine on high affinity glutamate uptake from rat striatum. Life Sciences 30 (1982) 1165-1172.

NIEOULLON A., KERKERIAN L., DUSTICIER N. Presynaptic dopaminergic control of high affinity glutamate uptake in the striatum. Neurosci.Lett., 43 (1983) 191-196.

NIEOULLON A., KERKERIAN L., DUSTICIER N. Presynaptic controls in the neostriatum : reciprocal interactions between the nigrostriatal dopaminergic neurons and the corticostriatal glutamatergic pathway. Exp.Brain Res., suppl.7 (1983) 54-65.

KERKERIAN L., DUSTICIER N., NIEOULLON A. Modulatory influence of dopaminergic neurons on high affinity glutamate uptake in the rat striatum (submitted).

Ionic Modulation of Glutamate Receptor in Rat Neuronal Tissues

Y. Yoneda, K. Ogita, H. Nakamuta and M. Koida

INTRODUCTION

L-Glutamic acid (GLU) has been believed to play an excitatory neurotransmitter role in the mammalian central nervous system through the induction of a transient increase in the postsynaptic membrane conductance of Na^+ and Ca^{2+}, respectively (Takeuchi and Onodera, 1973; Chang and Michaelis, 1980). The exact coupling mechanisms of this GLU receptor to ion channels, however, are not fully clarified at present. In this study, we have examined the effect of various ions on ^{3}H-GLU binding to diverse neural tissues including the brain, retina, adrenal and pituitary using a filtration assay method (Yoneda and Kuriyama, 1980) in order to elucidate these coupling mechanisms.

RESULTS AND DISCUSSION

Ammonium chloride was found to abolish the sodium acetate-induced augmentation of the binding, while itself elicited a significantly stimulatory action on the binding in a temperature-dependent manner. Similar raise of the binding exclusively resulted from the addition of anions known to penetrate the anion channel. In addition, calcium acetate (0.1-2.5mM) exerted a chloride-dependent enhancement of the binding. The Ca-stimulated as well as Cl-dependent binding was significantly deteriorated by an antagonist for anion channel (picrotoxinin, 10^{-3}M) and inhibitors of anion transport (ethacrynic acid, 10^{-3}M; DIDS, 10^{-4}-10^{-3}M), respectively. These results all support the proposal that central GLU receptor may be coupled to anion channel in addition to cation channel and that Ca may stimulate the binding through the action on the anion channel rather than affect the binding site directly.

---
*This work was supported in part by a grant (No. 59770150) to Y. Y. from the Ministry of Education, Science and Culture, Japan.

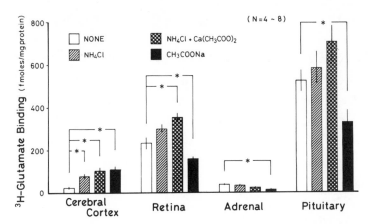

Fig.1 Comparison of ³H-glutamate binding in rat cerebral cortex, retina, adrenal and pituitary.

A significant basal binding activity of ³H-GLU was also detected in the adrenal and pituitary in addition to various central structures such as cerebral cortex and retina (Fig.1). Chloride and calcium ions exhibited a profound enhancement of the binding to central regions without significantly altering that to peripheral tissues. Sodium ion, however, induced a significant deterioration of the binding to retina, adrenal and pituitary despite the occurrence of a remarkable augmentation of the binding to other central structures (Fig.1).

CONCLUSION

GLU receptor may be present in various peripheral neural tissues such as adrenal and pituitary in addition to central structures and coupling mechanisms of the peripheral receptor to ion channels may be different from those in central tissues.

REFERENCES

Chang, H. H. and Michaelis, E. K. (1980). Effect of L-glutamic acid on synaptosomal and synaptic membrane Na⁺ fluxes and (Na⁺-K⁺)-ATPase. J. biol.Chem., 255, 2411-2417.
Takeuchi, A. and Onodera, K. (1973). Reversal potentials of the excitatory transmitter and L-glutamate at the crayfish neuromuscular junction. Nature (New Biol.), 242, 124-126.
Yoneda, Y. and Kuriyama, K. (1980). Presence of a low molecular weight endogenous inhibitor on ³H-muscimol binding in synaptic membranes. Nature, 285, 670-673.

Excitatory Amino Acid Binding Sites in Rat Brain Membranes

M. Recasens, J.-P. Pin, J.F. Rumigny and J. Bockaert

INTRODUCTION

The excitatory action of acidic amino acids are thought to be mediated through separate receptors, termed N-methyl-D-aspartate, quisqualate and kainate as primarily determined from electrophysiological studies. Ligand binding techniques, however, did not permit a correlation between the pharmacological specificity of binding sites and receptors. Although electrophysiological techniques have their own problems, the aim of the present report is to examine the various methodological aspects of excitatory amino acid binding techniques and to propose a classification of the various sites, taking into account the technical problems observed in such a study and using $L(^3H)$-glutamate ($L(^3H)GLU$), $L(^3H)$-aspartate ($L(^3H)ASP$) and $L(^3H)$-cysteine sulfinate ($L(^3H)CSA$) as ligands.

METHODOLOGICAL ASPECTS

Recent reports (Fagg et al., 1982) have shed some light on the discrepancies described in the literature, concerning the number of acidic amino acid binding sites, their characteristics and pharmacological specificities. These authors demonstrated that Cl^- ions, widely used in binding assay buffers, strongly stimulate GLU binding, resulting in a change of the pharmacological characteristics. Along these lines, our recent observations (Pin et al., 1984) suggest that this Ca^{2+}/Cl^- induced change in the pharmacological specificity of GLU binding sites may, in fact, result from a Cl^--stimulated accumulation into membrane vesicles, present in membranes prepared for binding assays. These observations are the following : 1) The time course of the Ca^{2+}/Cl^- dependent $L(^3H)$-GLU binding followed the pattern of excitatory amino acid uptake into synaptic vesicles and not a typical Michaelis-Menten curve, 2) Low incubation temperature (0-4°C) or freezing the membrane, completely inhibits the Ca^{2+}/Cl^- effects on the L(^{3}H)-GLU binding, 3) As the osmolarity of the incubation medium was increased by sucrose addition, the Ca^{2+}/Cl^- dependent enhancement of L(^{3}H)-GLU binding was decreased, 4) The cation and anion specificity coincide with the ability of these ions to create a membrane potential (interior negative) in membrane vesicles, 5) The density of the Ca^{2+}/Cl^- dependent GLU binding sites 80-100 pmoles/mg protein would fit better with an uptake capacity than with a number of receptor sites, 6) Electron micrograph revealed the presence of some membrane vesicles in extensively water washed membranes and 7) Chloride channel blockers (Furosemide, DIDS...) inhibit the Ca^{2+}/Cl^- induced increase in L(^{3}H)-GLU binding (unpublished data). From these observations, it can be deduced that three experimental variables (a : nature of the buffer, b : membrane preparation procedure and c : incubation temperature) are of primary importance to avoid L(^{3}H)-GLU accumulation which represents about 80% of the GLU "binding sites". a : Nature of the buffer : The incubation buffer must be free of permeant anions (Cl^-, Br^-). Under these conditions the driving force, which is generated supposedly by the relative permeability of the anion in relation to that of the cation through the membrane of the vesicles, is removed. b : Membrane preparation procedure. The membrane preparation procedure will be decisive in determining the amount of membrane vesicles which if possible, must be removed (i.e. how extensively the membranes were washed and "disrupted" by homogeneisation, hypotonic shocks, freezing and thawing, detergents or enzymatic digestion). One must remain aware of the possible conformational changes occurring during such treatments. Incubation temperature : Low temperature inhibits the Ca^{2+}/Cl^- augmented GLU binding, probably by blocking Cl^- entrance in vesicles. However, low temperature may also modify the conformation of the real GLU receptors. Thus, the choice of the incubation temperature remains controversial.

CHARACTERISATION OF ACIDIC AMINO ACID BINDING SITES

Taking the above observations into account, we re-examined the binding of three endogenous excitatory acidic amino acids, L(^{3}H)-GLU, L(^{3}H)-ASP and L(^{3}H)-CSA to crude rat brain synaptic membranes. Binding assays were performed at 0°C in 50 mM Tris-Citrate buffer, pH 7.4. Data analysis of saturation and displacement curves revealed the existence of two sites (high and low affinity) for both L-GLU and L-CSA and one site for L-ASP.

Ligand	K_d (µM)		% B_{max} (pmoles/mg protein)
L(^{3}H)GLU	K_d^H	0.14 ± 0.01	3.4 ± 1.9
	K_d^L	5.70 ± 1.90	74.0 ± 12.0
L(^{3}H)CSA	K_d^H	0.07 ± 0.03	2.2 ± 1.0
	K_d^L	12.40 ± 4.90	49.0 ± 27.0
L(^{3}H)ASP	K_d	5.8 ± 3.10	31.0 ± 11.0

The pharamcological profiles of the GLU, CSA and ASP high affinity sites respectively, are the following :

GLU sites : L-GLU = L-CSA > HOS (Hydroxylamine-O-sulfate > L-ASP > L- α amino adipate = sulphobromophthaleine >> D-CSA, QA = homocysteate = cysteate > Kynurenate
CSA sites : L-CSA > L-GLU > hydroxylamine-O-sulfate > L-ASP > D-CSA > Kyurenate = sulphobromophtaleine = Lα amino adipate > homocysteate = cysteate > QA
ASP sites : L-APS > L-CSA > L-GLU > Cysteate = sulphobromopthaleine > QA >> D-CSA = L α Amino adipate = homocysteate = Kynurenate.

From our experiments, we suggest that these endogenous excitatory amino acids interact at three distinct binding sites, - a high affinity site where CSA and GLU are equivalent - a CSA preferring site - an ASP preferring site. It must be emphasised that none of the classical agonists (NDMA, QA, KA) and antagonists (2-amino-5-phosphonovalerate, 2-amino-4-phosphonobutyrate) are efficient at these sites. These characteristics of specificity do coincide well with those reported by Michaelis et al. (1983) for the GLU binding protein purified from rat brain, and not with the pharmacology previously reported in this field with binding techniques (for reasons described in Methodological Aspects). These data also indicate that the GLU binding protein may represent the major population in crude synaptic membranes, masking the QA, KA and NMDA receptors. Further purification of the membranes is essential in order to study these receptors by binding techniques as recently demonstrated by Fagg et al., 1984. Nevertheless, the presence of a major GLU binding protein (and possibly CSA and ASP binding proteins) not directly linked to ion channels (unobservable by electrophysiological studies), raise the question of its (their) role(s). The interaction of excitatory amino acids with this protein may lead to intracellular biochemical events such as cyclic GMP formation or phosphatidyl inositol hydrolysis.

Fagg, E., Foster, A.C., Mena, E.E. and Cotman, C.W. (1982). Chloride and calcium ions reveal a pharmacologically distinct population of L-Glutamate binding sites in synaptic membranes : Correspondence between biochemical and electrophysiological data. J. Neurosci., 2, 958-965.
Fagg, E. and Matus, A. (1984). Selective association of N-methyl-aspartate and quisqualate types of L-glutamate receptor with brain postsynaptic densities. Proc. Natl. Acad. Sci. USA, 81, 6876-6880.
Michaelis, E.K., Michaelis, M.L., Stormann, T.M., Chittenden, W.L. and Grubbs, R.D. (1983). Purification and Molecular characterization of the brain synaptic membrane glutamate binding protein. J. Neurochem., 40, 1742-1753.
Pin, J.P., Bockaert, J. and Récasens, M. (1984). The Ca^{2+}/Cl^- dependent $L(^3H)$-glutamate binding : a new receptor or a particular transport process? F.E.B.S. Lett., 175, 31-36.

Na$^+$-Dependent Binding of D-Aspartate: Inhibition by K$^+$

N.C. Danbolt and J. Storm-Mathisen

Recently 3 groups including ourselves (Kramer and Baudry, 1984; Danbolt and Storm-Mathisen, 1984; Mena et al., 1985) have studied Na$^+$-dependent "binding" of ^{3}H-labelled glutamate or aspartate (Asp) in brain synaptic membranes and have reported that it is strongly inhibited by K$^+$ (IC$_{50}$ about 1 mM), which competes with Na$^+$. We now report that most of the specific "binding" may represent uptake into membrane bounded structures rather than binding to the transporter, but show that inhibition of this uptake by K$^+$ does not depend on ion gradients.

Rapid filtration through 0.45 um cellulose ester filters was used to separate free D-[^{3}H]Asp from "bound". Most of the specific "binding" was lost when the membranes were washed with hypotonic buffer (0.5 mM Tris acetate pH 7.4 on the filters) rather than with the 50 mM Tris acetate buffer containing 200 mM NaCl that was used for incubation (Table 1). The loss was not due to the removal of Na$^+$, since the "binding" was retained after washing with isotonic sucrose or with buffer containing 200 mM LiCl.

Uptake was present, but reduced, in membrane preparations treated to dissipate ion gradients [resuspended, sonicated, frozen (-80°C, thawed 5 min, 30°C) and/or left for extended periods of time in media of the same composition as the incubation medium, and/or treated with gramicidin "D"] (Table 1). The sensitivity of the "binding" to inhibition by gramicidin "D", hypotonic wash and freezing varied among membrane preparations (Table 1), whole tissue membranes (Foster and Roberts, 1978) always being more susceptible than synaptic membranes (Zukin et al., 1974; except washing in 50 mM Tris acetate rather than in H$_2$O). The latter type of preparation probably entails a more extensive disruption of membrane bounded particles than the former.

*Supported by the Norwegian Research Council for Science and the Humanities.

Table 1.   "Binding" of D-[³H]Asp in rat brain membranes

Membrane preparation	Washing with hypotonic buffer	Freezing/ thawing before incubation	Incubation with gramicidin (100ug/ml)	Incubation with both gramicidin and KCl	Incubation with KCl (3 mM)
Whole tissue	15±2(5)	48±7(4)	25±2(4)	6±1(3)	29±2(3)
Synaptic	30±5(3)	81±4(4)	60±10(3)	14±2(3)	29±1(3)

Membranes (6-16 ug prot. in 500 ul) were incubated (15 min, 30 °C) with 100 nM D-[³H]Asp. Filtration and 2 rinses took < 10 s. The results are expressed as % of control (appr. 90 and 20 pmol/mg prot. for whole tissue and synaptic membranes, respectively) and represent mean values ± SEM of n (in parentheses) separate experiments.

The inhibitory effect of K^+ was not changed by treatment affecting ion gradients and was independent of the type of membrane preparation (Table 1), suggesting that the inhibition does not reflect a reversed K^+ gradient (cf. Kanner, 1983; Roskoski et al., 1981). This was corroborated by experiments in which synaptosomes were lysed and the membranes washed in hypotonic media containing the same KCl concentration (3 mM) as the incubation medium.

CONCLUSIONS

(1) The "binding" of excitatory amino acids in membrane preparations may to a large extent represent transport into membrane bounded saccules (cf. Pin et al., 1984 for Na^+-independent "binding"). (2) Dissipation of the ion gradients is not sufficient to prevent carrier mediated uptake. (3) Na^+-dependent transport of this type is blocked by K^+ present at similar, low concentrations on both sides of the membrane. This indicates that K^+ interacts with the binding of D-Asp to the transporter or with the movement of the transporter between the two sides of the membrane. (4) The interaction of K^+ with the transporter may be important for the uptake mechanism.

REFERENCES

Danbolt, N.C. and Storm-Mathisen, J. (1984). Na^+-dependent binding of D-aspartate (Asp) and L-glutamate (Glu) to brain synaptic membranes is strongly inhibited by K^+. Neurosci. Lett. Suppl. 18, S431.

Foster, A.C. and Roberts, P.J. (1978). High affinity L-[^{3}H]glutamate binding to postsynaptic receptor sites on rat cerebellar membranes. J. Neurochem., 31, 1467-1477.

Kanner, B.I. (1983). Bioenergetics of neurotransmitter transport. Biochim. Biophys. Acta, 726, 293-316.

Kramer, K. and Baudry, M. (1984). Low concentrations of potassium inhibit the Na-dependent [^{3}H]glutamate binding to rat hippocampal membranes. Eur. J. Pharmac., 102, 155-158.

Mena, E.E., Monaghan, D.T., Whittemore, S.R. and Cotman, C.W. (1985). Cations differentially affect subpopulations of L-glutamate receptors in rat synaptic plasma membranes. Brain Res., 329, 319-322.

Pin, J.P., Bockaert, J. and Recasens M. (1984). The Ca^{2+}/Cl$^-$ dependent L-[^{3}H]glutamate binding: a new receptor or a particular transport process? FEBS Lett., 175, 31-36.

Roskoski, R. jr., Rauch, N. and Roskoski, L.M. (1981). Glutamate, Aspartate and gammaaminobutyrate transport by membrane vesicles prepared from rat brain. Arch. Biochem. Biophys., 207, 407-415.

Zukin, S.R., Young, A.B and Snyder, S.H. (1974). Gamma-aminobutyric acid binding to receptor sites in the rat central nervous system. Proc. Nat. Acad. Sci. USA, 71, 4802-4807.

Activation of Rat Brain Cortex Calcium Channels by Excitatory Amino Acids

N. Riveros and F. Orrego

INTRODUCTION

Calcium channels in the in vitro rat brain cortex are activated by all the excitatory amino acids tested, particularly by those acting at N-methyl-D-aspartate (NMDA) receptors, i.e. ibotenate and N-methyl-DL-aspartate (NMA), while the non-excitatory substances α-methylaspartate, N-acetylaspartate, or N-acetylaspartylglutamate were inactive (Berdichevsky et al., 1983; Riveros and Orrego, 1984). In the present work we have studied the sensitivity of these calcium channels to a number of calcium channel blockers, to tetrodotoxin, and to the endogenous inhibitory substances, γ-aminobutyrate (GABA) and adenosine.

METHODS

Thin rat brain cortex slices, obtained with blade and blade - guide, were incubated in vitro as described by Berdichevsky et al. (1983), except that the slices were pre-incubated for 30 min before adding the excitatory amino acids and ^3H-inulin plus ^{45}Ca. This lowered the control uptake of ^{45}Ca into the slice non-inulin space, as well as experimental variation.

RESULTS AND DISCUSSION

NMA (150 μM), kainate (KA) (2.5 mM), glutamate (5 mM), and DL-homocysteate (DLH) (5 mM), all increased both the initial rate of ^{45}Ca influx into the non-inulin space, as well as the ^{45}Ca content attained at equilibrium, thus indicating that the increased labelling of the cellular (non-inulin) compartments is primarily due to an increased calcium influx, and not to a decreased efflux, which would only affect the equilibrium level, but not the initial rate of influx.

The NMA-enhanced calcium influx, studied in Tris-buffered medium, was selectively blocked by 3 mM Cd or Co, but was completely insensitive to the dihydropyridine Ca channel "blocker" nifedipine (1 μM). This is characteristic of voltage-sensitive Ca channels of brain. Surprisingly, nifedipine, by itself, increased the initial rate of 45Ca influx, as well as the equilibrium level of 45Ca in the non-inulin space. This may be due to an agonist effect of nifedipine on some Ca channels or, else, to an indirect effect by which it blocks a system that normally inhibits these channels.

Tetrodotoxin (TTX) (1 μM) was unable to block NMA-stimulated Ca channels. This indicates that the opening of these Ca channels is not a consequence of depolarization mediated through voltage-sensitive sodium channels, which are highly sensitive to TTX.

GABA at 2.5, 5 or 10 mM increased Ca uptake by the cells from about 11 nmol/mg protein/10 min, to about 15 nmol/mg protein/10 min, but reduced NMA-stimulated Ca uptake from about 19 nmol/mg protein/10 min to exactly the same level (15 nmol/mg protein/10 min) found with GABA alone. We interpret this finding as due to a neuronal GABA-induced depolarization, mediated by a chloride-dependent mechanism, i.e. by a reduction of Cl_i. This depolarization partially opens the neuronal Ca channels, but also fixes the plasma membrane potential at the Cl equilibrium potential and, thus, NMA is unable to overcome this "clamping" of the membrane potential, and to depolarize it to the usual level. Adenosine, on the other hand, at 1 to 100 μM, was unable to modify either basal or NMA-activated Ca uptake.

In conclusion, the Ca channels opened by NMA or other excitatory amino acids are very similar, or identical, to neuronal voltage-sensitive Ca channels. It seems possible that the ionophore directly coupled to acidic amino acid receptors, is the TTX-insensitive cation channel described by Luini et al. (1981), the activation of which leads to neuronal depolarization, which in turn, opens the voltage-sensitive Ca channels. GABA is able to indirectly control the opening of these Ca channels, also by acting on the membrane potential.

Acknowledgements. Supported by Project B-1590 of the University of Chile, and 1081/84 of Fondo Nacional de Ciencias.

REFERENCES

Berdichevsky, E., Riveros, N., Sánchez-Armáss, S. and Orrego, F. (1983). Kainate, N-methylaspartate and other excitatory amino acids increase calcium influx into rat brain cortex cells in vitro. Neurosci. Lett. 36, 75-80.
Luini, A., Goldberg, O. and Teichberg, V.I. (1981). Distinct pharmacological properties of excitatory amino acid receptors in the rat striatum: study by Na+ efflux assay. Proc. natl. Acad. Sci. U.S.A. 78, 3250-3254.
Riveros, N. and Orrego, F. (1984). A study of possible excitatory effects of N-acetylaspartylglutamate in different in vivo and in vitro brain preparations. Brain Res. 299, 393-396.

Intrastriatal Quinolinic Acid Injection Results in Differential Monoaminergic Nerve Terminal Degeneration

S. Mazzari, C. Aldinio and M. Beccaro

INTRODUCTION

Intrastriatal injection of the endogenous excitotoxin quinolinic acid (QUIN) causes an axon-sparing lesion, thus providing an animal model to study the human neurodegenerative disorder-Huntington's disease (Schwarcz et al., 1983). In the acute and subacute phase of the QUIN lesion, striatal dopaminergic and serotonergic nerve terminals react to the loss of postsynaptic targets by increasing their activities (Mazzari et al., 1984). Secondary changes, including axonal atrophy and alterations of neurochemical characteristics of remaining neuronal processes, are expected to occur after the subacute stage of the QUIN lesion. We here report the modifications of dopaminergic and serotoninergic parameters 4, 60 and 120 days following the unilateral intrastriatal injection of QUIN. Data obtained provide evidences of a differential axonal regression of striatal monoaminergic afferents.

METHODS

Male Sprague-Dawley rats (200-250 gr) were used throughout the experiments. Unilateral microinjections of 300 n moles of QUIN were made into striatum as described by Schwarcz et al. (1983). Measurements of tissue levels of dopamine (DA), serotonin (5HT) and their metabolites, namely homovanillic acid (HVA), dihidroxyphenylacetic acid (DOPAC) and 5-hydroxyindoleacetic acid (5HIAA), were performed by high performance liquid chromatographic analysis with electrochemical detection.

RESULTS

At fourth day after QUIN injection, both dopaminergic and serotonergic striatal afferents in the injected area enhance their activities in the lesioned side as reflected by the marked in-

Table 1. Striatal monoaminergic parameters at different time following QUIN injection.

Days	DA	HVA	DOPAC	5HT	5HIAA
4	+ 17*	+113*	+169*	+ 17	+126*
60	ND	− 40*	− 20°	+ 23	+ 29
120	− 31*	− 66*	− 47*	ND	ND

Modifications are listed as percent changes with respect to the homolateral site of PBS injected controls (ND = no difference *p < 0.01; °p < 0.05 Student's test).

crease of HVA, DOPAC and 5HIAA levels (Table 1). The chronic phase of the lesion is associated with striatal atrophy and the two monoaminergic systems undergo changes which are qualitatively different from each other. (i) Serotonergic parameters are apparently unmodified, probably because the progressive degeneration of serotonergic nerve terminals parallels the striatal atrophy. (ii) Dopaminergic parameters are significantly reduced suggesting that the regression of striatal dopaminergic nerve terminals is more prominent than the striatal atrophy. No significant modifications were observed either in the contralateral striatum or substantia nigra.

CONCLUSION

Striatum chronically lesioned by local injection of QUIN undergoes atrophy and secondary ventricular enlargement. In addition a differential reactions of innervating dopaminergic and serotonergic systems occur. Similar findings were reported in brains from patients suffering from Huntington's disease. However the possible link of QUIN to the pathogenesis of this neurodegenerative disease is still hypothetical and deserves further investigations.

References

Mazzari, S., Aldinio, C., Toffano, G. and Schwarcz, R. (1984). Effects of intrastriatal quinolinic acid injections on serotonergic and dopaminergic neuronal systems. 14th Annual Meeting of the Society for Neuroscience, Anaheim, October 10-15, Abstract 277.4.

Schwarcz, R., Whetsell, W.D. and Mangano, R.M. (1983). Quinolinic acid on endogenous metabolite that produces axon-sparing lesions in rat brain. Science, 219, 316-318.

Kainic Acid Damage to Hippocampal Pyramidal Cells and Interneurons: Influence of Hypoglycemia

F.F. Johansen and N.H. Diemer

Hippocampal interneurones in stratum oriens are resistant to cerebral ischemia, whereas pyramidal neurones are damaged (Johansen et al. 1983). Since release of neurotransmitters (glutamate and aspartate) in dendrotoxic amounts seems to play an important role in the pathogenesis of ischemic cell damage and since it has been shown that kainic acid (KA) also induces both hippocampal damage and glutamate release, a similar pattern of selective cell vulnerability should be anticipated after KA administration.

Fasted male Wistar rats were given 11mg/kg KA intraperitoneally (i.p.). A normoglycemic group consisted of 15 rats (plasma glucose level, 6-8 mmol/l)and 6 rats were rendered hypoglycemic with i.p. injection of 2-4 IU/kg crystalline insulin (glucose level 3-4 mmol/l). Three weeks later the surviving rats (6 normoglycemic and 6 hypoglycemic) were perfusion-fixed, the brains embedded and 7 um coronal sections cut and stained with cresylviolet. In CA-1 all pyramidal cells and all interneurons were counted (ocular grid, x 4oo). From our earlier studies on hippocampus interneurons in standard and Golgi sections, it seems resonable to operate with at least three types of CA-1 interneurons. Thus the alveus close interneuron, the bigger stratum oriens situated interneuron (basket cell) and the stratum radiatum located interneuron were counted separately.

CA-1 pyramidal cell countings revealed significant cell loss in all surviving rats and divided them into a group with moderate CA-1 cell loss (< 50%) and a group with severe CA-1 cell loss (> 90%). Loss of interneurons was only significant in normoglycaemic and hypoglycaemic

animals with severe CA-1 pyramidal cell damage. Normoglycaemic rats had a significant loss of all three types
of interneurons while hypoglycaemic rats only had significant loss of the stratum oriens interneurons.

This study indicate a correlation between loss of CA-1
interneurons and pronounced CA-1 pyramidal cell death
and furthermore that hypoglycaemia protects some CA-1 interneurons against KA damage. Presumably KA-receptors
work in synergy with a glutamatergic input. In the hippocampus CA-1 region only the striatum oriens interneurons are known to be innervated by (glutamatergic) collaterals from the CA-1 pyramidal cells, while the two
other interneuron types investigated by us probably receive extrinsic inputs and mediate feed-forward inhibition. The distribution of different glutamate receptors
(e.g. KA-receptors) on interneurons might explain their
vulnerability to pathological insults. It is tentatively
suggested that the general loss of interneurons in the
normoglycaemic animals partly is due to lactic acidosis,
and the protective effect of hypoglycaemia to the alveus
and stratum radiatum situated interneurons reflect absence of KA-receptors on these interneurons.

Hypoglycemic Brain Damage is Mediated by Excitotoxins

T. Wieloch, R.N. Auer, E. Westerberg, U. Tossman, U. Ungerstedt

and B. Engelsen

Insulin-induced hypoglycemic encephalopathy in the rat is characterized by selective neuronal lesions to specific areas of the brain such as the crest of the denate gyrus and the striatum (Auer et al., 1984). Although the primary factor for the development of hypoglycemic brain damage is the excessively decreased blood glucose levels, recent experimental evidence suggests that excitatory amino acids or related "excitotoxins", may be important extracellular mediators of irreversible neuronal damage.

When blood glucose levels decrease to below 1 umol/g following an insulin injection, the electrical activity of the brain ceases (see Siesjö and Agardh, 1983) and at the onset of isoelectricity the plasma membranes are depolarized concomitant with a decrease in high energy phosphates and activation of catabolic reactions such as lipolysis (Wieloch et al., 1984). In addition, the tissue ratio of aspartate/glutamate is increased (Siesjö and Agardh, 1983), which is also reflected in the extracellular levels of amino acids. Using a microdialysis method we have measured the extracellular levels of aspartate and glutamate in the striatum during hypoglycemia, Fig. 1. Ten minutes following the onset of isoelectricity the extracellular concentration of glutamate is increased 9-fold and the concentrations of aspartate 15-fold. It is thus evident that conditions prevail during hypoglycemia where the levels of toxic excitatory amino acids are elevated.

The cortical glutamatergic projections to the striatum can be transected by cortical ablations. Thus unilateral removal of the anterior neocortex leads to a decrease in the striatal levels of glutamate ipsilateral to the lesion by 11 % (Wieloch et al., 1985). The lesion also protects the striatum against hypoglycemic neuronal damage, Table 1. Thus in control animals and in the striatum contralateral to the lesion, 60-90% of the neurons were necrotic 1 week following 30 minutes of hypoglycemic coma, while no damage was noted in the ipsilateral striatum subjacent to the lesion.

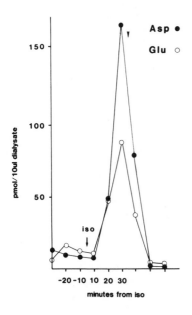

Fig. 1. The extracellular levels of glutamate and aspartate in the striatum of the rat prior to and during the isoelectric period (arrow), and following the onset of i.v. glucose infusion (arrowhead). The dialysis probe (0.4 mm in diameter, 4 mm active dialysis length) was placed horizontally into the striatum of the rat, approximately 60 min. prior to the onset of isoelectricity.

This protective effect of decortication is similar to that observed following injection of kainate into deafferented striata (Biziere and Coyle, 1979), suggesting that synaptic events such as release of excitatory amino acids could participate in the induction of hypoglycemic brain damage.

The NMDA-receptor antagonist 2-amino-7-phosphonoheptanoic acid (APH) which is protective against excitotoxin-induced neuronal damage in the striatum (Schwarcz et al., 1983), is also protective against hypoglycemia-induced neuronal damage. Thus unilateral intrastiatal injections of 40 ug of APH, mitigated the neuronal necrosis in an area surrounding the injection site, Table 2. This

Table 1. Neuronal damage in percent of total neuronal population in the dorsal striatum in unilaterally decorticated rats, subjected to 30 minutes of hypoglycemic coma and one week of recovery, and the striatal glutamate levels in unilaterally decorticated normoglycemic animals (from Wieloch et al., 1985).

| | Dorsal striatum | |
	contralateral	ipsilateral
Neuronal necrosis	79+3	0
Glutamate*	11.5 +0.3	10.4 +0.4 (p 0.05)

*Values are given as $umol \cdot g^{-1}$ wet weight $\pm$ S.E.M.

Table 2. The effect of intrastriatal injections of 2-amino-7-phosphonoheptanoic acid[*] on the density of neuronal necrosis in the striatum of rats subjected to 30 minutes of hypoglycemia induced isoelectricity and one week recovery.

	Neuronal necrosis in the striatum (%)	
	contralateral	ipsilateral
APH injection (n=6)	68 +6	5+3 (p 0.05)
NaCl injection (n=5)	61 +8	56+7

*Forty ug APH dissolved in 2 ul distilled water pH 7.0, or 2 ul saline were injected into the striatum over a period of 10 minutes, 30-50 minutes prior to the onset of isoelectricity.

demonstrates that APH sensitive receptors, such as the NMDA-receptor, are involved in the development of hypoglycemic brain damage.

The granule cells at the crest of the denate gyrus, are particular vulnerable to hypoglycemia, and neuronal damage is noted following a coma period of 10 minutes duration (Auer et al., 1984). An electron microscopical investigation of the ultrastructural changes associated with 10 minutes hypoglycemic coma revealed a dendritic axon-sparing lesion, Fig 2, similar to that observed following systemic or intracerebral injections of excitotoxins (Olney et al., 1983; Schwarcz et al., 1983).

In conclusion, we have shown that hypoglycemia-induced neuronal damage, is associated with increased extracellular levels of excitatory amino acids, is ameliorated by lesions of the glutamatergic projections, is mitigated by an NMDA-receptor antagonist, and have ultrastructural features of an excitotoxin-induced lesion. We thus conclude that hypoglycemic brain damage is an excitotoxin-mediated phenomenon.

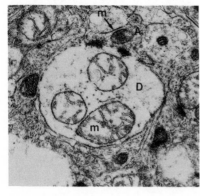

Fig. 2. The ultrastructural appearance of the outer layer of the dentate granule cell layer, showing dilated dendrites (D) with swollen mitochondria (m), and preserved axonal endings (A) with synaptic vesicles and normal mitochondria, in rats subjected to 10 minutes of hypoglycemic coma. Bar = 1 um.

REFERENCES

Auer, R.N., Wieloch, T., Olsson, Y., and Siesjö, B.K. (1984). The distribution of hypoglycemic brain damage. Acta Neuropath. (Berl.), 64, 177-191.

Biziere, K., and Coyle, J.T. (1979). Effects of cortical ablations on the neurotoxicity and receptor binding of kainic acid in striatum. J. Neurosci. Res., 4,383-398.

Olney, J.W. (1983) Excitotoxins: an overview. In Excitotoxins. (eds. K. Fuxe, P. Roberts, R. Schwarcz). McMillan Press, London. p.82-96.

Schwarcz, R., Whetsell Jr., W.O., and Foster, A.C. (1983). The neurodegenerative properties of intracerebral quinolinic acid and its structural analog cis-2,3piperidine dicarboxylic acid. In Excitotoxins. (eds. K. Fuxe, P. Roberts, R. Schwarcz). McMillan Press, London. p. 122-138.

Schwarcz, R., Whetsell Jr, W.O., and Mangano, R.M. (1983). Quinolinic acid: an endogenous metabolite that produces axon-sparing lesions in rat brain. Science,219, 316-318.

Siesjö, B.K. and Agardh, C.-D. Hypoglycemia. In Handbook of Neurochemistry 3rd ed. (ed. A. Lathja) Plenum Press, New York. vol 3. p. 353-376.

Wieloch, T., Harris, R., Symon, L., and Siesjö, B.K. (1984). Influence of severe hypoglycemia on brain extracellular calcium and potassium activities, energy charge, and phospholipid metabolism. J. Neurochem., 43, 160-168.

Wieloch T., Engelsen, B., Westerberg, E., and Auer R. (1985). Lesions of the glutamatergic cortico-striatal projections in the rat ameliorate hypoglycemic brain damage in the striatum. Neursci. Lett., in press.

Gliotoxic Effects of Neuroexcitatory Amino Acids on Brain Tumor Cell Lines

G. Le M. Campbell

There are no chemotherapeutic agents for the treatment and management of human gliomas. Some alkylating agents have been tried, but with little or no success. Agents developed by traditional screening protocols in general rely on the differential proliferative activity of normal and tumor cells as a basis for the therapeutic index. Agents developed by these procedures have little or no efficacy, either because they do not cross the blood-brain barrier or the side effects are too severe. An approach which utilizes differentially expressed biosynthetic pathways in brain cells provides a conceptually different basis for new chemotherapeutic agents. Such an approach may involve the concept of metabolic compartmentalization; a key metabolic enzyme for this concept is glutamine synthetase. Any strategy therefore which involves glutamine synthesis or utilization may therefore provide the basis for a new approach.

Evaluations of pathologic effects of neuroexcitatory amino acids and the analogues generated a classification - neurotoxic, gliotoxic, or no effect. Five analogs in particular were identified as being gliotoxic: amino adipic acid, phosphonobutyric acid, phosphonopropionic acid, α-methyl glutamic acid and α-methyl aspartic acid. Evaluations were performed on weanling rats, injected intraperitoneally. Extent of the gliotoxicity was limited, animals suffered no side effects (Olney et al, 1971). There was no obvious rationale for these effects; however it was of interest for us to determine whether aberrant or malignant glia were more or less susceptible to these analogs. We decided to test this hypothesis using a test panel of cell lines. In conjunction with Dr. Bigner and the Nervous System Neoplasia Group at Duke University, we selected six human cell lines; one osteosarcoma cell line, one medullablastoma cell line and four glioma cell lines. Selection criteria included their in vitro growth characteristics, their capacity for utilization in colony growth inhibition assays, and in vivo growth as solid tumors in nude mice. Experimental design was to evaluate the effect of these five gliotoxic analogs in addition

486

to two neurotoxic analogs on the growth of these six cell lines as reflected by changes in total protein and uptake of radiolabeled amino acids (Campbell et al, 1985). Experimental results demonstrated a heterogeneity of response. The osteosarcoma cell line showed no response to any of the analogs. The growth of two cell lines (U118 and D54MG) were affected by amino adipic acid and homocysteic acid, whereas the other three cell lines (TE 671, U373, and U251) were affected by phosphonobutyric acid. The other three gliotoxic analogs showed no effect on any of the cell lines. The medium in which these cell lines were grown contained 4 mM glutamine. We repeated these experiments in the presence and absence of glutamine in the medium. All cell lines except the osteosarcoma were now sensitive to amino adipic acid, homocysteic acid and phosphonobutyric acid. Such results clearly indicate that glutamine synthesis or utilization underlies the efficacy of these analogs. Dose response experiments indicated that 10-20 mM. concentrations of these analogs are effective, little or no effect was detectable at doses of 1 mM. Such a limited range of dose effectiveness may have advantages for localized drug delivery.

Further studies examined the effects of glutamate and glutamine on the growth of these cell lines with two types of response (Dranoff et al, 1985a). One group of cell lines such as TE-671 grows equally well in glutamine (0-4mM) and glutamate (0-5 mM), although the initial growth was somewhat slower in the lowest concentrations. The other group, e.g. D-54MG. did not grow in glutamate and the degree of growth in glutamine was dependent on the concentration in the medium. Correlations with the phenotypic expression of glutamine synthetase (GS), are consistent with the growth curve results. These analogues may have different mechanisms of action, as indicated by the phenotypic heterogeneity of GS in the cell lines examined. They may act as glutamine antagonists, as demonstrated by their effect on cell lines without glutamine synthetase. However, they may also interfere with glutamine synthetase. Both mechanisms may yield a similar end result - a reduction in the effective concentration of intracellular glutamine. What needs to be resolved at this point is what is the resultant effect of this reduction on energy metabolism and DNA synthesis. Bigner and his colleagues have examined multiple combinations of different drug regimens to maximize the potential effectiveness of this approach (Dranoff et al, 1985b). A better understanding of astrocyte metabolic pathways in human gliomas and medulloblastoma is essential to maximize the potential utilization of these kinds of agents in a multi-modal therapeutic regimen.

References:

Bigner, D.D. Bigner, S.H. Ponten, J. Westermark, B. Mahaley, M.S. Ruoshati, E. Hershmann, H. Eng, L.F. and Wikstrand, C.J. (1981). Heterogeneity of genotypic and phenotypic characteristics of fifteen permanent cell lines derived from human gliomas. J. Neuropath. Exp. Neurol., 40, 201-229.

Campbell, G.LeM. Bartel, R. Friedman, H.S. and Bigner, D.D. (1985).
Effects of glutamate analogs on brain tumor cell lines.
J. Neurochem. (in press).

Dranoff, G. Elion, G.B. Friedman, H.S. Campbell, G.LeM. and
Bigner, D.D. (1985a). Influence of Glutamine on the Growth of Human
Glioma and Medulloblastoma. Cancer Res. (in press).

Dranoff, G. Elion, G.B. Friedman, H.S. and Bigner, D.D. (1985b).
Combination Chemotherapy in Vitro exploiting Glutamine Metabolism of
Human Glioma and Medulloblastoma. Cancer Res., (in press).

Friedman, H.S. Bigner, S.H. McComb, R.D. Schold, S.C.
Pasternak, J.F. Groothuis, D.R. and Bigner, D.D. (1983). A Model
for human medulloblastoma – growth, morphology, and chromosomal
analysis in vitro and in athymic mice. J. Neuropathol., Exp.
Neurol., 42:485–503.

Olney, J.W. Lan Ho, O. and Rhee, V. (1971). Cytotoxic Effects of
Acidic and Sulphur containing Amino Acids on the Infant Mouse
Central Nervous System. Exp. Brain Res., 14, 61–78.

β-Kainic Acid as an Anticonvulsant and Possible NMDA Antagonist

A.G. Chapman, J.F. Collins, L. Turski and B.S. Meldrum

INTRODUCTION

β-kainic acid (β-KA) is a stereoisomer of α-KA with the D
configuration at the C2 carbon (Collins et al. 1984). α-KA is well
known as a potent convulsant and excitotoxin that interacts with a
specific population of receptors whose localization is distinct
from that of quisqualate or NMDA receptors. The configuration of
the C2 carbon in the β-KA molecule resembles that of the C2 carbon
in AP5 or AP7. This configuration seems to be important for the
pharmacological actions of the NMDA antagonists. We have therefore
investigated whether β-KA exhibits anticonvulsant action in two
animal models of seizures, i.e. in audiogenic convulsions in DBA/2
mice and in myoclinic seizures in Swiss S mice. Our second aim was
to chain elongate both in the α and β-series of compounds to yield
compounds (see Fig.1) which are analogous to γ-glutamylglycine and
γ-glutamylaminomethylphosphonate (Croucher et al, 1984).

α -KAINATE SERIES   β-KAINATE SERIES

R: -CH$_2$COOH,   KAINIC ACID

R: -CH$_2$CNHCH$_2$COOH,   KAINYL GLYCINE

R: -CH$_2$CNHCH$_2$P(OH)$_2$,   KAINYL AMINOMETHYLPHOSPHONATE

Fig. 1. α- and β-kainic acid and kainic acid derivatives.

MATERIAL AND METHODS
    1)   DBA/2 mice of mixed sexes and body weight 6-12 g, 21-28
days old, were exposed to auditory stimulation either: (1) 45 min
after i.c.v. injection of 10 μl of 67 mM sodium phosphate buffer
(pH 7.4) alone or containing: β-KA, α-kainyl glycine, β-kainyl
glycine, α-kainylaminomethylphosphonate, or β-kainyl aminomethylphosphonate, or (II) 45 min following i.p. administration of saline
(0.1 ml, pH 7) or β-KA. The seizure response (SR) to auditory
stimulation (109 dB for 60 s or until tonic extension occurred) was
assessed.
    2)   To characterize the anticonvulsant action of β-KA we
investigated its antagonistic properties against the convulsive
activity of excitatory amino acids in Swiss S mice. To compare the
anticonvulsant action of β-KA with excitatory amino acid antagonists
of a known pharmacological profile, we also tested 2-amino-7-
phosphonoheptanoate (APH), kynurenate (KYN) and γ-D-glutamylaminomethylsulphonate (GAMS). All drugs were injected i.c.v. CD50 values
and their 95% confidence limits were estimated for myoclonic
seizures from probit-log dosage regression curves (Turski et al,
1985).

RESULTS
    1)   In order of decreasing relative potency, β-KA, β-kainyl
glycine, α-kainyl glycine, α-kainyl aminomethylphosphonate and β-
kainylaminomethylphosphonate inhibit the development of audiogenic
seizures following their i.c.v. administration.
    2)   β-KA displays preferential anticonvulsant activity against
NMDA and quinolinate in Swiss S mice. It is almost ineffective
against quisqualate and totally ineffective against L-glutamate.
β-KA is of intermediate potency as an α-KA and D-homocysteinesulphinate antagonist.
    3)   APH and KYN have almost identical profile and potency of
anticonvulsant activity as those of β-KA.
    4)   The anticonvulsant activity of β-KA differs from the
profile of anticonvulsant action of GAMS, a preferential kainate/
quisqualate antagonist.

CONCLUSIONS
    The most significant novel finding is the demonstration of an
anticonvulsant action of β-KA. Coinjection experiments with
selective agonists indicate that β-KA is not acting as a preferential α-KA antagonist. Surprisingly it shows the same spectrum of
activity as known selective NMDA antagonists. The structural
requirements for α-kainate antagonism remain to be defined.

REFERENCES
Collins, J.F., Dixon, A.J., Badman, G., De Sarro, G., Chapman, A.G.,
Hart, G.P. and Meldrum, B.S. (1984). Neurosci. Lett., 51, 371-376.
Croucher, M.J., Meldrum, B.S., Jones, A.W. and Watkins, J.C. (1984).
Brain Res., 322, 111-114.
Turski, L., Meldrum, B.S. and Collins, J.F. (1985). Brain Res.,
336, 162-166.

Effects of the Two Convulsant β-Carbolines, DMCM and β-CCM on Excitatory Amino Acid Transmitter Systems in Rodents

A.G. Chapman

INTRODUCTION

The two β-carbolines, DMCM (methyl-6-7-dimethoxy-4-ethyl-β-carboline-3-carboxylate) and β-CCM (methyl-β-carboline-3-carboxylate), display similar interactions with the GABA-benzodiazepine-receptor complex, exhibit proconvulsant and convulsant actions in different seizure models (Braestrup et al, 1982, Meldrum & Braestrup, 1984), and affect GABA-mediated inhibition of neuronal firing: DMCM reduces the GABA-inhibition, while β-CCM reverses the diazepam-induced enhancement of the GABA response (Skerritt & Macdonald, 1984). The benzodiazepine receptors consist of 2 sub-classes, BZ_1 and BZ_2 which differ in their affinity for some of the ligands (triazolopyridazines and some of the β-carbolines; selective for BZ_1). The physiological role of the two receptor subclasses is not known, but they exhibit different regional distribution, and perhaps different pre- and post-synaptic location (Lo et al, 1983; Pan et al, 1984; Tietz et al, 1985). Although DMCM and β-CCM are both considered to induce convulsions by acting on the GABA/benzodiazepine receptor complex, the convulsions differ in several pharmacological and biochemical respects, especially with respect to the interaction of the β-carbolines with the excitatory amino acid transmitter system (Chapman et al, 1985).

METHODS AND RESULTS

Seizures were induced in adult DBA/2 mice by the ip administration of DMCM (1-4 mg/kg) or β-CCM (0.5-8 mg/kg) alone or following pretreatment with 2-APH (0.5 mmole/kg, 45 min, diazepam (1 mg/kg, 15 min) or quazepam (1 mg/kg, 15 min).

For biochemical studies fully convulsant doses of DMCM (15 mg/kg, ip) and β-CCM (10 mg/kg, ip) were used. The mice were sacrificed by microwave irradiation after 15 min of seizure activity, their brains dissected and extracted for regional amino acid determination (HPLC analysis following OPA derivatization).

[3]H-D-aspartate was released from superfused, preloaded rat cortical or hippocampal slices (McIlwain tissue chopper) by two

potassium (50 mM) pulses at 37°. Experimental (+ β-carbolines) and control release studies were run in parallel.

Pretreatment with the excitatory amino acid receptor antagonist, 2-APH significantly increased the ED_{50} for DMCM induced clonic seizures in DBA/2 mice two fold (4.7 μmoles/kg vs. 9.4 μmoles/kg) but did not significantly alter the ED_{50} for β-CCM (4.2 μmoles/kg vs. 6.6 μmoles/kg. Diazepam pretreatment had similar protective effect against β-CCM and DMCM seizures, while quazepam (selective) for BZ_1) selectively inhibited β-CCM seizures (4 fold increase in ED_{50} value; Chapman & De Sarro, unpublished).

Potassium-induced release of preloaded ^3H-D-aspartate from rat cortical or hippocampal minislices was enhanced in the presence of DMCM (100 μM), but not altered in the presence of β-CCM (100 μM).

When clonic or clonic/tonic seizures were induced in Swiss mice by the systemic administration of DMCM, regional aspartate levels fell 20-30% (in cortex and cerebellum) after DMCM, but were unchanged in all regions after β-CCM. Glutamate levels fell in cortex after β-CCM (23%) and in striatum after DMCM (15%).

GABA levels almost doubled in all regions after β-CCM seizures, whereas smaller GABA increases were seen after DMCM seizures.

DISCUSSION

Interaction with the excitatory amino acid transmission system appears to play a more important role during DMCM seizures than during β-CCM seizures. Thus: 1) DMCM seizures are inhibited by 2-APH, 2) DMCM enhances aspartate release, 3) regional cerebral aspartate levels are decreased during DMCM seizures. These parameters are not affected by β-CCM seizures. DMCM and β-CCM also differ in their selectivity for BZ receptor subtypes. DMCM appears relatively nonselective, whereas β-CCM appears selective for BZ_1 receptors (based on regional binding data and preferential inhibition of β-CCM seizures by quazepam). However, more information about pre- and post-synaptic location of BZ_1 and BZ_2 receptors and their interaction with excitatory amino acid terminals is required before the differences in β-CCM and DMCM effects can be attributed to their different interactions with BZ subtypes.

REFERENCES

Braestrup, C., Schmiechen, R., Neef, G., Nielsen M., and Petersen, E.N. (1982). Science 216, 1241-1243.

Chapman, A.G., Cheetham, S.C., Hart, G.P., Meldrum, B.S. and Westerberg, E. (1985). J. Neurochem. 45 in press.

Lo, M.M.S., Niehoff, D.L., Kuhar, M.J. and Snyder, S.H. (1983). Nature 306, 57-60.

Meldrum, B., Braestrup, C. (1964). In: Actions and Interactions of GABA and Benzodiazepines (Bowery N, eds). New York, Raven Press, pp.133-153.

Pan, H.S., Penney, J.B. and Young, A.B. (1984). J. Pharm. Exp. Therap. 230, 768-775.

Skerritt, J.H. and Macdonald, R.L. (1984). Europ. J. Pharmacol. 101 135-141.

Tietz, E.I., Chin T,H. and Rosenberg H.C. (1985). J. Neurochem. 44 1524-1534.

Glutamate, Aspartate and Related Amino Acids in Human Ventricular Cerebrospinal Fluid (vCSF)

B. Engelsen, V.M. Fosse and E. Myrseth

INTRODUCTION

Glutamate (Glu) and aspartate (Asp) are excitatory neuro-transmitters (Fonnum, 1984) as well as intermediates of brain metabolism. They may be important for the etiology and patho-genesis of epilepsy, various disorders of the basal ganglia and of cerebral ischemia (Meldrum, 1985). However, further studies in clinical neurology are required to elucidate the exact role of Glu and Asp in these disorders.

We have measured free amino acid concentrations in the vCSF of 16 patients with different types of hydrocephalus.

PATIENTS AND METHODS

The vCSF samples were stored at -80 C and deproteinized by means of heat inactivation. The amino acid analysis was performed by high performance liquid chromatography (HPLC).

Of the control patients (Pts.) with obstructive hydrocephalus 5 had various intracranial tumors. 2 Pts. had hydrocephalus of unknown etiology and 3 newborn babies had myelomeningocoele with hydrocephalus. 2 Pts. had highly increased CSF pressure and clinical symptoms of an impaired brain tissue circulation.

In Pt. BI (male, 30 years) a 4th ventricle ependymoma was removed and a post operative radiation therapy performed 8 years before the present operation. There has been no signs of a tumor recurrence as judged by clinical symptoms, CT-scans and recently by intraoperative inspection. He was admitted with various neurological deficits of a pontine and cerebellar origin. A cerebral CT-scan showed a distension of the 4th ventricle. An auditive evoked response showed signs of a pontine lesion.

The clinical symptoms disappeared or improved immediately after the opening of the 4th ventricle and a communicating cyst, in which the CSF was under high pressure.

Pt. RM (female, 68 years) was admitted with signs of severely increased intracranial pressure. A CT-scan showed a pineal tumor, large distensions of the 3rd and both lateral ventricles, brain oedema and no signs of cortical relief.

The clinical condition improved immediately after a ventricu-
loperitoneal shunt was inserted.

RESULTS

The mean concentrations (± SD) of vCSF Glu and Asp in the
control group were 3.0 ± 1.8 (n=14) and $\overline{0}$.2 ± 0.2 (n=9) µM
respectively. The Glu and Asp concentrations did not differ
between Pts. with and Pts. without intracranial tumors.

In Pt. BI Glu and ASP concentrations were increased by 1150%
and 1000% as compared to their respective mean control values.
Taurine was increased by 95% of the mean control value and iso-
leucine and α-aminobutyric acid were in the upper range of the
controls.

In Pt. RM the Glu and Asp concentrations were increased by
650% and 200% as compared to their respective mean control values.
Serine was in the upper range of the controls. There were no
other significant changes in amino acid concentrations, including
GABA.

DISCUSSION

Increased concentrations of vCSF Glu and Asp were the only
consistent changes in our 2 patients with symptoms which could be
due to an impaired brain circulation. This increase could be due
to either an increased release or a decreased re-uptake of Glu and
Asp in the cerebral tissue surrounding the ventricles. Increased
extracellular levels of Glu and Asp has recently been found in the
hippocampus of rats exposed to transient cerebral ischemia
(Benveniste et al., 1984).

However a defect in the normally saturable and net outward
transport of Glu and Asp through the blood-brain-CSF-barrier could
explain the changes in Glu and Asp, but not the increase in taurine
in Pt. BI. Our patients had no other conditions known to be
associated with increased Glu or Asp in lumbar CSF such as epilepsy
and hepatic encephalopathy.

The results of the present study may support the hypothesis of
Glu and Asp as important etiological factors in the development of
ischemic brain damage, however further clinical investigations are
needed to establish its validity.

REFERENCES

Benveniste, H., Drejer, J., Schousboe, A. and Diemer, N.H. (1984).
Elevation of the extracellular concentration of glutamate and
aspartate in rat hippocampus during transient cerebral ischemia
monitored by intracerebral microdialysis. J. Neurochem. 43, 1169-
1174.

Fonnum, F. (1984). Glutamate: A transmitter in mammalian brain.
J. Neurochem. 42, 1-11.

Meldrum, B. (1985). Possible therapeutic applications of
antagonists of excitatory amino acid neurotransmitters. Clinical
Science. 68, 113-122.

Quinolinic Acid, Tryptophan Availability to the Brain and Neurological Disorders

F. Moroni, G. Lombardi and V. Carlà

Quinolinic acid (QUIN) has been considered for many years a metabolic intermediate in the "de novo" biosynthesis of NAD from tryptophan. Recently it has been demonstrated that QUIN is an NMDA receptor agonist (Stone and Perkins, 1981) and when locally applied causes axon-sparing post synaptic degeneration of the neurons (Schwarcz et al., 1981). This prompted us to investigate whether or not QUIN is present in the brain and what the mechanisms are which regulate the concentration of this molecule in the CNS.

The regional brain content of QUIN was measured using a modification of a previously described method (Moroni et al., 1984).

Table 1 reports the concentrations of QUIN in various brain areas of rat, guinea-pig and rabbit. The table shows also that the concentrations of QUIN and those of 5HT and 5HIAA in the brain are in the same range. No relationship is present between the concentration of QUIN and those 5HT and of 5HIAA. In every animal species examined (including man) the cortex is the area containing the

Table 1. The distribution of QUIN 5HT and 5HIAA in the CNS

	Q U I N			5HT	5HIAA
	Guinea-pig	Rabbit	Rat	Rat	
CORTEX	1.9±0.2	1.6±0.1	1.8±0.2	1.2±0.1	1.3±0.11
STRIATUM	0.5±0.1	0.4±0.1	0.6±0.1	1.6±0.2	0.9±0.1
HIPPOCAMPUS	0.8±0.2	0.5±0.1	1.0±0.2	1.3±0.1	1.6±0.14
DIENCEPHALON	-	0.5±0.1	0.7±0.2	3.4±0.3	4.2±0.4
BRAIN-STEM	-	-	0.9±0.3	3.5±0.2	3.4±0.09
CEREBELLUM	1.4±0.2	1.1±0.2	1.4±0.2	0.5±0.08	0.68±0.03

Values are nmol/g w.w. and are the mean ± S.E. of at least 6 animals.

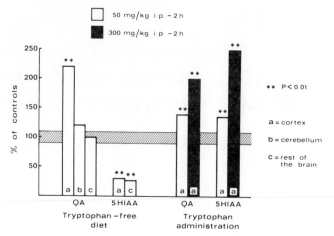

Fig.1. The brain content of QUIN and of 5HIAA after trypto
phan administration or after a tryptophan-free diet.

highest concentration of QUIN, while tha brain-stem is the area containing the highest concentration of 5HT and 5HIAA.

When tryptophan was administered to the animals the brain content of 5HT, of 5HIAA and of QUIN increased in a dose-dependent manner. This supports che concept that the concentration of these molecules in the brain are in some way dependent upon tryptophan availability. Surprisingly (fig. 1) the cortical content of QUIN increased
also in rats fed with a diet lacking in tryptophan. This sharply
contrasts with the large decrease of the content of 5HT an 5HIAA in
various brain areas of these animals. It seems, therefore, that a
metabolic pathway, not related with tryptophan and leading to QUIN
synthesis, is present in the brain.

A tryptophan-free diet leads to dementia (pellagra) both in man
and in animals and the possibility that an imbalance between the
high brain content of QUIN and the decrease of other tryptophan metabolites could be toxic for the neurons is now under investigation.
In particular, an imbalance between the concentration of QUIN and
those of kynurenic acid in the brain could be one of the mechanisms
leading to loss of neurons in the CNS.

REFERENCES

Moroni, F., Lombardi, G., Moneti, G., Carlà, V., (1984). The excitotoxin quinolinic acid is present and unevenly distributed in the rat
brain. Brain Res., 255, 352-355.

Schwarcz, A.C., Whetsell, W.C., Mangano, R.N., (1983). Quinolinic
acid: an endogenous metabolite that produce axon-sparing lesion in
rat brain. Science, 219, 316-318.

Stone, T.W., Perkins, M.N., (1981). Quinolinic acid: a potent endogenous excitant at amino acid receptors in CNS. European J. Pharmacol., 72, 411-412, 1981.

Chemical and Biochemical Stability of Ibotenic Acid and 7-HPCA and Attempts to Develop Irreversible QUIS Receptor Agonists

E.Ø. Nielsen, A. Schousboe, S.H. Hansen, J. Lauridsen

and P. Krogsgaard-Larsen

Ibotenic acid is an excitatory amino acid showing a complex pharmacological profile (excitation/inhibition) (MacDonald and Nistri, 1978) on single neurones in the mammalian CNS. This biphasic effect of ibotenic acid was confirmed by Curtis et al. (1979), and the prolonged depressant phase of the action of ibotenic acid was almost identical with the effect of muscimol. While it is generally agreed that the initial excitatory effect of ibotenic acid is mediated primarily by NMDA receptors (Watkins and Evans, 1981; McLennan, 1983; Foster and Fagg, 1984), the mechanisms underlying the inhibitory phase have not been clarified yet. This biphasic effect may be an inherent property of ibotenic acid (MacDonald and Nistri,1978) or, alternatively, the result of the formation of muscimol by decarboxylation of ibotenic acid within the tissue (Curtis et al., 1979).

Although ibotenic acid is known to be converted into muscimol in the dried Amanita muscaria mushrooms (Eugster, 1969), these aspects have not been studied in detail. In an attempt to elucidate the ability of ibotenic acid to undergo decarboxylation, we have studied the chemical and biochemical stability of ibotenic acid under different in vitro conditions in the presence or absence of brain or liver enzymes.

While ibotenic acid was shown to undergo decarboxylation at 100°C, the rates of decomposition being dependent on pH, the compound was stable at physiological temperatures and pH values in the absence of L-glutamate-decarboxylase (GAD). The rates of decomposition of ibotenic acid were strongly accelerated in the presence of GAD or liver homogenates (Nielsen et al., 1985).

Analogous studies on the highly selective QUIS receptor agonists AMPA and 7-HPCA (Krogsgaard-Larsen et al., 1984) showed that AMPA did not undergo detectable decarboxylation in the absence or presence of GAD. 7-HPCA was stable at 37°C at different pH values, whereas at 100°C the compound decomposed at rates slightly higher than those observed for ibotenic acid. Like AMPA, 7-HPCA was stable in the presence of GAD at physiological temperatures and pH values.

Ibotenic acid   7-HPCA   AMPA

ABPA   4-AHCP

In an attempt to develop agonists capable of binding irreversibly to the QUIS receptors, ABPA and 4-AHCP were designed and synthesized. Both of these compounds proved to be potent and highly selective agonists at QUIS receptors. However, neither the bromomethyl
group of ABPA nor the conjugated system in 4-AHCP seem to be sufficiently reactive or appropriately located for reactions with the QUIS
receptor macromolecule(s).

Curtis,D.R., Lodge,D., and McLennan,H. (1979). The Excitation and De-
    pression of Spinal Neurones by Ibotenic Acid. J.Physiol. 291,
    19–28.
Eugster,C.H. (1969). Chemie der Wirkstoffe aus dem Fliegenpilz.
    Fortschr.Chem.Org.Naturst. 27, 261–321.
Foster,A.C. and Fagg,G.E. (1984). Acidic Amino Acid Binding Sites in
    Mammalian Neuronal Membranes: Their Characteristics and Relation-
    ship to Synaptic Receptors. Brain Res.Rev. 7, 103–164.
Krogsgaard-Larsen,P., Nielsen,E.Ø., and Curtis,D.R. (1984). Ibotenic
    Acid Analogs. Synthesis and Biological and in Vitro Activity of
    Conformationally Restricted Agonists at Central Excitatory Amino
    Acid Receptors. J.Med.Chem. 27, 585–591.
MacDonald,J.F. and Nistri,A. (1978). A Comparison of the Action of
    Glutamate, Ibotenate and Other Related Amino Acids on Feline Spi-
    nal Interneurones. J.Physiol. 275, 449–465.
McLennan,H. (1983). Receptors for Excitatory Amino Acids in the Mam-
    malian Central Nervous System. Prog.Neurobiol. 20, 251–271.
Nielsen,E.Ø., Schousboe,A., Hansen,S.H., and Krogsgaard-Larsen,P.
    (1985). Excitatory Amino Acids. Studies on the Biochemical and Che-
    mical Stability of Ibotenic Acid and Related Compounds. J.Neuro-
    chem. (in press).
Watkins,J.C. and Evans,R.H. (1981). Excitatory Amino Acid Transmit-
    ters. Ann.Rev.Pharmacol.Toxicol. 21, 165–204.

Structural and Conformational Requirements for Activation of the QUIS and NMDA Receptors

U. Madsen, D.R. Curtis, L. Brehm, E.Ø. Nielsen and P. Krogsgaard-Larsen

The QUIS receptors probably represent the postsynaptic glutamic acid receptors (Monaghan et al., 1984). AMPA, 5- and 7-HPCA, synthesized using ibotenic acid as lead, are potent and highly selective QUIS receptor agonists (Krogsgaard-Larsen et al., 1984; 1985). 5-HPCA, which is a conformationally restricted analogue of AMPA, is likely to reflect the receptor-active conformation of AMPA. Although the structures of 5- and 7-HPCA look quite different, a comparison of the structures of the fully ionized forms of these compounds, which probably are recognized by the receptor, makes the similar agonist profiles of 5- and 7-HPCA understandable. Whilst the "glutamic acid structure elements" of AMPA and 7-HPCA adopt dissimilar conformations in the crystalline states, as shown by X-ray crystallography, the energy barrier for conversion of this conformation of AMPA into that reflected by 7-HPCA is low.

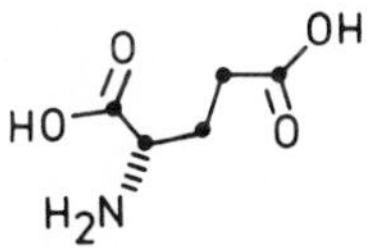

(S)-(+)-Glutamic acid (GLU)

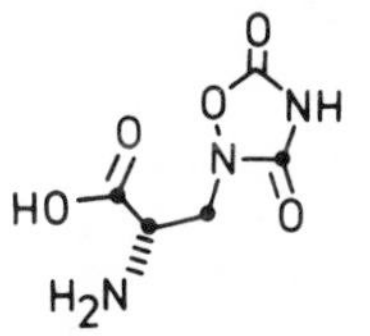

Quisqualic acid (QUIS)

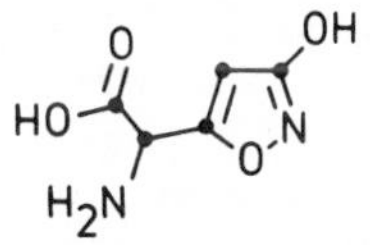

Ibotenic acid (IBO)

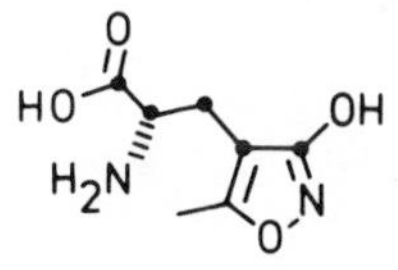

(S)-(+)-AMPA

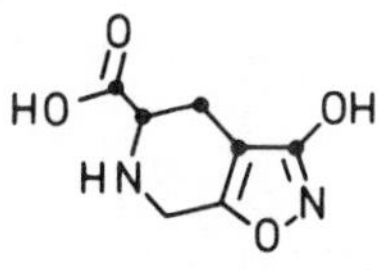

5-HPCA

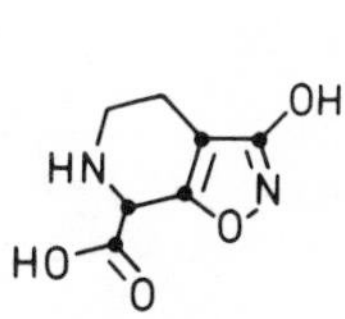

7-HPCA

Very little is known about the conformational requirements for activation of the NMDA receptors (Watkins, 1984). Besides NMDA itself, relatively few selective NMDA agonists are known, such as cis-1,3-ADCP and AMAA. In an attempt to shed light on the "receptor-active conformation" of these compounds, the conformationally restricted analogue 4-HPCA was tested (Madsen et al., 1985). This compound was, however, shown to be biologically inactive, indicating that it is unlikely to reflect the active conformation of NMDA, cis-1,3-ADCP or AMAA.

NMDA   Cis-1,3-ADCP   AMAA   4-HPCA

AP5   α-AA   6-HPCA

Similarly, in order to elucidate the conformation in which AP5 and α-AA bind to and block the NMDA receptors, structure-activity studies were made on the conformationally restricted analogue 6-HPCA (Madsen et al., 1985). This compound did not significantly reduce neuronal excitation induced by NMDA. Thus, the partially folded conformation of AP5 and α-AA, represented by 6-HPCA, does not seem to reflect the antagonist conformations of these compounds.

Krogsgaard-Larsen,P., Brehm,L., Johansen,S., Vinzents,P., Lauridsen, J., and Curtis,D.R. (1985). Synthesis and Structure-Activity Studies on Excitatory Amino Acids Structurally Related to Ibotenic Acid. J.Med.Chem. 28, 673-679.
Krogsgaard-Larsen,P., Nielsen,E.Ø., and Curtis,D.R. (1984). Ibotenic Acid Analogues. Synthesis and Biological and in Vitro Activity of Conformationally Restricted Agonists at Central Excitatory Amino Acid Receptors. J.Med.Chem. 27, 585-591.
Madsen,U., Schaumburg,K., Brehm,L., Curtis,D.R., and Krogsgaard-Larsen,P. (1985). Ibotenic Acid Analogues. Synthesis and Biological Testing of Two Bicyclic 3-Isoxazolol Amino Acids. Acta Chem.Scand. B39, in press.
Monaghan,D.T., Yao,D., and Cotman,C.W. (1984). Distribution of [3H]-AMPA Binding Sites in Rat Brain as Determined by Quantitative Autoradiography. Brain Res. 324, 160-164.
Watkins,J.C. (1984). Excitatory Amino Acids and Central Synaptic Transmission. Trends Pharmacol.Sci. 5, 373-376.

Index